REGENERATIVE MEDICINE TECHNOLOGY

ON-A-CHIP APPLICATIONS FOR DISEASE MODELING, DRUG DISCOVERY AND PERSONALIZED MEDICINE

GENE AND CELL THERAPY SERIES

Series Editors

Anthony Atala & Graça Almeida-Porada

PUBLISHED TITLES

Regenerative Medicine Technology: On-a-Chip Applications for Disease Modeling, Drug Discovery and Personalized Medicine
Sean V. Murphy and Anthony Atala

Therapeutic Applications of Adenoviruses
Philip Ng and Nicola Brunetti-Pierri

Cellular Therapy for Neurological Injury
Charles S. Cox, Jr

Placenta: The Tree of Life
Ornella Parolini

GENE AND CELL THERAPY

REGENERATIVE MEDICINE TECHNOLOGY

ON-A-CHIP APPLICATIONS FOR DISEASE MODELING, DRUG DISCOVERY AND PERSONALIZED MEDICINE

Edited by
Sean V. Murphy
Anthony Atala

CRC Press
Taylor & Francis Group
Boca Raton London New York

CRC Press is an imprint of the
Taylor & Francis Group, an **informa** business

CRC Press
Taylor & Francis Group
6000 Broken Sound Parkway NW, Suite 300
Boca Raton, FL 33487-2742

Printed on acid-free paper
Version Date: 20161010

International Standard Book Number-13: 978-1-4987-1191-3 (Hardback)

Library of Congress Cataloging-in-Publication Data

Names: Murphy, Sean V., editor. | Atala, Anthony, 1958- editor.
Title: Regenerative medicine technology : on-a-chip applications for disease modeling, drug discovery and personalized medicine / [edited by] Sean V. Murphy and Anthony Atala.
Other titles: Gene and cell therapy series.
Description: Boca Raton : Taylor & Francis, 2017. | Series: Gene and cell therapy series | Includes bibliographical references and index.
Identifiers: LCCN 2016027009| ISBN 9781498711913 (hardback : alk. paper) | ISBN 9781498711920 (e-book)
Subjects: | MESH: Lab-On-A-Chip Devices | Artificial Organs | Personalized Medicine | Printing, Three-Dimensional | Bioprinting | Regenerative Medicine--methods
Classification: LCC R856 | NLM QT 26 | DDC 610.28/4--dc23
LC record available at https://lccn.loc.gov/2016027009

Visit the Taylor & Francis Web site at
http://www.taylorandfrancis.com

and the CRC Press Web site at
http://www.crcpress.com

Printed and bound in the United States of America by Publishers Graphics, LLC on sustainably sourced paper.

Dedication

To my wife, Jess

Sean V. Murphy

To my family—Katherine, Christopher and Zachary

Anthony Atala

Contents

List of Figures

Series Preface

Gene and cell therapies have evolved in the past several decades from a conceptual promise to a new paradigm of therapeutics, able to provide effective treatments for a broad range of diseases and disorders that previously had no possibility of cure.

The fast pace of advances in the cutting-edge science of gene and cell therapy, and supporting disciplines ranging from basic research discoveries to clinical applications, requires an in-depth coverage of information in a timely fashion. Each book in this series is designed to provide the reader with the latest scientific developments in the specialized fields of gene and cell therapy, delivered directly from experts who are pushing forward the boundaries of science.

In this volume of the Gene and Cell Therapy book series, *Regenerative Medicine Technology: On-a-chip Applications for Disease Modeling, Drug Discovery and Personalized Medicine,* a remarkable group of authors comprehensively cover the latest developments in microfabrication, 3D bioprinting, 3D cell culture techniques, microfluidics, biosensor design and microelectronics, data collection, and predictive analysis. In the second part of this book, the authors present the most advanced concepts in body-on-a chip systems to study disease modeling and drug discovery, using single or combined arrays of tissues such as lung, liver, heart, skin, and kidney. The last section of this book series will report on current and future application of body-on-a-chip technologies, including successes in modeling tissue-specific cancers, metastasis, and tumor microenvironments, as well as applications for evaluating drug efficacy and toxicity, and for predicting outcomes of various disease treatment strategies.

We would like to thank the volume editors, Sean Murphy and Anthony Atala, and all the authors, all of whom are remarkable experts, for their valuable contributions. We would also like to thank our senior acquisitions editor, Dr. C.R. Crumly, and the CRC Press staff for all their efforts and dedication to the Gene and Cell Therapy book series.

Anthony Atala
Wake Forest Institute for Regenerative Medicine

Graça Almeida-Porada
Wake Forest Institute for Regenerative Medicine

Preface

Prior to the emergence of laboratory medicine, medical science was guided primarily by clinical observations and various observational theories. The invention of the microscope brought about a renaissance in the field of medical diagnostics and ushered in the era of laboratory medicine almost a century later. It led to novel discoveries, theories, and practices, including evidence-based approaches to medical decision-making. Like the microscope, many innovative ideas, devices, and technologies have had major impacts on modern medicine. One such example is the concept of a model system that provides a platform for studying and testing scientific hypotheses in a controlled and replicable environment.

Animals and animal models have been an integral part of scientific discovery, as models of normal and disease physiology as well as for the evaluation of therapies destined for human application. Animal models still find widespread use in medical research and scientific study and have become ever more sophisticated, such as genetically modified animal models that may better represent human disease and physiological responses to intervention. However, despite the impact animal models have had on medicine, they have failed to replicate many important human diseases. Therapies evaluated in these models often fail to predict therapeutic effectiveness, potential toxicity, or side effects upon translation into human trials.

Considerable advances have been made in the development of human cell culture models of tissue and organ function and physiological responses to external stimuli from various compounds, drugs, and toxins. Over the past 60 years, the culture and expansion of human cells for transplantation and tissue engineering applications has become routine in the laboratory. However, the common strategy for cell culture, namely expansion of a monolayer of cells on flat and hard plastic or glass 2D substrates, represents a reductionist approach where important components of tissues or organs are lacking. Specifically, cell–cell interactions, 3D architecture, and mechanical and biochemical cues may be lacking under these simplified conditions. This approach produces results that may not be predictive of the *in vivo* physiology and contributes to the unacceptably high failure rates for therapy development.

Recent advances and convergence of multiple areas of biology and engineering have facilitated the development of improved *in vitro* models of tissue and organ function. Specifically, progress in stem cell and cell biology has resulted in improved 3D culture conditions that have enabled greater levels of cell differentiation and physiological function. Additionally, various microfabrication technologies such as lithography and 3D bioprinting have provided opportunities to fabricate increasingly complex biological components into 3D functional tissues. The improvement in the functional output of these fabricated 3D tissues can now be evaluated using sophisticated biosensors and can be maintained *in vitro* for long periods within a system of microfluidic devices that house these sensors, provide nutrients, and remove waste from the functioning tissues. Biochemical diagnostics and assays have been miniaturized and automated, allowing the real-time evaluation of tissue function, while the large amount of data generated from these systems can be analyzed. These systems can be highly predictive and consistent with validated mechanisms.

Currently, multiple types of tissue organoids have been developed, comprising various cell types, fabrication strategies, microfluidic systems, biosensors, and functional output evaluation. Some examples include, but are not limited to, *in vitro* tissue models of the lung, liver, heart, skin, and kidney. These systems all closely recapitulate the tissue cellular components, structure, and organization of native tissues, thus reproducing tissue function at a more physiologically relevant level. Additionally, some progress had been made toward the development of a "body-on-a-chip" system, which consists of an interconnected network of multiple tissue-type organoids, connected through a biomimetic circulatory system, analogous to the vascular system of the human body. While still early in development, multiple applications of these systems have already been demonstrated, including the modeling of various diseases and disorders, such as inflammation, fibrosis, and cancer, as well as modeling potential therapeutic drug effects, toxicity, or side effects. Finally, these systems are being developed as platforms for personalized medicine where various medical interventions or treatments can be evaluated *in vitro* prior to delivery to the patient. By recapitulating not only the form but also the rudiments of function of their *in vivo* counterparts, these constructs have the potential to provide a key missing link between 2D culture systems or animal models and clinical translation.

This book is divided into three major sections, with a total of 18 chapters, each contributed by international experts to provide comprehensive coverage of the field of organ-on-a-chip technologies and their current and future applications. Section I provides detailed descriptions of the various technologies that have been developed, adapted, and applied in organ-on-a-chip systems, including microfabrication, 3D bioprinting, 3D cell culture techniques, biosensor design and microelectronics, microfluidics, and data collection and predictive analysis. Section II details specific tissue types that have been developed for disease modeling and drug discovery applications, including lung, liver, heart, skin, and kidney "on-a-chip" as well as recent progress in designing an entire "body-on-a-chip" system. Section III covers current and potential future applications of these systems, including achievements in modeling tissue-specific cancers, metastasis, and tumor microenvironments as well as applications for the evaluation of drug efficacy and toxicity and for the prediction of the outcomes of various disease treatment strategies.

SECTION I: TECHNOLOGIES

In Chapter 1, Prafulla Chandra, Carlos Kengla, and Sang Jin Lee present the current status of microfabrication strategies and techniques used in the emerging organ-on-a-chip field. Microfabrication techniques discussed include photolithography, soft lithography, etching (dry etching and wet etching), thin-film coating, two-photon excitation, 3D laser microfabrication, microdispensing, and 3D bioprinting. Strategies to fabricate increasingly complex biological components into 3D functional tissue are discussed as well as the challenges that need to be overcome to facilitate the widespread application of these technologies.

In Chapter 2, Ivy L. Mead and Colin E. Bishop provide an overview of the 3D cell culture field, discussing the utility and limitations of 2D cultures before going in depth about 3D cell culture strategies and the challenges associated with the future of cell culture. The authors highlight the importance of the microenvironment for the 3D cell cultures and provide a detailed description about the several critical components of 3D culture such as scaffolds, hydrogels, and spheroids. The authors also touch on the use of bioreactors and microfluidic chips for 3D culture, highlighting the specific methods and challenges associated with such supporting technologies.

Suichi Takayama and co-workers have contributed Chapter 3, dedicated to the development and application of electrochemical sensors that provide a sensitive and long-term measurement capability within organ-on-a-chip systems. Microfluidic incorporation of sensors that can accurately detect physiological states, such as temperature, pH, oxygen, carbon dioxide, nitric oxide, and glucose concentrations, are covered in this review of the field. Development of these sensor technologies are facilitating the precise control of fluidic flow and improved fabrication of micro- and nanostructures as well as more accurate representations of *in vivo* physiological states for the purpose of medical diagnostics.

Chapter 4, contributed by Panupong Jaipan and Roger Narayan, is a detailed and comprehensive description of the field of microfluidics and their impact on current development of biomedical research. Beginning with a general overview of microfluidic technology, the chapter covers important aspects of microfluidic design and control, specifically discussing transport, fluid flow, dispersion, mixing, separation, and electrokinetic and electro-osmosis phenomena. The chapter also highlights the essential components of microfluidic devices, including various types of microvalves and micropumps. Finally, the chapter provides commentary on multiple applications of microfluidics for organ-on-a-chip technology as well as in the fields of tissue engineering, biosensing, and drug delivery.

Chapter 5 by Andre Kleensang, Alexandra Maertens, and Thomas Hartung provides an interesting overview of the use of "Big Data" that can now be generated from biological systems. The chapter describes the applications and challenges involved in the various "-omics" technologies, as well as strategies to ensure the conclusions based on these platforms are predictive and consistent with validated mechanisms.

Chapter 6, contributed by Peter Ertl, covers the use of "lab-on-a-chip" systems for biomedical applications. Topics covered include the various materials, components, and sensing strategies used in biomedical microfluidic applications, and the use of 2D and 3D live cell microarray technologies. Some examples of lab-on-a-chip technologies discussed include single cell manipulation, immunoassays, stem cell cultivation, and applications.

SECTION II: ORGANS-ON-CHIPS

In Chapter 7, Joan E. Nichols and co-workers have compiled an extensive overview of the background, progress, and future directions of using microchip models of the respiratory tract and lung for disease modeling, drug discovery, and personalized medicine. This chapter covers the construction and scaling of respiratory models and includes discussion of the various cell sources and scaffolds used in these models. The authors provide a comprehensive discussion of current 2D and 3D models, the use of microfluidics to support these models, and some specific applications of this technology, including for modeling infectious disease, lung cancer, and for developing personalized therapies.

Chapter 8, contributed by Aleksander Skardal, describes the current state of liver- and liver cancer-on-a-chip systems and their potential for disease modeling and drug screening and evaluation. The chapter discusses the various concepts, approaches, and challenges involved in the development of 3D liver organoid models and provides multiple examples of 3D *in vitro* liver models and their applications. The incorporation of these models into microfluidic systems containing various fluidic support mechanisms and biochemical sensors is also covered in depth. Finally, Skardal reviews the current strategies to model liver cancer *in vitro,* including investigation of metastasis, epithelial-to-mesenchymal transition, tumor microenvironment, and pharmaceutical screening approaches.

Chapter 9 by Megan L. McCain describes the development of human-relevant, biomimetic, heart-on-a-chip platforms to improve our ability to study and predict the function of human heart tissue. The chapter describes the architecture, biological components, and function of heart tissue and highlights the need to mimic the essential structural and functional features of human heart tissue to provide new tools for disease modeling and cardiotoxicity screening. The various design parameters required for heart-on-a-chip platforms, including cell sources, biomaterials, and scaffolds and accessible readouts of electrical and contractile function, are covered along with some of the remaining challenges and future directions of this field.

Chapter 10, contributed by Claire G. Jeong, provides an interesting overview of the current efforts to develop complex 3D tissue–engineered skin equivalents achieving both functional and cosmetic satisfaction. The chapter highlights the importance of the structure and function of our skin as the most complex and largest organ serving as the primary protective physical barrier against the external environment. The current state of the art for engineering skin equivalents is described along with the use of novel biofabrication technologies, such as 3D bioprinting, spheroids, and microfluidic systems for producing improved *in vitro* skin models that recapitulate critical features and responses of normal or diseased human skin.

Chapter 11, contributed by Erica P. Kimmerling and David L. Kaplan, provides an overview of the functional components and anatomy of the kidney as well as descriptions of various cell sources available for generating *in vitro* kidney tissue models. Models discussed include glomerular tissue models and microfluidic renal cell culture systems with functional metrics for disease models and nephrotoxicity modeling. Applications of these systems such as 3D organoid models of polycystic kidney disease and evaluation of treatments suspected of halting cyst progression are also covered, as well as 3D tissue culture systems specifically designed for testing drug-induced nephrotoxicity.

Mahesh Devarasetty and co-workers introduce readers to the concept of the "body-on-a-chip" in Chapter 12, where multiple tissue-type organoids are connected through a biomimetic circulatory system, analogous to the vascular system of the human body. These systems bring together the technologies of 3D biofabrication, stem cells, tissue organoids, and biosensors and functional evaluation to recapitulate the physiological inter-organ metabolic relationships of the human body in both normal and disease tissue states. Multiple applications of this approach are discussed, including evaluation of how drug metabolites produced by one organoid can affect the function of another, as well as modeling of inter-organ tumor metastasis.

SECTION III: APPLICATIONS

In Chapter 13, Aleksander Skardal and co-workers describe their cutting-edge research to combine multiple organs within the same microfluidic device to model a simple organism-on-a-chip for drug and therapeutic studies. This chapter describes several examples of how integrated multi-organoid systems can model complex multi-organ interactions, including for applications such as cancer metastasis, drug testing and toxicology, and disease modeling. The authors discuss the importance of considering deployment of multi-organoid platforms over single organoid systems and describe strategies for organoid biofabrication, microfluidic hardware integration, common media development, and miniaturization and high-throughput platform development.

Chapter 14 by Ran Li, Michelle B. Chen, and Roger D. Kamm describes the current applications of microfluidic technologies to study processes and mechanisms relating to cancer diagnosis, progression, and treatment. Detailed discussion is provided on the ability to use microfluidics control factors such as growth factor, cytokine and extracellular matrix environmental cues, oxygen tension, cellular and mechanical cues, stiffness, and flow. Detailed examples of how various microfluidic systems can be applied to study the metastatic cascade, tumor angiogenesis, tumor-endothelial interactions and intravasation and extravasation are presented. The chapter concludes with an interesting discussion of the potential and future challenges of these systems to deepen our fundamental understanding of cancer biology and enable the discovery of new drug targets to combat metastatic progression.

In Chapter 15, Pierre-Alexandre Vidi and Sophie A. Lelièvre present a thorough description of cancer-on-a-chip applications. They describe how on-a-chip models that recapitulate normal and neoplastic breast tissues are ideal to study diseases such as breast cancer. The chapter also explores a central application of these on-a-chip models for drug screening, discussing the advantages of these models compared to other 3D culture systems as well as their limitations. Various cancer-on-a-chip models are described and important parameters involved in studying clinically relevant mechanisms such as cancer metastasis are highlighted. Finally, the possibility of breast-on-a-chip systems to serve the purpose of precision medicine and the role of biosensors integration in screening improvement are assessed.

In Chapter 16, Uta Grieshammer and Kelly A. Shepard discuss the current approaches used for *in vitro* disease modeling and for drug discovery and development. The chapter focuses on the application of pluripotent stem cell populations to explore disease mechanisms of pathogenesis, with specific examples from monogenic, complex, and infectious diseases. The authors detail several examples of the use of *in vitro* models for drug discovery and development and highlight some of the limitations and challenges in the current approaches. The chapter also provides some insight into how novel techniques and strategies can overcome these approaches, including the use of multiple functional cell types, use of 3D organoid culture, and incorporation of microfluidic devices to produce organ-on-a-chip systems for disease modeling. Several examples of currently used organ-on-a-chip systems for disease modeling and drug discovery are detailed.

In Chapter 17, Harry Salem and co-workers provide an interesting chapter discussing the various technologies that are currently being applied for predicting human pharmacology and toxicology. This chapter provides a timeline of how pharmacology and toxicology techniques have evolved over the course of decades, describing *in vivo* and recent *in vitro* and stem cell technology applications in the field. These programs are applied to study metabolic components, drug effects, and drug metabolism, to cite a few of the many examples provided in this chapter.

This book concludes with Chapter 18, written by Elisa Cimetta and co-workers, which provides an in-depth analysis of the developing field of "personalized medicine" and how organ-on-a-chip systems are contributing to this new approach to medicine. This chapter covers recent progress in the development of personalized medicine approaches, as well as technical, business, regulatory, ethical, and legal considerations. Various examples of personalized therapies are discussed, including those currently used clinically as well as those in development. Some examples include personalized cell and tissue therapies and personalized drug testing platforms.

Editors

Sean V. Murphy, PhD, is assistant professor of regenerative medicine at the Wake Forest Institute for Regenerative Medicine, Winston-Salem, North Carolina. He earned his bachelors' degree from the University of Western Australia and PhD from Monash University, Melbourne, Australia. His research focuses on developing regenerative medicine and tissue engineering strategies to establish and improve clinical treatments for lung disease. These strategies include cell therapies to restore normal function to lung tissue and minimize inflammation and scarring associated with disease, use of 3D bioprinting to fabricate new airway tissues for transplantation, and lung-on-a-chip technologies for disease modeling and drug discovery. Dr. Murphy is currently the associate editor of the journal *Bioprinting*, is on the editorial boards of multiple journals, including *Stem Cells Translational Medicine*, and is director, secretary, and founder of the Perinatal Stem Cell Society. Dr. Murphy has published over 30 peer-reviewed journal articles, multiple book chapters and reviews, and has received numerous awards and fellowships, most notably from the American Lung Association and the American Australian Association.

Anthony Atala, MD, is director of the Wake Forest Institute for Regenerative Medicine, Winston-Salem, North Carolina. He led the team that grew the first lab-grown organ to be implanted into a human. In addition to being involved in many of the field's top journals and books, he has devoted the last several decades of his career to the development of novel stem cells derived from the amnion, and to sustainable organs grown from a patient's own cells. Dr. Atala is a recipient of many awards, including the U.S. Congress–funded Christopher Columbus Foundation Award, bestowed on a living American who is currently working on a discovery that will significantly affect society; the World Technology Award in Health and Medicine; and the Edison Science/Medical Award. In 2011, he was elected to the Institute of Medicine of the National Academy of Sciences, and in 2014 was inducted to the National Academy of Inventors as a Charter Fellow. His work has been listed twice by *Time Magazine* as the top medical breakthrough of the year, and he was named by *Scientific American* in 2015 as one of the world's most influential people in biotechnology.

Contributors

Lissenya B. Argueta
Department of Microbiology and Immunology
University of Texas Medical Branch at Galveston
Galveston, Texas

Anthony Atala
Wake Forest Institute for Regenerative Medicine
Wake Forest School of Medicine
Winston-Salem, North Carolina

Sarindr Bhumiratana
EpiBone Inc.
Brooklyn, New York

Colin E. Bishop
Wake Forest Institute for Regenerative Medicine
Wake Forest School of Medicine
Winston-Salem, North Carolina

David Brown
School of Medicine
University of Texas Medical Branch at Galveston
Galveston, Texas

Daniel Carmany
Excet, Inc.
Edgewood Chemical Biological Center
Aberdeen Proving Ground, Maryland

Prafulla Chandra
Wake Forest Institute for Regenerative Medicine
Wake Forest School of Medicine
Winston-Salem, North Carolina

Michelle B. Chen
Department of Mechanical Engineering
Massachusetts Institute of Technology
Cambridge, Massachusetts

Joyce Han-Ching Chiu
Department of Biomedical Engineering
University of Michigan
and
Biointerfaces Institute
University of Michigan
Ann Arbor, Michigan

Elisa Cimetta
Assistant Professor
Department of Industrial Engineering, Padua University, Padova, Italy

Joaquin Cortiella
Department of Anesthesiology
University of Texas Medical Branch at Galveston
Galveston, Texas

Rodney Daniels
Department of Pediatrics and Communicable Diseases
University of Michigan
and
Michigan Center for Integrative Research in Critical Care
University of Michigan
Ann Arbor, Michigan

Mahesh Devarasetty
Wake Forest Institute for Regenerative Medicine
Wake Forest School of Medicine
Winston-Salem, North Carolina

Russell Dorsey
Department of the Army, RDECOM
Edgewood Chemical Biological Center
Aberdeen Proving Ground, Maryland

Adrienne Eastaway
School of Medicine
University of Texas Medical Branch at Galveston
Galveston, Texas

Peter Ertl
Faculty of Technical Chemistry
Vienna University of Technology
Vienna Austria

Steven D. Forsythe
Wake Forest Institute for Regenerative Medicine
Wake Forest School of Medicine
Winston-Salem, North Carolina

Uta Grieshammer
University of California, San Francisco
San Francisco, California

Thomas Hartung
Center for Alternatives to Animal Testing (CAAT)
Johns Hopkins Bloomberg School of Public Health
Baltimore, Maryland
and
University of Konstanz
Konstanz, Germany

Panupong Jaipan
Department of Materials Science and Engineering
North Carolina State University
Raleigh, North Carolina

Claire G. Jeong
Wake Forest Institute for Regenerative Medicine
Wake Forest School of Medicine
Winston-Salem, North Carolina
and
GlaxoSmithKline (GSK)
Complex In Vitro Models (Organoid Strategy Team)
Integrated Biological Platform Sciences
R&D Platform Technology & Science
King of Prussia, PA

Roger D. Kamm
Departments of Mechanical Engineering and Biological Engineering
Massachusetts Institute of Technology
Cambridge, Massachusetts

David L. Kaplan
Department of Biomedical Engineering
Tufts University
Medford, Massachusetts

Carlos Kengla
Wake Forest Institute for Regenerative Medicine
Wake Forest School of Medicine
Winston-Salem, North Carolina

Ge-Ah Kim
Biointerfaces Institute
University of Michigan
and
Department of Materials Science and Engineering
University of Michigan
Ann Arbor, Michigan

Erica Palma Kimmerling
Department of Biomedical Engineering
Tufts University
Medford, Massachusetts

Andre Kleensang
Centers for Alternatives to Animal Testing (CAAT)
Johns Hopkins Bloomberg School of Public Health
Baltimore, Maryland

Michael Lamprecht
EpiBone Inc.
Brooklyn, New York

Sang Jin Lee
Wake Forest Institute for Regenerative Medicine
Wake Forest School of Medicine
Winston-Salem, North Carolina

Sophie A. Lelièvre
Department of Basic Medical Sciences
Purdue University
West Lafayette, Indiana

Ran Li
Department of Biological Engineering
Massachusetts Institute of Technology
Cambridge, Massachusetts

Alexandra Maertens
Centers for Alternatives to Animal Testing (CAAT)
Johns Hopkins Bloomberg School of Public Health
Baltimore, Maryland

Megan L. McCain
Department of Biomedical Engineering
Viterbi School of Engineering
University of Southern California
Los Angeles, California

Ivy L. Mead
Maine Molecular Quality Controls
SACO, Maine

Sean Murphy
Wake Forest Institute for Regenerative Medicine
Wake Forest School of Medicine
Winston-Salem, North Carolina

Roger Narayan
Department of Biomedical Engineering
North Carolina State University
Raleigh, North Carolina

Joan E. Nichols
Departments of Internal Medicine, Infectious Disease, Microbiology and Immunology
University of Texas Medical Branch at Galveston,
Galveston Texas

Jean A. Niles
Departments of Internal Medicine, Infectious Disease
University of Texas Medical Branch at Galveston
Galveston, Texas

Mario Rothbauer
Faculty of Technical Chemistry
Vienna University of Technology
Vienna, Austria

Harry Salem
Department of the Army, RDECOM
Edgewood Chemical Biological Center
Aberdeen Proving Ground, Maryland

Kelly A. Shepard
California Institute for Regenerative Medicine
San Francisco, California

Thomas D. Shupe
Wake Forest Institute for Regenerative Medicine
Wake Forest School of Medicine
Winston-Salem, North Carolina

Aleksander Skardal
Wake Forest Institute for Regenerative Medicine
Wake Forest School of Medicine
Winston-Salem, North Carolina

Michael Smith
School of Medicine
University of Texas Medical Branch at Galveston
Galveston, Texas

Shay Soker
Wake Forest Institute for Regenerative Medicine
Wake Forest School of Medicine
Winston-Salem, North Carolina

Shuichi Takayama
Department of Biomedical Engineering
University of Michigan
and
Biointerfaces Institute
University of Michigan
and
Michigan Center for Integrative Research in Critical Care
University of Michigan
and
Department of Macromolecular Science and Engineering
University of Michigan
Ann Arbor, Michigan

Nina Tandon
EpiBone Inc.
Brooklyn, New York

Stephanie P. Vega
Departments of Internal Medicine, Infectious Disease, Microbiology and Immunology
Galveston, Texas

Pierre-Alexandre Vidi
Department of Cancer Biology
Wake Forest School of Medicine
Winston-Salem, North Carolina

David Wartmann
Faculty of Technical Chemistry
Vienna University of Technology
Vienna Austria

Nafissa Yakubova
EpiBone Inc.
Brooklyn, New York

Section I

Technologies

1 Microfabrication and 3D Bioprinting of Organ-on-a-Chip

Prafulla Chandra, Carlos Kengla and Sang Jin Lee

CONTENTS

1.1 INTRODUCTION

Organ-on-a-chip devices are microfabricated, biomimetic systems that are designed to model physiological functions of living tissues and organs *in vitro* [1–3]. These biomimetic systems, which would contain cells and extracellular matrix (ECM) components from human tissues, will recapitulate the microarchitecture and functions of the living tissues or organs. They will also be a valuable tool for multiple applications such as testing the effects of drugs on human organs, toxicity testing, disease modeling and could also revolutionize new drug discovery if adopted by the pharmaceutical industry. The paradigm used by pharmaceutical companies for new drug discovery and development is becoming obsolete due to the high cost, huge investment of money, and time (typically 15–20 years for a single drug discovery, development, and testing). Additionally, the preclinical animal testing process often fails to closely predict drug/toxicity responses in humans. One of the purposes of developing organ-on-a-chip devices is to overcome the current limitations.

Development of organ-on-a-chip devices relies on two core techniques: microfabrication, which can create microstructures for controlled cell and tissue organization and function, and microfluidics, which allows for delivery of nutrients, chemicals, biological factors, and so on in small amounts of fluids to cells and tissues, in a controlled microenvironment. A typical design for an organ-on-a-chip device can include a clear flexible polymer-based framework, or "chip," in which microfluidic channels are embedded and lined by cells derived from various human tissues. Other designs would have multiple microtissue structures (organoids) housed in the chip and linked by microfluidic channels. To mimic human-like physiology, this organ-on-a-chip can be integrated with an automated instrument, which can control fluid flow to the cells/tissues, including nutrients and test compounds, and permit real-time monitoring of biochemical functions of each tissue component.

Microfabrication, which includes several techniques such as photolithography, soft lithography, 3D printing, and so on, can be used to create micrometer-scale features in a variety of materials [4]. The ability to control surface topography and miniaturization of features are the novelties of microfabrication, which enables fabrication of miniaturized structures and devices with biologically-relevant scaled features. Implants and devices produced using these methods are able to fit into small, tissue-relevant dimensions and microenvironments that cells experience *in vitro* and *in vivo*. Recent advancements in microfabrication for *in vitro* applications include formation of static structures, such as wells and flow channels; moving parts, such as diaphragms and cantilevers; electrical devices, such as transistors and resistors, and biologically active surfaces, such as coated proteins and cells. Biosensing microdevices have been fabricated that can be implanted *in vivo* to detect a variety of biomolecules and conditions through mechanisms such as enzymatic-based electrochemical reactions [5] and sensor displacement from mechanical loading [6].

It is well known that micro- and nanoscale structures can lead to increased cellular adhesion, proliferation, differentiation, organization, and overall integration with the host [7,8]. Moreover, mechanical cues on the micro- and nanoscale are known to influence cellular phenotypes [9,10]. A more recent microfabrication technique called electron beam (E-beam) photolithography can be used to achieve nanometer-scale resolution of features [11,12]. 3D bioprinting is an emerging microfabrication technology being used for creating biological scaffolds, 3D cellular microenvironments, tissues, and even organs using biomaterials, ECM components, living cells, and bioactive factors [13–15]. Potential applications of 3D bioprinting include development of *in vitro* tissues, organoids, or biomimetic systems for toxicity testing, drug evaluation, drug discovery, and disease modeling. However, the most important clinical use of 3D bioprinting would be for regenerative medicine, including development of tissue constructs that can repair or replace injured or diseased body parts [16]. In this chapter, an overview of common microfabrication methods is provided, along with ways in which microfabrication is being applied to develop organ-on-a-chip platforms. 3D bioprinting is discussed in detail, with particular reference to its application for organ-on-a-chip applications.

1.2 MICROFABRICATION: FABRICATION METHODS

The most common microfabrication methods include photolithography, soft lithography, etching, microcontact printing, and thin film coating. More recent methods that are still evolving are two-photon excitation, 3D laser microfabrication, and 3D printing. In the field of tissue engineering and regenerative medicine, 3D bioprinting is becoming a valuable tool for fabricating clinically applicable tissue constructs, in shape and size, for reconstruction of damaged or diseased body parts [13]. Figure 1.1 provides an overview of some microfabrication methods relevant to on-chip platform development.

1.2.1 PHOTOLITHOGRAPHY

Photolithography is a popular microfabrication technique because of the high resolution and variety of patterns that can be created [17]. In this technique, a substrate material such as glass or silicone is coated with a layer of a photosensitive organic polymer, called a photoresist, followed by creation of a photomask, using glass or other transparent material. The desired pattern (to be transferred to the substrate) is created using an opaque material. The photomask is then placed on top of the photoresist and irradiated with UV light, thus exposing the transparent regions of the photoresist. In the case of a positive photoresist, the exposed regions break down on exposure to UV light and can be solubilized (and removed) in a developing solution; while with a negative photoresist, the photoresist polymer will become cross-linked upon exposure to UV light, thereby rendering it insoluble in the developing solution, while the rest of the photoresist polymer can be washed away. Depending on the application, the resulting photoresist patterns protects the covered substrate from subsequent etching or deposition of compounds or biomolecules on its surface. Using this method, both high resolution and a variety in patterns can be created.

1.2.2 SOFT LITHOGRAPHY

Soft lithography is a microelectromechanical system (MEMS) derivative technology that requires fabrication of a master pattern, which is then used to form subsequent molds. The initial patterns can be fabricated using several methods, such as glass etching, photo masks, photoresist film processing, or micromachined parts [18]. Poly(dimethyl siloxane) (PDMS) is a popular material for making the molds, particularly for creating fine feature resolution that is biologically relevant, such as for patterning of proteins or cells [19]. Microstamping, microfluidic patterning, and stencil patterning are the different types of soft lithography techniques. In stencil patterning, a master photoresist pattern (master template) is created and PDMS is added to it. However, the PDMS is prevented from covering the entire master template by adding the PDMS to a thickness that is lower than that of the features in the master template or by placing a barrier on the master template to prevent full coverage by PDMS. The end result is a PDMS mold that has holes in the pattern of the master template [64].

In microstamping [20], a soft stamp is made from a polymer (such as PDMS). The stamp is then inked with alkanethiols, silanes, alkylsilanes, or ECM proteins (similar to a traditional stamp and ink) and placed in contact with a surface.

	Photolithography	Wet etching	Dry etching	Physical vapor deposition	Chemical vapor deposition	STL/SLS/2PE	3D bioprinting
Advantages	• Robust and mature process • Re-usable masks • Repeatable • Techniques available for various length scales	• Cost-effective • Batch process • Atmospheric conditions	• Isotropic etch capable • High aspect ratios	• Deposit thin films of metals • Incorporate reactive gases for deposition of ceramic films • Batch processes	• Deposit dielectric materials • Various dopants • Batch processes depending on system	• Many material options including metals • Various high-resolution options depending on system • High-quality commercial systems	• Spatially place cells and biomaterials • Use of multiple cell types • Use of multiple biomaterials
Limitations	• 2D patterns • Transfer to 3D can be difficult especially with high aspect ratios • Not a biocompatible process	• Low control • Anisotropic • Hazardous reagents • Not a biocompatible process	• Costly equipment • No or small batch • Hazardous reagents • Not a biocompatible process	• Costly equipment • Difficult for curved substrates • Not a biocompatible process	• Costly equipment • Hazardous reagents • Not a biocompatible process	• Generally cell incorporation is by manual cell seeding • Typically not a biocompatible process	• Most systems are custom • Limited number of biomaterials for printing • Different modalities have resolution limitations
Applications	• Pattern generation for: • Integrated circuits • Optical circuits • Microfluidics • Specialized topography	• Pattern transfer • Surface preparation • Electronics	• Pattern transfer • Surface preparation • Steep wall fabrication • Microfluidics • Electronics • Optical waveguides • Destructive testing	• Electrical components for integrated circuits • Optical components like mirrors or thin films • Surface coating for hardware	• Dielectric components for integrated circuits • Optical cladding	• Scaffold fabrication for tissue engineering • Implant fabrication • Medical model production	• Tissue engineering • Complex 3D culture modeling • Body-on-a-Chip modeling • Scaffold fabrication

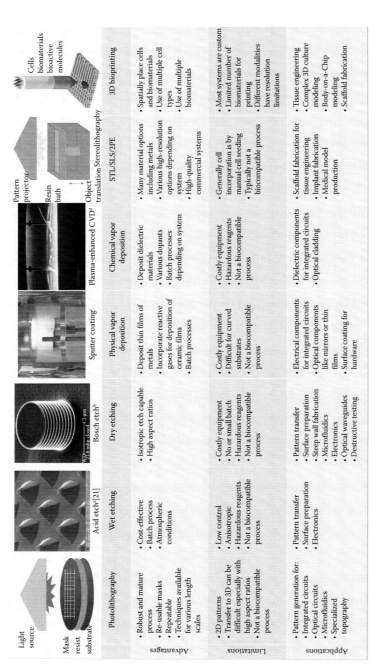

FIGURE 1.1 Processes used for microfabrication relevant to on-chip platform development. Each fabrication process has specific uses, advantages, and limitations depending on the design features needed in the device. (a) from Chen, Y.-C. et al. (2012). (b) https://en.wikipedia.org/wiki/Deep_reactive-ion_etching#/media/File:Bosch_process_PILLAR.jpg. (c) http://www.mtixtl.com/images/products/detail/PlasmarDetail01.jpg. and (d) http://www.mse.ncsu.edu/research/nanotobio/htm/DSCF2095.jpg.

Microcontact printing is a powerful method for patterning cells on scaffolds, creating medical devices or diagnostic platforms, and also for studying cell–ECM interactions [22].

Microfluidic-type soft lithography patterning utilizes a PDMS mold to create microchannels against a substrate. These microchannels are then used to pattern fluid materials onto a substrate. One unique feature of these microchannels is that they can maintain separate fluid streams through a single channel because of laminar flow. Patterning of cells for tissue engineering applications has also been carried out using this technique [23].

1.2.3 Etching (Dry Etching and Wet Etching)

Etching is a process by which topographical features are created on a surface by selective removal of material through physical or chemical means. Microfabrication through etching can be divided into two types: dry etching, which is carried out via gas-phase chemistry, and wet etching, which is done using liquid chemicals. In general, wet etching is more selective than dry etching. Materials such as silicon can be etched up to depths ranging from sub-micrometer to about 10 μm using dry etching. An advanced type of dry etching technology called deep-reactive–ion etching (DRIE) is revolutionizing microfabrication as it has the capability to make very deep and narrow structures in silicon, which will be valuable for many biological applications [24]. Common materials used for wet etching include silicone, glass, and thin films, while solutions of hydrofluoric acid, sulfuric acid, and nitric acid are commonly used for etching silicone. Wet etching of glass or other silicates is performed using a hydrofluoric acid-based chemistry.

1.2.4 Thin Film Coating

Thin film coating is a microfabrication method that can be divided into two separate categories: physical vapor deposition (PVD) and chemical vapor deposition (CVD) [25]. E-beam evaporation and sputter coating are the two primary technologies used for PVD. In E-beam evaporation, the target is placed in the system which uses magnetic fields to guide the E-beam to hit the target. The substrate is placed in a holder that typically rotates or orbits in order to increase film thickness uniformity, while sensors located near the substrate measure coating rates and pressure conditions as the process is conducted under vacuum. The E-beam energy and low pressure conditions atomize the target material, which flies off in all directions but is directed toward the substrate using a controllable shutter. The coating is achieved by controlling the rate of deposition and total time exposure to the material spray.

Sputter coating is similar to E-beam coating in that there is a target and a substrate, along with shutter control; however, the difference is the atmosphere and the method of atomizing. Sputtering takes place in a low-pressure system flooded with a noble gas, typically argon, where strong electromagnetic fields are used to form plasma in front of the target, blasting the surface with the noble gas nuclei. This bombardment dislodges atoms of the target material, which fly across the chamber to deposit on the substrate. Other gases can also be used, allowing atoms like oxygen or nitrogen to be incorporated into the film.

CVD is achieved by controlled exposure to vapors of liquid chemicals, typically hydrocarbons and silanes. The liquids are vaporized in the deposition chamber and energized to generate conditions for film growth. Silicon, diamond, and other materials can be deposited via CVD processes.

1.2.5 Two-Photon Excitation and 3D Laser Microfabrication

Two-photon excitation (TPE) microfabrication is a type of 3D microfabrication technique that is becoming popular for making prototypes [26,27]. In this method, a light curable resin is polymerized by simultaneously absorbing two photons at longer wavelength, usually in the red-infrared (IR) spectral region. The TPE process has many advantages, including better fabrication accuracy (around 100-nm lateral spatial resolution) compared to the single photon process, deeper penetration of the laser into the materials to induce polymerization at the desired depth without affecting other parts of the material, and fine control of 3D spatial resolution. Recently, TPE has also found application in fabrication of microfluidic devices [28], cell adhesive surfaces [29], and tissue scaffolds [30].

Another related technique, 3D lithographic microfabrication, is being used for biomedical applications. This method is based on a technique called depth-resolved widefield illumination (DRWFI) [31]. Multibeam 3D laser microfabrication is a method in which a laser is used for polymerization of a material (similar to stereolithography), but here the laser-assisted polymerization can start anywhere in the material (not necessarily from the surface layer) and is completed using multiphoton pinpoint addressing [32]. The lasers used as an irradiation source in the polymerization process range from extreme UV to near-infrared (NIR) wavelengths and are either pulsed (at nanosecond, picosecond, or femtosecond widths) or continuous wave.

1.2.6 Microdispensing

Microdispensing refers to the generation of small drops of solutions, typically in the nanoliter range [33]. These droplets are dispensed using a printing system that has nozzles capable of generating these droplets and then depositing them on a target surface without the nozzle ever contacting the surface (noncontact).

One of the methods to create these micron-sized droplets is by using piezoelectrically actuated inkjet dispensers. This process is called inkjet dispensing or inkjet printing. Inkjet printing technology can be used to dispense fluid droplets with diameters of about 20 to 200 μm (approximately 5 pl–5 nl, depending on the material) at very high rates. The high uniformity and small size of the droplet created makes this technology desirable for creating micron or submicron-sized features for fabricating precision microstructures, depositing materials in MEMS and BioMEMS devices, creating biomolecular arrays for high throughput analysis, and so forth. Inkjet-based printing is also flexible, requires no tooling, and is data-driven, as the structures can be directly printed using computer-aided design (CAD) models. For biological applications, piezoelectric-based inkjet printing technology can offer accurate and high-throughput deposition of biomaterials, biopolymers, bioactive fluids, and cells for applications such as combinatorial chemistry, toxicology, microarrays (genomics, gene expression, and proteomics), pharmaceutical (drug discovery/development), clinical (diagnostics and biosensors), coatings on medical devices, and regenerative medicine (cell patterning and tissue engineering).

1.2.7 3D Bioprinting

Three-dimensional printing, also referred to as additive manufacturing (AM) or solid free-form fabrication (SFF), is a manufacturing technique in which objects are created by depositing and fusing materials, such as plastics, powders, ceramics, or metals, in a layer-by-layer process to produce a 3D object [34]. In 1986, Charles W. Hull described a technology for 3D fabrication of objects that he named stereolithography, which is now one of the several fabrication modalities called 3D printing. Hull sequentially printed thin layers of a material that can be cured with UV light to form a multilayered solid 3D structure. This process was later adapted to create biologically-relevant scaffolds, incorporate cells in hydrogels, and create *in vivo*-like tissue constructs using methods popularly referred to as 3D bioprinting [13,35].

The ultimate goal in tissue engineering and regenerative medicine is to repair or replace dysfunctional tissues and organs with bioengineered tissues and organs for restoring structure and function in the body. In 3D bioprinting, biological components such as cells, ECM proteins, and bioactive molecules are printed layer-by-layer, with great spatial control to form the required 3D biological structures. Researchers are now aiming to develop living 3D functional human tissues and organs by replicating the tissue or organ structure in the body, a process called biomimicry. Examples of biomimicry include replicating the vasculature and branching patterns of the vascular tree in a tissue construct, and recreating physiologically accurate biomolecular gradients as seen in differentiating tissues. A detailed understanding of the target tissue microenvironment, including ECM organization, cell types and arrangement, gradients of soluble or insoluble factors, and so forth, will be valuable in designing the 3D bioprinting strategy. Such constructs will have greater clinical relevance for restoration of structure and function in diseases or damaged tissues and organs.

There are several types of 3D bioprinting technologies; the more commonly used ones are inkjet bioprinting, direct-write bioprinting, and laser-assisted printing [13]. Inkjet bioprinting is carried out using inkjet printers, also known as drop-on-demand printers [36]. Here, controlled volumes of liquid are delivered to predefined locations based on digital designs (2D or 3D) created on a computer. The first inkjet-based bioprinters were actually modified, commercially available, 2D ink-based printers in which the ink in the cartridge was replaced with a biological material and printing along a z axis (along with x and y axes) was added to the printer function.

Today's inkjet-based 3D bioprinters are custom-designed machines that can print synthetic polymers and biological components, including cells, at high precision, speed, and resolution. In this printing mechanism, the ink dispensing technologies are based on acoustic and thermal systems that force microsized drops of liquid out of the nozzle and onto a substrate [37,38]. 3D objects are constructed by addition of multiple layers. Using thermal inkjet printers, a variety of biological components, such as DNA and cells, have been printed. Some of the disadvantages of inkjet printing technology include thermal and mechanical stress to the printing components, nonuniform droplet size, and frequent nozzle clogging. Piezoelectric-based inkjet printers are free from many of these problems and are preferred for 3D printing of biological components [39].

The biological components, such as ECM proteins, cells, hydrogels, or polymers used in the printers for fabrication are often referred to as *bioinks*. The bioink cartridges in piezoelectric-based inkjet printers are fitted with a piezoelectric crystal, which expands and relaxes on application of an electrical signal (waveform), forcing droplets of fluid out of the nozzle tip. A variety of biological materials in solutions (including cells) and low-viscosity materials (such as some polymer solutions and hydrogels) can be printed using this technology.

Direct-write (extrusion-based) bioprinting technology seems to be the most versatile 3D printing technology, where finely controlled extruded materials are deposited onto a substrate by physical contact, using a microextrusion head [13,16,40]. Direct-write bioprinters have a temperature-controlled material-handling and dispensing system, as well as a stage capable of movement along the x, y, and z axes. Some systems also have multiple bioink cartridges that facilitate serial or parallel dispensing of multiple materials without the need for changing the print head. As the name suggests, extrusion-based direct-write bioprinters dispense materials continuously, and the newly dispensed material has to physically contact the previous layer in order to build the 3D structure. The print pattern is directed by the CAD file, which is converted into instructions for printing using a computer-aided manufacturing (CAM) and printer-specific software. The main advantage of extrusion-based bioprinting is the ability to dispense high viscosity materials (such as polymer melts and thick hydrogels) and also deposit cells at very high densities. In addition to cells, spheroids containing multiple cell types can also be used in extrusion-based 3D bioprinting where these spheroids would be allowed to self-assemble into higher-order 3D tissue structure in the printed construct [41].

A major requirement for 3D bioprinting is to find suitable materials that are not only compatible with the printing process and the biological components, particularly cells, but that can provide the desired mechanical and functional properties to the structure being printed. However, the effects of the bioprinted material and material processing on cell viability must be investigated. Studies ought to be conducted to ensure biocompatibility of material components and processing conditions. Material components may include cross-linking or gelling agents, while process conditions may include humidity, operating temperature, or ultraviolet light exposure for crosslinking.

Currently, the structural materials that are most commonly used for 3D bioprinting are synthetic polymers, such as poly(lactic acid) (PLA), poly(lactic-co-glycolic acid) (PLGA), poly(ε-caprolactone) (PCL), poly(ethylene glycol) (PEG), and naturally-derived polymers, such as collagen, fibrin, alginate, and hyaluronic acid. The naturally-derived polymers are mostly ECM components that are present in body tissues, which have inherent biological properties. Synthetic polymers can be tailored with specific physical and mechanical properties to suit particular requirements for 3D bioprinting. For bioprinting of cells, hydrogels made of collagen type I, alginate, agarose, gelatin, and fibrin have been successfully used to build 3D tissue structures with microscale patterns [14,42].

During 3D bioprinting, precise deposition with a high level of spatial and temporal control is considered an important property of a material. Thus, the viscoelastic and thermal properties of printing materials are some of the most important parameters for printing live cells. The choice of printing materials could be significantly influenced by the material's properties to protect cell viability during the printing process, as most of printing methods involve high-shear stress to the material while passing through a fine nozzle for printing [43,44]. Additional considerations for 3D bioprinting include printing parameters, such as material flow rate, nozzle diameter, light intensity, distance of print head from the substrate, and linear write speed. The total time required for each printing process is another variable to consider while designing and printing larger structures containing live cells. Extended periods of exposure to nonoptimal temperatures or humidity conditions can negatively affect both material properties and cell viability [45,46].

Laser-assisted bioprinting (LAB) is based on the principles of laser-assisted forward transfer (LAFT), which was originally developed for transfer of metals [47,48]. This technology has now been successfully applied to printing biological materials, such as DNA, peptides, and cells [47,49,50]. A typical LAB device consists of a source of pulsed laser beam, a focusing system, a "ribbon," usually made from glass that is covered with a laser-absorbing layer (either titanium or gold), a layer of biological material to be printed, and a substrate. Focused laser pulses on the absorbing layer of the ribbon generate a high-pressure bubble that propels the biological material toward the collector substrate.

Another approach for creating clinically relevant 3D tissues is creation of micro- or mini-tissues/organs (organoids). The concept of organoids is relevant because many tissues and organs are comprised of smaller structural and functional building blocks [51]. Using design and 3D bioprinting technologies, organoids can be fabricated and assembled into larger constructs which would resemble the target tissue or organ. The organoids can also be created by promoting self-assembling of cell units, which are then assembled into a macrotissue [41,52,53]. These organoids are particularly useful for *in vitro* applications or for developing diagnostic devices, such as an "organ-on-a-chip," where the organoids representing different organs are placed on a single platform and are connected using microfluidic channels [54,55]. A higher-order of fabricating a tissue for *in vitro* testing, such as toxicity testing, would be to create organs using 3D bioprinting [56]. Organ-on-a-chip technology would be a valuable tool not only as a screening platform for drugs, toxic compounds, or vaccines; it can also serve as *in vitro* models for understanding the pathophysiology of diseases [57,58].

1.3 MICROFABRICATION FOR *IN VITRO* BIOLOGICAL MICROENVIRONMENTS

Over the past half century, the semiconductor industry has developed powerful manufacturing tools that have enabled development and parallel fabrication of highly efficient electronic devices with large numbers of integrated components. Biomedical researchers have been inspired by the power of this manufacturing approach and have been attempting to bring the same level of scaling to biology and medicine. Photolithography, soft lithography, etching, and film deposition are the most important microfabrication techniques that have been adapted for biomedical applications [17,18,24,25]. Also, higher resolution photolithographic techniques, such as x-ray and e-beam lithography, are being used to create smaller feature size and more intricate topography that are biologically relevant.

The major advantages of microfabrication are miniaturization, high surface-to-volume ratio, small sample volumes, high throughput, geometric control, and integration with electronics for real-time analysis. In addition, miniaturized devices are portable and can be placed in constrained spaces. Examples include active implantable medical devices (AIMDs) such as cardiac pacemakers, defibrillators, stimulators (for bladder, diaphragm, and sphincter), drug administration devices, sensors, cochlear implants, and implantable active monitoring devices. Decreasing sample volumes can also be beneficial for many applications. In drug discovery applications, use of smaller volumes can reduce reagent volume, assay waste, and consequently result in huge cost savings. Miniaturization can lead to fabrication of high-throughput devices, particularly in the fields of drug discovery and genomic research. Microfabricated channels for capillary electrophoresis and nucleic acid arrays are some examples where miniaturization has increased throughput several fold [59,60].

Microfabrication can offer geometrical control of microstructures where patterns of varying geometries can be built in the same space with micrometer dimensional accuracy using techniques such as photolithography. Bhatia et al. used microfabrication to precisely control the spatial organization of fibroblasts and hepatocytes cells in co-cultures and investigated cell–cell interactions [61]. This system allowed them to discover that albumin production was localized to hepatocytes at the heterotypic interface and liver-specific function was dependent on the amount of heterotypic interface in the co-culture. It is also possible to integrate electronic components into microfabricated systems because the processes for conventional semiconductor fabrication and microfabrication of biomedical devices are very similar [62]. The integrated circuits on the device would then amplify the recorded signals, process the data, and give an output of signals coming from cortical neurons. Microfabrication techniques allow for the integration of optics and fluidics as well [63].

Microfabrication has a great deal of relevance for organ-on-a-chip platforms. The microscale channels can direct fluid flow through the chip to areas of interest for biosensing or for chemical reaction, while passive layers can protect active layers or create openings for interaction within the whole structure wherever necessary. Integrating electronic and optical probes with a microfluidic sample delivery system can be carried out using advanced microfabrication techniques and will be useful in the design and manufacturing of organ-on-a-chip devices.

1.4 APPLICATIONS OF MICROFABRICATION IN ORGAN-ON-A-CHIP

Microfabrication techniques have not only allowed for miniaturization of already existing medical devices, but have also aided in the creation of completely new devices for biomedical applications, including organ-on-a-chip platforms. While photolithographic etching methods (which are used to manufacture computer microchips) are now being used to create biological "chip" platforms, a combination of microfabrication techniques are being used to control sizes, shapes, and patterns of surface feature in these devices on a scale that cells sense in their natural microenvironments. The ultimate goal of using this approach is to build a miniaturized tissue or organ system that recapitulates the structure and functions of the full size native tissues or organs (Table 1.1). For example, a microfluidic chamber that is lined with a single type of cultured cell such as hepatocytes or lung epithelial cells and performs specific functions of that tissue type would represent a single organ-on-a-chip [54].

A higher level of design would involve two or more microfluidic channels which are separated by porous membranes and where each channel is lined by different cell types as present at the interfaces between different tissues. For a multiple organ-on-a-chip design, specific chambers are lined with cells from different organs and each chamber is linked using channels. A specially created, miniaturized organ housed in a separate chamber and connected by microfluidic channels to mimic *in vivo* physiological structure and interactions between different organs is known as a body-on-a-chip [2,3,64,65]. To replicate tissue and organ more closely, recent organ-on-a-chip designs have incorporated cells embedded in ECM gels to create a 3D microenvironment [68], and multicellular constructs have been created using different types of tissue engineering techniques, including 3D bioprinting [54,65,66]. For example, it has been possible to create epithelial or endothelial tissues that stably express differentiated functions in models of kidney [67], liver [68], brain [69], heart [70], skeletal muscle [71], and intestine [71]. On a higher order of function, microfabrication has allowed for integrating polarized epithelium with living vascular endothelium or stromal cell–containing connective tissue in 3D microfluidic devices that represent the eye [73] and brain [74].

Organs in the human body are complex systems, with different types of cells and tissues that form complex tissue–tissue interfaces. They also have multimodular structures consisting of smaller, repeating functional units that individually perform major characteristic functions of the whole organ (e.g., gas exchange in the alveoli of the lung, absorption in the villi of the gut, etc.). Therefore, to develop a useful tissue- or organ-level device for *in vitro* analysis of complex human physiology, it is necessary to mimic the complex physical microenvironment in which cells are normally situated and also reproduce the physiological process.

Replica molding techniques have been used to build a PDMS microdevice that mimics the complex structure of the endothelial–epithelial interface in liver sinusoid [69]. Similarly, two stacked microfabricated PDMS chambers that were separated using a thin porous membrane were used to culture rat renal tubular epithelial

TABLE 1.1
Applications of Microfabrication Methods for Development of *In Vitro* Biological Systems and Organs-on-a-Chip

Tissue/Organ	Fabrication Methods	Platform Design	Applications
Bone marrow	Lithography, spin-coating, and laser cutting [75]	Bone marrow-on-a-chip	Radiation-induced toxicity effects [75]
Brain	Multilayer lithography [76]; lithography and microfluidics [77]; photolithography, spin-coating, and PDMS molding [69,78]	Chip-based microfluidic device mimicking cerebral vasculatures; blood-brain barrier-on-a-chip	Drug permeability testing through the blood brain barrier [76]; analysis of microfluidic blood-brain barrier device to replicate permeability and cell characteristics [77]; use of TEER measurements to test functionality of blood-brain barrier-on-a-chip [69]; axon diodes [78]
Cancer/tumor	PDMA molding [79]; soft lithography, PDMS molding, and oxygen plasma treatment [80]; replica molding and optical lithography [81]; soft lithography, PDMA molding, and coating [82]	Cells-on-a-chip; microfluidic device	Studying chemotherapy resistance using a lung cancer microfluidic model [79]; observing the effects of various drugs on breast cancer cells through a droplet microfluidic device [80]; EIS biosensor for prostate cancer screening [81]; extravasation of cancer cell sthrough an endothelial barrier [82]
Heart	Microcontact printing and soft lithography/PDMS [83]; spin-coating [70]; PDMS molding and oxygen plasma treatment [84]	Heart-on-a-chip; cardiac muscle-on-a-chip	Reconstituting Barth's syndrome through a functional heart-on-a-chip [83]; structure–function relationships in laminar cardiac muscle [70]; ischemia/reperfusion injury analysis [84]
Intestine	Multilayer lithography [85]; lithography, spin-coating, and etching [72]	Cells in a microfluidic device; microfabricated membranes	Testing drug permeability in the intestinal epithelial cell membrane through use of microhole trapping [85]; polymer membranes for body-on-a-chip devices and functional model for potential integration in a body-on-a-chip design [72]

(Continued)

TABLE 1.1 *(Continued)*
Applications of Microfabrication Methods for Development of *In Vitro* Biological Systems and Organs-on-a-Chip

Tissue/Organ	Fabrication Methods	Platform Design	Applications
Kidney	Photolithography, replica molding, and PDMS bonding [86]; photolithography, PDMA molding, and oxygen plasma treatment [67]; PDMS molding [87]	Kidney-on-a-chip; cells in a microfluidic device	Testing the effects of drugs on the proliferation of a dynamic kidney microfluidic chip [86]; collecting duct-on-a-chip to investigate the effects of changes within a renal environment [67]; modeling epithelial-to-mesenchymal transition in proximal tubules [87]
Liver	Photolithography and spin-coating [88]; PDMS molding [89]; photolithography, spin-coating, PDMS molding, and oxygen plasma treatment [90]; hhotolithography and PDMS molding [91]	Cells in a microfluidic device; microtissue	3D HepaTox chip for hepatotoxicity testing [88]; integrated insert in a dynamic microfluidic platform to evaluate dynamic organ-organ interactions of the liver and intestine [89]; artificial liver sinusoid for primary hepatocyte culture [90]; making 3D hepatic microtissues [91]
Lung	Photolithography, PDMS casting, spin-coating, and chemical etching [92]; PDMS molding, and chemical etching [93]	Lung-on-a-chip	Modeling disease functions through a chip undergoing cyclic mechanical strain [92]; testing the transport properties with physiological "breathing" chip design [93]
Spleen	Photolithography [94]	Spleen-on-on-a-chip	Fast-flow/slow-flow device for testing deformability of RBC in the spleen model [94]
Vasculature	Micromilling, PDMS molding, and oxygen plasms treatment [95]	Microfluidic chip with tubular cell structures	Functional analysis: using tissue-engineered microenvironment systems to replicate vasculature environment [95]

(Continued)

TABLE 1.1 *(Continued)*
Applications of Microfabrication Methods for Development of *In Vitro* Biological Systems and Organs-on-a-Chip

Tissue/Organ	Fabrication Methods	Platform Design	Applications
Skeletal muscle	Photolithography, spin-coating, deep reactive ion etching, and plasma enhanced chemical vapor deposition [71]	Muscle on silicon cantilever	Drug/toxicity testing: toxin detection [71]; muscle performance enhancement [71]
Multiple organ	Replica molding, plasma oxidation [96]	Multiple cells-on-a-chip	Drug/toxicity testing: human-on-a-chip fabrication for drug testing [96]
Multiple organ	3D bioprinting [53,56]	Body-on-a-chip	Drug/toxicity testing: organoid-based body-on-a-chip [53,56]

cells on the upper surface of the membrane, and represented a simplified kidney model for analysis of kidney transport barrier functions [68].

As another example, Huh et al. recently created a human lung-on-a-chip platform with alveolar-capillary–like interface of the lung air sac which could recapitulate the mechanism of breathing in the human lung [57]. This lung-on-a-chip consisted of a 3D microfluidic device with two parallel microchannels, separated by a thin porous ECM coated PDMS membrane [97]. The flexible PDMS membrane in the central channel could be mechanically stretched and relaxed by application of cyclic suction to the two hollow chambers that enclosed the microchannels. Cyclic vacuum was applied to the side hollow chambers to induce outward bending or inward recoil of the elastic walls of the central PDMS microchannels and repeating this actuation cycle to produce ~10% cyclic strain. This way, it was possible to generate dynamic mechanical distortions similar to those observed in the air sac of the living human lung. This is an interesting example of using microfabrication to produce a functional lung-on-a-chip.

Recently, it has been demonstrated that the rational use of microfabrication methods is a brain-on-a-chip system that has asymmetric microchannels to mimic the complex, oriented neuronal networks and tissue–tissue interfaces seen in the brain [78]. Here, two cell culture chambers were separated by an array of gradually narrowing funnel-shaped microchannels that formed selective filters to enable unidirectional axonal growth.

Another chip device, microfabricated elastomeric valves, were used to allow precise positioning and co-culture of hippocampal neurons and glial cells for several weeks [98]. By culturing primary rabbit corneal epithelial cells on a thin collagen gel membrane suspended across a PDMS microchannel, a microfluidic cornea model was developed to create a fully biological tissue–tissue interface [73].

This microengineered, microfluidics-based cornea-on-a-chip device can be used for screening ocular irritants, as was demonstrated by permeability assays to detect corneal epithelial damage induced by sodium hydroxide.

Cells and tissues are exposed to different kinds of biochemical gradients in their native microenvironment and attempts to model this system in an *in vitro* environment can only be possible because of novel uses of microfluidics. Kim et al. have created such a system, a human gut-on-a-chip [99]. The microfluidics system was used to exert cyclic mechanical strain and fluid flow onto cultured human Caco-2 intestinal epithelial cells, and this stimulus results in spontaneous formation of villi-like undulated structures that exhibited normal intestinal barrier functions. This gut-on-a-chip was also used for long-term co-culture of the bacteria *Lactobacillus* sp. (a normal gut microflora) along with host epithelial cells, and it was found that intestinal barrier function was enhanced and was similar to that seen *in vivo*.

An example of generating mechanical cues through microfluidics is mechanical distortion of cells in a lung-on-a-chip to reproduce ventilator-induced lung injury seen in many patients [100]. Jang and Suh used a kidney-on-a-chip device to demonstrate that in addition to inducing morphological polarization, application of the shear stress also resulted in the formation of normal transport functions in the epithelium, as seen in healthy human kidneys [67].

In additional to mechanical cues, microfabrication has been used to apply electrical stimulation to cultured cells and analyze cellular responses. In devising a heart-on-a-chip, Grosberg et al. cultured neonatal rat ventricular cardiomyocytes on micropatterned (with ECM proteins) thin, flexible PDMS films that induced myogenesis [70]. Electrical stimulation was then applied to this tissue, which generated contractile stresses. This heart-on-a-chip system can provide a robust platform to measure the effects of drugs and their cardiotoxicity in response to pharmacological stimulation.

Engineering of biochemical microenvironments is possible using microfabrication methods, which can create biomimetic microsystems *in vitro*. In a study by Allen et al., oxygen gradients generated in a perfused bioreactor caused regional variations in hepatocyte function along the liver sinusoids—a characteristic of normal liver zonation *in vivo* [101]. Microengineered perfusion systems could enhance hepatic transport [102] and maintain viability of 3D liver tissues [103]. Unlike 3D static cultures, the microfluidic-based chip platform enables design and control over many system parameters by integrating microsized sensors into the system. This would allow real-time analysis of a broad array of physiological processes in the cultured cells or tissues in the chip platform. Some examples where microsensor-based chips are being used to analyze physiological processes include measurement of tissue barrier integrity [104], cell migration [105], oxygen, glucose, pH [106], and fluid pressure [107] measurements. Microchannel geometry can also be optimized to enhance oxygen and nutrient delivery to cells in organoids within the organ-on-a-chip, thereby increasing survival and functionality.

Another use of microfabrication for controlling cellular interactions and response in a device is by creating nanoporous membranes [108] and posts [109], which would separate the cells from the fluid flow path, thereby independently controlling exposure of chemical gradients to individual cell types and also minimizing fluid shear stress on cells [90]. In a microfabricated liver-on-a-chip, the natural endothelial barrier of the liver sinusoid was recreated by culturing primary human hepatocytes in an area that was separated from the active flow channel by microchannels with around 30-μm inner diameters. This device was then used for analyzing the hepatoxicity of the drug diclofenac [90].

More importantly, microfabricated organ-on-a-chip platforms can be used to model disease pathologies for understanding mechanisms and testing therapies. Torisawa et al. developed a multilayered microfluidic device in which cancer cells were patterned in a microchannel at spatially defined positions relative to cells that secreted chemoattractants (source cell) and cells that scavenged the chemokines (sink cells) [109]. This system allowed modeling of chemotaxis of cancer cells during metastasis. Renal interstitial fibrosis, which occurs during development of proteinuria nephropathy, was recently recreated using a kidney-on-a-chip platform, where exposure of renal epithelial cells to complement C3a containing culture media induced apoptosis in these cells [87]. Heart ischemia has been recreated using an organ-on-chip device, using a dynamic variation of oxygen tension [84]. Such devices and pathological models can be useful for screening drug candidates capable of treating this life-threatening condition.

1.5 3D BIOPRINTING OF ORGAN-ON-A-CHIP

To build any physiologically relevant organ-on-a-chip platform, it is important to first create small tissue- or organ-like structures (organoids) that have structural and functional characteristics similar to the native tissue or organ being represented. Different types of systems have been developed as organs-on-a-chip. These include some simple ones containing a perfused microfluidic chamber with a single type of cultured cell (e.g., cardiomyocytes or hepatocytes), and some complex ones that have two or more cell types lined along porous membranes and connected with microchannels. All these systems have been useful in studying some aspects of human tissues *in vitro*, but they still do not truly represent the complex structure and function of native tissues and organs. To develop these platform systems, advanced tissue engineering techniques are needed, and 3D bioprinting represents one such highly capable platform.

The advantages of the 3D bioprinting technologies (mostly direct-write bioprinting) are that multiple biological components (biomaterials, cells, or bioactive factors) can be printed simultaneously in a precise manner to mimic the native tissue or organ structures, and subsequently its specific functions [13–15]. These technologies would better control the geometric and compositional structure of the tissue- or organ-specific organoids to provide more anatomical and functional similarity to human tissues or organs (Figure 1.2).

In addition, 3D bioprinting technologies would be used to deliver multiple organoids-in-a-chip to mimic the human body's response to harmful chemical and biological agents and to develop potential treatments for such exposure [110]. One can envision that in future multi-organ physiological systems using 3D bioprinting capacity, it might be possible to use the heart-on-a-chip to actively and naturally

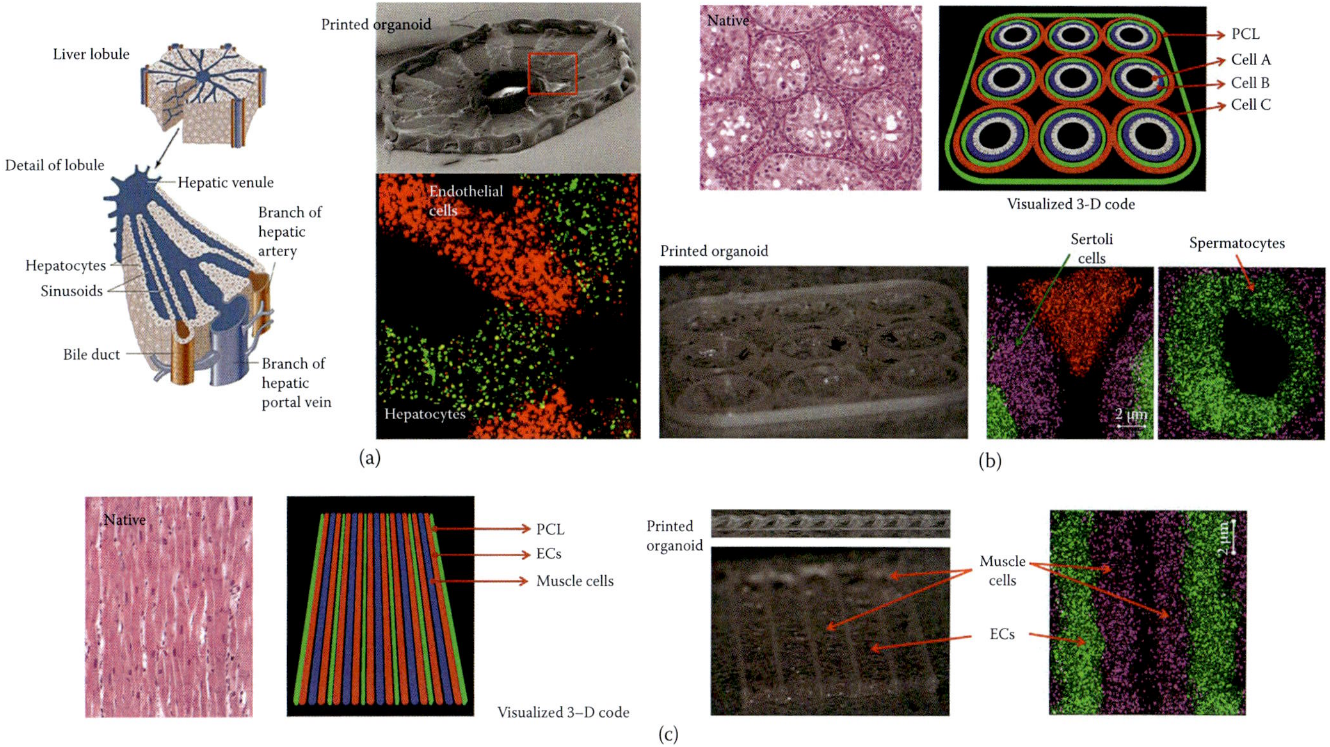

FIGURE 1.2 Bioprinting of tissue- or organ-specific structures: (a) liver, (b) testis, and (c) cardiac muscle organoids.

pump fluid though the whole system, while using the lung-on-a-chip to oxygenate, liver-on-a-chip to metabolize, kidney chip to cleanse the circulating blood substitute, and so on. In the near-term, it is likely that external pumps would be used to drive fluid flow and control variations in fluid flow and dynamics that are required to maintain viability of many different cells linked within a single, integrated, chip-based, multi-organ physiological test system.

1.6 CURRENT CHALLENGES

Several challenges need to be overcome before organ-on-a-chip platforms can find widespread use in clinical laboratories, medical facilities, and pharmaceutical and biopharmaceutical companies. Fabrication of organs-on-a-chip requires precise microengineering capabilities, such as uniformly seeding cells in the microfluidic channels and controlling cell–ECM and cell–cell interactions to achieve optimal relationships between tissue structure and function. Also, continuous perfusion of the system may erode (or degrade) thin ECM coatings over time, thereby affecting cell survival, and formation of bubbles in the microfluidic channels can damage cell layers and hamper the function of the chip.

Another technical challenge is the availability of human-derived cells in enough numbers, their incorporation into the chip, and maintaining their viability for the duration of the chip's use. In the case of macroscale 3D culture systems, it would be a challenge to carry out high-resolution imaging or analyze processes of inner parts of the tissue in the chip. Possible solutions to this include integrating these chips with on-line analytical tools such as microfluorimetry, electrode arrays, or even a fluorescence microscope.

Organ–organ interactions add another layer of complexity to multiple organ-on-a-chip platforms, mainly in scaling of the results from *in vitro* testing using cells or miniaturized tissues to the original size of the human tissue or organ *in vivo* [111,112]. Hence, the practical usefulness of organ-on-a-chip systems would depend on addressing this scaling effect and devising a way to accurately correlate the analysis data.

Another challenge for creating a clinically useful organ-on-a-chip is technical robustness of the system. Various factors, such as cells, transport of nutrients and biological factors, gradient maintenance, and fluidic control must work together to achieve optimal function. Improvements in chip design should address factors that are relevant to the function of the organs, such as recreating the microscale ecosystem for cells, inducing appropriate mechanical forces, efficient permeation rates, and so forth. For a body-on-a-chip device, where tissues representing different organs of the human body are linked together to function as a single system, a "universal blood substitute" is also required that can support all tissues, similar to what blood does in the human body. A significant hurdle here is that culture media is usually optimized for specific tissues, and some media are also selective in the use of externally added serum, growth factors, and other additives. For example, in a study examining the use of growth factors in an organ-on-a-chip device that included cells from the kidney, lung, and liver, it was found that addition of TGF-β enhanced the function of one cell type but inhibited the function of another [113].

Creating a system for time-release of growth factors (or drugs) or devising a universal culture medium to support multiple cell types can be a promising approach to integrating multiple tissue types on a single chip platform. However, recreating some *in vivo* tissue-specific functions, such as capturing the sustained production of metabolic enzymes or directional biliary ductal clearance might still be difficult in a liver-on-a-chip that is made today. Finally, the different types of organ-on-a-chip being currently developed (or those that will be developed in the future) will require reproducibility and reliability as an analytical platform. Simplification of microfabrication methods for mass production and automation of device operation are requirements so that they can be used reliably as commercial analytical platforms and point-of-care diagnostic devices.

1.7 FUTURE OUTLOOK

Microfabricated organs-on-a-chip are emerging as a powerful platform not only for *in vitro* testing of chemicals, toxic compounds, and biological agents, but also for the study of human pathophysiology and tissue development. Many pioneering efforts have developed devices which mimic *in vivo*–like microenvironments to model human tissue structures and physiology outside the body.

Future improvements in organ-on-a-chip designs would include incorporation of nanoscale sensors and molecular reporters that would enable extension of in-line and time-lapse monitoring of multiple tissues or organs for parameters such as temperature, pressure, fluid flow, glucose, pH, oxygen, and so on [111,113]. Simultaneous analysis and scaling-up of sample numbers would increase the statistical significance of results obtained from these chips. Controlling long-term survival and enhancing differentiation and function of cells in the chip can be achieved by controlling fluid flow, and this would enable testing or analysis to be carried out on clinically relevant time scales for chronic pathophysiological conditions which are usually for a duration of several weeks. Controlling fluid flow in organs-on-a-chip can enable modeling of absorption, distribution, metabolism, and excretion (ADMET) properties of drugs, which can predict their pharmacokinetic/pharmacodynamic properties more reliably. A combination of mathematical scaling and clearance dynamics data is already making this possible [114]. Pharmaceutical and biotechnology companies can derive the greatest value from organ-on-a-chip platforms if clinically relevant pharmacokinetic/pharmacodynamic modeling protocols can be developed for these systems [114]. In addition to validation and prioritization of lead drug candidates, these companies can use the organ-on-a-chip platforms to determine dosing and safety margins of drugs for clinical trials and also for analyzing drug toxicities and alternate effects. Identification of new biomarkers of disease response, drug efficacy, and toxicity are other advantages of using organs-on-a-chip during clinical testing of pharmaceutical compounds.

The development of organs-on-a-chip is still in its infancy. New avenues for drug discovery and personalized medicine can open up if this technology can effectively recapitulate a broad range of responses to drugs and toxins that organs in the body generate. Other advancements that could popularize the use of organ-on-a-chip devices include automation of sample collection and processing, feedback controls, programmable monitoring, and so on.

Fundamental processes that can be addressed and analyzed using this platform system include tissue development and regeneration, differentiation, disease process, immune response, infection, and so on. However, complex physiological systems such as the nervous system, adaptive immune response, and pathophysiology of chronic diseases might still be difficult to evaluate using the currently available organ-on-a-chip devices. For example, in the case of liver-on-a-chip, it has still not been possible to integrate a biliary outflow tract into these devices or to sustain physiological function of the liver in these chips for a long-term period. Only simple physiological processes, such as production of albumin, clotting factors (liver-specific proteins), effects of ischemia, regional hypoxia and nutritional deprivation, and energy metabolism, can currently be determined. However, with the rapid advancement taking place in this field, analysis of complex physiological and pathophysiological phenomena may soon be a possibility. If multiple parameters such as precise position of cells, molecular and oxygen gradients, and mechanical forcing can be applied independently and simultaneously with real-time analysis of molecular-scale events in a chip platform, this will represent the true power of microfabrication and microsystem engineering.

Synthetic *in vitro* models of the whole body are being created by generating functional units of organs with functional interfaces and placing them in a single, integrated chip platform. A useful advancement in such a system would be to replace inert channels with an endothelium-lined, microfluidic vasculature to create a true human "body-on-a-chip" which can be integrated with automated, time-lapse microscopic imaging and different types of sensors for analyzing multiple parameters simultaneously. It is highly likely that future organ-on-a-chip systems will have the capabilities to produce testing outcomes that are equivalent to or better than those obtained from animal testing. Organ-on-a-chip devices incorporating human cell-derived functional structure would offer real-time, quantitative analysis of the physiological response to drugs, toxins, and biological agents.

ACKNOWLEDGMENT

This study is supported by Armed Forces Institute of Regenerative Medicine (X81XWH-08-2-0032).

REFERENCES

1. Huh D, Hamilton GA, Ingber DE. From 3D cell culture to organs-on-chips. *Trends Cell Biol* 2011; 21: 745–754.
2. Huh D, Torisawa, YS, Hamilton GA, et al. Microengineered physiological biomimicry: Organs-on-chips. *Lab Chip* 2012; 12: 2156–2164.
3. Van der Meer AD, van den Berg A. Organs-on-chips: Breaking the *in vitro* impasse. *Integr Biol* 2012; 4: 461–470.
4. Betancourt T, Brannon-Peppas L. Micro- and nanofabrication methods in nanotechnological medical and pharmaceutical devices. *Int J Nanomed* 2006; 1(4): 483–495.

5. Hasan A, Nurunnabi M, Morshed M, et al. Recent advances in application of biosensors in tissue engineering. *Biomed Res Int* 2014; 2014: 307519.
6. Ainslie KM, Desai TA. Microfabricated implants for applications in therapeutic delivery, tissue engineering and biosensing. *Lab Chip* 2008; 8(11): 1864–1878.
7. Khademhosseini A, Langer R, Borenstein J, Vacanti JP. Microscale technologies for tissue engineering and biology. *Proc Nat Acad Sci U S A* 2006; 103: 2480.
8. Jung DR, Kapur R, Adams T, et al. Topographical and physicochemical modification of material surface to enable patterning of living cells. *Crit Rev Biotech* 2001; 21(2): 111–154.
9. Lopez CA, Fleischman AJ, Roy S, Desai TA. Evaluation of silicon nanoporous membranes and ECM-based microenvironments on neurosecretory cells. *Biomaterials* 2006; 27(16): 3075–3083.
10. Saha K, Pollock JF, Schaffer DV, Healy KE. Designing synthetic materials to control stem cell phenotype. *Curr Opin Chem Biol* 2007; 11: 381.
11. Kolodzie CM, Maynard HD. Electron-beam lithography for patterning biomolecules at the micron and nanometer scale. *Chem Mater* 2012; 24(5): 774–780.
12. Vigneswaran N, Samsuri F, Ranganathan B, Padmapriya. Recent advances in nano patterning and nano imprint lithography for biological applications. *Procedia Engineer* 2014; 97: 1387–1398.
13. Murphy SV, Atala A. 3D bioprinting of tissues and organs. *Nat Biotechnol* 2014; 32(8): 773–785.
14. Seol YJ, Kang HW, Lee SJ, et al. Bioprinting technology and its applications. *Eur J Cardiothorac Surg* 2014; 46(3): 342–348.
15. Seol YJ, Yoo JJ, Atala A. Bioprinting of three-dimensional tissues and organ constructs In: Atala A, Yoo JJ, eds. *Essentials of 3D Biofabrication and Translation*. London, UK: Elsevier/Academic Press; 2015; pp. 283–292.
16. Kang H-W, Kengla C, Lee SJ, et al. 3-D organ printing technologies for tissue engineering applications. In: Narayan R, ed. *Rapid Prototyping of Biomaterials: Principles and Applications*. Oxford: Woodhead; 2014; pp. 236–253.
17. Leuschner R, Pawlowski G. Photolithography. *Mat Science Technol* 2013; Available at: http://onlinelibrary.wiley.com/doi/10.1002/9783527603978.mst0258/abstract?.
18. Xia Y, Whitesides GM. Soft lithography. *Angew Chem Int Ed Engl* 1998; 37(5): 550–575.
19. Kane RS, Takayama S, Ostuni E, et al. Patterning proteins and cells using soft lithography. *Biomaterials* 1999; 20(23–24): 2363–2376.
20. Ruiz SA, Chen CS. Microcontact printing: A tool to pattern. *Soft Matter* 2007; 3: 168–177.
21. Chen, Y.-C. et al. The Formation and the Plane Indices of Etched Facets of Wet Etching Patterned Sapphire Substrate. *J. Electrochem. Soc.* 2012; 159: D362–D366.
22. Quist AP, Pavlovic E, Oscarsson S. Recent advances in microcontact printing. *Anal Bioanal Chem* 2005; 381(3): 591–600.
23. D'Arcangelo E, McGuigan AP. Micropatterning strategies to engineer controlled cell and tissue architecture in vitro. *BioTechniques* 2015; 58(1): 13–23.
24. Fu YQ, Collin A, Fasoli A, et al. Deep reactive ion etching as a tool for nanostructure fabrication. *J Vac Sci Technol* 2009; B 27: 1520.
25. Madou M. *Fundamentals of Microfabrication: The Science of Miniaturization*. 2nd ed. New York, NY: CRC Press; 2002; p. 723.
26. Mauro S, Fourkas JT. Recent progress in multiphoton microfabrication. *Laser Photon Rev* 2008; 2: 100–111.
27. Sun HB, Kawata S. Two-photon photopolymerization and 3D lithographic microfabrication. In *NMR-3D Analysis-Photopolymerization*, Vol 170; Advances in Polymer Science Series. Heidelberg, Germany: Springer-Verlag Berlin 2004, pp. 169–273.

28. Stoneman M, Fox M, Zeng CY, et al. Real-time monitoring of two-photon photopolymerization for use in fabrication of microfluidic devices. *Lab Chip* 2009; 9(6): 819–827.
29. Pins GD, Bush KA, Cunningham LP, et al. Multiphoton excited fabricated nano and micro patterned extracellular matrix proteins direct cellular morphology. *J Biomed Mat Res Part A* 2006; 78A(1): 194–204.
30. Claeyssens F, Hasan EA, Gaidukeviciute A, et al. Three-dimensional biodegradable structures fabricated by two-photon polymerization. *Langmuir* 2009; 25(5): 3219–3223.
31. So PT, Kim D. Depth resolved wide field illumination for biomedical imaging and fabrication. *Conf Proc IEEE Eng Med Biol Soc* 2009; 2009: 3234–3235.
32. Applications of femtosecond lasers in 3D machining. In Misawa H, Juodkazis S, eds. *3D Laser Microfabrication: Principles and Applications.* Wiley-VCH Verlag GmbH & Co. KGaA, Weinheim, FRG: 2006; pp. 341–378.
33. Bammesberger S, Ernst A, Losleben N, Tanguy L, Zengerle R, Koltay P. Quantitative characterization of non-contact microdispensing technologies for the sub-microliter range. *Drug Discov Today* 2013; 18(9–10): 435–446.
34. Schubert C, van Langeveld MC, Donoso LA. Innovations in 3D printing: A 3D overview from optics to organs. *Br J Ophthalmol* 2014; 98(2): 159–161.
35. Atala A, Yoo JJ, eds. *Essentials of 3D Biofabrication and Translation.* Academic Press, London, UK, 2015; pp. 1–440.
36. Boland T, Xu T, Damon B, Cui X. Application of inkjet printing to tissue engineering. *Biotechnol J* 2006; 1: 910–917.
37. Derby B. Inkjet printing of functional and structural materials: Fluid property requirements, feature stability, and resolution. In Clarke DR, Ruhle M, Zok F, eds., *Annual Review of Material Research*, 2010; Vol. 40; pp. 395–414.
38. Cui X, Boland T, D'Lima DD, Lotz, MK. Thermal inkjet printing in tissue engineering and regenerative medicine. *Recent Pat Drug Deliv Formul* 2012; 6: 149–155.
39. Sumerel J, Lewis J, Doraiswamy A, et al. Piezoelectric ink jet processing of materials for medical and biological applications. *Biotechnol J* 2006; 1: 976–987.
40. Chang CC, Boland ED, Williams SK, Hoying JB. Direct-write bioprinting three-dimensional biohybrid systems for future regenerative therapies. *J Biomed Mater Res B Appl Biomater* 2011; 98(1): 160–170.
41. Mironov V, Visconti RP, Kasyanov V, Forgacs G, Drake CJ, Markwald RR. Organ printing: Tissue spheroids as building blocks. *Biomaterials* 2009; 30: 2164–2174.
42. Tasoglu S, Demirci U. Bioprinting for stem cell research. *Trends Biotechnol* 2013; 31(1): 10–19.
43. Talbot EL, Berson A, Brown PS, Bain CD. Evaporation of picoliter droplets on surfaces with a range of wettabilities and thermal conductivities. *Phys Re. E* 2012; 85: 061604.
44. Sperling LH. *Introduction to Physical Polymer Science,* Vol XXVII. New York: Wiley; 1992; p. 594.
45. Smith CM, Christian JJ, Warren WL, Williams SK. Characterizing environmental factors that impact the viability of tissue-engineered constructs fabricated by a direct-write bioassembly tool. *Tissue Eng* 2007; 13(2): 373–383.
46. Pepper ME, Cass CA, Mattimore JP, et al. Post-bioprinting processing methods to improve cell viability and pattern fidelity in heterogeneous tissue test systems. *Conf Proc IEEE Eng Med Biol Soc* 2010; 1: 259–262.
47. Guillemot F, Souquet A, Catros S, et al. High-throughput laser printing of cells and biomaterials for tissue engineering. *Acta Biomater* 2010; 6: 2494–2500.
48. Guillotin B, Souquet A, Catros S, et al. Laser assisted bioprinting of engineered tissue with high cell density and microscale organization. *Biomaterials* 2010; 31: 7250–7256.
49. Colina M, Serra P, Fernandez-Pradas JM, et al. DNA deposition through laser induced forward transfer. *Biosens Bioelectron* 2005; 20: 1638–1642.

50. Dinca V, Kasotakis E, Catherine J, et al. Directed three-dimensional patterning of self-assembled peptide fibrils. *Nano Lett* 2008; 8: 538–543.
51. Campbell PK, Jones KE, Huber RJ, Horch KW, Normann RA. A silicon-based, three-dimensional neural interface: manufacturing processes for an intracortical electrode array. *IEEE Trans Biomed Eng*. 1991; 38(8): 758–768.
52. Kelm JM, Lorber V, Snedeker JG, et al. A novel concept for scaffold-free vessel tissue engineering: Self-assembly of microtissue building blocks. *J Biotechnol* 2010; 148: 46–55.
53. Kengla C, Atala A, Lee SJ. Bioprinting of organoids. In: Atala A, Yoo JJ, eds. *Essentials of 3D Biofabrication and Translation*. London, UK: Elsevier; 2015; pp. 271–282.
54. Bhatia SN, Ingber DE. Microfluidic organs-on-chips. *Nat Biotechnol* 2014; 32: 760–772.
55. Caplin JD, Granados NG, James MR, et al. Microfluidic organ-on-a-chip technology for advancement of drug development and toxicology. *Adv Healthc Mater* 2015; 4(10): 1426–1450.
56. Kang HW, Atala A, Yoo JJ. Bioprinting of organs for toxicology testing. In: Atala A, Yoo JJ, eds. *Essentials of 3D Biofabrication and Translation*. London, UK: Elsevier; 2015; pp. 179–186.
57. Huh D, Matthews BD, Mammoto A, Montoya-Zavala M, Hsin HY, Ingber DE. Reconstituting organ-level lung functions on a chip. *Science* 2010; 328: 1662–1668.
58. Sonntag F, Schilling N, Mader K, et al. Design and prototyping of a chip-based multi-micro-organoid culture system for substance testing, predictive to human (substance) exposure. *J Biotechnol* 2010; 148: 70–75.
59. Pei J, Dishinger JF, Roman DL, et al. Microfabricated channel array electrophoresis for rapid characterization and screening of enzymes using RGS-G protein interactions as a model system. *Anal Chem* 2008; 80(13): 5225–5231.
60. Obeid PJ, Christopoulos TK. Microfabricated systems for nucleic acid analysis. *Crit Rev Clin Lab Sci* 2004; 41(5–6): 429–465.
61. Bhatia SN, Balis UJ, Yarmush ML, Toner M. Probing heterotypic cell interactions: Hepatocyte function in microfabricated co-cultures. *J Biomater Sci Polym Ed* 1998; 9(11): 1137–1160.
62. Najafi K. Solid-state microsensors for cortical nerve recordings. *IEEE Eng Med Biol Mag* 1994; 13(3): 375–387.
63. Zhang D, Men L, Chen Q. Microfabrication and applications of opto-microfluidic sensors. *Sensors (Basel)* 2011; 11(5): 5360–5382.
64. Sung JH, Esch MB, Prot JM, et al. Microfabricated mammalian organ systems and their integration into models of whole animals and humans. *Lab Chip* 2013; 13: 1201–1212.
65. Li N, Tourovskaia A, Folch A. Biology on a chip: Microfabrication for studying the behavior of cultured cells. *Crit Rev Biomed Eng* 2003; 31: 423–488.
66. Andersson H, van den Berg A. Microfabrication and microfluidics for tissue engineering: State of the art and future opportunities. *Lab Chip* 2004; 4: 98–103.
67. Jang KJ, Suh KY. A multi-layer microfluidic device for efficient culture and analysis of renal tubular cells. *Lab Chip* 2010; 10: 36–42.
68. Nakao Y, Kimura H, Sakai Y, Fujii T. Bile canaliculi formation by aligning rat primary hepatocytes in a microfluidic device. *Biomicrofluidics* 2011; 5: 022212-1–022212-7.
69. Griep LM, Wolbers F, de Wagenaar B, et al. BBB ON CHIP: Microfluidic platform to mechanically and biochemically modulate blood-brain barrier function. *Biomed Microdevices* 2013; 15: 145–150.
70. Grosberg A, Alford PW. McCain ML, Parker KK. Ensembles of engineered cardiac tissues for physiological and pharmacological study: Heart on a chip. *Lab Chip* 2011; 11: 4165–4173.
71. Wilson K, Das M, Wahl KJ, et al. Measurement of contractile stress generated by cultured rat muscle on silicon cantilevers for toxin detection and muscle performance enhancement. *PLoS One* 2010; 5: e11042.

72. Esch MB, Sung JH, Yang J, et al. On chip porous polymer membranes for integration of gastrointestinal tract epithelium with microfluidic 'body-on-a-chip' devices. *Biomed Microdevices* 2012; 14: 895–906.
73. Puleo CM, McIntosh Ambrose W, Takezawa T, et al. Integration and application of vitrified collagen in multilayered microfluidic devices for corneal microtissue culture. *Lab Chip* 2009; 9: 3221–3227.
74. Park J, Koito H, Li J, Han A. Multi-compartment neuron-glia co-culture platform for localized CNS axon-glia interaction study. *Lab Chip* 2012; 12: 3296–3304.
75. Torisawa YS, Spina CS, Mammoto T, et al. Bone marrow-on-a-chip replicates hematopoietic niche physiology in vitro. *Nat Methods* 2014; 11(6): 663–669.
76. Yeon JH, Na D, Choi K, et al. Reliable permeability assay system in a microfluidic device mimicking cerebral vasculatures. *Biomed Microdevices* 2012; 14(6): 1141–1148.
77. Booth R, Kim H. Characterization of a microfluidic in vitro model of the blood-brain barrier (μBBB). *Lab Chip* 2012; 12(10): 1784–1792.
78. Peyrin JM, Deleglise B, Saias L, et al. Axon diodes for the reconstruction of oriented neuronal networks in microfluidic chambers. *Lab Chip* 2011; 11: 3663–3673.
79. Zhang L, Wang J, Zhao L, et al. Analysis of chemoresistance in lung cancer with a simple microfluidic device. *Electrophoresis* 2010; 31(22): 3763–3770.
80. Gong Z, Zhao H, Zhang T, et al. Drug effects analysis on cells using a high throughput microfluidic chip. *Biomed Microdevices* 2011; 13(1): 215–219.
81. Chiriacò MS, Primiceri E, Montanaro A, et al. On-chip screening for prostate cancer: An EIS microfluidic platform for contemporary detection of free and total PSA. *Analyst* 2013; 138(18): 5404–5410.
82. Jeon JS, Zervantonakis IK, Chung S, et al. In vitro model of tumor cell extravasation. *PLoS One* 2013; 8(2): e56910.
83. Wang G, McCain ML, Yang L, et al. Modeling the mitochondrial cardiomyopathy of Barth syndrome with induced pluripotent stem cell and heart-on-chip technologies. *Nat Med* 2014; 20(6): 616–623.
84. Khanal G, Chung K, Solis-Wever X, et al. Ischemia/reperfusion injury of primary porcine cardiomyocytes in a lowshear microfluidic culture and analysis device. *Analyst (Lond)* 2011; 136: 3519–3526.
85. Yeon JH, Park JK. Drug permeability assay using microhole-trapped cells in a microfluidic device. *Anal Chem* 2009; 81(5): 1944–1951.
86. Baudoin R, Griscom L, Monge M, et al. Development of a renal microchip for in vitro distal tubule models. *Biotechnol Prog* 2007; 23(5): 1245–1253.
87. Zhou M, Ma H, Lin H, Qin J. Induction of epithelial-to-mesenchymal transition in proximal tubular epithelial cells on microfluidic devices. *Biomaterials* 2014; 35: 1390–1401.
88. Toh YC, Lim TC, Tai D, et al. A microfluidic 3D hepatocyte chip for drug toxicity testing. *Lab Chip* 2009; 9(14): 2026–2035.
89. Bricks T, Paullier P, Legendre A, et al. Development of a new microfluidic platform integrating co-cultures of intestinal and liver cell lines. *Toxicol In Vitro* 2014; 28(5): 885–895.
90. Lee PJ, Hung PJ, Lee LP. An artificial liver sinusoid with a microfluidic endothelial-like barrier for primary hepatocyte culture. *Biotechnol Bioeng* 2007; 97: 1340–1346.
91. Li CY, Stevens KR, Schwartz RE, Alejandro BS, Huang JH, Bhatia SN. Micropatterned cell-cell interactions enable functional encapsulation of primary hepatocytes in hydrogel microtissues. *Tissue Eng Part A* 2014; 20(15–16): 2200–2212.
92. Huh D, Leslie DC, Matthews BD, et al. A human disease model of drug toxicity-induced pulmonary edema in a lung-on-a-chip microdevice. *Sci Transl Med* 2012; 4(159): 159ra147.

93. Huh D, Matthews BD, Mammoto A, et al. Reconstituting organ-level lung functions on a chip. *Science* 2010; 328(5986): 1662–1668.
94. Rigat-Brugarolas LG, Elizalde-Torrent A, Bernabeu M, et al. A functional microengineered model of the human spleen-on-on-a-chip. *Lab Chip* 2014; 14(10): 1715–1724.
95. Tourovskaia A, Fauver M, Kramer G, et al. Tissue-engineered microenvironment systems for modeling human vasculature. *Exp Biol Med (Maywood)* 2014; 239(9): 1264–1271.
96. Zhang C, Zhao Z, Abdul Rahim NA, van Noort D, Yu H. Towards a human-on-chip: Culturing multiple cell types on a chip with compartmentalized microenvironments. *Lab Chip* 2009; 9: 3185–3192.
97. Ochs M, Nyengaard JR, Jung A, et al. The number of alveoli in the human lung. *Am J Respir Crit Care Med* 2004; 169: 120–124.
98. Majumdar D, Gao Y, Li D, Webb DJ. Co-culture of neurons and glia in a novel microfluidic platform. *J Neurosci Methods* 2011; 196: 38–44.
99. Kim HJ, Huh D, Hamilton GA, Ingber DE. Microengineered physiological biomimicry: organs-on-chips. *Lab Chip* 2012; 12(12): 2156–2164.
100. Douville NJ, Zamankhan P, Tung YC, et al. Combination of fluid and solid mechanical stresses contribute to cell death and detachment in a microfluidic alveolar model. *Lab Chip* 2011; 11: 609–619.
101. Allen JW, Khetani SR, Bhatia SN. In Vitro Zonation and Toxicity in a Hepatocyte Bioreactor. *Toxicol Sci* 2005; 84: 110–119.
102. Goral VN, Hsieh YC, Petzold ON, et al. Perfusion-based microfluidic device for three-dimensional dynamic primary human hepatocyte cell culture in the absence of biological or synthetic matrices or coagulants. *Lab Chip* 2010; 10: 3380–3386.
103. Domansky K, Inman W, Serdy JA. et al. Perfused multiwell plate for 3D liver tissue engineering. *Lab Chip* 2010; 10: 51–58.
104. Douville NJ, Tung YC, Li R, Wang JD, El-Sayed ME, Takayama S. Fabrication of two-layered channel system with embedded electrodes to measure resistance across epithelial and endothelial barriers. *Anal Chem* 2010; 82: 2505–2511.
105. Nguyen TA, Yin TI, Reyes D, Urban GA. Microfluidic chip with integrated electrical cell-impedance sensing for monitoring single cancer cell migration in three dimensional matrixes. *Anal Chem* 2013; 85: 11068–11076.
106. Eklund SE, Thompson RG, Snider RM, et al. Metabolic discrimination of select list agents by monitoring cellular responses in a multianalyte micro-physiometer. *Sensors (Basel)* 2009; 9: 2117–2133.
107. Liu MC, Shih HC, Wu JG, et al. Electrofluidic pressure sensor embedded microfluidic device: A study of endothelial cells under hydrostatic pressure and shear stress combinations. *Lab Chip* 2013; 13: 1743–1753.
108. Carraro A, Hsu WM, Kulig KM, et al. In vitro analysis of a hepatic device with intrinsic microvascular-based channels. *Biomed Microdevices* 2008; 10: 795–805.
109. Torisawa YS, Mosadegh B, Bersano-Begey T, et al. Microfluidic platform for chemotaxis in gradients formed by CXCL12 source-sink cells. *Integr Biol* 2010; 2: 680–686.
110. Hsu J. Tiny 3-D-printed organs aim for "body on a chip." *Scientific American*; September 16, 2013; Accessed: September 12, 2016. http://www.scientificamerican.com/article/tiny-3d-printed-organs-ai/.
111. Esch MB, Smith AS, Prot JM, et al. How multi-organ microdevices can help foster drug development. *Adv. Drug Delivery Rev* 2014; 6: 158–169.
112. Wikswo JP, Curtis EL, Eagleton ZE, et al. Scaling and systems biology for integrating multiple organs-on-a-chip. *Lab Chip* 2013; 13: 3496–3511.
113. Wikswo JP, Block FE 3rd, Cliffel DE, et al. Engineering challenges for instrumenting and controlling integrated organ-on-chip systems. *IEEE Trans Biomed Eng.* 2013; 60: 682–690.
114. Sung JH, Esch MB, Shuler ML. Integration of in silico and in vitro platforms for pharmacokinetic-pharmacodynamic modeling. *Expert Opin Drug Metab Toxicol* 2010; 6: 1063–1081.

2 Three-Dimensional Cell Culture

Ivy L. Mead and Colin E. Bishop

CONTENTS

2.1 SUMMARY OF THREE-DIMENSIONAL CELL CULTURE: A BRIEF OVERVIEW OF THE FIELD

2.1.1 Legacy and Utility of Two-Dimensional Cell Culture

Two-dimensional cell culture refers to the practice of growing homogenous cell populations, in monolayer, on a plastic culture surface (Figure 2.1). This method of maintaining and expanding cells has been instrumental to expanding our understanding of how cells interact under a wide variety of conditions. This technique has been so successful,

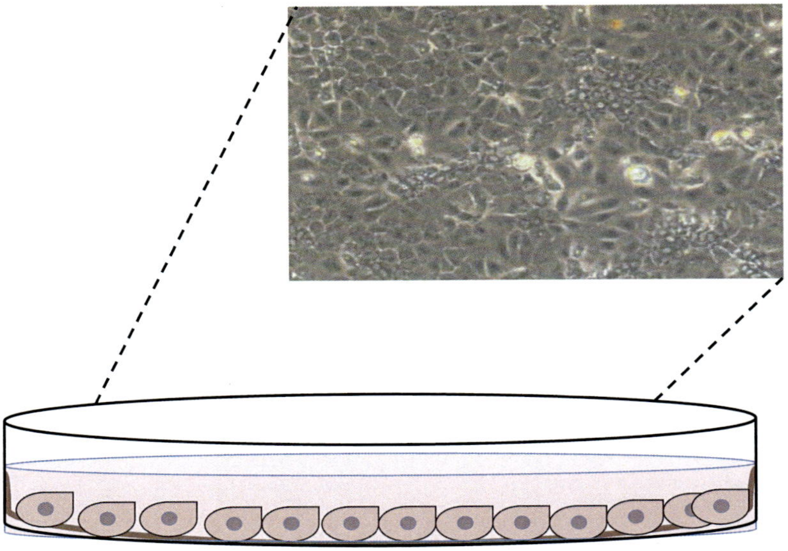

FIGURE 2.1 Illustration of traditional monolayer cell culture. Cells are often seeded in a specific density according to their growth dynamics, particularly rate of proliferation. Monolayer culture can be further supported by the addition of an ECM coating on the surface of the culture plastic. Differentiated HepaRG cell line grown in monolayer is shown in light microscopy image above. Note that visible cell morphology is often visibly different in cells cultured in two dimensions, particularly over time in culture.

in fact, that it is now the bedrock that underlies decades of life sciences research (Tibbitt and Anseth 2009). Two-dimensional cell culture has shaped the way cellular assays have been developed and optimized, and allows for repeatable research. In some cases, the cellular assays can even be automated, making high-throughput testing possible. Cells in this format are also simple to manipulate, allowing for ease of training. In combination with other supportive technologies, such as improved culture medium compositions and methods involving thin surface coatings of extracellular matrix materials over the plastic surface of the culture vessel, monolayer culture has become more sophisticated (Zhang et al. 2009). Some cell types behave similarly to *in vivo* when cultured on traditional, flat surfaces; for example, keratinocytes and corneal epithelial cells are able to self-organize to form complex structures (Suuronen et al. 2005).

2.1.2 Limitations of Two-Dimensional Cell Culture

However, a majority of cell types do not adapt naturally to two-dimensional culture conditions, presenting challenges when attempting to use them as accurate and meaningful models. It is not an ideal configuration for most cell and tissue types, because it is not representative of how those cells would thrive *in vivo*. In monolayer culture, cells are forced to adapt a rigid, flattened morphology as they adhere to the planar surface rather than other cells or extracellular matrix as they would in a native, three-dimensional microenvironment. This tends to cause cells to dedifferentiate and display altered proliferation similar to a cancerous phenotype (von der Mark et al. 1977; Petersen et al. 1992). This alteration is associated with reorganization of the cell cytoskeleton forming stress fibers, as well as changes in how cells interact physically with each other, such as by intracellular junctions (Wong and Gotlieb 1986; Harris et al. 2012). The phenotypic alterations go beyond morphological characteristics, with gene expression changes occurring when the cells are not receiving the signals they would in their native tissue, further altering cell behavior and response (Ghosh et al. 2005).

These significant differences between cells cultured in two dimensions compared to *in vivo* have been a major limiting factor in preclinical drug testing. In this context, monolayer cultures are commonly used to predict drug toxicity to provide a representation of human tissue response that is lacking in animal models (Paul et al. 2010). Primary human hepatocytes serve as an example of a cell type that, when standardly cultured in monolayer on a thin coating of Matrigel or collagen I, does not maintain viability or function sufficient for most applications past a few days (Knobeloch et al. 2012). However, when cultured in three dimensions either by embedding the cells in hydrogel or by the formation of spherical aggregates, hepatocytes are able to retain valuable drug detoxification functionality and viability for longer time points (Berthiaume et al. 1996; Messner et al. 2013). The major distinction is that by providing hepatocytes with a three-dimensional culture environment, they are able to form important cell-cell and cell-extracellular matrix connections as they would in the native liver, leading to polarization and long-term differentiation, preventing cell death (Lee et al. 2013). Two-dimensional cell culture has been a reliable method for studying a variety of aspects of cell biology; however, it may not be ideal for replicating most *in vivo* cell states.

2.1.3 Characteristics of Three-Dimensional Cell Culture

Three-dimensional cell culture provides a vastly improved microenvironment when compared to traditional monolayer methods. In this setting, cells are better able to interact closely with other cells and the surrounding extracellular matrix. This causes considerable changes to cell phenotype and behavior, opening up a whole new context for research. In three dimensions, cells are able to more easily polarize and self-organize to form functional structures similar to *in vivo* (Chen 1997; Abu-Absi et al. 2002). When cell interaction is facilitated in three dimensions it leads to the formation of solute gradients and also allows for intracellular signaling, both of which are more challenging to achieve in monolayer. In addition, cells are able to move more freely, allowing for self-organization, and are able to interact directly with the surrounding extracellular matrix, allowing for mechanisms relying on adhesion, cell contraction, and extracellular matrix remodeling (Huh, Hamilton, and Ingber 2011).

2.1.4 Three Dimensions, The Future of Cell Culture

The possible applications for three-dimensional cell culture are wide-reaching, and these models have already been adopted in many different fields of study.

Three-dimensional cell culture can also be used to model of tissues, specific aspects with potential for use in exploring mechanistic hypotheses. Historically, cancer biology has pioneered and developed many three-dimensional cell culture methods for the study of cancer development; for example, the use of aggregates to simulate aspects of tumor development (Mueller-Klieser 1987). There is a need for *in vitro* models allowing for the stable culture of human cells. Animal models are able to provide invaluable insight into many clinical challenges, however they are not comparable to human responses in all respects, particularly when studying pathogen-specific or drug-specific effects (McGonigle and Ruggeri 2014). In addition, there is a commercial need for less expensive and more high-throughput models for human tissues. Development of models that provide human tissue specificity, tissue complexity and functionality, and high-throughput screening capabilities could aid in decreasing the high cost of new drug development by making the preclinical process more efficient and predictive. Three-dimensional cell culture models also have great potential for clinical contexts by allowing for the production of more complex and functional tissues, which may successfully be used in the replacement or repair of damaged regions. Three-dimensional cell culture models are an invaluable tool for researchers in multiple fields and help bridge the gap between traditional monolayer cell culture and the complexity of animal models.

2.1.5 Challenges Associated with 3D Cell Culture

Several ongoing issues remain in three-dimensional cell culture, some more challenging than others, depending on the culture format. These challenges range from maintaining cell functionality and viability with the addition of more complexity to a lack of techniques capable of analyzing these cultures. Frequently, oxygen availability is a limiting factor in three-dimensional cell culture. Increase in size and complexity of the model, an important feature of this method, must be carefully controlled if necrosis is to be avoided (Asthana and Kisaalita 2012). Another significant set of challenges for three-dimensional cell culture is a lack of optimized methods for analyzing more complicated structures. In particular, it is difficult to analyze using methods requiring fluorescent or chemiluminescent signals due to limitation of sample thickness or layering (Graf and Boppart 2010). Difficulty in validation, which allows for a more confident comparison between these complex models and *in vivo* tissues, prevents more widespread adoption of three-dimensional cell culture techniques.

2.2 THREE-DIMENSIONAL CELL CULTURE MICROENVIRONMENT

In cell culture, the microenvironment provides important cues that are crucial for influencing many aspects of cell behavior, including cell adhesion, proliferation, differentiation, morphology, and gene expression (Tibbitt and Anseth 2009; Asthana and Kisaalita 2013). One example is the process of differentiating stem cells into specific niches with the assistance of growth factor gradients (Petersen et al. 1992). The degree of complexity provided by extracellular matrix and molecular gradients allows three-dimensional cell culture models to effectively mimic native tissues to a much greater degree than in traditional monolayer culture.

2.2.1 Extracellular Matrix is Important for Cell Behavior

Important aspects of microenvironment include composition and derivation of the extracellular matrix. Defined matrix content varies by tissue type, as well as by disease state. Extracellular matrix is composed of a network of collagen and other support fibers punctuated by hydrated glycosaminoglycans, proteoglycans, and glycoproteins (Lee, Cuddihy, and Kotov 2008). Fibrous proteins, such as collagens, fibronectin, and laminin, provide the basis of mechanical action for cells. Other proteins provide sites for integrin binding and other factors allowing for communication between cells and extracellular matrix (Cukierman et al. 2001). Stiffness of extracellular matrix provides specific messaging to cells, leading to distinct phenotypic changes (Engler et al. 2006; Gieni and Hendzel 2008). For example, some cells may manipulate the extracellular matrix by contraction, providing cues to other cells, i.e., the degree to which they contract is controlled by the degree of stiffness (Grinnell et al. 1999). In addition, stiffness affects how easily cells can migrate through their microenvironment, with a soft matrix facilitating migration more than a stiff matrix (Zaman et al. 2005). By binding growth factors and other proteins, such as TGFβ, VEGF, and HGF, the extracellular matrix also plays a part in controlling local gradients of solutes (Davis 1988; Ruhrberg et al. 2002). Extracellular matrix components bind these factors, slowing their diffusion or storing them until cells trigger their release during extracellular matrix modulation (Griffith and Swartz 2006). In many disease states, extracellular matrix remodeling is a key part of the pathology. In liver fibrosis, for example, the extracellular matrix is dramatically altered with a replacement of the normal basement membrane-like extracellular matrix with a continuous, stiff, collagen-rich scar, perpetuating the local inflammatory response and altering cell behavior (Bataller and Brenner 2005). Changes in the extracellular matrix composition, and thus its characteristics, affect how the local cells behave.

2.2.2 Molecular Gradients in Three-Dimensional Cell Culture

In three-dimensional cell culture, molecular gradients of components such as oxygen, nutrients, and effector molecules influence cell viability and differentiation. Gradients of molecules can be altered by a variety of factors, such as rate of diffusion or perfusion of tissue, degree of cell consumption or production of materials, and whether or not there is a source of flow surrounding the culture (Derda et al. 2009). These gradients can be important because they may directly affect local cell behavior, i.e., a stronger concentration of a solute in one area of the culture may lead to a different cell phenotype than it would in an area with a weaker concentration. One of the critical limiting factors of many forms of three-dimensional cell culture is the oxygen gradients. Oxygen is poorly soluble in culture medium, which can lead to decreased viability in some regions of the culture (Glicklis, Merchuk, and Cohen 2004). The problem of oxygen deprivation is sometimes overcome either by providing movement of culture medium or by generating a vascular network within the tissue to ease oxygen perfusion into the center of the tissue (Asthana and Kisaalita 2012). However, in some models, such as in tumor development, the lack of oxygen may be advantageous.

2.3 SCAFFOLDS

Scaffolds, critical to tissue engineering, are three-dimensional structures that support the growth of a culture, allowing cells to self-organize as they shape or remodel their environment (Lee, Cuddihy, and Kotov 2008). The properties of individual scaffolds are dictated by the characteristics of the biomaterial from which they are made, the selection of which varies according to the application. Scaffolds are used for both experimental cell culture and clinical applications. There are several important aspects that should be examined when designing scaffolds for three-dimensional cell culture. Scaffolds are designed to be inert, biocompatible, or supportive, particularly when used for clinical applications. Porosity and pore size of the scaffold influence the way that the cells adhere, navigate, and interact with the material, and also how nutrients perfuse through the structure. Optimal porosity and pore size allow for better solute perfusion within a structure and in some cases can even be used to promote formation of vascular structures (Bramfeldt et al. 2010). As with native extracellular matrix, the stiffness and other mechanical properties of the biomaterial can alter cell phenotype, as well as provide simple architectural support. Specific properties can be incorporated into the scaffold depending on the cell culture requirements. Many scaffold materials can be coated with extracellular matrix components or pre-seeded with growth factors or other biochemical factors, which can affect cell populations in various ways (Mann and West 2002). Some scaffolds are designed to be biodegradable to allow for remodeling of the microenvironment by the cells. This is also desirable in clinical applications where a permanent structure is not needed once the engineered tissue is stable.

2.3.1 Methods and Materials for Scaffolds

A range of methods and materials have been developed to produce functional scaffolds to support cell culture for a variety of applications. Synthetic materials commonly used to produce more solid structures include polylactic acid, polyglycolic acid, and polycaprolactone. Design of scaffolds provides the opportunity for creativity, with many configurations possible. One method uses paper augmented with hydrogel, stacked to form a supportive scaffold allowing for construction of oxygen and nutrient gradients (Derda et al. 2009). A description of the full range of methods for manufacturing scaffolds is outside the scope of this chapter but can be found in a number of specific reviews (Carletti, Motta, and Migliaresi 2011). It is important to note that scaffold production methods differ in their ability to reliably reproduce a final product. Commonly used methods for scaffold manufacturing and manipulation include solvent casting and particulate leaching (Mikos et al. 1994), emulsion casting, fiber bonding, electrospinning (Pham, Sharma, and Mikos 2006), and solid free-form fabrication (Hollister 2005). Each of these methods has its own advantages and challenges. Cells can be implanted onto the scaffold during or after the manufacturing process, as in three-dimensional bioprinting (Mironov et al. 2003). Hydrogels, a type of scaffold, are important enough to merit their own section, to follow.

2.3.2 Challenges Associated with Scaffolds

There are challenges associated with the use of scaffolds for three-dimensional cell culture, although many are related to the specific method used. As previously mentioned, one issue that can affect three-dimensional scaffold cultures is inadequate oxygen perfusion throughout the entire structure. This can cause aberrant cell phenotypes and decreased viability. Fabrication techniques can also be labor-intensive and difficult to automate. Combined with the aforementioned analysis challenges, this limits high-throughput applications.

2.4 HYDROGELS

Hydrogels have long been used as a tool for three-dimensional cell culture. They are broadly defined as a network of interlacing, hydrated polymer chains forming a three-dimensional matrix encapsulating cells, and can be considered a type of scaffold (Cushing and Anseth 2007). Hydrogels can be made in a variety of formats depending on the type of material used and the method of fabrication. Depending on the cell type, the hydrogel microenvironment may be tailored to provide any mixture of traits, such as matrix stiffness, degradation, porosity, or availability of endogenous signals. For example, the way that a cell migrates through a hydrogel depends on the hydrogel's material characteristics (Gobin and West 2002). A major area of focus for hydrogel development has involved increasing mechanical versatility, allowing for greater manipulation required for applications such as bioprinting, and reactivity to biological stimuli (Kopeček 2007). Hydrogels have been designed that are responsive to specific thresholds of pH, temperature, and light (Miyata, Asami, and Uragami 1999). Furthermore, hydrogels can be biodegradable, which allows for support of the cells while allowing them to remodel their structure as needed (Ulbrich, Strohalm, and Kopeček 1982). Biodegradability of hydrogels can also be used in order to provide controlled release of solutes, such as small molecules and growth factors (Li, Rodrigues, and Tomás 2012). Thus, they can be helpful for differentiating cells or as part of clinical tissue repair technologies.

2.4.1 Methods and Materials for Hydrogel Culture

Hydrogel characteristics can be changed by modulating a variety of different factors, including concentrations of the ingredients (e.g., amounts of specific extracellular matrix components), crosslinking to increase stiffness, and providing different numbers of adhesion ligands for cell-extracellular interaction. Hydrogels can be made from a variety of biomaterials, each with their own characteristics to consider. They fall into two main categories: naturally-derived and synthetic, but can also be mixed in hybrid configurations.

Natural hydrogels include Matrigel, which is made out of native extracellular matrix proteins collected from a cell line, as well as collagen and alginate. Another form of natural hydrogel uses decellularized tissue extracts.

Extracellular matrix is collected from a specific tissue, then combined with a hydrogel material to be used to support cells of that tissue type (Skardal et al. 2012). Chitosan hydrogel is an example of a naturally-derived substrate that is degradable and supportive for several different cell types (Moura et al. 2011). Hyaluronic acid hydrogels are also commonly used because they are compatible with crosslinking materials.

Synthetic gels can be produced from a variety of inert materials and using many fabrication techniques. Poly-(ethylene glycol) is an example of a material commonly used in three-dimensional culture. In contrast to natural gels, synthetic hydrogels can be produced uniformly and are easily reproducible and characterized. Simple synthetic materials lack some functional signals for cells, like the active sites found in natural extracellular matrix, limiting their potential to support cells (Mahoney and Anseth 2006). Hybrid natural-synthetic materials offer a compromise and allow for more control over the ability to reconstruct a specific microenvironment. By combining natural components, such as extracellular matrix molecules, with defined synthetic gels, a more easily reproducible and functional scaffold can be produced (Salinas et al. 2007).

2.4.2 Challenges Associated with Hydrogel Cell Culture

One challenge with natural hydrogels is that their components can be variable and not well characterized, which can make it difficult to isolate specific cellular mechanisms responsible for resulting aspects of the culture. Hydrogels also carry the previously mentioned challenges of oxygen availability and analysis limitations associated with most three-dimensional cell culture. They can also be difficult to manipulate, although specific mechanisms for selectively solidifying gels (such as changes to pH or temperature, and UV light treatment) have been designed to make them more user-friendly.

2.5 SPHEROIDS

Spheroids, scaffold-free aggregates of cells, have been used for a variety of purposes over the decades (Holtfreter 1943; Moscona and Moscona 1952). Spheroid culture is particularly useful for co-culture in which cells can easily organize themselves into distinct layers. This flexibility in microenvironment also allows for natural neovascularization (Figure 2.2) (Fennema et al. 2013). Spheroids (in the form of embryoid bodies) have been used to aid in the process of differentiating stem cells, as well as to better understand the dynamics of tumor development (Kurosawa 2007). However, recently their uses have been expanded to include a broad array of research applications. They are especially useful in drug toxicity studies, as their spheroid form allows for the culture of human cells in a more stable environment, supporting differentiation and viability (Kelm et al. 2003). In addition, being composed of multiple layers of cells rather than a single one, they provide a greater physical barrier to perfusion of drug compounds than in a traditional monolayer culture. Most importantly, spheroids have the potential for high

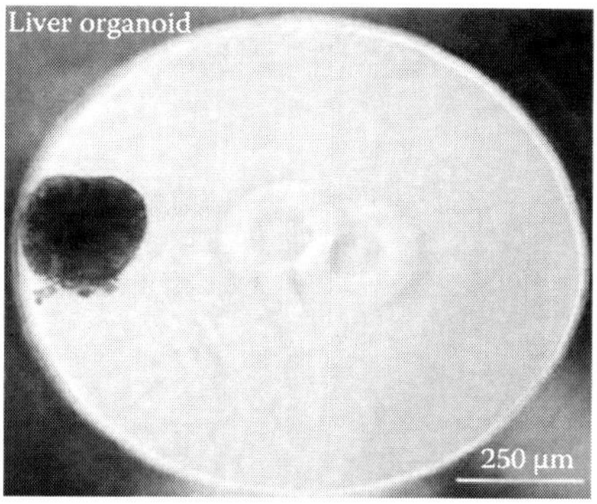

FIGURE 2.2 Spheroid aggregation process. First, cells are seeded in the aggregating environment (e.g., hanging drop plate) along with basal medium and ECM components to aid in formation. Cells begin to aggregate and self-organize rapidly, forming a tight aggregate. A whole-mount fluorescence image shows a cardiac spheroid expressing neovascularization marker VEGF (left); a light microscopy image shows a liver spheroid cultured for long-term characterization (center); and an H&E stained section of a liver spheroid shows its tight structure (right).

throughput screening applications (Hsiao et al. 2012). The process of producing spheroids and many of the aspects of testing required for drug toxicity studies can be easily automated (Kelm and Fussenegger 2004). Many methods for spheroid production produce tissues of controllable and uniform size, avoiding size-related oxygen limitations. Due to their relatively small size, they are also more easily analyzed using common techniques such as microscopy.

2.5.1 Methods and Materials for Spheroid Culture

Spheroids can be produced by a large variety of methods, so only the most common will be discussed here. All of these methods of spheroid formation rely on a single principle of cell biology, inherent cadherin and integrin binding abilities of cells (Casey et al. 2001; Robinson, Foty, and Corbett 2004; Shimazui et al. 2004).

One widely used method is hanging drop aggregation, which allows the cells to form a compact aggregate in the bottom of a droplet hanging from the underside of a surface (Foty 2011). This method is technically simple and is able to produce uniform sizes of tissue based on the number of cells seeded in the drop (Mehta et al. 2012). Spinner flasks, or similarly designed bioreactors, can be used to rapidly form spheroids and provide the added benefit of medium movement (Engelberg, Ropella, and Hunt 2008). However, many cell types do not readily form aggregates in the spinner flask system and aggregate size cannot be controlled. Micropatterned or treated materials providing an ultra-low adherence surface can be used to form spheroids by making it easier for the cells to adhere to each other rather than the surface, although this method may not provide compact spheroids using all cell types (Yuhas et al. 1977).

Synthetic molds, frequently fabricated out of polydimethylsiloxane, can also be used to produce uniform aggregates, and can be made more efficient and supportive when combined with microfluidic systems (Ungrin et al. 2008; Torisawa et al. 2007).

2.5.2 Challenges Associated with Spheroid Culture

The primary challenge with the spheroid culture format (when these tissues are considered as single units), is that the degree of tissue-specific complexity possible is limited. This limitation is due to their small size and their lack of the architecture necessary to support complexity. However, using spheroids as functional building blocks has been explored to overcome this problem (Mattix et al. 2014). In addition, some cell types do not form spheroids as readily as others. Particularly slow are differentiated adult cells such as hepatocytes. These cells can benefit from being co-cultured with support cells such as human umbilical vein endothelial cells (HUVECs).

2.6 SUPPORTING TECHNOLOGIES—BIOREACTORS AND MICROFLUIDIC CHIPS

Complexity can be added to cultures created with three-dimensional cell culture methods by utilizing one of the several supporting technologies. This section will feature bioreactors and microfluidic chips, technologies that have been used to provide the property of shear flow, the movement of medium around a tissue.

2.6.1 Background, Methods, and Challenges with Bioreactors

Bioreactors in the context of three-dimensional cell culture are machines used to support cell cultures in a specific physiological environment. Bioreactors can be used for cell expansion in large quantities as well as for tissue fabrication (Hansmann et al. 2013). They have been used to produce spherical aggregates, as described in the previous section, and are useful for producing larger tissues as well (Peshwa et al. 1993). In many cases, medium movement is desired in cell culture, mimicking the fluid shear stress forces found *in vivo*. One important advantage of bioreactors is that they allow for dynamic movement of culture medium, which provides more consistent perfusion of nutrients and waste materials. Bioreactors can also be adjusted to provide ideal temperature, humidity, oxygen, carbon dioxide, and other relevant concentrations for cell growth.

There are several different types of bioreactors: spinner flasks, which are simple containers typically combined with stir bars to provide medium movement (Ismadi et al. 2014). Rotating wall vessels which are made up of two cylinders—the inner one rotates containing the tissues and medium (Hammond and Hammond 2001). Perfusion bioreactors push medium through scaffolds using a pump, allowing for medium flow over samples in a chamber (Yan, Bergstrom, and Chen 2012). The primary challenge with using bioreactors in tissue manufacturing is that the results are largely uncontrolled, leading to nonuniform sizes and shapes. Even with medium flow around them, larger structures will be prone to necrosis due to oxygen starvation in the center of the tissue.

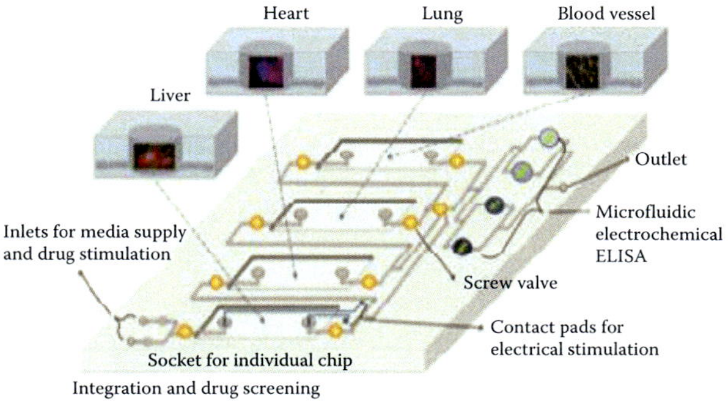

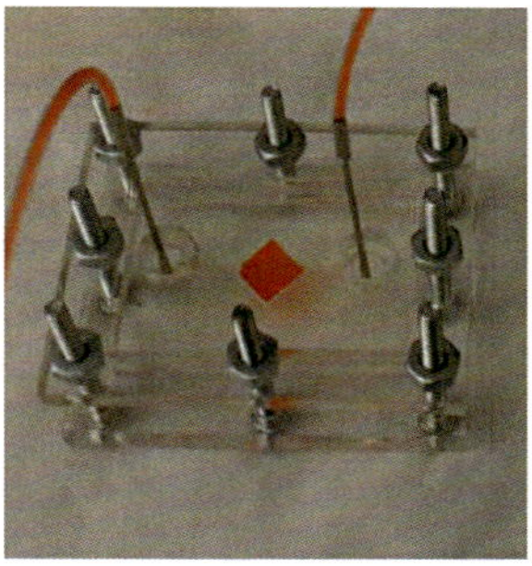

FIGURE 2.3 A figure showing a body-on-a-chip microfluidic system, which provides on-board, real-time monitoring of liver, heart, lung, and blood vessel organoids in tandem. (Courtesy of the Wake Forest Institute for Regenerative Medicine, Winston-Salem, North Carolina.)

2.6.2 Background, Methods, and Challenges with Microfluidic Chips

Microfluidic chips, also referred to as "organs-on-a-chip," frequently combine three-dimensional culture methods, such as scaffolds or spheroids, with microfluidic techniques to form an integrated system designed to provide a more complex culture environment (Figure 2.3) (Bhatia and Ingber 2014). As with bioreactors, microfluidic chips can provide controlled movement of medium around the tissue or culture (Lee et al. 2011). They are also frequently tailored to provide specific mechanical, architectural, or biochemical characteristics to aid in better recapitulating the environment of native tissue (Mark et al. 2010). There are a number of fabrication methods yielding structures that provide many different possibilities for mechanical design or cell interaction. These systems can be designed for ease of testing, allowing for integration of sensors, or live imaging by interfacing with microscopes. The main challenge associated with microfluidic chips is with their use in high-throughput applications. Manufacturing is challenging and often expensive, especially for models with high complexity.

2.7 CONCLUSION

Three-dimensional cell culture methods are flexible enough to be used for any desired cell or tissue type and for a variety of applications. Unlike most two-dimensional cell culture methods, three-dimensional culture provides an environment in which cells can interact with each other and the surrounding extracellular matrix, resulting in improved cell viability and functionality. In addition, it supports the formation of more complex microenvironmental effects, more closely replicating *in vivo* contexts. These methods are already revolutionizing what is possible both academically and clinically. However, more broad applicability of three-dimensional cell culture depends on further development and optimization of assays capable of adequately characterizing and testing these complicated structures, a task that is being pursued both by academia and industry.

REFERENCES

Abu-Absi, Susan Fugett, Julie R. Friend, Linda K. Hansen, and Wei-Shou Hu. 2002. Structural Polarity and Functional Bile Canaliculi in Rat Hepatocyte Spheroids. *Experimental Cell Research* 274 (1): 56–67. doi:10.1006/excr.2001.5467.

Asthana, Amish, and William S. Kisaalita. 2012. Microtissue Size and Hypoxia in HTS with 3D Cultures. *Drug Discovery Today* 17 (15–16): 810–17. doi:10.1016/j.drudis.2012.03.004.

Asthana, Amish, and William S. Kisaalita. 2013. Biophysical Microenvironment and 3D Culture Physiological Relevance. *Drug Discovery Today* 18 (11–12): 533–40. doi:10.1016/j.drudis.2012.12.005.

Bataller, Ramón, and David A. Brenner. 2005. Liver Fibrosis. *The Journal of Clinical Investigation* 115 (2): 209–18. doi:10.1172/JCI24282.

Bhatia, Sangeeta N., and Donald E. Ingber. 2014. Microfluidic Organs-on-Chips. *Nature Biotechnology* 32 (8): 760–72. doi:10.1038/nbt.2989.

Bramfeldt, H., G. Sabra, V. Centis, and P. Vermette. 2010. Scaffold Vascularization: A Challenge for Three-Dimensional Tissue Engineering. *Current Medicinal Chemistry* 17 (33): 3944–67.

Carletti, Eleonora, Antonella Motta, and Claudio Migliaresi. 2011. Scaffolds for Tissue Engineering and 3D Cell Culture. *Methods in Molecular Biology (Clifton, N.J.)* 695: 17–39. doi:10.1007/978-1-60761-984-0_2.

Casey, R.C., K.M. Burleson, K.M. Skubitz, S.E. Pambuccian, T.R. Oegema, L.E. Ruff, and A.P. Skubitz. 2001. Beta 1-Integrins Regulate the Formation and Adhesion of Ovarian Carcinoma Multicellular Spheroids. *The American Journal of Pathology* 159 (6): 2071–80.

Chen, C.S. 1997. Geometric Control of Cell Life and Death. *Science* 276 (5317): 1425–28. doi:10.1126/science.276.5317.1425.

Cukierman, E., R. Pankov, D.R. Stevens, and K.M. Yamada. 2001. Taking Cell-Matrix Adhesions to the Third Dimension. *Science (New York, N.Y.)* 294 (5547): 1708–12. doi:10.1126/science.1064829.

Cushing, Melinda C., and Kristi S. Anseth. 2007. Materials Science. Hydrogel Cell Cultures. *Science (New York, N.Y.)* 316 (5828): 1133–34. doi:10.1126/science.1140171.

Davis, B.H. 1988. Transforming Growth Factor Beta Responsiveness Is Modulated by the Extracellular Collagen Matrix during Hepatic Ito Cell Culture. *Journal of Cellular Physiology* 136 (3): 547–53. doi:10.1002/jcp.1041360323.

Derda, R., A. Laromaine, A. Mammoto, S.K.Y. Tang, T. Mammoto, D.E. Ingber, and G.M. Whitesides. 2009. Paper-Supported 3D Cell Culture for Tissue-Based Bioassays. *Proceedings of the National Academy of Sciences of the United States of America* 106 (44): 18457–62. doi:10.1073/pnas.0910666106.

Engelberg, Jesse A., Glen E.P. Ropella, and C. Anthony Hunt. 2008. Essential Operating Principles for Tumor Spheroid Growth. *BMC Systems Biology* 2 (1): 110. doi: 10.1186/1752-0509-2-110.

Engler, Adam J., Shamik Sen, H. Lee Sweeney, and Dennis E. Discher. 2006. Matrix Elasticity Directs Stem Cell Lineage Specification. *Cell* 126 (4): 677–89. doi:10.1016/j.cell.2006.06.044.

Fennema, Eelco, Nicolas Rivron, Jeroen Rouwkema, Clemens van Blitterswijk, and Jan de Boer. 2013. Spheroid Culture as a Tool for Creating 3D Complex Tissues. *Trends in Biotechnology* 31 (2): 108–15. doi:10.1016/j.tibtech.2012.12.003.

Foty, Ramsey. 2011. A Simple Hanging Drop Cell Culture Protocol for Generation of 3D Spheroids. *Journal of Visualized Experiments: J Vis Exp* (51): e2720. doi:10.3791/2720.

Francois Berthiaume, Prabhas V. Moghe, Mehmet Toner, and Martin L. Yarmush. 1996. Effect of Extracellular Matrix Topology on Cell Structure, Function, and Physiological Responsiveness: Hepatocytes Cultured in a Sandwich Configuration. *FASEB Journal: Official Publication of the Federation of American Societies for Experimental Biology* 10 (13): 1471–84.

Ghosh, Sourabh, Giulio C. Spagnoli, Ivan Martin, Sabine Ploegert, Philippe Demougin, Michael Heberer, and Anca Reschner. 2005. Three-Dimensional Culture of Melanoma Cells Profoundly Affects Gene Expression Profile: A High Density Oligonucleotide Array Study. *Journal of Cellular Physiology* 204 (2): 522–31. doi:10.1002/jcp.20320.

Gieni, Randall S., and Michael J. Hendzel. 2008. Mechanotransduction from the ECM to the Genome: Are the Pieces Now in Place? *Journal of Cellular Biochemistry* 104 (6): 1964–87. doi:10.1002/jcb.21364.

Glicklis, Rachel, Jose C. Merchuk, and Smadar Cohen. 2004. Modeling Mass Transfer in Hepatocyte Spheroids via Cell Viability, Spheroid Size, and Hepatocellular Functions. *Biotechnology and Bioengineering* 86 (6): 672–80. doi:10.1002/bit.20086.

Gobin, Andrea S., and Jennifer L. West. 2002. Cell Migration through Defined, Synthetic ECM Analogs. *FASEB Journal: Official Publication of the Federation of American Societies for Experimental Biology* 16 (7): 751–53. doi:10.1096/fj.01-0759fje.

Graf, Benedikt W., and Stephen A. Boppart. 2010. Imaging and Analysis of Three-Dimensional Cell Culture Models. *Methods in Molecular Biology (Clifton, N.J.)* 591: 211–27. doi:10.1007/978-1-60761-404-3_13.

Griffith, Linda G., and Melody A. Swartz. 2006. Capturing Complex 3D Tissue Physiology in vitro. *Nature Reviews. Molecular Cell Biology* 7 (3): 211–24. doi:10.1038/nrm1858.

Grinnell, F., C.H. Ho, Y.C. Lin, and G. Skuta. 1999. Differences in the Regulation of Fibroblast Contraction of Floating versus Stressed Collagen Matrices. *The Journal of Biological Chemistry* 274 (2): 918–23.

Hammond, T.G., and J.M. Hammond. 2001. Optimized Suspension Culture: The Rotating-Wall Vessel. *American Journal of Physiology. Renal Physiology* 281 (1): F12–25.

Hansmann, Jan, Florian Groeber, Alexander Kahlig, Claudia Kleinhans, and Heike Walles. 2013. Bioreactors in Tissue Engineering-Principles, Applications and Commercial Constraints. *Biotechnology Journal* 8 (3): 298–307. doi:10.1002/biot.201200162.

Harris, Andrew R., Loic Peter, Julien Bellis, Buzz Baum, Alexandre J. Kabla, and Guillaume T. Charras. 2012. Characterizing the Mechanics of Cultured Cell Monolayers. *Proceedings of the National Academy of Sciences of the United States of America* 109 (41): 16449–54. doi:10.1073/pnas.1213301109.

Hollister, Scott J. 2005. Porous Scaffold Design for Tissue Engineering. *Nature Materials* 4 (7): 518–24. doi:10.1038/nmat1421.

Holtfreter, Johannes. 1943. A Study of the Mechanics of Gastrulation. Part I. *Journal of Experimental Zoology* 94 (3): 261–318. doi:10.1002/jez.1400940302.

Hsiao, Amy Y., Yi-Chung Tung, Xianggui Qu, Lalit R. Patel, Kenneth J. Pienta, and Shuichi Takayama. 2012. 384 Hanging Drop Arrays Give Excellent Z-Factors and Allow Versatile Formation of Co-Culture Spheroids. *Biotechnology and Bioengineering* 109 (5): 1293–1304. doi:10.1002/bit.24399.

Huh, Dongeun, Geraldine A. Hamilton, and Donald E. Ingber. 2011. From 3D Cell Culture to Organs-on-Chips. *Trends in Cell Biology* 21 (12): 745–54. doi:10.1016/j.tcb.2011.09.005.

Ismadi, Mohd-Zulhilmi, Priyanka Gupta, Andreas Fouras, Paul Verma, Sameer Jadhav, Jayesh Bellare, and Kerry Hourigan. 2014. Flow Characterization of a Spinner Flask for Induced Pluripotent Stem Cell Culture Application. Edited by Katriina Aalto-Setala. *PLoS One* 9 (10): e106493. doi:10.1371/journal.pone.0106493.

Kelm, Jens M., and Martin Fussenegger. 2004. Microscale Tissue Engineering Using Gravity-Enforced Cell Assembly. *Trends in Biotechnology* 22 (4): 195–202. doi:10.1016/j.tibtech.2004.02.002.

Kelm, Jens M., Nicholas E. Timmins, Catherine J. Brown, Martin Fussenegger, and Lars K. Nielsen. 2003. Method for Generation of Homogeneous Multicellular Tumor Spheroids Applicable to a Wide Variety of Cell Types. *Biotechnology and Bioengineering* 83 (2): 173–80. doi:10.1002/bit.10655.

Knobeloch, Daniel, Sabrina Ehnert, Lilianna Schyschka, Peter Büchler, Michael Schoenberg, Jörg Kleeff, Wolfgang E. Thasler, et al. 2012. Human Hepatocytes: Isolation, Culture, and Quality Procedures. *Methods in Molecular Biology (Clifton, N.J.)* 806: 99–120. doi:10.1007/978-1-61779-367-78.

Kopecek, Jindrich. 2007. Hydrogel Biomaterials: A Smart Future? *Biomaterials* 28 (34): 5185–92. doi:10.1016/j.biomaterials.2j007.07.044.

Kurosawa, Hiroshi. 2007. Methods for Inducing Embryoid Body Formation: In vitro Differentiation System of Embryonic Stem Cells. *Journal of Bioscience and Bioengineering* 103 (5): 389–98. doi:10.1263/jbb.103.389.

Lee, Chia-Yen, Chin-Lung Chang, Yao-Nan Wang, and Lung-Ming Fu. 2011. Microfluidic Mixing: A Review. *International Journal of Molecular Sciences* 12 (12): 3263–87. doi:10.3390/ijms12053263.

Lee, Jungwoo, Meghan J. Cuddihy, and Nicholas A. Kotov. 2008. Three-Dimensional Cell Culture Matrices: State of the Art. *Tissue Engineering. Part B, Reviews* 14 (1): 61–86. doi:10.1089/teb.2007.0150.

Lee, Seung-A, Da Yoon No, Edward Kang, Jongil Ju, Dong-Sik Kim, and Sang-Hoon Lee. 2013. Spheroid-Based Three-Dimensional Liver-on-a-Chip to Investigate Hepatocyte-Hepatic Stellate Cell Interactions and Flow Effects. *Lab on a Chip* 13 (18): 3529–37. doi:10.1039/c3lc50197c.

Li, Yulin, João Rodrigues, and Helena Tomás. 2012. Injectable and Biodegradable Hydrogels: Gelation, Biodegradation and Biomedical Applications. *Chemical Society Reviews* 41 (6): 2193–2221. doi:10.1039/c1cs15203c.

Mahoney, Melissa J., and Kristi S. Anseth. 2006. Three-Dimensional Growth and Function of Neural Tissue in Degradable Polyethylene Glycol Hydrogels. *Biomaterials* 27 (10): 2265–74. doi:10.1016/j.biomaterials.2005.11.007.

Mann, Brenda K., and Jennifer L. West. 2002. Cell Adhesion Peptides Alter Smooth Muscle Cell Adhesion, Proliferation, Migration, and Matrix Protein Synthesis on Modified Surfaces and in Polymer Scaffolds. *Journal of Biomedical Materials Research* 60 (1): 86–93.

Mark, Daniel, Stefan Haeberle, Günter Roth, Felix von Stetten, and Roland Zengerle. 2010. Microfluidic Lab-on-a-Chip Platforms: Requirements, Characteristics and Applications. *Chemical Society Reviews* 39 (3): 1153. doi:10.1039/b820557b.

Mattix, Brandon, Timothy R. Olsen, Yu Gu, Megan Casco, Austin Herbst, Dan T. Simionescu, Richard P. Visconti, Konstantin G. Kornev, and Frank Alexis. 2014. Biological Magnetic Cellular Spheroids as Building Blocks for Tissue Engineering. *Acta Biomaterialia* 10 (2): 623–29. doi:10.1016/j.actbio.2013.10.021.

McGonigle, Paul, and Bruce Ruggeri. 2014. Animal Models of Human Disease: Challenges in Enabling Translation. *Biochemical Pharmacology* 87: 162–71. doi:10.1016/j.bcp.2013.08.006.

Mehta, Geeta, Amy Y. Hsiao, Marylou Ingram, Gary D. Luker, and Shuichi Takayama. 2012. Opportunities and Challenges for Use of Tumor Spheroids as Models to Test Drug Delivery and Efficacy. *Journal of Controlled Release* 164 (2): 192–204. doi:10.1016/j.jconrel.2012.04.045.

Messner, S., I. Agarkova, W. Moritz, and J.M. Kelm. 2013. Multi-Cell Type Human Liver Microtissues for Hepatotoxicity Testing. *Archives of Toxicology* 87 (1): 209–13. doi:10.1007/s00204-012-0968-2.

Mikos, Antonios G., Amy J. Thorsen, Lisa A. Czerwonka, Yuan Bao, Robert Langer, Douglas N. Winslow, and Joseph P. Vacanti. 1994. Preparation and Characterization of Poly(l-Lactic Acid) Foams. *Polymer* 35 (5): 1068–77. doi:10.1016/0032-3861(94)90953-9.

Mironov, Vladimir, Thomas Boland, Thomas Trusk, Gabor Forgacs, and Roger R. Markwald. 2003. Organ Printing: Computer-Aided Jet-Based 3D Tissue Engineering. *Trends in Biotechnology* 21 (4): 157–61. doi:10.1016/S0167-7799(03)00033-7.

Miyata, T., N. Asami, and T. Uragami. 1999. A Reversibly Antigen-Responsive Hydrogel. *Nature* 399 (6738): 766–69. doi:10.1038/21619.

Moscona, A., and H. Moscona. 1952. The Dissociation and Aggregation of Cells from Organ Rudiments of the Early Chick Embryo. *Journal of Anatomy* 86 (3): 287–301.

Moura, M. José, H. Faneca, M. Pedroso Lima, M. Helena Gil, and M. Margarida Figueiredo. 2011. In Situ Forming Chitosan Hydrogels Prepared via Ionic/Covalent Co-Cross-Linking. *Biomacromolecules* 12 (9): 3275–84. doi:10.1021/bm200731x.

Mueller-Klieser, W. 1987. Multicellular Spheroids. A Review on Cellular Aggregates in Cancer Research. *Journal of Cancer Research and Clinical Oncology* 113 (2): 101–22.

Paul, Steven M., Daniel S. Mytelka, Christopher T. Dunwiddie, Charles C. Persinger, Bernard H. Munos, Stacy R. Lindborg, and Aaron L. Schacht. 2010. How to Improve R&D Productivity: The Pharmaceutical Industry's Grand Challenge. *Nature Reviews Drug Discovery* 9: 203–214.

Peshwa, M.V., Y.S. Kyung, D.B. McClure, and W.S. Hu. 1993. Cultivation of Mammalian Cells as Aggregates in Bioreactors: Effect of Calcium Concentration of Spatial Distribution of Viability. *Biotechnology and Bioengineering* 41 (2): 179–87. doi:10.1002/bit.260410203.

Petersen, O.W., L. Rønnov-Jessen, A.R. Howlett, and M.J. Bissell. 1992. Interaction with Basement Membrane Serves to Rapidly Distinguish Growth and Differentiation Pattern of Normal and Malignant Human Breast Epithelial Cells. *Proceedings of the National Academy of Sciences of the United States of America* 89 (19): 9064–68.

Pham, Quynh P., Upma Sharma, and Antonios G. Mikos. 2006. Electrospinning of Polymeric Nanofibers for Tissue Engineering Applications: A Review. *Tissue Engineering* 12 (5): 1197–211. doi:10.1089/ten.2006.12.1197.

Robinson, Elizabeth E., Ramsey A. Foty, and Siobhan A. Corbett. 2004. Fibronectin Matrix Assembly Regulates alpha5beta1-Mediated Cell Cohesion. *Molecular Biology of the Cell* 15 (3): 973–81. doi:10.1091/mbc.E03-07-0528.

Ruhrberg, Christiana, Holger Gerhardt, Matthew Golding, Rose Watson, Sofia Ioannidou, Hajime Fujisawa, Christer Betsholtz, and David T. Shima. 2002. Spatially Restricted Patterning Cues Provided by Heparin-Binding VEGF-A Control Blood Vessel Branching Morphogenesis. *Genes & Development* 16 (20): 2684–98. doi:10.1101/gad.242002.

Salinas, Chelsea N., Brook B. Cole, Andrea M. Kasko, and Kristi S. Anseth. 2007. Chondrogenic Differentiation Potential of Human Mesenchymal Stem Cells Photoencapsulated within Poly(Ethylene Glycol)–Arginine-Glycine-Aspartic Acid-Serine Thiol-Methacrylate Mixed-Mode Networks. *Tissue Engineering* 13 (5): 1025–34. doi:10.1089/ten.2006.0126.

Shimazui, Toru, Jack A. Schalken, Koji Kawai, Risa Kawamoto, Adrie van Bockhoven, Egbert Oosterwijk, and Hideyuki Akaza. 2004. Role of Complex Cadherins in Cell-Cell Adhesion Evaluated by Spheroid Formation in Renal Cell Carcinoma Cell Lines. *Oncology Reports* 11 (2): 357–60.

Skardal, Aleksander, Leona Smith, Shantaram Bharadwaj, Anthony Atala, Shay Soker, and Yuanyuan Zhang. 2012. Tissue Specific Synthetic ECM Hydrogels for 3-D in vitro Maintenance of Hepatocyte Function. *Biomaterials* 33 (18): 4565–75. doi:10.1016/j.biomaterials.2012.03.034.

Suuronen, Erik J., Heather Sheardown, Kimberley D. Newman, Christopher R. McLaughlin, and May Griffith. 2005. Building in vitro Models of Organs. *International Review of Cytology* 244: 137–73. doi:10.1016/S0074-7696(05)44004-8.

Tibbitt, Mark W., and Kristi S. Anseth. 2009. Hydrogels as Extracellular Matrix Mimics for 3D Cell Culture. *Biotechnology and Bioengineering* 103 (4): 655–63. doi:10.1002/bit.22361.

Torisawa, Yu-suke, Bor-han Chueh, Dongeun Huh, Poornapriya Ramamurthy, Therese M. Roth, Kate F. Barald, and Shuichi Takayama. 2007. Efficient Formation of Uniform-Sized Embryoid Bodies Using a Compartmentalized Microchannel Device. *Lab on a Chip* 7 (6): 770–76. doi:10.1039/b618439a.

Ulbrich, Karel, Jiří Strohalm, and Jindřich Kopeček. 1982. Polymers Containing Enzymatically Degradable Bonds. VI.Hydrophilic Gels Cleavable by Chymotrypsin. *Biomaterials* 3 (3): 150–54. doi:10.1016/0142-9612(82)90004-7.

Ungrin, Mark D., Chirag Joshi, Andra Nica, Céline Bauwens, and Peter W. Zandstra. 2008. Reproducible, Ultra High-Throughput Formation of Multicellular Organization from Single Cell Suspension-Derived Human Embryonic Stem Cell Aggregates. Edited by Patrick Callaerts. *PLoS One* 3 (2): e1565. doi:10.1371/journal.pone.0001565.

von der Mark, K., V. Gauss, H. von der Mark, and P. Müller. 1977. Relationship between Cell Shape and Type of Collagen Synthesised as Chondrocytes Lose Their Cartilage Phenotype in Culture. *Nature* 267 (5611): 531–32.

Wong, M.K., and A.I. Gotlieb. 1986. Endothelial Cell Monolayer Integrity. I. Characterization of Dense Peripheral Band of Microfilaments. *Arteriosclerosis (Dallas, Tex.)* 6 (2): 212–19.

Yan, X., D.J. Bergstrom, and X.B. Chen. 2012. Modeling of Cell Cultures in Perfusion Bioreactors. *IEEE Transactions on Bio-Medical Engineering* 59 (9): 2568–75. doi:10.1109/TBME.2012.2206077.

Yuhas, J.M., A.P. Li, A.O. Martinez, and A.J. Ladman. 1977. A Simplified Method for Production and Growth of Multicellular Tumor Spheroids. *Cancer Research* 37 (10): 3639–43.

Zaman, Muhammad H., Roger D. Kamm, Paul Matsudaira, and Douglas A. Lauffenburger. 2005. Computational Model for Cell Migration in Three-Dimensional Matrices. *Biophysical Journal* 89 (2): 1389–97. doi:10.1529/biophysj.105.060723.

Zhang, Yuanyuan, Yujiang He, Shantaram Bharadwaj, Nevin Hammam, Kristen Carnagey, Regina Myers, Anthony Atala, and Mark Van Dyke. 2009. Tissue-Specific Extracellular Matrix Coatings for the Promotion of Cell Proliferation and Maintenance of Cell Phenotype. *Biomaterials* 30 (23–24): 4021–28. doi:10.1016/j.biomaterials.2009.04.005.

3 Electrochemical Sensors for Organs-on-a-Chip

Joyce Han-Ching Chiu, Ge-Ah Kim, Rodney Daniels and Shuichi Takayama

CONTENTS

3.1 INTRODUCTION

An *in vitro* analytical system capable of accurate representation of *in vivo* physiological states for the purpose of medical diagnostics is the goal of human-on-a-chip–based technology. Microfluidics provides the platform on which this goal can be achieved. The precise control of fluidic flow and improved fabrication of micro- and nanostructures allow for optimal control over cell culture environment. The innate fluidic flow system not only provides recirculating fluids to save reagent use and enhance autocrine signaling, but also links the culture area with downstream sampling for *in situ* real-time analysis. Such platforms are useful for drug testing and cell culture where detection of target analytes provide information on cell response or cell quality. Flexible choices in biocompatible substrate facilitate a custom environment for individual needs. Furthermore, microfluidic devices allow for better oxygenation and perfusion capability due to its small dimensions compared to the rigid and static system of conventional cell culture.

With the advance in human-on-a-chip, where physiological systems are interconnected by metabolic reactions, it is necessary to closely monitor operations by measuring pH, oxygen, nitric oxide concentration, temperature changes, and glucose consumption, all of which can shed light on the contributions of these factors on several diseases. These parameters also provide indications on the dynamic state of a human-on-a-chip system.

Electrochemical-based measurements have long been utilized in cell culture systems. The sensitivity and long-term measurement capability are ideal for *in situ* setup. Amperometric electrodes are readily integrated onto device structures with breakthroughs in fabrication techniques. Besides metabolic studies, impedance measurements for cell viability is another noninvasive technique which utilizes the change in impedance due to cell adhesion and proliferation to monitor cell growth.[1,2]

Electrochemical sensing is not without competition, however. The main competing method is optical chemical sensing. Optical transparency is readily available in most lab-on-a-chip substrates. The use of fluorescent-based detection boasts a faster response time than electrochemical systems in most cases, and optical sensing is usually preferred over electrochemical methods[3,4] in the measurement of pH or CO_2 levels. The introduction of pH-sensitive dye combined with an LED excitation light source was the basis for the pH sensor proposed by Kostov et al.[5] The fluorescent detection of dissolved oxygen is facilitated by use of ruthenium fluorophore via a frequency dependent experiment where the lifetime of ruthenium is measured by phase shift between excitation and emission. Applications of optical sensors can be found in microbioreactor array designed with pH and O_2 sensing capabilities.[6,7] However, optical sensing suffers from potential optical cross-talk due to the dye used for different analytes, and generally has a narrower detection range. Optical sensors in general provide better resolution in the low oxygen concentration range, while the electrochemical methods show superior performance at higher oxygen levels.[8] In a human-on-a-chip system, electrochemical sensing could be used for interstitial small molecules, such as glucose, O_2, etc., therefore freeing up bandwidth for optical measurement of biomarkers using fluorescent tags, thus combining the two techniques to allow for a more robust system.

3.2 ELECTRODES

Electrochemical sensors utilize redox reactions of electroactive species for measurements of target molecules. In general, electrodes may be categorized into two types. The more conventional electrochemical electrode will have both the working and counter electrodes directly immersed in target solution, while the Clark type electrode, often found in oxygen sensors, is characterized by an electrode surrounded by buffer solution with an outer sleeve coated with a gas permeable membrane so that the electrodes, which are not in direct contact with the aqueous system, will be dependent on the diffusion process of the target molecule, but less influenced by other components of the system. The Clark type electrode is usually more bulky and difficult for miniaturization. Despite its use in culture system, the traditional Clark type electrodes are not suitable for organs-on-a-chip system.

Ion-sensitive filed-effect transistor (ISFET) is an alternative electrode for potentiometric devices. Similar to the metal oxide semiconductor field-effect transistor, ISFET requires source, drain, and gate terminals. Current is generated when applying a voltage between the source and drain with an n-type channel, which is formed when the holes in the channels are repelled by applying a positive voltage higher than the gate voltage. In a biosensor, the application of voltage at the gate terminal is replaced by modification at the gate terminal using an ion-selective membrane. Ion accumulation at the gate terminals will result in the change of electrical charges similar to the current generation. ISFET's native integration with electronics gives it an edge over optical systems, and advancements in the silicon industry have made fabrication for miniaturization of on-chip sensors a standard process. Other modifications of gate channels such as immune-ISFET can be accomplished by utilizing antibodies, and such sensors are investigated for use in biomolecule identification. However, we will focus on traditional ISFET sensors that are used in pH sensors due to their high resistance to corrosion, tolerance, and measurement capability in extreme pH ranges.

Nanomaterials introduce non-conventional materials into the world of biosensors. Carbon nanotubes are investigated extensively for the possibility for use in electrochemistry.[9,10] Banks et al. examined the redox reaction of NADH using multiwalled carbon nanotube modified electrodes and compared it to traditional carbon electrodes.[11–13] The low signal-to-noise ratio and reduction in over-potential, termed the *electro-catalytic effect*, accomplished through the implementation of carbon nanotubes, is particularly attractive to researchers throughout the field.

Modern innovations in manipulation of graphene sheets have generated a storm of interest in its application. As a zero-band gap material, it provides a unique opportunity to act as a transistor and a conductor. It is also chemically inert, therefore ideally suited as an electrode for electrochemical measurement. Compared to traditional electrodes, which use either transition metals such as Fe and Cu, heavy inert metals such as Au and Ag, or metals found in oxides such as Sn and In, graphene possesses higher thermal conductivity and better biocompatibility, making it ideal for biosensing applications. Similar to carbon nanotubes, it also exhibits high electrocatalytic activity.[14,15]

Biofouling of electrodes is one of the key considerations when dealing with biological samples. Adsorption of proteins onto the electrodes will progressively degrade sensor readouts. The primary approach in preventing adsorption involves surface modification of electrodes. Patel et al. proposed an alternative in tackling the biofouling issue.[16] Instead of preventing adsorption, the focus lies on circumventing the effect of adsorption by use of nanopores. Small active analytes can pass through the nanopores of the electrodes so that redox reactions can freely proceed. Although large proteins will still adsorb on the electrodes as usual, electron transfer and the exchange of redox species proceeds with minimal interference and with little impact on redox performance. At the same time, the porous electrode possesses a larger surface area, which provides higher active sites for redox reactions to proceed compared to planar electrodes (Figure 3.1a). Electron transfer of ferricyanide using nanoporous gold electrodes in the presence of biofouling agents, namely bovine serum albumin and bovine fibrinogen, were evaluated with cyclic voltammetry (Figure 3.1b) and showed little deviation from the peak shape at initial condition after 22 hours, whereas planar gold electrodes demonstrate peak flattening and significant reduction in signal within approximately 5 to 10 minutes.

Another approach for laying electrodes in human-on-a-chip devices may come from syringe-injectable electrodes.[17] Traditionally, fabrication of microfluidic based electrochemical analysis was built from a bottom-up approach. Electrodes are plated on substrates using deposition methods such as electrochemical deposition, sputtering, or metal evaporation, followed by modification of electrodes before bonding a polymer-based microfluidic channel substrate on top of the electrodes. However, Liu et al. (2015) successfully fabricated a mesh-like scaffold capable of fitting into syringe needles that can unfold and relax into the cavities of target locations (Figure 3.1c, d). The scaffold was designed for *in vivo* measurement but can fit the microfluidic based human-on-a-chip device structure as well. Field electric

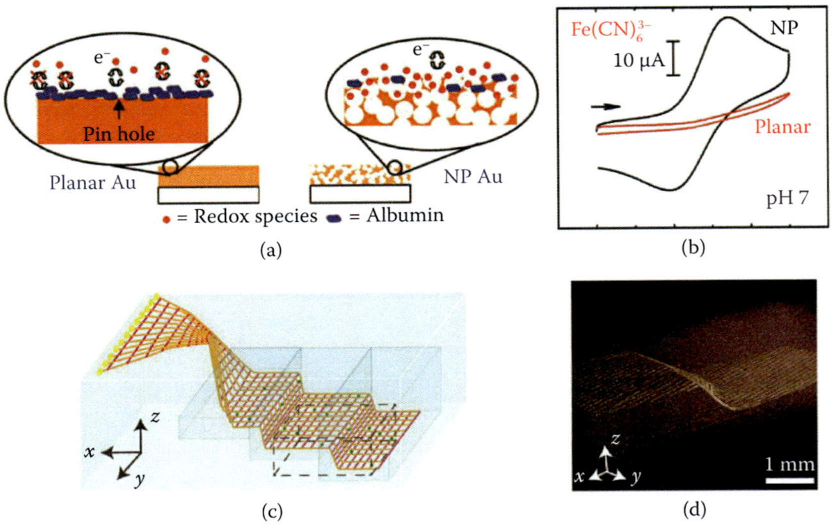

FIGURE 3.1 (a) Cartoon representation of planar versus nanoporous gold surface. Redox species were able to pass through with porous Au while hindered at planar Au, thereby inhibiting electron transfer. (Reprinted with permission from Patel, J., et al., *Anal Chem*, 85(23), 11610–18, 2013. Copyright [2013] American Chemical Society.) (b) Cylic voltammetry showed that nanoporous gold was able to maintain hysteresis peak shape while planar Au could not after 22 hours (planar flattening occurred within approximately 10 minutes). (c, d) Mesh scaffold relaxation within cavities. (Reprinted with permission from Liu, J., et al., *Nat Nanotechnology*, 10, 629–639. Copyright 2015, Macmillan Publishers Ltd.)

transistor type sensors can be built on top of the mesh and used to investigate the pH changes of vascular smooth muscles cells.[18]

3.3 OXYGEN

Oxygen is a critical ingredient for the aerobic metabolism of cells, and its concentration must be maintained at the optimal range to prevent the inhibition of cell growth and metabolism. Oxygen uptake rate, which can be obtained from dissolved oxygen (DO) sensing, not only offers cell population estimates but also provides information on the metabolism pathway. When combined with additional biochemical sensing, such as pH, glucose, or carbon dioxide, it is possible to determine the metabolism pathway. For example, the extracellular chemical environment regulates glucose metabolism pathways, and simultaneous measurements of O_2 and pH can detect this influence.[19]

Oxygen sensors in biological environments can be largely categorized into two types: optical and electrochemical. However, these two types exhibit quite contrasting characteristics. Optical sensors enable remote and noninvasive sensing[20] and are based on the photoluminescence quenching effect of O_2, in which the introduction of O_2 decreases the lifetime of luminophores according to the Stern-Volmer equation.

On the other hand, electrochemical sensors require online sensing and detect electric potential or current flow between cathode and anode where the electric signal is generated from the redox reaction:

$$O_2 + 4e^- + 2H_2O \rightarrow 4OH^- \tag{3.1}$$

Clark et al. first presented the idea of membrane probes, which became the most commonly used concept for the electrochemical DO sensors. The sensor had an Ag reference anode immersed in saturated potassium chloride, and a Pt cathode which was separated from the electrolyte by an insulted reservoir and a polyethylene membrane.[21] The concentration of O_2 diffused through the membrane was proportional to the DO partial pressure surrounding the outside of the membrane. The electric potential difference formed between the electrodes exhibited linear change in response to the O_2 concentration change. Other studies on Clark-type sensors have been done with different membrane materials: polyvinyl chloride, Teflon, Nafionz.[19,22]

However, these sensors involve O_2 consumption and influence the gas concentration surrounding the cathode. In addition, the Clark-type apparatus is difficult to be miniaturized for *in situ* O_2 sensing of lab-on-a-chip application. Due to development in microfabrication techniques, researchers have found ways to address these issues. The effect of cathode size reduction on blood O_2 partial pressure sensing behavior was studied. The Suzuki group constructed a strip consisting of a μsquare Ag cathode and Ag/AgCl anode which is immersed in the electrolyte solution (Figure 3.2a).[23] Here, silicon rubber works as an O_2-permeating membrane (Figure 3.2b and c). For the blood p_{O2} measurement, the cathode portion of the strip is first immersed in the calibration buffer solution followed by the vinyl-covered blood sample–filled container (Figure 3.2d). Cathode size reduction results in two competing effects on sensing accuracy: lower flow dependency and lower sensitivity. If the cathode is sufficiently small—in this case 50 μm × 50 μm and 25 μm × 25 μm—decreased diffusion effect and flow dependency have stronger influence.

An even smaller sensor was achieved by Krommenhoek et al., who fabricated ultra-microelectrode arrays (UMEA) on the surface of oxidized silicon substrates to fit in the 96-well microreactor for yeast cultivation.[24] Using photolithography, a set of Pt macrostrips and patterned polyimide layers create recessed arrays of UMEAs with a radius of 2 μm (Figure 3.2e). Compared to the conventional Clark-type sensor, UMEA self-consumes relatively negligible portion of the electrode-surrounding DO concentration owing to its reduced effective surface area. however, this is a tradeoff with regard to signal-to-noise ratio. Optimization of electrode surface area is crucial to integrate the UMEA-type sensor to the lab-on-a-chip devices.

The concept of a disposable wristwatch-type blood analyzer with multiple electrochemical biosensor arrays was suggested as an application for miniaturized DO-sensing biochips.[25] Including the oxygen sensor, the electrodes for the sensor arrays are covered with a gel-type electrolyte, which will prevent the electrolyte from being displaced but allow faster ion transfer at operating temperature. The blood loading to the sensor reservoir is initiated by injecting the blood sample to the middle layer of a five-layer structured device. Afterwards it is pushed upward to the multiplexer layer by an air-bursting on-chip power source, followed by distribution to the reservoir and the portable detection module attached.

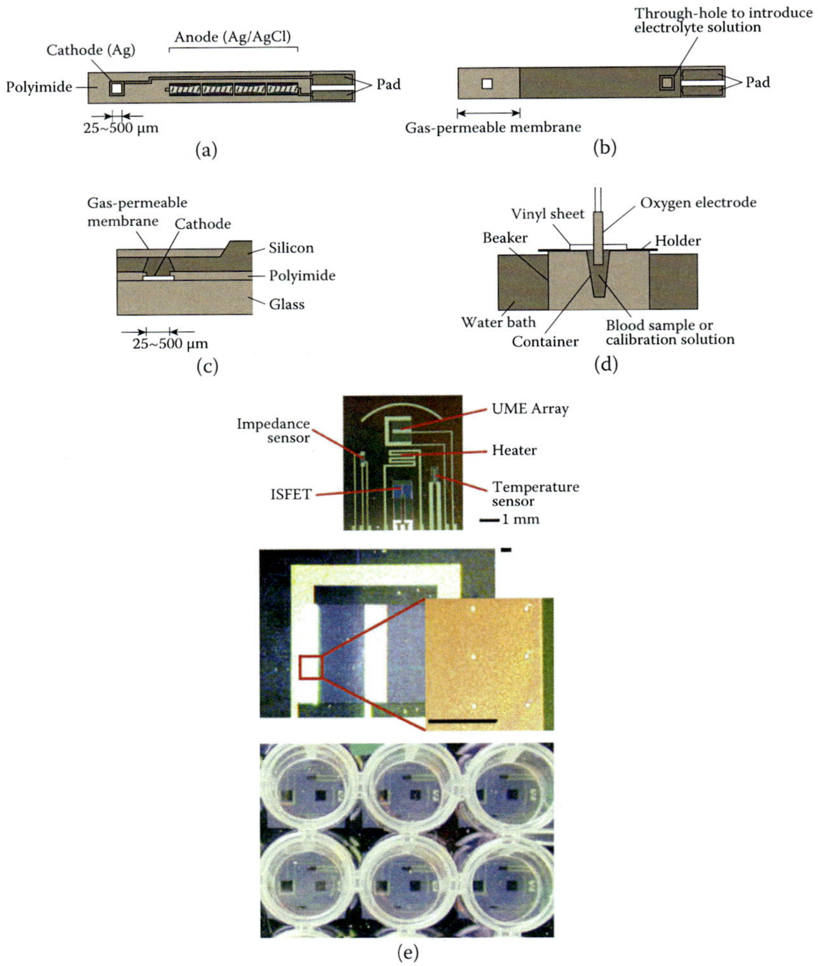

FIGURE 3.2 Schematics of miniaturized oxygen sensor: (a) the electrodes (cathode and anode) patterned on the glass substrate, (b) the layer for gas permeable membrane and the electrolyte solution inlet, (c) the vertical cross-section of the working electrode, and (d) the experimental setup for sensing O_2 partial pressure in the blood sample covered with vinyl to prevent dissolution of O_2 from contact with air. (Reprinted from Determination of blood pO_2 using a micromachined Clark-type oxygen electrode, 431, Suzuki, H.,et al., *Anal. Chim. Acta*, 249–259, Copyright (2001), with permission from Elsevier.) (e) UMEA for dissolved oxygen sensing sensor arrays put under 96-well plate to demonstrate their dimension. (Krommenhoek, E.E., et al.: Lab-scale fermentation tests of microchip with integrated electrochemical sensors for pH, temperature, dissolved oxygen and viable biomass concentration. *Biotechnol Bioeng*. 2008. 99. 884–92. Copyright Wiley Periodicals, Inc. Reproduced with permission.)

3.4 CARBON DIOXIDE/pH

It is not just cell growth that depends heavily on the pH environment. pH also influences metabolic activity and plays a role in synthesis of matrix macromolecules.[26] The production of CO_2 occurs during glucose consumption, and acidic chemicals are often generated as byproducts of metabolic reactions. Carbon dioxide levels also act as an indicator for respiratory system health and epithelial function of the alveoli. Therefore, detection of CO_2 level is important for lung-on-a-chip studies.[27,28] The dissolved CO_2 concentration can be related to the atmospheric partial pressure via Henry's Law[29]: $[CO_2]aq = K_H pCO_2$. For culture that maintains a constant pH, it is usually assumed that the exchange between atmospheric CO_2 and dissolved CO_2 within the culture media is in equilibrium. In addition, changes in pH levels may reflect toxic effects from drugs or toxins, thus making the ability to measure pH in organs-on-a-chip more imperative.

One of the earliest pH and CO_2 electrode meters of biological materials was produced by Severinghaus.[30] The electrode estimates the blood CO_2 level via the reaction of CO_2 and sodium bicarbonate and uses the Henderson-Hasselbalch equation to estimate the dissociation of hydrogen. A linear relationship exists between measured CO_2 and pH within the range tested for this electrode. The electrode is designed with immobilized H_2CO_3 and bicarbonate buffer solution encased in a permeable membrane and is capable of comparable measurement in air and aqueous solutions. However, the pH-dependent behavior is not ideal for direct CO_2 measurements[31] because the pH level depends on other acidic species, such as lactic acid, and not simply the changes in CO_2.[32]

The need for direct *in situ* measurement of CO_2 concentration is of particular interest for cultivation of certain bacterial types for fermentation due to its dual capability for both promotion and inhibition of growth. Cell growth also differs at various levels of CO_2 generation among a variety of O_2 flows, and the ability to vary pulse duration to observe the effect on cell culture can be beneficial in microorganism studies.[33] The electrode proposed by Puhar et al. may be calibrated *in situ* by both gaseous CO_2 and aqueous buffer solution with known pH values so that the voltage measurement may be converted to partial pressure of CO_2 via the Nerst equation.[34] The change in CO_2 concentration in the liquid phase could be correlated with the measured partial pressure of CO_2 in the gas phase.

An *in situ* electrode for CO_2/pH sensing was proposed by Shoda et al. which relies on the diffusion of CO_2 across the membrane of the electrode and its subsequent dissolution in the electrolyte solution.[35] The equilibrium reaction, $CO_2 + H_2O \Leftrightarrow H_2CO_3 \Leftrightarrow HCO_3^- + H^+$, provides the pH of the system and is proportional to the concentration of dissolved CO_2 within the culture media. The electrode was used to monitor growth of *Escherichia coli* over a period of 8 hours and was able to measure the dissolved CO_2 generated by metabolism after glucose consumption. The dissolved CO_2 was shown to be five times more concentrated than the partial CO_2 pressure. Simultaneous monitoring of CO_2 in liquid and gas phase was recommended for cultivation of microbial organisms.

The glass Severingaus type electrodes are still considered to be state of the art in terms of pH measurement, but due to their bulkiness, they are difficult to be integrated onto chip-based devices. ISFET provides the solution for miniaturization. These silicon-based electrochemical sensors are equipped with a gate oxide that is exposed to the hydrogen ions available within the solution. The pH value is defined

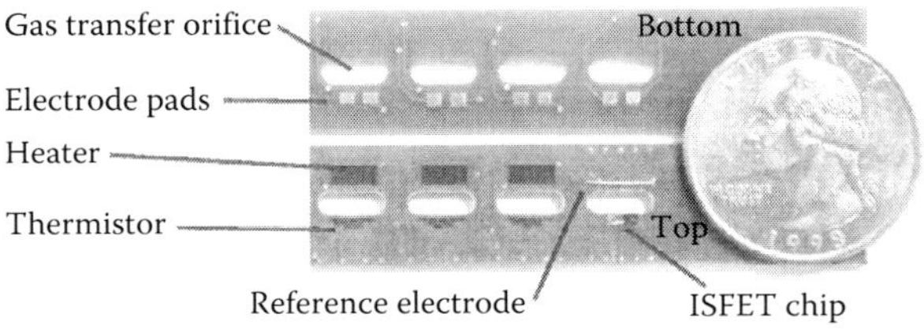

FIGURE 3.3 Assembled ISFET chip. (Maharbiz, M.M., et al.: Microbioreactor arrays with parametric control for highthroughput experimentation. *Biotechnol. Bioeng.* 2004. 85. 376–81. Copyright Wiley Periodicals, Inc. Reproduced with permission.)

by the surface potential, which is measured as voltage and can be calculated using a modified Nernst equation. Maharbiz et al. designed a microreactor (Figure 3.3), to control growth of *E. coli* while continuously monitoring pH level.[36] Similarly, Krommenhoek et al. fabricated a microchip equipped with capabilities for measuring temperature, O_2 concentration, and an ISFET type Ta_2O_5-gated electrode for pH measurement with accuracy below 0.1 pH unit.[24,37] Automated titration to control pH level is included in the bioreactor to control growth condition, and *Candida utilis* was cultured up to 10 hours. Drift is often associated with the ISFET-type pH electrode but will stabilize after 1 to 2 hours of run time, so the measurement can be corrected effectively.

3.5 NITRIC OXIDE

The measurement of nitric oxide (NO) gained a great deal of momentum in research after it was identified as a key player in many physiological processes. It acts as a messenger to control vascular tone[38] and neurotransmission,[39,40] and may also cause detrimental effects on DNA integrity[41] and protein structure, including the potential to cause tissue and organ damage.[42] NO exists at nM levels in physiological systems. Therefore detection methods must possess high sensitivity and ideally be able to perform continuous measurement. As NO is sensitive to light and temperature variations, stability of measurement is also imperative in NO electrode design.

Electrochemistry provides an elegant solution to the aforementioned problems and is suitable for direct detection in biological samples, whereas competing methods in NO analysis such as chemiluminescence, spectrophotometry, bioassay, and other spectroscopic methods are often incapable of continuous monitoring. Commercial NO electrodes are readily available, and one such electrode is based on the Clark's type sensor, which relies on the redox reaction between NO and oxygen contents.[43] However, the size and low sensitivity of the electrode renders the incorporation onto chip-based systems impossible. Another type uses surface modified platinum electrodes, which prevents the generation of harmful or interfering species by NO oxidation (Figure 3.4a). Pariente et al. used a Nafion-coated electrode to eliminate interference from nitrite, ascorbic acids, and negatively charged species, but positively charged molecules resulting from the redox reaction will eventually inhibit accurate detection of NO over time.[44]

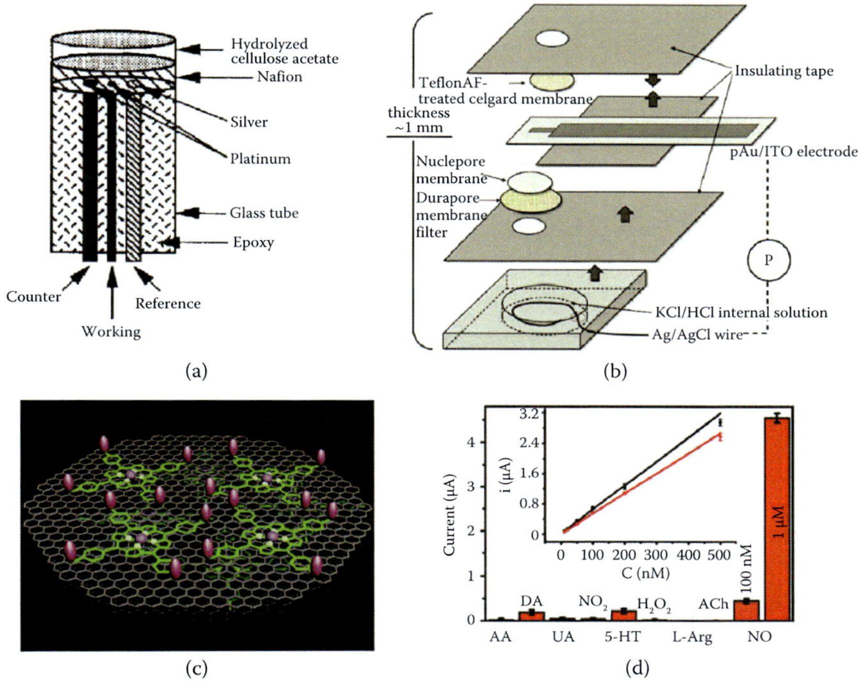

FIGURE 3.4 (a) Schematics of integrated electrodes for NO sensing. (Reprinted from Pariente, F., et al., Chemically modified electrode for the selective and sensitive determination of nitric oxide (NO) *in vitro* and in biological systems, *J. Electroananl. Chem.*, 379: 191–197, Copyright [1994], with permission from Elsevier.) (b) Schematics of multilayered microfluidic system for real-time detection of NO during cell culture. (Reprinted with permission from Cha, W., et al., *Anal. Chem.* 82, 3300–3305, 2010. Copyright [2010] American Chemical Society.) (c) Schematic representation of functionalized graphene-based NO analysis system. (From Liu, Y.-L., et al., *Chem. Sci.*, 6, 1853–1858, 2015.) (d) Assay of nitrogen derived species and charged species showed interference from positively charged species. (Liu, J., et al., Syringe-injectable electronics. *Nat Nanotechnol*, 2015; 10(7): 629–36. Reproduced by permission of The Royal Society of Chemistry.)

In an attempt to clarify the exact mechanism of NO delivery to vascular systems, Kotsis et al. utilized a carbon electrode coated with Nafion in a microbore tubing flow cell to monitor NO content. Endothelial cells are introduced into the flow cell with a fibronectin coating on the tubing and the cells are incubated and cultured to confluence. This microsystem mimicked the environment, including shear forces experienced by endothelial cells in a vascular system, and enabled elucidation of the role of NO in the process of vasodilation. The function of NO was concluded to be the same when the system was modified to a Polydimethylsiloxane (PDMS) microchannel device. The microchannels were fabricated using photolithography and the electrode was deposited with a carbon ink electrode method with Nafion post-modification. The same system was later found to produce NO readings without statistical differences, whether the Nafion coating was present or not. This microchannel design is more robust than its initial microbore tubing and has the potential to be part of a human-on-a-chip device.[45–47]

Nafion-modified carbon electrodes were used to monitor NO production in a human glioblastoma cell culture system in conjunction with a gold electrode for simultaneous detection of superoxide free radicals by Chang et al. Despite the high reactivity between both molecules, the electrodes were shown not to have signal cross-talk by introducing inhibitory reagents of either species into the culture system, which may prove useful in drug discovery.[48]

The interference of cationic molecules during NO oxidation processes or native positively charged molecules within solutions may be prevented by use of NO-selective membrane, WPIss-10, as coating for electrodes. This carbon microelectrode introduced by Zhang et al. is designed for measurement in biological samples. The electrode reported detection of NO as low as 0.3 nM and boasted a sensitivity of 17 pA/nM NO for the amperometric response. The temperature effect on the electrode is characterized to be 10 nM NO/degree C, and the researchers claimed that such deviation may be easily corrected.[49,50]

Despite the versatility of PDMS in microfluidic and human-on-a-chip devices, it is notoriously gas permeable and hydrophobic.[51] Different curing agent ratios for the PDMS matrix were used to circumvent this behavior, but a different substrate may be more suitable for electrochemical measurements on a chip. Teflon was chosen as the substrate for the design of a microchannel culture system by Cha et al.[52] Sputtered gold/indium tin oxide electrode on top of a polymer membrane, Nuclepore, coated with Au-HCF served as the working electrode with a Celgard membrane as the gas-permeable membrane (Figure 3.4b). This combination was chosen for its low over-potential for NO oxidation. Nuclepore provides a porous structure where the electrode may contact the solution and also provide flexibility to the working electrode. The setup was shown to have a sensitivity of ~10 pA/nM NO with a detection limit of approximately 1nM, and has a lifetime of 4 weeks. Biological studies were performed with macrophage-type cells. Endotoxin was introduced into the device to simulate the response of bacterial intrusion or immune response, and NO upregulation was observed as expected. The performance of the electrode in Dulbecco's Modified Eagle's Medium decreased more rapidly than the control Phosphate buffered saline (PBS) studies, but demonstrated another success and method for electrode integration on chip.

Electrodes made from layered nanosheets with a grapheme-based device were demonstrated for NO analysis in a cell culture system.[53] Fe(III) meso-tetra (4-carboxyphenyl) porphyrin combined with graphene oxide was deposited onto an indium tin oxide microelectrode array to create the nanosheets (Figure 3.4c). The surface was further functionalized by 3-aminophenylboronic acid, which serves as the adhesion layer so that endothelial cells may attach onto the surface and multiply. The sensor was able to achieve 55 pM and 90 pM limits of detection in PBS and culture media, respectively. The electrode also performed well for most species in selectivity tests where interfering superoxide and nitrogen-derived molecules were added to the system; however, positively charged molecules demonstrated slight interferences with detection (Figure 3.4d). The device was also found to be reusable and was postulated to have potential *in vivo* capability.

3.6 GLUCOSE

Glucose consumption in metabolic processes during cell culture and within human-on-a-chip devices presents its own niche of importance and interest, but the primary drive for the need of biosensor technologies for glucose measurement stemmed from the management and control of diabetes mellitus in which glucose must be closely monitored and tightly controlled. In this condition, high glucose concentrations are problematic and can cause diabetic ketoacidosis, a potentially life-threatening condition requiring continuous insulin therapy and monitoring in the hospital setting. In addition, due to insulin injections utilized for diabetic treatment, among other factors, deviation from the normal range to a lower than normal range can also occur, and if severe may cause seizures, shock, or progress to a life-threatening condition in some circumstances. Therefore, accurate identification and tight control of glucose levels remain important objectives for future research with considerable market value.

Glucose oxidase is the basis for amperometric electrochemical sensors, and the first glucose electrode was developed in 1962 utilizing this enzyme-catalyzed reaction:

$$\text{Glucose} + O_2 \xrightarrow[\text{glucose oxidase}]{} \text{gluconic acid} + H_2O_2 \tag{3.2}$$

with the platinum cathode monitoring the oxygen consumption reaction.[54] Later, different glucose oxidase-based sensors using other reduction reactions were proposed as well.[55] As interest in biosensors continues to grow, great strides are being made in design, device size, sensitivity, and detection limit. Synthetic electron acceptors, called mediators, eventually replaced oxygen to transport electrons from the active sites of glucose oxidase to electrodes.[56] Ohara et al. introduces polymer films as bridges between glucose oxidase redox sites and electrode surfaces to reduce the response time and provide high current output.[57]

Despite the widespread use of electrochemical glucose sensors for *in vivo* or *ex vivo* measurements in biomedical applications, the incorporation of glucose sensors into cell culture chips took place much later. Miniaturization and fabrication of thin-film electrochemical glucose sensors enables simultaneous analysis on a microfluidic cell culture chip. Pereira Rodrigues et al. fabricated a PDMS biochip for human hepatoblastoma cells with oxygen and glucose sensors on the inlet and outlet of the fluid flow (Figure 3.5a). Nafion modified gold electrode was chosen as the working electrode with a glucose oxidase coating. Chronoamperometric measurement of intermittent injections of glucose and blank PBS solution showed repeatable and linear response corresponding to different concentrations. Cell culture experiments were only recorded over a period of 12 hours, but the electrodes were stable under microfluidic conditions for over 12 days.[58]

The landscape of biosensors also changed with the emergence of nanotechnology. With nanofabrication, even smaller electrodes were made possible. Carbon nanotubes became the new platform for electrochemical biosensors by coupling glucose oxidase to dangling tips of carbon nanotubes[59,60] (Figure 3.5b). Despite the advantages provided by carbon nanotubes over traditional electrode materials, the cost of fabrication is too high for mass production. Relatively affordable graphene sheets present an alternative. Boero et al. demonstrated a carbon nanotube glucose/lactate sensor system for cell culture systems.[61,62]

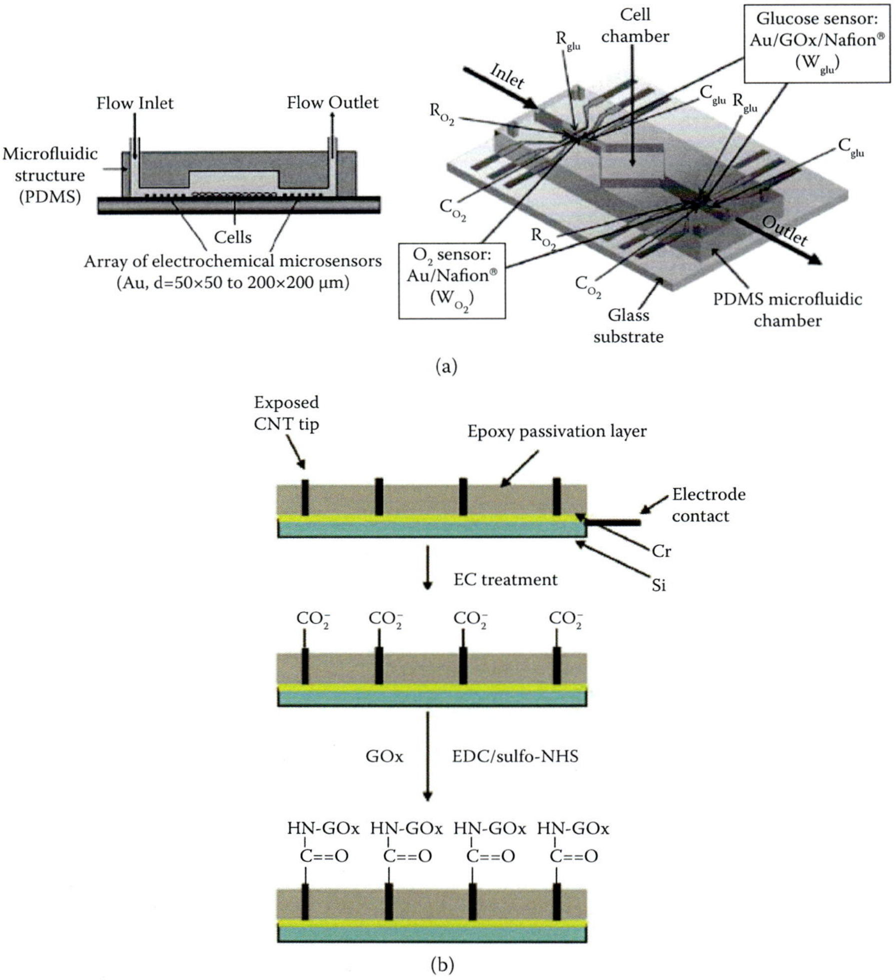

FIGURE 3.5 (a) Schematic representation of glucose and oxygen electrochemical microsensor. (Reprinted from Cellbased microfluidic biochip for the electrochemical realtime monitoring of glucose and oxygen, 132(2), N. Pereira Rodrigues, Y. Sakai, T. Fujii, *Sensors Actuators B Chem.*, 608–613, Copyright [2008], with permission from Elsevier.) (b) Cartoon representation of fabrication process for carbon nanoelectrodes. (Reprinted with permission from Lin, Y., et al., *Nano Letters*, 4, 191–195, 2004. Copyright [2004] by the American Physical Society.)

Many studies have realized glucose sensors based on graphene sheets. Nafion was previously shown to provide higher electrocatalytic efficiency and antifouling properties with carbon nanotubes[63] and was used by Lu et al. in their fabrication of grapheme-based glucose sensors.[64] Eliminating mediators simplifies the number of reactions involved, therefore reducing the redox potential. This can be realized by graphene sheets, as demonstrated by Kang et al., who used immobilized glucose oxidase on chitosan-modified graphene.[65] Results from cyclic voltammograms showed anodic and cathodic peak potential close to standard electrode potential, which suggested near-direct electron transfer. Similar direct electron transfer has been reported using different graphene modifications, such as polyvinylpyrrolidone,[66] gold nanoparticle-chitosan,[67] and 3-aminopropyltriethoxysilane.[68] Despite the widespread interest in graphene as glucose-sensing electrodes, there is yet no report on incorporation of such electrodes with cell based organs-on-a-chip or human-on-a-chip devices.

Electrochemical glucose sensors for organ-on-a-chip devices are still rare, while fluorescent-based glucose-sensing technology is relatively easier to find.[69–71] Glucose metabolism is a proposed indicator for the quality of embryo development, and Heo et al. designed a microfluidic device capable of embryo culture and glucose measurement. The system consists of an on-chip braille display fluidic control with *in situ* indium tin oxide heating element, and the depletion of glucose concentration is measured by fluorescent quantification via glucose oxidase reaction.

3.7 TEMPERATURE

The kinetics and thermodynamics of chemical reactions is dependent on reaction temperature, including redox reactions that are used in the sensors for molecular detection, and the activity of enzymes and cells are also highly temperature dependent. Therefore, control and sensing of temperature in an organ-on-a-chip system is a necessary feature. However, compared to oxygen, pH, glucose, or NO sensing, temperature measurements are complicated in microreactor systems due to small fluid volumes and complex thermal transfer processes associated with a variety of device material properties and structures. This means that large thermal couples or contact-type thermal devices may disturb the distribution of heat and change the dynamic of temperature transfer in a microreactor. As a result, on-chip miniaturization of temperature sensors is desirable.

Analysis of samples collected from organs-on-a-chip can be analyzed using conventional techniques off-chip, such as western blots, enzyme-linked immunoassays, or polyermerase chain reactions (PCRs).[72] Microfluidic-based PCR for DNA amplification has been achieved and is one technique that would benefit from *in situ* temperature control due to the repeated thermal cycles necessary for the procedure. Taking advantage of the small reaction chamber and consequently lower heat capacity of a microreactor, thermal cycles may become more efficient and enable the capability to undergo more cycles before losing polymerase enzyme activity. High-throughput microfluidic devices equipped with Pt thin-film thermal heating and sensor controls were demonstrated to be capable of 30 cycles in 3 minutes, compared to the minimum 30 minutes for conventional PCR machines.[73] However, due to the comparatively large surface area-to-volume ratio of the microchip, quantification of DNA amplification using on-chip PCR is shown to be less optimal than conventional PCR machines. This is postulated to be caused by biomolecules being adsorbed onto the surface of microchips.

The silicon microchip proposed by Yoon[73] requires fabrication processes that are generally more expensive than the soft lithography processes of polymer-based

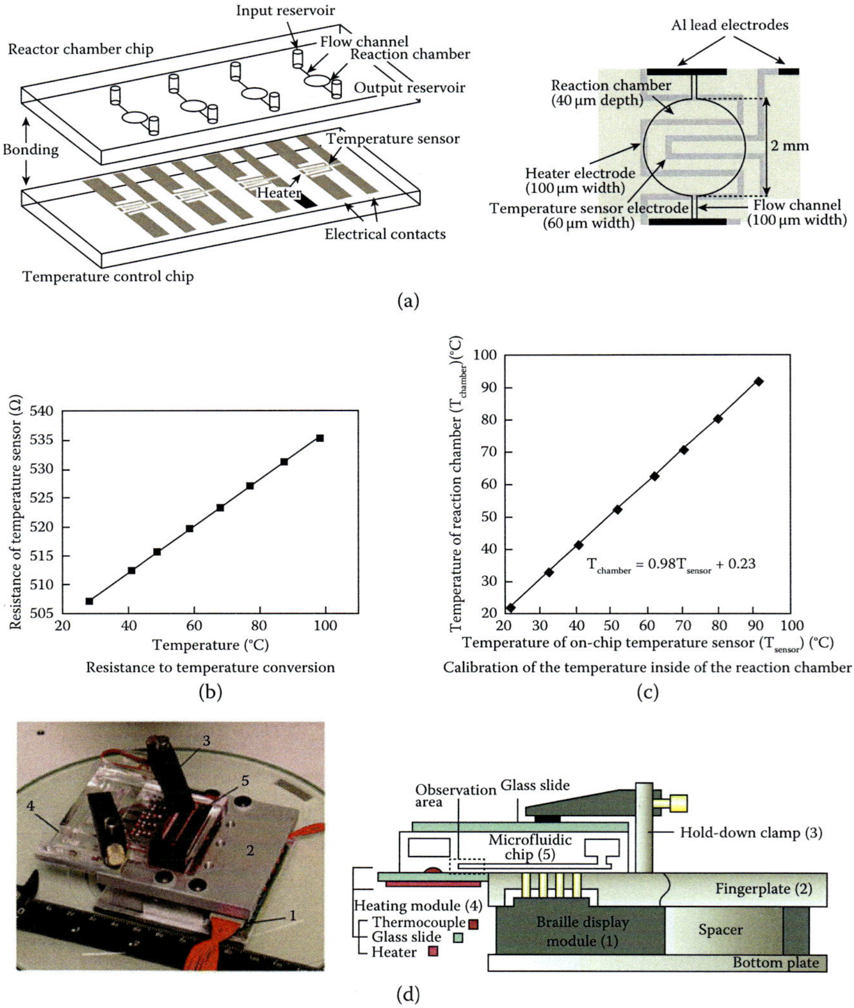

FIGURE 3.6 (a) Schematics of a temperature control chip showing the zigzag resistive heaters and sensors. (From Yamamoto, T., T. Fujii, and T. Nojima, PDMS-glass hybrid microreactor array with embedded temperature control device. Application to cell-free protein synthesis. *Lab Chip*, 2002; 2(4): 197–202. Reproduced by permission of The Royal Society of Chemistry.) (b, c) Linear calibration curves for resistance readout temperature. (From Yamamoto, T., T. Fujii, and T. Nojima, PDMS-glass hybrid microreactor array with embedded temperature control device. Application to cell-free protein synthesis. *Lab Chip*, 2002; 2(4): 197–202. Reproduced by permission of The Royal Society of Chemistry.) (d) Thermocouple-based temperature sensor with integrated Braille display module for portable long-term culture. (From Futai, N., et al., Handheld recirculation system and customized media for microfluidic cell culture. *Lab Chip*, 2006; 6(1): 149–54. Reproduced by permission of The Royal Society of Chemistry.)

microchip systems commonly used in organs-on-a-chip. El-Ali et al. presented an SU-8 PCR microchip equipped with Pt-resistive heaters and sensors. The surface of SU-8 was treated using dichlorodimethylsilane to decrease the aforementioned biomolecule adsorption, and the microchip yielded two-thirds the quantity of amplified products compared to conventional PCR.[74] On-chip analysis for analyzing organ-on-a-chip products can thus be realized.

Protein synthesis in microreactors is another procedure that requires temperature control. Although external heating and cooling systems are available, *in situ* elements that provide more precise control seem to be preferred. Zigzag-shaped resistive heaters and temperature sensors (Figure 3.6a) made from indium tin oxide were patterned on a temperature control chip for a PDMS microreactor.[75] Photolithography allows precise placement of sensors for accurate readout of reactor temperature, and the temperature profile is shown to be quite uniform across the reaction chamber. Linear resistance-temperature calibration curves can be obtained to convert the resistance readout to temperature during experiments (Figure 3.6b, c).

Recently, Ren et al. demonstrated an integrated microelectrochemical system for biological analysis using a capacitive temperature sensor.[76] Digital capacitance measurements were used to infer the dielectric constant of temperature-sensing PDMS fluids. Temperature can be determined from the linear relationship between the dielectric constant and temperature. Besides using bottom-up fabrication of temperature sensors, wired thermocouples can be attached to the external surface of lab-on-chip devices (Figure 3.6d). Futai et al. combined a Braille display module with heaters and temperature sensors to create a portable cell culture microfluidic device. The Braille-driven chip with accompanying media reservoir provides dynamic control, while the temperature is controlled by a digital controller with a thermocouple affixed on the device to monitor *in situ* temperature in the cell culture area.[77]

3.8 CONCLUSION

The organ-on-chip field has developed sufficiently such that the question at hand is less "can we create useful organs-on-a-chip," but rather "how can we efficiently extract useful information from the organs-on-a-chip?" Initial efforts to evaluate cell function in organs-on-a-chip were based on microscope visualizations and end-point evaluations, and often after fixing and staining cells within devices, or after extracting nucleic acids from the cells. As culture durations in organs-on-a-chip increase and the cell function evaluation goes beyond live/dead responses to more subtle non-lethal cell function differences, the ability to perform minimally-invasive, real-time, continuous or semi-continuous monitoring of microfluidic cell culture becomes increasingly important. Along with this ability exists the need for miniaturization and multiplexing of sensing capabilities. Therefore electrochemical sensors integrated with small electrodes and utilizing existing technologies and infrastructures from the microelectronics industry promise to provide an attractive direction for organ-on-a-chip sensor development.

REFERENCES

1. Yang, L., et al., Interdigitated microelectrode (IME) impedance sensor for the detection of viable *Salmonella typhimurium*. *Biosens Bioelectron*, 2004; **19**(10): p. 1139–47.
2. Suehiro, J., et al., Selective detection of viable bacteria using dielectrophoretic impedance measurement method. *J Electrostat*, 2003; **57**(2): p. 157–68.
3. Pattison, R.N., et al., Measurement and control of dissolved carbon dioxide in mammalian cell culture processes using an in situ fiber optic chemical sensor. *Biotechnol Prog*, 2000; **16**(5): p. 769–74.
4. Wolfbeis, O.S., et al., Fluorimetric analysis. *Fresenius' Z Anal Chem*, 1983; **314**(2): p. 119–24.
5. Kostov, Y., et al., Low-cost microbioreactor for high-throughput bioprocessing. *Biotechnol Bioeng*, 2001; **72**(3): p. 346–52.
6. Wu, M.H., et al., Development of high throughput optical sensor array for on-line pH monitoring in micro-scale cell culture environment. *Biomed Microdevices*, 2009; **11**(1): p. 265–73.
7. Lee, H.L., et al., Microbioreactor arrays with integrated mixers and fluid injectors for high-throughput experimentation with pH and dissolved oxygen control. *Lab Chip*, 2006; **6**(9): p. 1229–35.
8. Harms, P., Y. Kostov, and G. Rao, Bioprocess monitoring. *Curr Opin Biotechnol*, 2002; **13**(2): p. 124–27.
9. Luo, H., et al., Investigation of the electrochemical and electrocatalytic behavior of single-wall carbon nanotube film on a glassy carbon electrode. *Anal Chem*, 2001; **73**(5): p. 915–20.
10. Sherigara, B.S., W. Kutner, and F. D'Souza, Electrocatalytic Properties and Sensor Applications of Fullerenes and Carbon Nanotubes. *Electroanalysis*, 2003; **15**(9): p. 753–72.
11. Banks, C.E. and R.G. Compton, Exploring the electrocatalytic sites of carbon nanotubes for NADH detection: An edge plane pyrolytic graphite electrode study. *Analyst*, 2005; **130**(9): p. 1232–9.
12. Banks, C.E., et al., Electrocatalysis at graphite and carbon nanotube modified electrodes: Edge-plane sites and tube ends are the reactive sites. *Chem Commun (Camb)*, 2005; **7**: p. 829–41.
13. Banks, C.E., et al., Investigation of modified basal plane pyrolytic graphite electrodes: Definitive evidence for the electrocatalytic properties of the ends of carbon nanotubes. *Chem Commun (Camb)*, 2004; **16**: p. 1804–5.
14. Shao, Y., et al., Graphene Based Electrochemical Sensors and Biosensors: A Review. *Electroanalysis*, 2010; **22**(10): p. 1027–36.
15. Zhu, Z., et al., A critical review of glucose biosensors based on carbon nanomaterials: Carbon nanotubes and graphene. *Sensors (Basel)*, 2012; **12**(5): p. 5996–6022.
16. Patel, J., et al., Electrochemical properties of nanostructured porous gold electrodes in biofouling solutions. *Anal Chem*, 2013; **85**(23): p. 11610–18.
17. Liu, J., et al., Syringe-injectable electronics. *Nat Nanotechnol*, 2015; **10**(7): p. 629–36.
18. Tian, B., et al., Macroporous nanowire nanoelectronic scaffolds for synthetic tissues. *Nat Mater*, 2012; **11**(11): p. 986–94.
19. Eklund, S.E., et al., Modification of the Cytosensor™ microphysiometer to simultaneously measure extracellular acidification and oxygen consumption rates. *Anal Chim Acta*, 2003; **496**(1–2): p. 93–101.
20. Bambot, S.B., et al., Phase fluorometric sterilizable optical oxygen sensor. *Biotechnol Bioeng*, 1994; **43**(11): p. 1139–45.
21. Clark, L.C., Jr., Monitor and control of blood and tissue oxygen tensions. *ASAIO J*, 1956; **2**(1): p. 8.

22. Eklund, S.E., et al., A microphysiometer for simultaneous measurement of changes in extracellular glucose, lactate, oxygen, and acidification rate. *Anal Chem*, 2004; **76**(3): p. 519–27.
23. Suzuki, H., et al., Determination of blood pO_2 using a micromachined Clark-type oxygen electrode. *Anal Chim Acta*, 2001; **431**(2): p. 249–59.
24. Krommenhoek, E.E., et al., Lab-scale fermentation tests of microchip with integrated electrochemical sensors for pH, temperature, dissolved oxygen and viable biomass concentration. *Biotechnol Bioeng*, 2008; **99**(4): p. 884–92.
25. Ahn, C.H., et al., Disposable smart lab on a chip for point-of-care clinical diagnostics. *Proceedings of the IEEE*, 2004; **92**(1): p. 154–73.
26. Wu, M.H., et al., Effect of extracellular ph on matrix synthesis by chondrocytes in 3D agarose gel. *Biotechnol Prog*, 2007; **23**(2): p. 430–4.
27. Kumar Mahto, S., J. Tenenbaum-Katan, and J. Sznitman, Respiratory physiology on a chip. *Scientifica (Cairo)*, 2012; **2012**: p. 364054.
28. Briva, A., et al., High CO_2 levels impair alveolar epithelial function independently of pH. *PLoS One*, 2007; **2**(11): p. e1238.
29. Dixon, N.M. and D.B. Kell, The control and measurement of 'CO_2' during fermentations. *J Microbiol Methods*, 1989; **10**(3): p. 155–76.
30. Severinghaus, J.W., Electrodes for blood pO_2 and pCO_2 determination. *Anesthesiology*, 1959; **20**(1): p. 136.
31. Donaldson, T.L. and H.J. Palmer, Dynamic response of the carbon dioxide electrode. *AIChE J*, 1979; **25**(1): p. 143–51.
32. Nicholls, D.G. and P.B. Garland, Electrode measurements of carbon dioxide. *Methods Microbiol*, 1972; **6**: p. 55–63.
33. Shang, L., et al., Inhibitory effect of carbon dioxide on the fed-batch culture of Ralstonia eutropha: Evaluation by CO_2 pulse injection and autogenous CO_2 methods. *Biotechnol Bioeng*, 2003; **83**(3): p. 312–20.
34. Puhar, E., et al., Steam-sterilizable pCO_2 electrode. *Biotechnol Bioeng*, 1980; **22**(11): p. 2411–16.
35. Shoda, M. and Y. Ishikawa, Carbon dioxide sensor for fermentation systems. *Biotechnol Bioeng*, 1981; **23**(2): p. 461–6.
36. Maharbiz, M.M., et al., Microbioreactor arrays with parametric control for high–throughput experimentation. *Biotechnol Bioeng*, 2004; **85**(4): p. 376–81.
37. Krommenhoek, E.E., et al., Integrated electrochemical sensor array for on-line monitoring of yeast fermentations. *Anal Chem*, 2007; **79**(15): p. 5567–73.
38. Moncada, S., M.W. Radomski, and R.M.J. Palmer, Endothelium-derived relaxing factor. *Biochem Pharmacol*, 1988; **37**(13): p. 2495–501.
39. Bult, H., et al., Nitric oxide as an inhibitory non-adrenergic non-cholinergic neurotransmitter. *Nature*, 1990; **345**(6273): p. 346–7.
40. Rajfer, J., et al., Nitric oxide as a mediator of relaxation of the corpus cavernosum in response to nonadrenergic, noncholinergic neurotransmission. *Obstetrical & Gynecological Survey*, 1992; **47**(5): p. 350–51.
41. Wink, D.A. and J. Laval, The Fpg protein, a DNA repair enzyme, is inhibited by the biomediator nitric oxide in vitro and in vivo. *Carcinogenesis*, 1994; **15**(10): p. 2125–9.
42. Bonfoco, E., et al., Apoptosis and necrosis: Two distinct events induced, respectively, by mild and intense insults with N-methyl-D-aspartate or nitric oxide/superoxide in cortical cell cultures. *Proc Natl Acad Sci U S A*, 1995; **92**(16): p. 7162–6.
43. Liu, X., et al., Quantitative measurements of NO reaction kinetics with a Clark-type electrode. *Nitric Oxide*, 2005; **13**(1): p. 68–77.
44. Pariente, F., J.L. Alonso, and H.D. Abruña, Chemically modified electrode for the selective and sensitive determination of nitric oxide (NO) in vitro and in biological systems. *J Electroanal Chem*, 1994; **379**(1–2): p. 191–7.

45. Kotsis, D.H. and D.M. Spence, Detection of ATP-induced nitric oxide in a biomimetic circulatory vessel containing an immobilized endothelium. *Anal Chem*, 2003; **75**(1): p. 145–51.
46. Spence, D.M., et al., Amperometric determination of nitric oxide derived from pulmonary artery endothelial cells immobilized in a microchip channel. *Analyst*, 2004; **129**(11): p. 995–1000.
47. Hulvey, M.K. and R.S. Martin, A microchip-based endothelium mimic utilizing open reservoirs for cell immobilization and integrated carbon ink microelectrodes for detection. *Anal Bioanal Chem*, 2009; **393**(2): p. 599–605.
48. Chang, S.C., et al., An electrochemical sensor array system for the direct, simultaneous in vitro monitoring of nitric oxide and superoxide production by cultured cells. *Biosens Bioelectron*, 2005; **21**(6): p. 917–22.
49. Zhang, X., et al., A novel microchip nitric oxide sensor with sub-nM detection limit. *Electroanalysis*, 2002; **14**(10): p. 697.
50. Zhang, X., Real time and in vivo monitoring of nitric oxide by electrocehmical sensors- from dream to reality. *Front Biosci*, 2004; **9**(1–3): p. 3434.
51. Merkel, T.C., et al., Gas sorption, diffusion, and permeation in poly(dimethylsiloxane). *J Polym Sci B Polym Phys*, 2000; **38**(3): p. 415–34.
52. Cha, W., et al., Patterned electrode-based amperometric gas sensor for direct nitric oxide detection within microfluidic devices. *Anal Chem*, 2010; **82**(8): p. 3300–5.
53. Liu, Y.-L., et al., Functionalized graphene-based biomimetic microsensor interfacing with living cells to sensitively monitor nitric oxide release. *Chem Sci*, 2015; **6**(3): p. 1853–8.
54. Clark, L.C. and C. Lyons, Electrode systems for continuous monitoring in cardiovascular surgery. *Ann N Y Acad Sci*, 2006; **102**(1): p. 29–45.
55. Guilbault, G.G. and G.J. Lubrano, An enzyme electrode for the amperometric determination of glucose. *Anal Chim Acta*, 1973; **64**(3): p. 439–55.
56. Wang, J., Electrochemical glucose biosensors. *Chem Rev*, 2008; **108**(2): p. 814–25.
57. Ohara, T.J., R. Rajagopalan, and A. Heller, "Wired" enzyme electrodes for amperometric determination of glucose or lactate in the presence of interfering substances. *Anal Chem*, 1994; **66**(15): p. 2451–7.
58. Pereirarodrigues, N., Y. Sakai, and T. Fujii, Cell-based microfluidic biochip for the electrochemical real-time monitoring of glucose and oxygen. *Sens Actuators B Chem*, 2008; **132**(2): p. 608–13.
59. Patolsky, F., Y. Weizmann, and I. Willner, Long-range electrical contacting of redox enzymes by SWCNT connectors. *Angew Chem Int Ed Engl*, 2004; **43**(16): p. 2113–17.
60. Lin, Y., et al., Glucose biosensors based on carbon nanotube nanoelectrode ensembles. *Nano Lett*, 2004; **4**(2): p. 191–5.
61. Boero, C., et al., Design, development, and validation of an in-situ biosensor array for metabolite monitoring of cell cultures. *Biosens Bioelectron*, 2014; **61**: p. 251–9.
62. Boero, C., et al., Highly sensitive carbon nanotube-based sensing for lactate and glucose monitoring in cell culture. *IEEE Trans Nanobiosci*, 2011; **10**(1): p. 59–67.
63. Wang, J., M. Musameh, and Y. Lin, Solubilization of carbon nanotubes by Nafion toward the preparation of amperometric biosensors. *J Am Chem Soc*, 2003; **125**(9): p. 2408–9.
64. Lu, J., et al., Simple fabrication of a highly sensitive glucose biosensor using enzymes immobilized in exfoliated graphite nanoplatelets nafion membrane. *Chem Mater*, 2007; **19**(25): p. 6240–6.
65. Kang, X., et al., Glucose oxidase-graphene-chitosan modified electrode for direct electrochemistry and glucose sensing. *Biosens Bioelectron*, 2009; **25**(4): p. 901–5.
66. Shan, C., et al., Direct electrochemistry of glucose oxidase and biosensing for glucose based on graphene. *Anal Chem*, 2009; **81**(6): p. 2378–82.

67. Shan, C., et al., Graphene/AuNPs/chitosan nanocomposites film for glucose biosensing. *Biosens Bioelectron*, 2010; **25**(5): p. 1070–4.
68. Wang, Z., et al., Direct electrochemical reduction of single-layer graphene oxide and subsequent functionalization with glucose oxidase. *J Phys Chem C*, 2009; **113**(32): p. 14071–5.
69. Prill, S., M.S. Jaeger, and C. Duschl, Long-term microfluidic glucose and lactate monitoring in hepatic cell culture. *Biomicrofluidics*, 2014; **8**(3): p. 034102-1–034102-9.
70. Lin, Z., et al., In-situ measurement of cellular microenvironments in a microfluidic device. *Lab Chip*, 2009; **9**(2): p. 257–62.
71. Heo, Y.S., et al., Real time culture and analysis of embryo metabolism using a microfluidic device with deformation based actuation. *Lab Chip*, 2012; **12**(12): p. 2240–6.
72. Bhatia, S.N. and D.E. Ingber, Microfluidic organs-on-chips. *Nat Biotechnol*, 2014; **32**(8): p. 760–72.
73. Yoon, D.S., et al., Precise temperature control and rapid thermal cycling in a micromachined DNA polymerase chain reaction chip. *J Micromech Microeng*, 2002; **12**(6): p. 813–23.
74. El-Ali, J., et al., Simulation and experimental validation of a SU-8 based PCR thermocycler chip with integrated heaters and temperature sensor. *Sens Actuators A Phys*, 2004; **110**(1–3): p. 3–10.
75. Yamamoto, T., T. Fujii, and T. Nojima, PDMS-glass hybrid microreactor array with embedded temperature control device. Application to cell-free protein synthesis. *Lab Chip*, 2002; **2**(4): p. 197–202.
76. Ren, Q., et al., A novel capacitive temperature sensor for a lab-on-a-chip system. *IEEE SENSORS 2014 Proceedings*, 2014: p. 436–9.
77. Futai, N., et al., Handheld recirculation system and customized media for microfluidic cell culture. *Lab Chip*, 2006; **6**(1): p. 149–54.

4 Microfluidics

Panupong Jaipan and Roger Narayan

CONTENTS

4.1 OVERVIEW

In this chapter, we provide promising aspects of microfluidics to answer questions that why in the near future researchers would prefer to invest in microfluidics for integration with their medical technology, and how microfluidic devices have significant impact on current development of biomedical research fields.

4.2 MICROFLUIDICS

Microfluidics [1–6] is a multidisciplinary research field, and there are many diverse disciplines of its applications. For the ongoing development in the design and utilization of microfluidic devices for fluid transport and fluid manipulation, there is a wide range of practical applications from the healthcare industries for biomedicine (e.g., drug delivery, detection, and diagnostic devices) and pharmaceuticals. In the medical field, novel microfluidic devices are capable of offering an alternative method for noninvasive diagnostics and surgery [1]. Microfluidic devices consist of the smallest microscaled design features, including microchannels, micropumps, microvalves, micromixers, microfilters, and microsensors; these components can be used for dilution, metering, flow switching, incubation of reaction materials and reagents, mixing, pumping, particle separation, sample dispensing, and/or sample injection [1,2]. In addition, most practical applications of the microdevices are involved in liquid fluid transport. Microfluidic devices have many advantages in terms of flow manipulations and economics since the devices are able to regulate fluid flow with rapid response and relatively high accuracy, improve throughput of samples, handle small fluid volumes from microliter scales down to picoliter scales, and pragmatically reduce the cost of operating systems and analyses. Furthermore, microfluidic devices can provide the chance to minimize experiment operations onto single chips by using a smaller amount of reagents and shorter reaction times relative to the traditional milliliter-scaled devices [1–3,7,8].

Many concepts of microfluidics are therefore significant for the development of both lap-on-chip (LOC) (i.e., organ-on-chip [OOC]) devices and micro-total-analysis systems for medical fields in which the transport processes (e.g., mixing, flow manipulation of the fluids, reagents, and particles, separations, and reactions) have been applied on a relatively small scale compared to conventional engineering technologies [1,2].

The development of the health-care devices would significantly obtain many good points from more accurate, faster, and more highly precise diagnostic devices would greatly reduce the healthcare cost in hospitals in both developed and developing countries, as they can provide better epidemiological data from the infectious disease OOC models [3,9]. For instance, the microfluidic chip was used to study capillary-cell invasion for malignant breast tumor and brain tumor [4,10,11]. In addition, microfluidic devices have the capability to mimic complex organ physiology, which has the potential to be used for studying the development of the human-disease models in order to determine medical procedures to cure these diseases, provide better understanding of the pharmacokinetics of the drug release mechanism, and identify toxicity in patients.

4.2.1 Dimensionless Groups Involved in Microfluidics

There are two fundamental dimensionless groups relevant to the transport phenomena of the fluid; the Reynolds number (Re) and the Peclet number (Pe). The Reynolds number is typically used to indicate the type of the fluid flow—whether the flow is laminar, turbulent, or in transition between laminar and turbulent [3,12].

The Reynolds number is the ratio of inertia to viscous force densities, and we can determine the dimensionless group by the following equation [3]:

$$\mathrm{Re} = \frac{\rho v D_h}{\mu} \tag{4.1}$$

In this equation, D_h is the hydraulic diameter of the channel, ρ is the fluid density, v is the fluid velocity, and μ is the fluid viscosity. In terms of classifications of the types of fluid flow pattern, if the range of the Re number is less than 2100, the flow pattern is classified as a laminar flow; if the range of the Re number is greater than 4000, the fluid flow is turbulent, and if the Re number is in between 2100 and 4000, it indicates the transition region for both laminar and turbulent flows. For the microflow regime, the Re range is typically between 10^{-6} and 10 [3,13].

The other dimensionless group relevant to microfluidics is the Peclet number, which is related to the strength of the diffusive mixing. The Pe number represents the relative strength of the convection versus the diffusion. It can be calculated from the following equation:

$$Pe = \frac{vw}{D} \tag{4.2}$$

In this equation, D is the diffusion coefficient of the solute particles, v is the fluid velocity, and w is the microchannel width.

4.3 TRANSPORT PROCESSES IN MICROFLUIDICS

To design and optimize microfluidic-based devices for medical applications, it is necessary to understand transport processes [2,3] and interfacial phenomena at small scales (e.g., microscales, nanoscales, or smaller), and the design of microfluidics also depends upon the particular use. The transport processes of the microfluidics are fundamentally involved in laminar flow pattern and passive mixing by diffusion, which typically takes place in the microchannels, and the laminar flow generates easy flow patterns with very little diffusion, eliminating potential difficulties in the mixing processes [3]. Electrokinetic phenomena [2], including electro-osmosis, electrophoresis, and streaming potential, also have significant roles in many applications of microfluidics. The small dimensions of the microscaled channels lead to an increase in the surface area-to-volume (SAV), thereby enhancing the processes (e.g., capillary electrophoresis becomes more effective because of easier removal of heat), and also minimizing electrokinetic flow due to rapid diffusion of macromolecules and adsorption to the surfaces of the microchannels. More significantly, controlling electrokinetic phenomena in microfluidic devices for transporting fluids and particles is crucial for organ-on-chips applications [2].

The flow stability and the precise manipulation of the nanoliter down to picoliter volumes of fluids in the microfluidic devices are also important for many applications, including DNA detection in microfluidic devices, since microfluidics can prevent bubble formation inside the channels and chambers due to sensitive detection in the microscaled systems; the undesired bubbles are capable of negatively affecting the fluid flow and causing detection failures, especially in highly sensitive optical detection systems [3]. However, bubbles can be advantageous of the actuation mechanism in some applications [14].

4.3.1 Manipulation of Fluid Flow in Microfluidics

Many kinds of external driving forces [5] (e.g., pressure, electric, magnetic, capillary) can manipulate fluid flows in microfluidic systems, and the fluid flows can be successfully controlled by forces applied macroscopically at either inlets or outlets of the devices. Following is a list of the forces contributing to fluid flows in microfluidic devices discussed in this section (e.g., pressure differences, electric fields, capillary driving, and free-surface flow driven by gradients in interfacial tension) [1,3].

- **Pressure differences:** The velocity distribution is parabolic across the channel, and it is controlled by hydrodynamic systems, which measure the pressure differences in the microchannels via a low-pressure manometer. This pressure-driven approach usually depends upon the flow rates of two fluids.
- **Electric fields:** The velocity profile is commonly uniform, and the stresses are often present in charged (i.e., Debye double) layers near the boundaries drive the bulk fluid motion.
- **Capillary driving:** Forces from wetting of surfaces by the fluid, leading to pressure gradients in liquids, which is similar to applied pressure differences, but it is relatively more concerned with the shape of the interface.
- **Gradient in the interfacial tension (i.e., Marangoni flows):** This driving force also depends upon the dependence of surface tension on temperature or chemical concentration [15,16].

Alternatively, the fluid flow can be achieved by integrating the microchannels with the components, such as micropumps and microvalves. In particular, when a gas-liquid or liquid-liquid interface is present, controlling spatial variations of surface tension (i.e., Marangoni stresses) can generate the fluid motion; the thermal, chemical, or electrical gradients can cause the variations. In addition, using capillary pressure gradients allows transport of liquid/gas or liquid/liquid fluids in the channels. According to the Young-Laplace's law, the capillary pressure gradient is generated by changing wetting properties (e.g., contact angle, or surface tension) or geometrical features (channel diameter) [5]. Moreover, the number of the forms of electrokinetics have been of considerable interest for controlling the fluid flows in microchannels. For electro-osmosis, the fluid moves relative to stationary charged boundaries. The other form, dielectrophoresis, moves particles in a gradient of electric field; in electrowetting, there is an electric field modifying wetting properties of the fluids. The magnetic fields are also able to directly control the flow or regulate the dispersed magnetic particles.

4.3.2 Dispersion

The Taylor-Aris dispersion [1] occurs because of thermal fluctuations (i.e., molecular diffusion), which cause non-uniform velocity profiles of the streamlines. The dispersion is often prevented by creating electroosmotic flows, since the flows of the electroosmotic pattern have the uniform, or mostly uniform, velocity profiles. However, some designs (e.g., three-dimensional networks) for generating electroosmotic flows can still cause dispersion. To minimize hydrodynamic dispersion, it is necessary to understand the detailed fluid flow in all of the microchannels. Dispersion is controlled by the local features of velocity distribution in the channels in which it causes the speed of the fluid flow near the walls of the microchannels disparate from the flow speed in the center of the channels [17].

4.3.3 Mixing

Mixing [1,5] is one of the most important processes in microfluidic devices, but it is difficult for the small-scaled system to perform mixing well because of laminar flows (i.e., low Re number) in the systems and the absence of turbulent pattern. The flow speeds and channel lengths are not influential enough to provide the molecular diffusion, but the chaotic particle paths of the fluid itself can enhance the mixing of two fluids or a tracer in a fluid. Interestingly, the laminar flows have sufficient factors to contribute to chaotic mixing formation (i.e., laminar chaos). Currently, integration of microfluidic devices with active and passive pumping systems is capable of improving the mixing in the channels.

4.3.4 Separation

Separation [1] is undeniably still critical for some biomedical applications (e.g., separating biopolymers during electrophoresis [18]). Particularly, the techniques using cross fields (e.g., electric, thermal, etc.) can produce and enhance the spreading of an injected solute in the flow direction, and this axial spreading bolsters separation processes practically. Moreover, modifying surface properties of the microchannels is possibly able to enhance separations due to interactions between particles and surfaces. Integration of microfluidic devices with the AC operation can also improve the separation efficacy of the microsystems [5,19].

4.3.5 Electrokinetic Phenomena in Microfluidics

Electrokinetics [5] refers to the coupling between electric currents and fluid flow in liquid containing electrolytes. The Debye screening layer forming at charged interfaces typically generates these bulk effects. When the walls bounding the liquid are charged, there are two primary electro-hydrodynamic phenomena taking place, namely, electro-osmosis and streaming current and streaming potentials. First, electro-osmosis occurs as the result of generating a flow from applying the electric field along a liquid-filled channel. Streaming current and streaming potential occurs when a pressure-driven flow draws ions tangential to the surface and consequently creates the electric current. The current can be collected through an electric short-circuit, or recirculated within the electrolyte channel by conductivity; in the case of a steady-state electric potential, the difference can be measured between the ends of the microchannel. Additional phenomena occurring at the interface between an immersed charged object and liquid include electrophoresis, sedimentation potential, and electroviscous effects. When an electric field generates motion of the object relative to the fluid, it causes the electrophoresis effect. While a vertical potential difference is present in a sedimentation system, it engenders the sedimentation potential effect. The electroviscous effect takes place when the flow-induced distortion of the Debye layer has impact with the static and dynamic properties of the suspension.

The following section focuses on electro-osmosis rather than streaming potential, because the latter has not yet been clearly investigated for studies in the microfluidic systems, though it may be advantageous for some applications.

4.3.5.1 Electro-Osmosis Phenomenon

The chemical state of the surface of the microchannels is changed by ionization of covalently bound surface groups or by ion adsorption once the electrolyte is adjacent to the surface [5]. A charge from the surface is the net effect whereby counter-ions are released into the liquid; typically the surface made from glass, SiOH, in the water ionizes and generates SiO^-, charged surface and also releases one photon (H^+). The balance between electrostatic interactions and thermal agitation engenders the charge density profile at the equilibrium state. The liquid is electrically neutral except for a charged layer near the boundary, which exhibits a charge opposite in sign and equal in amplitude to the surface charge. The characteristic thickness of this Debye layer, λ_D, decreases as the inverse square root of the ion concentration in the bulk of the liquid, and typically has a magnitude of 1–100 nm in water [5].

When we apply electric field $\boldsymbol{E}_{ext}$ along the channel, the conductive current and the corresponding local electric field $\boldsymbol{E}$ are generated throughout the liquid. The bulk of the liquid still remains electrically neutral, and it is not affected by the net force. In contrast, there is a net electrical charge density in the Debye layer, and the local electric field $\boldsymbol{E}$ which is tangential to the surface of the microchannel causes the body force on the fluid, thereby inducing a shear. As a result, the fluid velocity increases from zero on the surface to a finite value $-\boldsymbol{m}_{EO}\boldsymbol{E}$ at the edge of the thin Debye layer, where $\boldsymbol{m}_{EO}$ is a local mobility characteristic of the surface and is related to the surface charge density σ_{el}. As the surface potential is small, it leads to

$$m_{EO} = \frac{\sigma_{el}\lambda_D}{\mu} = \frac{\zeta \in \in_0}{\mu} \tag{4.3}$$

where $\in$ is the dielectric constant, $\in_0$ is the permittivity of the vacuum, ζ is the potential of the surface, and μ is the shear viscosity. Note that there is one assumption that there is a no-slip boundary condition on the equation (4.3); however, if there is a finite slip length $\boldsymbol{a}$ on the surface, the mobility increases by a factor $\left(1+\frac{\boldsymbol{a}}{\lambda_D}\right)$.

For the electro-osmotic flow [20], the magnitude of the velocities is dependent on the effective slip phenomenon, but if the dimensions of the cross-section of the microchannel are greater than λ_D, the velocity is independent of the cross-sectional dimensions. Additionally, the system requires a high-voltage supply to generate the electrostatic potential drop between the two ends of the centimeter-long channel. Nevertheless, the very high voltage can cause unstable electro-osmosis; therefore, the electro-osmosis method integrated with the hydrodynamic system is used in microsystem in order to handle the problems of pressure-driven and electro-osmotic techniques [21,22].

4.3.5.2 Importance of Electrokinetics in Microsystems

When all electrokinetic effects [5] take place on the surface of the channels, their effects can provide many advantages to the microfluidic devices, as follows:

1. Electro-osmotic flow can help the microfluidic-based system drive the liquids. In the case of the narrow channel with thickness h and width w, for the given potential drop the volume flow rate of the electro-osmotic flow is proportional to hw. For the given pressure drop, the volume flow rate is proportional to h^3w.
2. Electro-osmotic flow is capable of maintaining high electric fields (greater than 100 volt/cm) with low currents due to the high electrical resistance to ionic presence in the small cross-sectional area of the fluid-filled channels.
3. Spontaneous convection, such as thermally driven convection, is weak because of viscous damping for the small systems (less than 100 microns).
4. Electrophoresis can effectively enhance separation processes in many analytical microsystems.
5. For the electro-osmotic flow in homogeneous channels, the plug flow allows the transport of samples due to hydrodynamic dispersion present in the pressure-driven flows.

4.3.6 Applications of Mixing and Dispersion in Microfluidics

Mixing and dispersion [5] of the reagents, including small molecules, macromolecules such as proteins, or particles, are both significant processes for medical research in the microfluidic system. Some examples of applications in microfluidic devices relevant to mixing and dispersion are as follows:

1. If the chemical transformation, including deposition onto the surface of the microchannel, needs to be performed at particular locations within the channels, minimal mixing of the solute between adjacent laminar streams is required since the transverse mixing occurs by diffusion only, and laminar flow is desired for this application.
2. In the case of initiation of the chemical reaction, the complete and rapid mixing of adjacent laminar streams is required to achieve this purpose, and the laminar flow is not suited for this application; hence, convective mixing has to be improved.

3. In the case of transporting the plugs of analytes for detection applications in microfluidic devices, uniform electro-osmotic flow is required to obtain satisfactory analytical results. A pressure-driven flow with a parabolic profile causes the solute (i.e., analyte) to be stretched out along the direction of the flow; as the result, the detector reads the values inconsistently.

Note that convective transport is typically faster than diffusive transport even in the microscales or smaller channels. In addition, laminar flow with a high Pe number is ideal for applications in delivery and confinement of reagents. In particular, laminar flow has been applied to deliver multiple reagents to a single cell with subcellular spatial resolution [5,23].

4.3.7 Improvement of the Mixing Process in Microfluidics

Typically, mixing between the laminar streams in the microchannel takes place by diffusion only. In order to improve mixing in the microscaled channels, first, the mixing length and mixing time should be decreased, and transverse flows should be generated. If this is not done, the solutions, such as protein solution, cannot be transported within the microchannel. For generating transverse flows, there are three common methods [5] to achieve this goal:

1. Passive method: the interaction of the externally-driven flow, including electro-osmotic flow and pressure-driven, with the fixed channel geometry can engender transverse flows.
2. Active method: the oscillatory forcing (e.g., mechanical or electrical) within the microchannel is capable of generating transverse flows.
3. Adding microstructural deformable elements to the liquids in the case of viscoelastic liquids: this technique generates elastic stresses, resulting in activating flow instabilities [24,25]; consequently, it significantly enhances mixing in a curved microchannel [26].

4.4 ESSENTIAL COMPONENTS OF THE MICROFLUIDICS

Microfluidic devices can be integrated with many kinds of functional components, such as micropumps, microvalves, micromixers, microfilters, and microsensors. This section focuses on a variety of microvalves and micropumps which have been used in medical applications. The following details illustrate how both microvalves and micropumps can contribute to the functionality of the microfluidic devices.

4.4.1 Microvalves

A microvalve [27] is one of the main significant components for designing microfluidic devices with complex functionality; it is used to control flowing, timing, and separating of fluids in the micro-scaled channel. This section interprets five major types of microvalves: electrokinetic, pneumatic, pinch, phase change, and burst microvalves.

4.4.1.1 Electrokinetic Microvalves

Electrokinetic valves [28–31] are typically integrated within microfluidic devices for operating in applications having a continuous flow system only, and performing as a fluid router which uses electro-osmotic flow to switch fluids from one channel to the other. However, there are limitations and drawbacks for this microvalve to be considered when designing the devices. First, for all electrokinetic transport processes, it strongly depends upon the surface properties of the microchannel, and the valve works reliably on glass surfaces only, which makes it difficult and expensive to do micromachining. Second, it is strongly dependent on the ionic composition of the buffer. Third, high-voltage sources and expensive switches are required to operate the system. In addition, the valve is limited to work on continuous-flow application only and is not suited for batch processes.

4.4.1.2 Pneumatic Microvalves

Pneumatic microvalves [27] typically utilize a flexible membrane (e.g., polydimethylsiloxane [PDMS]) to control the flow pattern in the microchannel. It works by applying pneumatic pressure on the membrane via the control channel [32–39]. Moreover, the pneumatic type can be easily integrated with standard soft lithography processes, and it can also be filled with fluid without air interrupting into the microchannel while the control channel is working by activating from external pressure source. Nonetheless, there is a concern that changing the pneumatic pressure can cause deflecting the valve membrane and sealing against the seat. The three typical types of the pneumatic valves are normally-open, normally-closed, and check microvalves.

Normally-open microvalves [33,36,37,39] (e.g., quake, plunger, and plunger microvalves) are used to stop the flow through the valve by deflecting the membrane from applying positive pressure on the control channel, but the valves do not fully close off the flow. Quake microvalves have been used in applications for protein separation [40] and sell sorting [41,42]. Plunger microvalves have been used for microinjection of fluid through a microneedle [43], and have recently been used for delivering a bolus of fluid orthogonal to the direction of the inlet of the fluid [44]. The later-deflection membrane microvalves integrated with a bifurcating channel junction have been successfully employed for high-speed cell sorting [45].

The check microvalves [47,48] enable fluid to flow in one direction only. To open the valve, the upstream pressure is increased in order to overcome restoring forces (e.g., gravity) and downstream pressure. This type of check microvalve is also capable of performing delivery of biocompatible materials to mouse embryo fibroblast cells [47].

4.4.1.3 Pinch Microvalves

Pinch microvalves [49–51] are operated using mechanical pressure for causing physical deformation of the bulk of the PDMS which forms the device rather than making deformation of the PDMS membrane next to the fluidic microchannel as the pneumatic microvalves do. In addition, pinch microvalve operation depends upon the localized pressure from mechanical generation instead of external generation of pressure supplying the wall as the pneumatic models do.

Pinch microvalves [52–54] have other aspects similar to pneumatic microvalves. For example, Braille pinch microvalves (one type of pinch microvalve) generate localized pressure by employing mechanical pins of Braille displays, which is easy to operate with the programmable valve control method [52,53]. Braille pin pinch microvalves have been used to perform cell seeding and compartmentalization [52], and culture media recirculation [53].

4.4.1.4 Phase-Change Microvalves

Phase-change microvalves [27] regulate the fluid flow by the phase transition of the valve materials (e.g., paraffin [55–57], hydrogel [58,59], and polymer [60]); thus, the microfluidic devices do not need a pneumatic connection to the valve seat for operation. However, the response time for actuation is slow, which is not crucial for some applications, but the phase-change microvalve is not suited for applications that are concerned with rapid fluidic switching. Paraffin microvalves have been used for disposable bioassay microchip designs because they are affordable and easy for actuation due to the low tunable melting point of the paraffin [55,59], but paraffin valves can typically be used only once for modulating fluid flow [57].

4.4.1.5 Burst Microvalves

Burst microvalves [61–63] are single-use passive microvalves which open irreversibly when the driving pressure overcomes the flow resistance of the valve. Because of their single-use and simple fabrication, burst microvalves are very attractive for applications of disposable microfluidics in which there is no off-chip controller available. The capillary burst microvalve, one of the types of burst designs, regulates fluid flow by increasing capillary resistance inside the microchannel as the result of a sudden change in the geometry or surface chemistry of the microchannel.

4.4.2 Micropumps

Micropumps [27] are also an important component for microfluidic devices, for regulating and modulating fluid flow. This section provides functional information of two practical types of the micropumps, active and passive pumps, which have been employed with the microfluidic systems.

4.4.2.1 Passive Micropumps

Fluid volume in contact with the microstructure surfaces moves immediately as a result of the interaction between the surface tension of liquid and the chemical composition of the surface. The fluid flow always goes in the direction that minimizes the free energy between the vapor, fluid, and solid interfaces. For instance, spontaneous wetting of microchannel or capillary can be used to pump fluids as the passive-pumping mechanism does. Delamarche fabricated a capillary pump that was capable of generating the flow rate of 220 nL/s, with an average flow speed of 55 mm/s [64].

4.4.2.2 Active Micropumps

Active micropumps depend on an external signal to initiate and stop pumping actions. The external signal also enhances the efficacy for regulating the flow rate and the transient action of the pump, thereby significantly increasing the complexity of organ-on-chip microfluidic operations.

4.4.2.2.1 Pneumatic Membrane Micropumps

This kind of micropump works when a fluid volume (i.e., bolus) is bound between the activated pumping membranes and moves through sequential activation of the pumping membranes. As the result, the bolus moves away from its initial position, thereby producing volume displacement in the microchannel. Pneumatic micropumps integrated with hydrogel check valves [65] can generate flow rates of up to 25 μL/minute [47], and displacement volumes of up to 120 nL/stroke [66].

4.4.2.2.2 Piezoelectric Micropumps

Piezoelectric pumps [27,67] consist of a piezoelectric disk attached on top of a diaphragm above the pumping chamber; the piezoelectric materials (e.g., lead zirconate titanate) change their shape significantly when applied with the electric current. The stress exerted by piezoelectric materials connected to the diaphragm is able to pump the fluids, and alternating the deflection of the piezo-coupled diaphragm from the pumping chamber can engender the pumping strokes.

4.4.2.2.3 Electrochemical Micropumps

Micropumps can operate from the electrolysis of the aqueous solution to pump fluid through the microchannel. Bohm [68,69] developed the electrochemical micropump by using gas generation from the electrolysis of the KNO_3 to pump the sample fluid into the microchannel, and this generated flow rates ranging from 0.8 nL/minute to 4 nL/minute. However, there was a concern about the cross-contamination between the sample fluid and the products from the electrolysis [27].

4.4.2.2.4 Electro-Osmotic Micropumps

The electro-osmotic pump consists of a microchannel with electrodes submerged in fluid reservoirs at one end [70–73]. As the DC electric field is applied across the electrodes, there is a high force taking place at the wall of the microchannel, thereby causing the bulk motion called "electro-osmotic flow," and leading to moving the charge and fluid through the microchannel. The advantages of the micropump are that there are no moving parts, and it has the capability to be compatible with fluids over a wide range of conductivities. In addition to employing the DC electric field, the electro-osmotic micropump with applying the AC electric field on the electrodes is capable of engendering a flow rate of 450 μm/second at a power supply of less than 5 VRMS of AC voltage [27,72].

4.5 APPLICATIONS OF MICROFLUIDICS

This section discusses currently promising aspects of microfluidics in practical applications—including organs-on-chips, tissue engineering, biosensors, and drug delivery—which could enhance our standard of living and improve quality of life to a large number of people in the near future.

4.5.1 Microfluidic Organs-on-Chips

Organ-on-a-chip [74] devices are microfluidic devices that are used to examine living cells (e.g., evaluating cell contributions to the physiological functions of tissues and organs). The purpose of these devices is to study particular units of organ or tissue, not to create a whole living organ; microfluidic organs-on-chips are also capable of fabricating tissue and organ functionalities, which are limited for conventional two-dimensional and three-dimensional culture systems known as organoids. Microfluidic organs-on-chips can also be used to investigate *in vitro* analysis of biochemical, genetic, and metabolic mechanisms of the living cells in tissues and organs, and to study tissue development, organ physiology, and disease etiology by observation with high-resolution microscope and real-time imaging. The two-dimensional culture system could previously not support differentiated functions of many cell types, and could not predict in vivo tissue functions and drug activities [75]. For the three-dimensional cell culture systems or organoids [74], it is difficult to maintain cells in positions within the structures for extended analysis; it is also difficult to perform functional genetic analysis of living cells within the three-dimensional models.

The novel models of microfluidic organs-on-chips [74] are capable of overcoming the limitations of the previous two systems. Microfluidic chips can simply fabricate the system to investigate physiological function of one cell type (e.g., hepatocytes or kidney tubular epithelial cells), and the chips can also fabricate more complex systems consisting of different cell types by connecting two or more microchannels with porous membranes to form interfaces among the disparate cell types, such as the blood-brain barrier. Microfluidic chips can also be integrated with microsensors used for analysis of the cultured cells or microenvironmental conditions (e.g., cell migrations [76] and fluid pressure [77]), which is not compatible with the previous three-dimensional models. Another advantage of the chips is their capability to independently control fluid shear stress by switching the flow rates or microchannel dimensions [78,79] and by separating cells from the flow path using the nanoporous membranes [78]; hence, the cultured living cells and their functionalities have a high tendency to live longer. The chips can also allow different cell types to be cultured at the same microchannel with the other living cells by designing complex microchannel paths for connecting with adhesive substrates [74], by microprinting extracellular matrix (ECM) in different positions within the microchannels [80–82].

Microfluidic organs-on-chips have also been used for the investigation of basic mechanisms of organ physiology and disease. Applying organs-on-chips integrated with dynamic variation of oxygen tension makes it possible to study the beginning of disease states, including heart ischemia [83] or vaso-occlusion in sickle-cell disease, because of polymerization of hemoglobin S in deoxygenated erythrocytes [84]; as a result, the chips allow researchers to determine the specific drug for the particular threatening disease.

An organ comprises two or more different tissues, which are themselves formed of the groups of disparate types of cells [74]; consequently, the organ is a hierarchical structure. A human blood-brain-barrier-on-a-chip was fabricated by lining porous, fibronectin-coated polycarbonate membrane with human brain microvascular endothelium on one side and human astrocytes on the other side [85]. The chip was also integrated with microelectrodes to measure transepithelial electrical resistance across the barrier, which may lead to investigating how drugs can be transferred across the blood-brain barrier.

For another example of a microfluidic chip, a human lung-on-a-chip [74], was fabricated to study the mechanism of breathing motion in lung disease. In the the lung-on-a-chip, human alveolar epithelial cells were positioned above a porous and flexible ECM-coated membrane; human capillary endothelial cells were positioned on the bottom. The air flow was then fed through the upper channel to generate the air–liquid interface with the alveolar epithelium; culture medium entered the vascular channel either with or without human immune cells [86,87]. Stimulation of the breathing mechanism was assessed by applying cyclic suction to the full height of the chambers, distorting and relaxing the flexible PDMS side-walls and attached porous membrane. Silica nanoparticle simulants of environmental airborne particulates were then introduced into the air channels, which could cause a higher significant amount of reactive oxygen species, cellular uptake of nanoparticles, and transport of nanoparticles across cell layers and into the vascular channel if the cells were experiencing cyclic breathing motions. Lung-on-a-chip may be utilized to understand how the mechanical forces (e.g., breathing motions) play a role in interleukin-2-induced pulmonary edema [87].

4.5.2 Microfluidics in Tissue Engineering

Costantini et al. [88] applied microfluidics with a high internal phase emulsion (HIPE) technique for fabricating a new class of scaffolds from dextran-methacrylate. The method is capable of precise tuning of all structural parameters of the matrices (e.g., porosity, pore size, the lumen of interconnections between the pores). The three steps for fabricating the scaffolds are as follows: first, producing the monodisperse oil-in-water emulsion; second, cross-linking the external phase under UV light at the temperature of 50° for 24 hours; and third, extracting the inner phase and purifications. Interestingly, during the cross-linking process, no coalescence occurred. As a consequence, highly ordered and uniformly spaced in all dimensions structures were obtained; in contrast, the scaffolds fabricated by the conventional technique [46] without microfluidics had highly polydisperse with standard deviations of the order of tens of percent of the mean sizes, and the conventional technique cannot control or tune the final morphology of the scaffolds.

The microfluidic-HIPE method [88] is able to generate narrow distribution of the pore diameters with standard deviations below 10% of the mean, since the microfluidic flow-focusing junction can be used to tune the mean diameter of the pores. In addition, the microfluidic-HIPE can form greater sizes of interconnects exhibited by scaffolds than the conventional technique, leading to advantages for cell culture experiments in tissue engineering with capability for supply of nutrients and oxygen, and disposal of waste throughout the scaffolds for the studies. Another advantage of the novel technique is that it is capable of independently varying the dimension of interconnects while retaining constant pore size. Microfluidics are also able to form monodispersed droplets within the confined space of the outlet channel [88].

4.5.3 Microfluidics for Biosensing Applications

To enhance the sensitivity of the signals for a sensing system using optical phenomena (e.g., surface plasmon resonance and surface-enhanced Raman scattering) for many biological applications, including analysis of DNA and cells. Microfluidics are a promising candidate to achieve highly efficient detection for sensing systems [89]. Microfluidics are capable of controlling and measuring small amounts of fluids, such as reagents, from microliters down to picoliters, for high-throughput systems effectively to generate a higher degree of control for separations and detections [89,90]; thus, the integration of sensing systems with microfluidic devices have a great potential to bolster analysis in the biomedical sensing field. Microfluidics can also decrease cost and time for analysis [91–93], making microfluidic methods more attractive over the conventional milliliter-scaled glassware [94–97].

Localized surface plasmon resonance (LSPR) is a resonance phenomenon of waves localized in a metal nanostructure (e.g., nanoparticles and nanowires); an LSPR sensor is also an attractive technique used as a detection system due to its high sensitivity and no pretreatment of sample required for fast, qualitative detection of enantiomers [89]. Guo et al. [98] fabricated a biosensor integrated with microfluidics and LSPR to detect enantiomers of chiral compounds rather than chromatographic methods. They successfully invented the LSPR sensor integrated with microfluidics by forming dense gold nanorods on a self-assembled monolayer of 3-aminopropyltriethoxysilane (APTES) inside the channel walls of the microfluidic device for drug–protein interactions. In addition, they fabricated microfluidic channel on a glass substrate in which light was induced through an optical fiber using a tungsten halogen light source into the channel-generated LSPR signal, and the signal was detected by UV spectrometer. Before injection of the solution of gold nanorods into the microchannel and incubation for 5 hours to form the gold nanorods surface, the self-assembled monolayers of APTES were immobilized. LSPR signals were detected by measuring peak shifts at resonance condition before and after binding interaction; consequently, this method can be used for detection of other enantiomers with high sensitivity.

Another example of biosensing applications integrated with microfluidics is the detection of reactive oxygen species (ROS) caused by cigarette smoke [99]. The novel technique for ROS detection, an electrophoresis device integrated with a microfluidic chip and using the LSPR phenomenon [89,99], is capable of using less cigarette smoke consumption and shorter time for analysis less than conventional detection of ROS, which required a significant amount of cigarette smoke sample and took a longer time for the detecting process.

4.5.4 Microfluidics for Drug Delivery Systems

Researchers have been attempting to improve effective drug delivery systems by integrating with microfluidic platforms, and microfluidics can precisely control and transport small quantities of liquid from microliters to picoliters [100]. Hence, microfluidics have recently been utilized in the fabrication of self-assembled drug carriers, droplet-based drug carriers, and non-spherical drug carriers, because microfluidic systems are able to generate monodispersed and multifunctional drug carriers with effectively controllable physical and chemical properties to bolster transport, release, distribution, and elimination of drugs during the period of treatment [101,102].

For self-assembly drug carriers integrated with microfluidics [100], two or multiple streams of reagents are interfaced and the carriers are created at the interfacial layer. Typically, this method has been also integrated with hydrodynamic flow focusing in order to successfully generate the self-assembly reactions of the carriers by controlling the mixing rates between different fluid streams depending upon the shape of the microchannel, the flow rates, and the diffusion coefficient of different miscible streams [100,103]. The size of self-assembled drug carriers from this fabrication technique is typically less than 1 μm, which can decrease the possibility of phagocytosis when the drug carriers transport across physiological barriers [101].

Another technique for fabricating drug carriers, droplet-based microfluidic method [104,105], is able to form homogeneous drug-loaded particles with relatively larger size than the carriers fabricated by self-assembly microfluidic carriers, thereby providing a high dose of the drug encapsulated in the particles and maintaining drug release for long periods of time. However, this method cannot fabricate very small drug-loaded carriers if nano-sized drug-loaded particles are required.

Recent studies have illustrated that the shape of the drug-loaded carriers has significant effect on in vivo biodistribution, their uptake mechanisms, and their blood circulation time in the human body [100,106]. With the previous two drug fabrication techniques, most of drug-loaded particles are spherical, so targeting of the carriers to the disease cells in the human body is still challenging due to low surface-to-volume ratio compared to complex shapes. Hence, the non-spherical drug carriers integrated with the microfluidics method is a promising solution, since the non-spherical shapes of the carriers (e.g., rod-like, cylinder-like, and toroid-like particles) have a relatively higher surface-to-volume ratio than that of the spherical shape [100]. As the result, those shapes are capable of improving the attachment of drugs to the cell surface.

REFERENCES

1. Stone HA, Kim S. Microfluidics: Basic issues, applications, and challenges. *AIChE*, 2001;47(6):1250–1254.
2. Saliterman SS. Education, bioMEMS, and the medical microdevice revolution. *Expert Rev Med Device* 2005;2(5):515–519.
3. Duste SW, Yusof NA. Microfluidics-based lab-on-chip systems in DNA-based biosensing: An overview. *Sensors* 2011;11:5754–5768.
4. Huh D, Hamilton GA, Ingber DE. From 3D cell culture to organs-on-chips. *Trends Cell Biol* 2011;21(12):745–754.
5. Stone HA, Stroock AD, Ajdari A. Engineering flows in small devices: Microfluidics toward a lap-on-a-chip. *Annu Rev Fluid Mech* 2004;36:381–411.
6. Gravesen P, Branebjerg J, Jensen OS. Microfluidics—A review. *J Micromech Microeng* 1993;3(1993):168–182.
7. David NB, Philip JL, Luke PL. Microfluidics-based systems biology. *Mol Biosyst* 2006;2:97–112.
8. Michael SL, Ron W, Laura FL. Automated design and programming of a microfluidic DNA computer. *Nat Comput* 2006;5:1–13.

9. Yager P, Edwards T, Fu E, et al. Microfluidic diagnostic technologies for global public health. *Nat Biotechnol* 2006;442:412–418.
10. Zervantonakis IK, et al. Microfluidic devices for studying heterotypic cell–cell interactions and tissue specimen cultures under controlled microenvironments. *Biomicrofluidics* 2011;5:13406.
11. Sung KE, et al. Transition to invasion in breast cancer: A microfluidic in vitro model enables examination of spatial and temporal effects. *Integr Biol (Camb)* 2011;3:439–450.
12. Jang KJ, et al. Fluid-shear-stress-induced translocation of aquaporin-2 and reorganization of actin cytoskeleton in renal tubular epithelial cells. *Integr Biol (Camb)* 2011;3:134–141.
13. McCabe WL, Smith JC, Harriot P. *Unit Operations of Chemical Engineering,* 5th edition, Fluid-flow phenomenal. Clark BJ and Castellano E (eds.), Singapore: McGraw-Hill, 1993;42–61.
14. Garstecki P, Fuerstman MJ, Fischbach MA, et al. Mixing with bubbles: A practical technology for use with portable microfluidic devices. *Lab Chip* 2006;6:207–212.
15. Jaipan P, Poonsiriseth A, Taweerojkulsri C, et al. Thermocapillary effects in foams: Thermal diffusion through a bubble staircase and impact on shear modulus. *Colloid Surface A Physicochem Engineer Aspects* 2013;434:339–348.
16. Gallardo BS, Gupta VK, Eagerton FD, et al. Electrochemical principles for active control of liquids on submillimeter scales. *Science* 1999;283:57.
17. Ismagilov RF, Stroock AD, Kenis PJA, et al. Experimental and theoretical scaling laws for transverse diffusive broadening in two-phase laminar flow in microchannels. *Appl Phys Lett* 2000;76:2376.
18. Duke TAJ, Austin RH. Microfabricated sieve for the continuous sorting of macromolecules. *Phys Rev Lett* 1998;80:1552.
19. Debesset S, Hayden CJ, Dalton C, et al. A circular AC electroosmotic micropump for chromatographic applications. *Micro TAS* 2002;2:655–657.
20. Stroock AD, Weck M, Chiu DT, et al. Patterning electroosmotic flow with patterned surface charge. *Phys Rev Lett* 2000;84:3314–3317. Erratum 2001. *Phys Rev Lett* 86:6050.
21. Wang C, Nguyen NT, Wong TN, et al. Investigation of active interface control of pressure driven two-fluid flow in microchannels. *Sens Actuat A* 2007;133:323–328.
22. Li H, Nguyen, NT, Wong TN, Ng SL. Microfluidic on-chip fluorescence-activated interface control system. *Biomicrofluidics* 2010;4:044109:1–044109:13.
23. Takayama S, Ostuni E, LeDuc P, et al. Laminar flows: subcellular positioning of small molecules. *Nature* 2001;411:1016.
24. Larson RG, Shaqfeh ESG, Muller SJ. A purely elastic instability in Taylor-Couette flow. *J Fluid Mech* 1990;218:573–600.
25. Byars JA, Öztekin A, Brown RA, McKinley GH. Spiral instabilities in the flow of highly elastic fluids between rotating parallel disks. *J Fluid Mech* 1994;271:173–218.
26. Groisman A, Steinberg V. Efficient mixing at low Reynolds numbers using polymer additives. *Nature* 2001;410:905–908.
27. Au AK, Lai H, Utela BR, Folch A. Microvalves and micropumps for bioMEMS. *Micromachines* 2011;2:179–220.
28. Lee DE, Soper S, Wang WJ. Design and fabrication of an electrochemically actuated microvalve. *Microsyst Technol* 2008;14:1751–1756.
29. Kaigala GV, Hoang VN, Backhouse CJ. Electrically controlled microvalves to integrate microchip polymerase chain reaction and capillary electrophoresis. *Lab Chip* 2008;8:1071–1078.
30. Jacobson SC, Ermakov SV, Ramsey JM. Minimizing the number of voltage sources and fluid reservoirs for electrokinetic valving in microfluidic devices. *Anal Chem* 1999;71:3273–3276.

31. Schasfoort RBM, Schlautmann S, Hendrikse L, van den Berg A. Field-effect flow control for microfabricated fluidic networks. *Science* 1999;286:942–945.
32. Sundararajan N, Kim D, Berlin AA. Microfluidic operations using deformable polymer membranes fabricated by single layer soft lithography. *Lab Chip* 2005;5:350–354.
33. Studer V, Hang G, Pandolfi A, et al. Scaling properties of a low-actuation pressure microfluidic valve. *J Appl Phys* 2004;95:393–398.
34. Hosokawa, K., Maeda, R. A pneumatically-actuated three-way microvalve fabricated with polydimethylsiloxane using the membrane transfer technique. *J Micromech Microeng* 2000;10:415–420.
35. Grover WH, Skelley AM, Liu CN, et al. Monolithic membrane valves and diaphragm pumps for practical large-scale integration into glass microfluidic devices. *Sens Actuat B* 2003;89:315–323.
36. Unger MA, Chou HP, Thorsen T, et al. Monolithic microfabricated valves and pumps by multilayer soft lithography. *Science* 2000;288:113–116.
37. Thorsen T, Maerkl SJ, Quake SR. Microfluidic large-scale integration. *Science* 2002;298:580–584.
38. Lee SJ, Chan JCY, Maung KJ, et al. Characterization of laterally deformable elastomer membranes for microfluidics. *J Micromech Microeng* 2007;17:843–851.
39. Pandolfi A, Ortiz M. Improved design of low-pressure fluidic microvalves. *J Micromech Microeng* 2007;17:1487–1493.
40. Wang YC, Choi MN, Han JY. Two-dimensional protein separation with advanced sample and buffer isolation using microfluidic valves. *Anal Chem* 2004;76:4426–4431.
41. Fu AY, Chou HP, Spence C, et al. An integrated microfabricated cell sorter. *Anal Chem* 2002;74:2451–2457.
42. Studer V, Jameson R, Pellereau E, et al. A microfluidic mammalian cell sorter based on fluorescence detection. *Microelectron Eng* 2004;73–74:852–857.
43. Lee S, Jeong W, Beebe DJ. Microfluidic valve with cored glass microneedle for microinjection. *Lab Chip* 2003;3:164–167.
44. Sip CG, Folch A. An open-surface micro-dispenser valve for the local stimulation of conventional tissue cultures. In *Proceedings of the 14th International Conference on Miniaturized Systems for Chemistry and Life Sciences (microTAS)*, Groningen, The Netherlands, 3–7 October 2010, pp. 1778–1780.
45. Abate AR, Agresti JJ, Weitz DA. Microfluidic sorting with high-speed single-layer membrane valves. *Appl Phys Lett* 2010;96:203509.
46. Baek JY, Park JY, Ju JI, Lee TS, Lee SH. A pneumatically controllable flexible and polymeric microfluidic valve fabricated via in situ development. *J Micromech Microeng* 2005;15:1015–1020.
47. Kim J, Baek J, Lee K, et al. Photopolymerized check valve and its integration into a pneumatic pumping system for biocompatible sample delivery. *Lab Chip* 2006;6:1091–1094.
48. Voldman, J, Gray ML, Schmidt MA. An integrated liquid mixer/valve. *J Microelectromech Syst* 2000;9:295–302.
49. Pemble CM, Towe BC. A miniature shape memory alloy pinch valve. *Sens Actuat A* 1999;77:145–148.
50. Weibel DB, Siegel AC, Lee A, et al. Pumping fluids in microfluidic systems using the elastic deformation of poly(dimethylsiloxane). *Lab Chip* 2007;7:1832–1836.
51. Weibel DB, Kruithof M, Potenta S, et al. Torque-actuated valves for microfluidics. *Anal Chem* 2005;77:4726–4733.
52. Gu W, Zhu XY, Futai N, et al. Computerized microfluidic cell culture using elastomeric channels and braille displays. *Proc Natl Acad Sci U S A* 2004;101:15861–15866.
53. Futai N, Gu W, Song JW, Takayama S. Handheld recirculation system and customized media for microfluidic cell culture. *Lab Chip* 2006;6:149–154.

54. Gu W, Chen H, Tung YC, Meiners JC, Takayama S. Multiplexed hydraulic valve actuation using ionic liquid filled soft channels and braille displays. *Appl Phys Lett* 2007;90:033505. doi: 10.1063/1.2431771.
55. Yang BZ, Lin Q. A latchable microvalve using phase change of paraffin wax. *Sens Actuat A* 2007;134:194–200.
56. Yoo JC, Choi YJ, Kang CJ, Kim YS. A novel polydimethylsiloxane microfluidic system including thermopneumatic-actuated micropump and paraffin-actuated microvalve. *Sens Actuat A* 2007;139:216–220.
57. Liu RH, Bonanno J, Yang JN, et al. Single-use, thermally actuated paraffin valves for microfluidic applications. *Sens Actuat B* 2004;98:328–336.
58. Yu Q, Bauer JM, Moore JS, Beebe DJ. Responsive biomimetic hydrogel valve for microfluidics. *Appl Phys Lett* 2001;78:2589–2591.
59. Liu CW, Park JY, Xu YG, Lee S. Arrayed ph-responsive microvalves controlled by multiphase laminar flow. *J Micromech Microeng* 2007;17:1985–1991.
60. Kaigala GV, Hoang VN, Backhouse CJ. Electrically controlled microvalves to integrate microchip polymerase chain reaction and capillary electrophoresis. *Lab Chip* 2008;8:1071–1078.
61. Cho H, Kim HY, Kang JY, Kim TS. How the capillary burst microvalve works. *J Colloid Interface Sci* 2007;306:379–385.
62. Chen JM, Huang PC, Lin MG. Analysis and experiment of capillary valves for microfluidics on a rotating disk. *Microfluid Nanofluid* 2008;4:427–437.
63. Riegger L, Mielnik MM, Gulliksen A, et al. Dye-based coatings for hydrophobic valves and their application to polymer labs-on-a-chip. *J Micromech Microeng* 2010;20:045021.
64. Juncker D, Schmid H, Drechsler U, et al. Autonomous microfluidic capillary system. *Anal Chem* 2002;74:6139–6144.
65. Futai N, Gu W, Takayama S. Rapid prototyping of microstructures with bell-shaped cross-sections and its application to deformation-based microfluidic valves. *Adv Mater* 2004;16:1320–1323.
66. Kim JY, Park H, Kwon KH, et al. A cell culturing system that integrates the cell loading function on a single platform and evaluation of the pulsatile pumping effect on cells. *Biomed Microdevices* 2008;10:11–20.
67. Tracey MC, Johnston ID, Davis JB, Tan CKL. Dual independent displacement-amplified micropumps with a single actuator. *J Micromech Microeng* 2006;16:1444–1452.
68. Bohm S, Olthuis W, Bergveld P. An integrated micromachined electrochemical pump and dosing system. *Biomed Microdevices* 1999;1:121–130.
69. Bohm S, Timmer B, Olthuis W, Bergveld P. A closed-loop controlled electrochemically actuated micro-dosing system. *J Micromech Microeng* 2000;10:498–504.
70. Jacobson SC, Hergenroder R, Koutny LB, Ramsey JM. Open-channel electrochromatography on a microchip. *Anal Chem* 1994;66:2369–2373.
71. Ramsey RS, Ramsey JM. Generating electrospray from microchip devices using electroosmotic pumping. *Anal Chem* 1997;69:1174–1178.
72. Mpholo M, Smith CG, Brown ABD. Low voltage plug flow pumping using anisotropic electrode arrays. *Sens Actuat B* 2003;92:262–268.
73. Debesset S, Hayden CJ, Dalton C, Eijkel JC, Manz A. An ac electroosmotic micropump for circular chromatographic applications. *Lab Chip* 2004;4:396–400.
74. Bhatia SN, Ingber DE. Microfluidic organs-on-chips. *Nature Biotechnol* 2014;32:8.
75. Greek R, Menache A. Systematic reviews of animal models: Methodology versus epistemology. *Int J Med Sci* 2013;10:206–221.
76. Nguyen TA, Yin TI, Reyes D, Urban GA. Microfluidic chip with integrated electrical cell-impedance sensing for monitoring single cancer cell migration in three-dimensional matrixes. *Anal Chem* 2013;85:11068–11076.

77. Liu MC, et al. Electrofluidic pressure sensor embedded microfluidic device: A study of endothelial cells under hydrostatic pressure and shear stress combinations. *Lab Chip* 2013;13:1743–1753.
78. Carraro A, et al. In vitro analysis of a hepatic device with intrinsic microvascular based channels. *Biomed Microdevices* 2008;10:795–805.
79. Griep LM, et al. BBB on chip: Microfluidic platform to mechanically and biochemically modulate blood-brain barrier function. *Biomed Microdevices* 2013;15:145–150.
80. Kane BJ, Zinner MJ, Yarmush ML, Toner M. Liver-specific functional studies in a microfluidic array of primary mammalian hepatocytes. *Anal Chem* 2006;78:4291–4298.
81. Tumarkin E, et al. High-throughput combinatorial cell co-culture using microfluidics. *Integr Biol (Camb)* 2011;3:653–662.
82. Agarwal A, Goss JA, Cho A, et al. Microfluidic heart on a chip for higher throughput pharmacological studies. *Lab Chip* 2013;13:3599–3608.
83. Khanal G, Chung K, Solis-Wever X, et al. Ischemia/reperfusion injury of primary porcine cardiomyocytes in a lowshear microfluidic culture and analysis device. *Analyst (Lond)* 2011;136:3519–3526.
84. Wood DK, et al. A biophysical indicator of vaso-occlusive risk in sickle cell disease. *Sci Transl Med* 2012;4:123–126.
85. Booth R, Kim H. Characterization of a microfluidic in vitro model of the blood-brain barrier (mBBB). *Lab Chip* 2012;12:1784–1792.
86. Huh D, et al. Reconstituting organ-level lung functions on a chip. *Science* 2010;328:1662–1668.
87. Huh D, et al. A human disease model of drug toxicity-induced pulmonary edema in a lung-on-a-chip microdevice. *Sci Transl Med* 2012;4:159ra147, pp. 1–8.
88. Costantini M, Colosi C, Guzowski J, et al. Highly ordered and tunable polyHIPEs by using microfluidics. *J Mater Chem B* 2014;2:2290.
89. Lee H, Xu L, Koh D, et al. Various on-chip sensors with microfluidics for biological applications. *Sensors* 2014;14(9):17008–36.
90. Daniele MA, Boyd DA, Mott DR, Ligler FS. 3D hydrodynamic focusing microfluidics for emerging sensing technologies. *Biosens Bioelectron* 2015;67:25–34.
91. Lee CC, Sui GD, Elizarov A, et al. Multistep synthesis of a radiolabeled imaging probe using integrated microfluidics. *Science* 2005;310:1793–1796.
92. Lee BS, Lee JN, Park JM, et al. A fully automated immunoassay from whole blood on a disc. *Lab Chip* 2009;9:1548–1555.
93. Kong J, Jiang L, Su XO, et al. Integrated microfluidic immunoassay for the rapid determination of clenbuterol. *Lab Chip* 2009;9:1541–1547.
94. Janasek D, Franzke J, Manz A. Scaling and the design of miniaturized chemical-analysis systems. *Nature* 2006;442:374–380.
95. Watts P, Haswell SJ. Microfluidic combinatorial chemistry. *Curr Opin Chem Biol* 2003;7:380–387.
96. Cullen CJ, Wootton RCR, de Mello AJ. Microfluidic systems for high-throughput and combinatorial chemistry. *Curr Opin Drug Disc* 2004;7:798–806.
97. Roberge DM, Ducry L, Bieler N, et al. Microreactor technology: A revolution for the fine chemical and pharmaceutical industries? *Chem Eng Technol* 2005;28:318–323.
98. Guo LH, Yin YC, Huang R, et al. Enantioselective analysis of melagatran via an LSPR biosensor integrated with a microfluidic chip. *Lab Chip* 2012;12:3901–3906.
99. Wang HS, Xiao FN, Li ZQ, et al. Sensitive determination of reactive oxygen species in cigarette smoke using microchip electrophoresis-localized surface plasmon resonance enhanced fluorescence detection. *Lab Chip* 2014;14:1123–1128.
100. Riahi R, Tamayol A, Shaegh SAM, et al. Microfluidics for advanced drug delivery systems. *Curr Opin Chem Engineer* 2015;7:101–112.

101. Owens DE III, Peppas NA. Opsonization, biodistribution, and pharmacokinetics of polymeric nanoparticles. *Int J Pharmaceut* 2006;307:93–102.
102. Gañán-Calvo AM, Montanero JM, Martín-Banderas L, Flores-Mosquera M. Building functional materials for health care and pharmacy from microfluidic principles and flow focusing. *Adv Drug Deliv Rev* 2013;65:1447–1469.
103. Lo CT, Jahn A, Locascio LE, Vreeland WN. Controlled self-assembly of monodisperse niosomes by microfluidic hydrodynamic focusing. *Langmuir* 2010;26:8559–8566.
104. Zhao C-X. Multiphase flow microfluidics for the production of single or multiple emulsions for drug delivery. *Adv Drug Deliv Rev* 2013;65:1420–1446.
105. Mou C-L, Ju X-J, Zhang L, et al. Monodisperse and fast-responsive poly(N-isopropylacrylamide) microgels with open-celled porous structure. *Langmuir* 2014;30:1455–1464.
106. Mathaes R, Winter G, Besheer A, Engert J. Non-spherical micro-and nanoparticles: Fabrication, characterization and drug delivery applications. *Expert Opin Drug Deliv* 2015;12:481–92.

5 From Big Data to Predictive Analysis from *In Vitro* Systems

Andre Kleensang, Alexandra Maertens and Thomas Hartung

CONTENTS

5.1 INTRODUCTION

A number of high-content technologies allow deriving what is nowadays often called "big data" from cellular models. The challenge is to make "big sense" from "big data." Toxicology with its challenges (Hartung 2009) is used as a prime example here. The main opportunity is to deduce the underlying adverse outcome pathways, i.e., the pathways of toxicity and disease manifestations. However, in order to understand aspects of these processes and possible perturbations in disease or due to toxic or traumatic insults, it is necessary to make such cell models available. It is not about "big data" but about "good big data" and this starts with good cell models. With the advent of stem cells, this field has started to boost a development of organotypic models, which promise to represent developmental aspects of organogenesis and the possible targets of disruption. Using patient-derived stem cells, the relevant genetic background can be studied and even combined with stressors, which might aggravate the manifestation of developmental disorders. Bioengineering offers additional tools for further optimizing cell cultures to make them organotypic ("microphysiological systems" or "organ-on-a-chip"), i.e., reproduce organ functionalities. However, the necessary complement of this is quality assurance of the cell systems.

The next step is the generation of big data by high-content technologies, especially the various omics technologies. They have again very different levels of standardization and quality assurance. The increasing level of commercialization and a number of quality assurance initiatives have helped a lot here. Taking especially the example of mass-spectroscopy-based metabolomics, some problems will be illustrated.

The challenge lies clearly, however, in making sense of the data, as these are technologies full of measurement noise, where a relatively small number of experiments are used to assess a multitude of variables. This requires a reduction in dimensionality: the most common approaches are based on significance of changes (when done well with false-discovery rate adjustment) and clustering of findings. The latter depends critically on prior knowledge on pathways, and here the enormous publication biases hit hard. But the emerging data-mining and bioinformatics tools give hope for a predictive interpretation of such data.

Bioinformatics plays a key role in mining the information-rich new technologies and making sense of the output by modeling. With interdisciplinary collaboration, toxicology can take advantage of such expert knowledge. The challenge and the opportunity lie in the transition from mode of action (MoA) models to pathway modeling (Hartung and McBride 2011; Kleensang et al. 2014; Bouhifd et al. 2014; 2015a), increasing the resolution of analysis, and then back to building Integrated Testing Strategies (ITS) on this understanding of adverse outcome pathways and, ultimately, a systems integration of this mechanistic knowledge (Hartung et al. 2012). Environmental contaminations do not present themselves in isolation but as mixtures with unknown "cocktail" effects. Traditional animal test approaches are not suitable for testing many combinations of doses and timing. New pathway-based tests, in contrast, could allow the identification of critical combinations and provide better environmental protection.

In the meantime, several newly emerging technologies have demonstrated capabilities for the development of more modern approaches for toxicology to replace the traditional "black-box" animal-based paradigms by providing mechanistic details of events at the cellular and molecular levels (Leist et al. 2014). Such high-content methods are the logical complement to sophisticated organotypic cultures where a maximum of information is obtained from the lower number of replicates because of duration of model preparation and technical effort for each and every parallel cell system.

5.2 ORGANOTYPIC CELL CULTURE

Cell culture is prone to artifacts. Cells live—and as any living being, they react to survive. Survival of the most adaptable one: a principle well known in the evolution of organisms holds true also for cell populations under the selection pressure of culture conditions. The closer the cellular environment mimics the physiological situation, the less cells need to move away from tissue-type differentiation.

We have first examples that show organotypic cultures (Andersen et al. 2014; Hartung 2014) to provide more relevant results (Marx et al. 2012, 2016; Alépée et al. 2014). They might help us to make better predictions of the organism's response to treatment, disease agents, and chemical exposure. We need to admit that the current systems are of limited use, e.g., to decide on developing agents with higher probability of success in clinical trials (Hartung 2013). In this context, 3D is a key aspect to get to organotypic cultures, as pursued by our group for brain models (Hogberg et al. 2013; Pamies et al. 2014) for developmental neurotoxicity (Smirnova et al. 2014). Traditional 2D cultures look like pan-fried eggs "sunny side up." Their environment: half plastic, half culture medium, and a little bit of other cells. They typically have less than 1% of both cell density per volume and cell-to-cell contacts when compared to native tissue. Intracellular communication is difficult, whether by contact or paracrine mediators, which are instantly diluted by cell culture medium. Most cultures do not result in polarization of cells, as especially epithelial cells show in the organism (Gordon et al. 2015).

A number of technical solutions to achieve 3D cultures yield benefits: cell differentiation, reduced variability, long-term stability, etc.—not all in each and every setup, but in many cases (Alépée et al. 2014). But everything comes with a price: more work, more costs, slower growth, and heterogeneity; cells on the surface of our 3D cultures are not the same as those in the inside. And there is a nutrition problem; if not combined with perfusion, medium supply for the inner cells is limited by diffusion and cell barriers. Some hundred micrometer diameter is a typical limit, before lack of oxygen and nutrients leads to necrosis at the center.

But 3D alone is not yet organotypic (Hartung 2014). The challenge starts with the choice of cell types. If tumor lines are used, 3D cannot restore their genetic make-up (Hartung 2013; Kleensang et al. 2016); thousands of point mutations, chromosomal multiplications, rearrangements, and losses cannot be turned back. There is tremendous hope that we can increasingly use stem cells to obtain quasi-primary human cells, but the differentiation protocols still have major limitations. So we will often have to use primary cells, but their supply is challenging if sourcing from humans. Often 3D culture will slow down dedifferentiation, the loss of specific cell functions typical for primary cells brought into culture, but again reliable protocols are only emerging. We are left with time windows between establishing the 3D culture and critical loss of differentiation. Adding the fourth dimension, time, to make long-term exposures and long-term reactions of our tissue equivalents possible is the next challenge.

There is more to do in order to make a 3D culture organotypic. Perfusion can make culture more homeostatic, if we are not recirculating the culture media. Each medium change is the most drastic change of environment for a cell that we can imagine. In an instant, all waste is gone and nutrients are replenished. To adapt, cells need to stay flexible, i.e., they have to avoid terminal differentiation. But an organ consists of many cell types, which we can model in co-cultures—and they are organized in structures, often form functional units. This is very challenging to recreate *in vitro*. Moreover, there is an extracellular matrix to add.

We are only starting to engineer all of this, including mimicking the impact of physical factors—stretch, pressure, peristaltic, and more. The current excitement for

3D in cell culture technology is fueled by the increasing awareness of how much we miss with traditional approaches. In a collaborative R&D funding initiative by the US agencies National Institutes of Health (NIH), Defense Advanced Research Projects Agency (DARPA) and the Food and Drug Administration (FDA) as well as a similar program by Defense Threat Reduction Agency (DTRA), $200 million over 5 years have been made available, aimed at creating 3D chips with living cells and tissues that model the structure and function of human organs (Hartung and Zurlo 2012). Such tools not only are expected to help develop medical countermeasures for chemical and biological warfare and terrorism but also are equally important in the general drug discovery and development area for predicting more accurately how effective a therapeutic candidate would be in clinical studies. European programs are more dispersed, less coordinated, but certainly not of much smaller dimension.

They bring bioengineers and the *in vitro* testing community together. We can learn from alternative methods and their validation, which have addressed quality assurance for *in vitro* tests over the last decades, most explicitly with the development of the good cell culture practice (GCCP) guidance (Coecke et al. 2005). If we want to become more predictive, we have to follow this avenue, and going 3D is among the first of many steps of our journey to meaningful models of tissues and organs.

5.3 SOURCES OF CELLS FOR 3D CULTURES

To provide human cellular material, which is increasingly seen as the gold standard in toxicology of the twenty-first century (NRC 2007; Leist et al. 2008), new technologies are emerging that allow differentiation of human embryonic stem cells (hESCs), human induced pluripotent stem cells (hiPSCs) or fetal cells such as fetal neural progenitor cells (NPCs). Although human cell systems are in the end desired for making predictions of human health, rodent systems can also be of high value for risk assessment. For instance, rodent models may be used in parallel to a human cell-based model. By using both cultures, toxicity observed in human cells *in vitro* can be extrapolated to the large database on toxicity in rodents *in vivo*. Such species extrapolations that use comparison of *in vitro* and *in vivo* data are called the "parallelogram approach of toxicology." For instance, species-specific differences in sensitivity observed between human and rat organoid cultures are most likely to reflect, and to predict, disparities in cellular toxicodynamics. These experimentally-derived toxicodynamic factors can then be employed for hazard and risk assessment. However, degrees of maturation between human and rodent systems have to be carefully assessed, as speed of maturation differs between species, not only *in vivo* but also *in vitro* (Baumann et al. 2014).

5.4 BETTER CELL SYSTEMS ALSO NEED BETTER QUALITY ASSURANCE

Our attempts to establish GCCP (Coecke et al. 2005) and publication guidance for *in vitro* studies (Leist et al. 2010) desperately await broader implementation. Earlier, we discussed the shortcomings of typical cell culture (Hartung 2007a, 2013). These articles summed up experiences gained from the validation of *in vitro* systems and in the course of developing GCCP guidance (Coecke et al. 2005). Cell cultures are prone to artifacts (Hartung, 2007a)—far too many artificially chosen and difficult-to-control conditions influence our experiments. Quality assurance is the gift from alternative methods to the life sciences. While good laboratory practice (GLP) (at least originally) addressed only regulatory *in vivo* studies, and International Organization for Standardization (ISO) guidance is not really specific for life science tools, neither addresses the key issue, i.e., the relevance of a test. This is the truly unique contribution of validation (Leist et al. 2012a), which is far too infrequently applied in other settings.

We do not obtain *in vivo*-like differentiation because we often start with tumor cells (ten thousands of mutations, loss, and duplications of chromosomes), over-passaging with selection of subpopulations, nonphysiologic culture conditions (hardly any cell contact, low cell density, no polarization, limited oxygen supply, non-homeostatic media exchange, temperature, and electrolyte concentrations reflective of humans, not rodents), forcing growth (fetal calf serum, growth factors), no demand on cell functions due to over-pampering, and no *in vitro* kinetics giving consideration to the fate of test substances in the culture and lack of cell type interactions. For most aspects there are technical solutions, but few are applied, and if they are applied, it occurs in isolation, solving some—but not all—of the problems. On top of this, we have a lack of quality control. If we take the following estimates, it is likely that only 60% of studies use the intended cells without mycoplasma infection. Misidentified cells are a threat to all *in vitro* work. The most impressive are HeLa cells. Since 1967, cell line contaminations have been evident, i.e., another cell type was accidentally introduced into a culture and slowly took over. They are the most promiscuous so far and HeLa cells actually were the first human tumor cell line. Recently, the HeLa genome has been sequenced (Landry et al. 2013) finding extra versions of most chromosomes (up to five copies) and many genes were duplicated even more extensively. Many chromosomes showed drastically altered arrangement of the genes. Do we really expect such a cell monster to show normal physiology? The cell line was found to be remarkably durable and prolific, as illustrated by its contamination of many other cell lines. It is assumed that, even today, 10–20% of cell lines are actually HeLa cells and, in total, 18–36% of all cell lines are wrongly identified. Even over the last decade, studies analyzing the problem of inauthenticity in cell banks range from 15–18% (Hughes et al. 2007) and a very useful list of such mistaken cell lines is available. A 2004 study (Buehring et al. 2004) showed that HeLa contaminants were used unknowingly by 9% of survey respondents, a likely underestimation of the problem, and only about a third of respondents were testing their lines for cell identity. It is a scandal that a large percentage of *in vitro* research is done on cells other than the supposed ones, and thus misinterpreted this way.

Another type of contamination that is astonishingly frequent and has a serious impact on *in vitro* results is microbial infection, especially with mycoplasma (Langdon, 2003). Screening by the FDA for more than three decades shows that, of 20,000 cell cultures examined, more than 3000 (15%) were contaminated with mycoplasma (Rottem and Barile 1993). Studies in Japan and Argentina reported mycoplasma contamination rates of 80% and 65%, respectively (Rottem and Barile 1993). An analysis by the German Collection of Microorganisms and Cell Cultures (DSMZ) of 440 leukemia-lymphoma cell lines showed that 28% were mycoplasma positive (Drexler and Uphoff 2002). Laboratory personnel are the main sources of

Mycoplasma orale, M. fermentans, and *M. hominis*. These species of mycoplasmas account for more than half of all mycoplasma infections in cell cultures and physiologically are found in the human oropharyngeal tract (Nikfarjam and Farzaneh 2012). *M. arginini* and *A. laidlawii* are two other mycoplasmas contaminating cell cultures that originate from fetal bovine serum or newborn bovine serum. Trypsin solutions derived from swine are a major source of *M. hyorhinis*. It is important to understand that the complete lack of a bacterial cell wall of mycoplasma implies resistance against penicillin (Bruchmüller et al. 2006), and they even pass 0.2 μm sterility filters, especially at higher-pressure rates (Hay et al. 1989). Mycoplasma can have diverse negative effects on cell cultures, and it is extremely difficult to eradicate this intracellular infection (Drexler and Uphoff 2002; Nikfarjam and Farzaneh 2012). While there is good understanding in the respective fields of biotechnology, this is much less the case in basic research, and mycoplasma testing is neither internationally harmonized with validated methods nor common practice in all laboratories on a regular basis. For a comparison of the different mycoplasma detection platforms see Lawrence et al. (2010) and Young et al. (2010).

The documentation practices in laboratories and publications are often subpar. There is some guidance available (GLP has been increasingly adapted, GCCP see below) but little is applied. The more recent mushrooming of cell culture protocol collections is an important step, but adherence is still uncommon and deviations are unclear in publications. We tend to toy around with the models until they work for us, and too often only for us. Such standardization forms the basis for formal validation, as developed by the European Union Reference Laboratory for alternatives to animal testing (EURL-ECVAM), adapted and expanded by Interagency Coordinating Committee on the Validation of Alternative Methods (ICCVAM) of the US National Toxicology Program and other validation bodies, and, finally, internationally harmonized by Organisation for Economic Co-operation and Development (OECD) (OECD 2005). Validation is the independent assessment of the scientific basis, the reproducibility, and the predictive capacity of a test. It was redefined in 2004 in the Modular Approach (Hartung et al. 2004) but needs to be seen as a continuous adaptation of the process to practical needs and a case-by-case assessment of what is feasible (Hartung 2007b; Leist et al. 2012a). The most important changes to the Modular Approach were the introduction of an applicability domain (borrowing the concept from quantitative structure activity relationships [QSAR]), the use of existing data (retrospective validation), and the independence of reproducibility and relevance assessment, which allows for leaner study designs and performance standards for similar tests to be considered equivalent to a validated one. The framework of evidence-based medicine is increasingly being translated to toxicology (Hoffmann and Hartung 2006), and it recently led to the creation of the Evidence-Based Toxicology Collaboration (Zurlo 2011) and increased use of its key tool, systematic reviews (Stephens et al. 2016).

The advent of human embryonic and induced pluripotent stem cells appears to be something of a game-changer. First, it promises to overcome the problems of availability of human primary cells, though a variety of commercial providers make almost all relevant human cells available in reasonable quality (but at costs that are challenging for academia). It is important to note, however, that we do not yet have protocols to achieve full differentiation of any cell type from stem cells.

This is probably only a matter of time, but many of the nonphysiologic conditions taken from traditional cell cultures contribute here. Stem cells have been praised for their genetic stability, which appears to be better than for other cell lines, but we have increasingly discovered their limitations in that respect, too (Mitalipova et al. 2005; Lund et al. 2012; Steinemann et al. 2013). The limitations experienced first are costs of culture and slow growth—many protocols require months, and labor, media, and supplement costs add up. The risk of infection increases unavoidably. We still do not obtain pure cultures and often require cell sorting, which, however, implies detachment of cells with the respective disruption of culture conditions and physiology.

5.5 THE USE OF OMICS TECHNOLOGIES FOR PATHWAY MAPPING AND TOXICOLOGICAL TESTING STRATEGIES

A living cell is an incredibly complex, dynamic system comprised of hundreds of thousands active genes, transcribed mRNA, proteins with all of their modifications, metabolites, and structural constituents from lipids and carbohydrates, to mention only a few. Even under homeostatic conditions all of this is undergoing continuous change and exchange regulated by complex interactions in networks resulting in rhythmic and chaotic patterns. This becomes even more complex if we see a population of cells, different cell types interacting or then the organ functions they form and their systemic interaction in the organism; virtually all toxic endpoints involve an emergent property of a population of cells. Even worse, life means reacting to the environment, which is constantly impacting on all levels of organization. It is illusive to fully describe such a system and model it. It is also naïve to take any component and expect it to reflect the whole system. The goal must be to know enough about a system to understand the major impacts, which is essentially what research into diseases or toxicology is about: understanding the impacts which make a lasting and severe change to the system.

To use an analogy (Smirnova et al. 2015), to understand the traffic in a larger city, we need to characterize a system of hundred thousands of pedestrians, cars, bicycles, etc. But we do not need and we cannot understand each and every element's behavior to understand the way flow is impaired—in fact, focusing on the micro understanding would distract from understanding the macro level. If there is a traffic accident, we see patterns of changes (traffic jam, redirection of flow, emergency forces deployed, etc.). If we take a snapshot photograph from a satellite of the situation, we might already see certain clusters or the appearance of ambulances. Even better if we can visualize fluxes and show where flow is hindered and see the direction of movement.

Omics technologies are these type of satellite photographs, usually just as a snapshot of the system. By comparison with the "normal" situation, we can start to identify the major derangements, especially when we have time series, replicates and dose-response analysis available, best if we can determine fluxes. We do not need to monitor every car; some of them suffice to characterize what happens on the main roads and places, and some of them, such as the ambulances, the police cars, or the fire trucks, are more telling. Different types of interferences can result in similar patterns (accident, construction work, a sport event) if hitting the same place/region. The stronger the disruption, the more easy to detect perturbation at places further away or whatever we measure (a traffic jam will impact little on pedestrians and bicyclists, but a roadblock does).

The analogy falls short when we see that our omics snapshot photographs are selective—they see either mRNA, proteins or metabolites, etc. This would be a camera seeing only cars but missing the anomalies of a marathon or a bicycle race taking place in the city. In order to understand these situations, we need to combine our monitoring.

A few lessons from our analogy:

- A dynamic system can hardly be understood from a single snapshot.
- Repeated and varied measurements, especially of different components, will give a more robust view of the system.
- The better we understand normal traffic and earlier perturbations, the better we know where and what to monitor and how to interpret it.
- Knowing the ambulance and police cars (the early and stress responses) is a good way to sense trouble, even without knowing why they were deployed.
- Simulation of traffic helps planning and can be done understanding only the major principles of the system.
- The stronger the hit to the system and the longer lasting the effect, the more likely we will see it and interpret it correctly.

Back to toxicology—let's spare the discussion, why we need to rethink the way we do our assessments by animal experiments (see OECD n.d.). We will only explore here what the availability of our new satellite photo cameras (the omics technologies) means for understanding perturbations of the organism by chemicals. It means first that we change the resolution—we see what happens in the city and do not need to wait until everything collapses (death of the animal). This will be likely much more relevant for low-dose and long-term exposures. We can compare different cities (species) to see whether the same perturbations occur, so whether they follow the same mode of action and can learn from each other.

Having set the scene for our expectations as to the promise of omics technologies, a few lessons learned over the last decade follow.

5.5.1 Each Omics is Different

Obviously, they measure different things and do so using different technologies (van Vliet 2011). These come with very different levels of standardization, maturation, detection limits and variability, experience in the toxicological user communities, quality assurance, analysis procedures and ways of expressing results, etc.

To contrast only two, i.e., gene array-based transcriptomics and mass-spectroscopy–based metabolomics: the most advanced omics technology is clearly the microarray, with a variety of commercial products available, detailed guidance for use and reporting, and a considerable use experience.

Microarrays have, over the past 20 years, become a mainstay of any systems-level approach. However, arriving at this point required years of developing statistical techniques to take advantage of the insights offered by the high-dimensional data while eliminating much of the noise intrinsic to the technology. On the other hand, other—omics technologies, such as metabolomics, are very promising (van Vliet et al. 2008) yet present some technical difficulties. Metabolomics (Bouhifd et al. 2013; Ramirez et al. 2013)—defined as measuring the concentration of "all" low molecular weight (<1500 Da) molecules in a system of interest—has yet to join transcriptomics and proteomics as an essential part of systems biology.

The fact that metabolomics is ultimately very close to the phenotype turns out to be a double-edged sword, as it means that metabolomics is extraordinarily sensitive to slight changes in experimental parameters, and it requires a scrupulous commitment to protocol and a long-term commitment to troubleshooting, as virtually any small change—different brands of food for animals, different plastic plates in tissue culture—can introduce artifacts (Bouhifd et al. 2015b). Additionally, sample preparation must be kept to a minimum, as every step has the potential to add artifacts.

In terms of analytical chemistry, metabolomics presents another challenge: the universe of metabolites consists of chemicals with a vast range of properties—there are approximately 2000 polar and natural lipids, 500 class-specific metabolites, 200 redox metabolites, and 800 primary metabolites—and the different biochemical properties precludes coverage with any one platform, e.g., HPLC will have different coverage than gas chromatography. Therefore, while untargeted metabolomics attempts to catch "all" the metabolites, the choice of platform will likely favor some over others. This is important to keep in mind for pathway analysis, as metabolites that are invisible to a specific platform but are heavily represented on a pathway of interest may skew the result, i.e., cells treated with estrogen may have steroid-specific pathways upregulated, but if a technology does not adequately capture large, nonpolar compounds, any impact on that pathway may be difficult to see. Furthermore, metabolomics, unlike transcriptomics, does not produce a list of unambiguously identified "features." Instead, it depends on several intricate steps of data analysis to go from a chromatogram to a list of metabolites with concentrations. Final metabolite identification is highly dependent on the accuracy of the prior analysis steps as well as the database used for compound identification; this is perhaps the most significant bottleneck for metabolomics to become a modality commonly employed by systems biologists. Metabolite identification is hampered by the fact that our knowledge of metabolic networks is still relatively incomplete, the databases still comparatively new, and the data infrastructure lacking, which present challenges for both metabolite identification and pathway analysis.

5.5.2 A Model Does Not Become Better by Adding Fancy Omics Endpoint

All models are wrong, some are useful

George Box
British Statistician

We cannot repeat this mantra often enough. We need to understand for each and every model how wrong/how useful it is. The limitations of *in vivo*, *in vitro*, and *in silico* models have been discussed elsewhere (Hartung 2007a, 2008, 2013; Hartung and

Hoffman, 2009). The important point to be made here is that a higher-level, composite endpoint does not overcome but amplifies the problems, as it adds noise and difficult-to-control variables to the model. On the upside, a broader phenotyping of the system, which comes hand in hand with the use of omics, shows exactly these shortcomings. If we can avoid the cherry-picking of results and the false discoveries due to multiple testing, but keep the broad outlook on what they tell us about the system, they are most valuable to understand what is really happening. However, this is no easy feat, and involves ensuring that conclusions based on one—omics platform are reproducible on other platforms (to limit technological artifacts) and are consistent with known and validated mechanisms. Additionally, this requires a commitment to quality assurance and reproducibility that has been generally lacking in—omics technologies as well as *in vitro* science as a whole.

5.5.3 The Problem is Not in Generating but Interpreting Data—The Signal vs. Noise Problem and Signatures of Toxicity

The amount of data available to biologists has shown an almost exponential explosion in the last few decades—National Center for Biotechnology Information's Gene Expression Omnibus (NCBI's GEO) platform boasts more than 3000 data sets and continues to grow. The embarrassment of riches, however, is not helpful unless the data can be turned into knowledge. This will require dimensionality reduction of the data so that the noise from both the biological variability and technical aspects does not overwhelm the signal, and that the derived Pathways of Toxicity are not the result of over-fitting to one limited set of data and are robust when compared with existing data.

Additionally, this will necessitate machine learning techniques that prune the information used in a model—something that will become even more important as ToxCast and other high-throughput data become available. It is a well-established fact in data mining that often additional data merely add noise or cause model overfitting. It is always a temptation to assume that using all available data will improve accuracy; however, the reality is that more descriptors may simply be adding more noise and not offering additional information. Finally, as bioinformatics methods generate more and more testable hypotheses, a smarter approach to exploring such proposed regulatory mechanisms is required. One possibility is to transform the genetic regulatory networks produced by—omics approaches into a systems biology markup language (SBML) model (Hucka et al. 2003).

5.5.4 Biomarkers of Mechanism Promise to "Clean" Signatures of Toxicity and Make Them Translatable between Model Systems

The closer a signature is to the MoA, the closer to the source—to draw upon our previous traffic metaphor, knowing the intersection of an accident is preferable to the 10,000 ft. view of traffic congestion—and the less likely it will be that the effects of biological variability and technical noise will muddy the signatures of toxicity. Biomarkers are experimental endpoints which reflect mechanisms (Blaauboer et al. 2013), which thus especially qualify to develop predictive test systems.

5.5.5 The Challenge of Pathway of Toxicity Identification and Annotation

Fortunately, Pathway of Toxicity (PoT) annotation can build upon the information architecture that has been extensively developed to describe biological pathways (Kleensang et al. 2014). SBML is a system designed to formally describe any biological entities that are linked by interactions or processes in a machine-readable format and (through graphical interpretation) human-readable diagrams. It is sufficiently flexible to specify genetic regulatory circuits, metabolic pathways, or cell-signaling pathways and can describe a system in as much or as little detail as necessary to capture the essential features. An additional benefit is that there are several SBML-compatible curated pathways (e.g., PANTHER Pathways) to help structure the data; therefore, SBML models often do not require starting from scratch, but usually only the far simpler task of adding the relevant information suggested by the high-throughput approach along with existing pathways.

Because SBML requires explicit, formally specified interactions, it often shows areas in proposed pathways that are poorly understood or characterized and, in some cases, conflicting. The standard diagrams employed by cell biologists to describe mechanisms in molecular biology generally involve a bunch of arrows and symbols. The arrows could mean anything—transcription, activation, phosphorylation, or merely a vague and unspecified interaction—and the symbols could be genes, proteins, small molecules, or, worse still, vague concepts such as "oxidative stress." Therefore, structuring proposed pathways using SBML or other standardized, controlled formats would not only help both prune and extend the network generated by "-omics" technologies, it would also help make the leap from basic pathway identification and hypothesis generation to models than can be used for more complex simulations, which can both weed out false positives and point to areas where proposed transcriptional regulatory mechanisms are clearly inadequate to describe the data. Because SBML requires explicit, formally specified interactions, it often shows areas in proposed pathways that are poorly understood or characterized and, in some cases, conflicting.

5.5.6 The Challenge of PoT Qualification

The temptation (and, owing to the complexity of interpretation, the comparative ease) of spinning high-throughput/high-dimensional data into a "good story" means that quality assurance is of critical importance to any alternative method based on such techniques. This will be of particular importance to metabolomics, owing to the sensitivity of the technique, the ambiguity of metabolite identification, and the high probability of artifacts—the temptation will always be there for researchers to treat a fluke as a profound finding, and the only guard against this is a culture of quality assurance and reproducibility.

5.5.7 The Challenge of Strategically Integrating Multiple Tests

Despite the fact that toxicology uses many stand-alone tests, a systematic combination of several information sources very often is required. Examples include when not all possible outcomes of interest (e.g., modes of action), classes of test substances (applicability domains), or severity classes of effect are covered in a single test; when the positive test result is rare (low prevalence leading to excessive false-positive results); and when the gold standard test is too costly or uses too many animals, creating a need for prioritization by screening. Similarly, tests are combined when the human predictivity of a single test is not satisfactory or when existing data and evidence from various tests will be integrated. Increasingly, kinetic information also will be integrated to make an *in vivo* extrapolation from *in vitro* data.

Integrated testing strategies (ITS) (Hartung et al. 2013) offer the solution to these problems. ITS have been discussed for more than a decade, and some attempts have been made in test guidance for regulations. Despite their obvious potential for revamping regulatory toxicology, however, we still have little guidance on the composition, validation, and adaptation of ITS for different purposes. Similarly, weight of evidence and evidence-based toxicology approaches require different pieces of evidence and test data to be weighed and combined.

ITS also represent the logical way of combining pathway-based tests, as suggested in the article "Toxicology for the Twenty-First Century." The state of the art of ITS and suggestions as to the definition, systematic combination, and quality assurance of ITS have been published (Hartung et al. 2013; Rovida et al. 2015).

5.5.8 The Goal of Systems Toxicology

Omics technologies provide a valuable opportunity to refine existing methods and provide information for so-called integrated testing strategies via the creation of signatures of toxicity. By mapping these signatures to underlying pathways of toxicity, some of which have been identified by toxicologists over the last few decades, and bringing them together with pathway information determined from biochemistry and molecular biology, a "systems toxicology" (Hartung et al. 2012) approach will enable virtual experiments to be conducted that can improve the prediction of hazard and the assessment of compound toxicity. It is important to define what is meant by "systems toxicology," which borrows heavily from "systems biology," i.e., attempts to model the (patho)physiology of the body with computational tools. It is proposed here that we will need such modeling, both to identify putative PoT (to enable their experimental validation) and ultimately to make sense of the data, as well as to make predictions about the effect of a substance in humans.

The underlying structure of systems biology and toxicology is a network. The move toward systems toxicology and massively parallel techniques opens new opportunities, but at the same time raises problems in deriving meaningful information out of the wealth of generated data. Such data are increasingly represented as networks in which the vertices (e.g., transcripts, proteins, or metabolites) are linked by edges (correlations, interactions, or reactions, respectively). Networks can vary in their functionalities. Some are undirected graphs that enable only the study of structure; others, like the biochemical network, are characterized by interactions of varying strengths, strongly nonlinear dynamics, and saturating response to inputs (Wagner, 1996).

Network analysis has evolved into a very active and interdisciplinary area of research encompassing biology, computer science, and social and information sciences.

Many studies are highly theoretical, but they may eventually help in identifying PoTs. The network research has three primary goals. First, it aims to understand statistical properties that characterize the network structure in order to suggest appropriate ways to measure these properties. This is very relevant to signature of toxicity identification. Second, it aims to create models of networks that can help us understand the meaning of these properties and how they interact with one another. Third, it aims to predict what the behavior of networked systems will be on the basis of measured structural properties. This direction can be very useful in elucidating pathways. Structural analysis of networks has already led to new insights into biological systems, and it is a helpful method for proposing new hypotheses. Several techniques for such structural analysis exist, such as the analysis of the global network structure, e.g., scale-free networks, network motifs (i.e., small subnetworks that occur significantly more often in the biological network than in random networks), network clustering (modularization of the network into parts) and network centralities. Network centralities are used to rank elements of a network according to a given importance concept (Koschutzki and Schreiber, 2008).

Biological systems are networked at many levels: hormone regulations, signaling cascades, gene regulation by transcription factors, or microRNA, for example. Increasingly, these systems can be modeled dynamically, though often in isolation—and they are difficult to combine. Fortunately, this challenge has been taken up by a broad scientific community in the life sciences, and toxicology can profit from many parallel or pioneering developments. This offers hope that solutions to understanding the interplay of network structure, function, and dynamics will emerge rapidly. Ultimately, virtual cells and organs need to be set up to integrate these networks, their interactions, and, for toxicology, their perturbation by exogenous substances (e.g., the German Virtual Liver Network [http://www.virtual-liver.de/] or similar activities for liver [http://epa.gov/ncct/virtual_liver/] and the virtual embryo at US EPA [http://epa.gov/ncct/v-Embryo/]). Such systems increasingly allow simulations and virtual experiments to be carried out. We have learned that such biological networks typically have critical nodes, which largely reflect the derangement of the network. Often these are the crossings of different pathways, which may enable the simplification of some of the models.

Systems toxicology, with its new datasets and large scale data integration, will enable the exploration of properties of biological systems beyond what is currently possible. It will provide the prospects of investigating natural variation and stochastic effects and their role in defining phenotypes and the transitions between them. To deliver on this potential, systems toxicology will need tools. There is an urgent need for novel tools for navigating, filtering, aggregating, visualizing, and assessing research content. In taking inspiration from systems biology, we grow more and more appreciative of collaborative, open-source tools to accelerate interoperability and to leverage resources available to scientists.

A number of challenges in modeling biological networks remain (Hartung et al. 2012):

- Few measurements are continuous, but we need dynamic/kinetic modeling; we have only snapshots of the dynamic system, which we need to combine with knowledge on (reaction) kinetics.

- Each of these networks has its own set of technologies to monitor them, and they are not necessarily compatible to measure in the same sample at the same time.
- Many systems are not completely known or measurable (only some network members can be followed).
- Many biological systems have a spatial (e.g., compartmentalization) or temporal (e.g., sequence and timing of events) aspect.
- Developmental aspects (establishment, maturation, [de-]differentiation, aging, etc.) take place in the models or represent measures of interest.
- Many relevant physiological processes involve interactions of different cell types or tissues, adding layers of complexity to the modeling.
- Inter-individual differences affect data acquisition.
- The multitude of parameters and conditions creates problems of multiple testing, over-fitting, noise/signal ratios.

And this is only to model the physiological process. It does not yet include the interference of substances. The opportunity lies in using the measurement endpoints as input parameters for the systems biology model and simulating the effect on the dynamics of the system. Ideally, this includes points of interaction with the foreign substance and effect data on nodes in the networks.

Systems toxicology is an exciting new prospect on which to base our studies on information-rich methods and bioinformatics, which reflect the dynamics and complexity of physiology. The fact that we have our "disease agent," i.e., the toxicant, at hand and can induce derangement by varying timing, conditions, concentrations, etc. as often as we want, distinguishes systems toxicology from similar approaches for clinical problems. The question arises how can such a systems toxicology approach be quality controlled and validated? Again, toxicology is more advanced than other medical fields here, with experience in such quality assurance schemes as GLP and formal validation. Toxicology, therefore, should not wait as a bystander to embrace the systems biology developed in other scientific disciplines but should instead bring its specific opportunities and experiences to the table. This promises, in return, to advance toxicology to a true reflection of human toxicity.

ACKNOWLEDGMENT

This article is based on our earlier publications (Hartung et al. 2012, 2013; Hartung 2013, 2015; Alépée et al. 2014); we would like to express our thanks to our coauthors and workshop partners for this input.

REFERENCES

Alépée N, Bahinski T, Daneshian M, De Wever B, Fritsche E, Goldberg A, Hansmann J, et al. State-of-the-art of 3D cultures (organs-on-a-chip) in safety testing and pathophysiology. *ALTEX* 2014;31:441–477.

Andersen M, Betts K, Dragan Y, Fitzpatrick S, Goodman JL, Hartung T, Himmelfarb J, et al. Developing microphysiological systems for use as regulatory tools—Challenges and opportunities. *ALTEX* 2014;31:364–367.

Baumann, J., Barenys, M., Gassmann, K, Fritsche E. Comparative human and rat 'Neurosphere Assay' for developmental neurotoxicity testing. *Current Protocols in Toxicology* 2014;59:12.21.11–12.21.24.

Blaauboer BJ, Boekelheide K, Clewell HJ, Daneshian M, Dingemans MM, Goldberg, AM, et al. The use of biomarkers of toxicity for integrating in vitro hazard estimates into risk assessment for humans. *ALTEX* 2012;29:411–425.

Bouhifd M, Andersen ME, Baghdikian C, Boekelheide K, Crofton KM, Fornace AJ Jr., Kleensang A, et al. The Human Toxome project. *ALTEX* 2015a;32:32:112–124.

Bouhifd M, Beger R, Flynn T, Guo L, Harris G, Hogberg HT, Kaddurah-Daouk R, et al. Quality assurance of metabolomics. *ALTEX* 2015b;32(4):319–26.

Bouhifd M, Hogberg HT, Kleensang A, Maertens A, Zhao L, Hartung T. Mapping the human toxome by systems toxicology. *Basic Clin Pharmacol Toxicol* 2014;115:1–8.

Bouhifd M, Hartung T, Hogberg HT, Kleensang A, Zhao L. Review: Toxicometabolomics. *J Appl Toxicol* 2013;33:1365–1383. doi: 10.1002/jat.2874.

Bruchmüller I, Pirkl E, Herrmann R, Stoermer M, Eichler H, Klüter H, Bugert P. Introduction of a validation concept for a PCR-based mycoplasma detection assay. *Cytotherapy* 2006;8:62–69.

Buehring GC, Eby EA, Eby MJ. Cell line cross-contamination: How aware are mammalian cell culturists of the problem and how to monitor it? *In Vitro Cell Devel Biol Animal* 2004;40:211–215.

Coecke S, Balls M, Bowe G, Davis J, Gstraunthaler G, Hartung T, Hay R, et al. Guidance on good cell culture practice. *ATLA* 2005;33:261–287.

Drexler HG, Uphoff CC. Mycoplasma contamination of cell cultures: Incidence, sources, effects, detection, elimination, prevention. *Cytotechnology* 2002;39:75–90.

Gordon S, Daneshian M, Bouwstra J, Caloni F, Constant S, Davies DE, Dandekar G, et al. Non-animal models of epithelial barriers (skin, intestine and lung) in research, industrial applications and regulatory toxicology. *ALTEX* 2015;32:327–378.

Hartung T. 3D—A new dimension of in vitro research. *Adv Drug Deliv Rev* 2014;69–70:vi.

Hartung T, Luechtefeld T, Maertens A, Kleensang A. Integrated testing strategies for safety assessments. *ALTEX* 2013;30:3–18.

Hartung T. Look back in anger—What clinical studies tell us about preclinical work. *ALTEX* 2013;30:275–291.

Hartung T, van Vliet E, Jaworska J, Bonilla L, Skinner N, Thomas R. Systems toxicology. *ALTEX* 2012;29:119–128.

Hartung T, Zurlo J. Alternative approaches for medical countermeasures to biological and chemical terrorism and warfare. *ALTEX* 2012;29:251–260.

Hartung T, McBride M. Food for thought … on mapping the human. *ALTEX* 2011;28:83–93.

Hartung T, Hoffmann S. Food for thought on … in silico methods in toxicology. *ALTEX* 2009;26:155–166.

Hartung T. Food for thought … on animal tests. *ALTEX* 2008;25:3–9.

Hartung T. Toxicology for the twenty-first century. *Nature* 2009;460:208–212.

Hartung, T. Food for thought … on cell culture. *ALTEX* 2007a;24:143–152.

Hartung, T. Food for thought … on validation. *ALTEX* 2007b;24:67–72.

Hartung T, Bremer S, Casati S, Coecke S, Corvi R, Fortaner S, Gribaldo L, et al. A modular approach to the ECVAM principles on test validity. *ATLA* 2004;32:467–472.

Hay RJ, Macy ML, Chen TR. Mycoplasma infection of cultured cells. *Nature* 1989;339:487–488.

Hoffmann S, Hartung T. Toward an evidence-based toxicology. *Hum Exper Toxicol* 2006;25:497–513.

Hogberg HT, Bressler J, Christian KM, Harris G, Makri G, O'Driscoll C, Pamies D, Smirnova L, Wen Z, Hartung T. Toward a 3D model of human brain development for studying gene/environment interactions. *Stem Cell Res Ther* 2013;4(Suppl 1):S4.

Hucka M, Finney A, Sauro HM, Bolouri H, Doyle JC, Kitano H, Arkin AP, et al. The systems biology markup language (SBML): A medium for representation and exchange of biochemical network models. *Bioinformatics* 2003;19(4):524–531.

Hughes P, Marshall D, Reid Y, Parkes H, Gelber C. The costs of using unauthenticated, over-passaged cell lines: how much more data do we need? *BioTechniques* 2007;43:575–584.

Kleensang A, Maertens A, Rosenberg M, Fitzpatrick S, Lamb J, Auerbach S, Brennan R, et al. Pathways of toxicity. *ALTEX* 2014;31:53–61.

Kleensang A, Vantangoli M, Odwin-DaCosta S, Andersen ME, Boekelheide K, Bouhifd M, Fornace AJ Jr, et al. Genetic variability in a frozen batch of MCF-7 cells invisible in routine authentication affecting cell function. *Scientific Reports* 2016;6:28994–28994.

Koschützki D, Schreiber F. Centrality analysis methods for biological networks and their application to gene regulatory networks. *Gene Regul Syst Bio* 2008;2:193–201.

Landry JJM, Pyl PT, Rausch T, Zichner T, Tekkedil MM, Stütz AM, Jauch A, et al. The genomic and transcriptomic landscape of a HeLa cell line. G3 (Bethesda) 2013;3(8):1213–1224. doi: 10.1534/g3.113.005777.

Langdon SP. Cell culture contamination: An overview. In: Langdon SP (Ed.). *Methods in Molecular Medicine, Vol. 88: Cancer Cell Culture: Methods and Protocols Cancer Cell Culture*. Totowa, NJ: Humana Press; 2003, pp. 309–318.

Lawrence B, Bashiri H, Dehghani H. Cross comparison of rapid mycoplasma detection platforms. *Biologicals* 2010;38:6–6.

Leist M, Hasiwa N, Rovida C, Daneshian M, Basketter D, Kimber I, Clewell H, et al. Consensus report on the future of animal-free systemic toxicity testing. *ALTEX* 2014;31:341–356.

Leist M, Hasiwa M, Daneshian M, Hartung T. Validation and quality control of replacement alternatives—Current status and future challenges. *Tox Res* 2012a;1:8–22. doi: 10.1039/C2TX20011B.

Leist M, Lidbury BA, Yang C, Hayden PJ, Kelm JM, Ringeissen S, Detroyer A, et al. Novel technologies and an overall strategy to allow hazard assessment and risk prediction of chemicals, cosmetics, and drugs with animal-free methods. *ALTEX* 2012b;29:373–388.

Leist M, Efremova L, Karreman C. Food for thought ... considerations and guidelines for basic test method descriptions in toxicology. *ALTEX* 2010;27:309–317.

Leist M, Hartung T, Nicotera P. The dawning of a new age of toxicology. *ALTEX* 2008b;25:103–114.

Lund RJ, Närvä E, Lahesmaa R. Genetic and epigenetic stability of human pluripotent stem cells. *Nat Rev Genet* 2012;13:732–744.

Marx U, Andersson TB, Bahinski A, Beilmann M, Beken S, Cassee FR, Cirit M, et al. Biology-inspired microphysiological system approaches to solve the prediction dilemma of substance testing using animals. *ALTEX* 2016;33:272–321.

Marx U, Walles H, Hoffmann S, Lindner G, Horland R, Sonntag F, Klotzbach U, et al. 'Human-on-a-chip' developments: A translational cuttingedge alternative to systemic safety assessment and efficiency evaluation of substances in laboratory animals and man? *Altern Lab Anim* 2012;40:235–257.

Mitalipova MM, Rao RR, Hoyer DM, Johnson JA, Meisner LF, Jones KL, Dalton S, Stice SL. Preserving the genetic integrity of human embryonic stem cells. *Nature Biotechnol* 2005;23:19–20.

Nikfarjam L, Farzaneh P. Prevention and detection of mycoplasma contamination in cell culture. *Cell J* (Yakhteh) 2012;13:203–212.

NRC. *Toxicity Testing in the 21st Century: A Vision and a Strategy*. Washington, DC: The National Academies Press; 2007.

OECD (2005). Guidance document on the validation and international acceptance of new or updated test methods for hazard assessment. OECD Series on Testing and Assessment No. 34. ENV/JM/MONO(2005)14.

OECD (n.d.) Guidelines for the testing of chemicals. Section 4, Health Effects, DOI: 10.1787/20745788.

Pamies D, Hartung T, Hogberg HT. Biological and medical applications of a brain-on-a-chip. *Exper Biol Med* 2014;239(9):1096–1107. doi: 10.1177/1535370214537738.

Ramirez T, Daneshian M, Kamp H, Bois FY, Clench MR, Coen M, Donley B, et al. Metabolomics in toxicology and preclinical research. *ALTEX* 2013;30:209–225.

Rottem S, Barile MF. Beware of mycoplasmas. *Trends Biotechnol* 1993;11:143–151.

Rovida C, Alépée N, Api AM, Basketter DA, Bois FY, Caloni F, Corsini E, et al. Integrated testing strategies (ITS) for safety assessment. *ALTEX* 2015;32:25–40.

Smirnova L, Harris G, Leist M, Hartung T. Cellular resilience. *ALTEX*, 2015;32(4):247–260.

Smirnova L, Hogberg HT, Leist M, Hartung T. Developmental neurotoxicity—Challenges in the 21st century and in vitro opportunities. *ALTEX* 2014;31:129–156.

Steinemann D, Göhring G, Schlegelberger B. Genetic instability of modified stem cells—A first step towards malignant transformation? *Am J Stem Cells* 2013;2:39–51.

Stephens ML, Betts K, Beck NB, Cogliano V, Dickersin K, Fitzpatrick S, Freeman J, et al. The emergence of systematic review in toxicology. *Toxicol Sci*, 2016;152(1):10–16. doi: 10.1093/toxsci/kfw059.

van Vliet E. Current standing and future prospects for the technologies proposed to transform toxicity testing in the 21st century. *ALTEX* 2011;28(1):17–44.

van Vliet E, Morath S, Eskes C, Linge J, Rappsilber J, Honegger P, Hartung T, Coecke S. A novel in vitro metabolomics approach for neurotoxicity testing, proof of principle for methyl mercury chloride and caffeine. *Neurotoxicology* 2008;29:1–12.

Wagner A. Does evolutionary plasticity evolve? *Evolution* 1996;50:1008–1023.

Young L, Sung J, Stacey G, Masters JR. Detection of Mycoplasma in cell cultures. *Nat Protoc* 2010;5:929–934.

Zurlo J. Evidence-based toxicology collaboration kick-off meeting. *ALTEX* 2011;28:152.

6 Lab-on-a-Chip Systems for Biomedical Applications

David Wartmann, Mario Rothbauer and Peter Ertl

CONTENTS

6.1 INTRODUCTION

The application of micromachining technologies for biomedical research has fostered the development of microscale technologies for advanced *in vitro* cell analysis such as microfluidic cellular microarrays, micro–cell culture systems, and specialized micro-analytical platforms. The greatest benefit of these miniaturized cell chip systems is the ability to provide quantitative data in real time to shed light on rapidly changing, dynamic biological systems. Additionally, lab-on-a-chip systems for biomedical applications have shown to exhibit high reliability, reproducibility, and robustness, while affording the opportunity to conduct measurements under physiologically-relevant conditions, which is considered one the key criteria for next-generation cell-based assays. This chapter presents a short overview on fabrication methods of microfluidic devices, fluid actuators, and biosensors for cell-based applications, describes microfluidic single-cell, multicell and three-dimensional (3D) cell culture arrays, and reflects on recent lab-on-a-chip advances and their applications for quality-control and stem cell biology.

6.2 MATERIALS, COMPONENTS, AND SENSING STRATEGIES

Common to all lab-on-a-chip systems for cell analysis is that they consist of micro–cell culture chambers that are connected to a microfluidic channel network in combination with various on-chip and off-chip optical imaging and/or electrical sensing strategies. Consequently advanced *in vitro* cell analysis systems need to autonomously perform an increasing number of operations such as reproducible cell seeding, reliable cell culture maintenance, and automated cell manipulation as well as on-chip analysis. This means that a number of issues concerning the integration of an appropriate fluid handling system, optimum cell culture biointerface, necessary cell actuation or stimulation, and biosensing strategies need to be carefully considered to enable *in vivo*-like cell culture conditions.

6.2.1 Materials and Fabrication Methods of Cell Chips

Fabrication methods used to build lab-on-a-chip systems for cell analysis are predominantly based on micromachining tools and microelectromechanical systems (MEMS) technology including soft lithography, hot embossing, injection molding, laser micromachining, and photolithography as well as 3D printing techniques (Fiorini et al. 2005; Dragone et al. 2013). The identification of the appropriate fabrication method is guided by a variety of parameters including available infrastructure (e.g., technical equipment), required fabrication speed and cost (e.g., multi-use and disposable devices), and necessary feature resolution (e.g., microstructures), as well as material properties (e.g., transparency, biocompatibility). Initially materials such as glass and silicon (Harrison et al. 1992; Manz et al. 1992) were used to cultivate cell cultures, because of existing manufacturing procedures and established biofunctionalization protocols that allowed the cultivation of adherent cell cultures. In recent years, however, plastics have become a dominant choice due to their improved material properties and compatibility with rapid prototyping technologies such as replica molding of PDMS (McDonald et al. 2002; Zhang et al. 2009; Zhang et al. 2010; Berthier et al. 2012), thermoset composites (Carlborg et al. 2011; Sollier et al. 2011) and thermoplastics (Duffy et al. 1998; Rudd et al. 1998; Fiorini et al. 2003; Fiorini et al. 2004; Golden et al. 2007; Novak et al. 2013). The main advantage of using replica molding (Xia et al. 1998) techniques is the elimination of cost-intensive clean room infrastructures, which made rapid prototyping accessible to a broader scientific community including bioengineers, medical researchers, and biological research groups. The increasing need for disposable high-throughput screening platforms has further fostered the introduction of hot embossing and injection molding technologies for biomedical applications. While hot embossing involves molding of thermoplastic sheets (e.g. polymethylmethacrylate [PMMA], polycarbonate [PC], cyclic olefin copolymer [COC], polystyrene [PS], polyvinylchloride [PVC], and polyethyleneterephthalate [PETG], etc.) (Becker et al. 2002; Novak et al. 2013; Ren et al. 2013) using metallic masters under heat and pressure (Locascio et al. 2006), injection molding allows for industrial scale-up production of biochips (Mair et al. 2006; Attia et al. 2009). Table 6.1 lists different fabrication methods, materials, and applications for cell analysis (Fiorini and Chiu 2005; Kim et al. 2008; Coltro et al. 2010; Sollier et al. 2011; Wu et al. 2011). Due to the complexity with respect to fabrication technology, most microdevices comprising a single material are predominantly produced by photolithography and soft lithography techniques. The majority of microdevices, however, are hybrids consisting of multiple materials including silicone, glass, photoresist (e.g., SU-8, TMMF, etc.), PDMS, and other polymers (e.g., PMMA, PC, PS, etc.) to gain synergistic material properties fit for the application.

6.2.2 Liquid Handling and Actuation Strategies for Cell Chips

The integration of fluid handling and actuation technology in lab-on-a-chip systems for cell analysis enables active nutrient supply and waste removal as well as the controlled addition of soluble factors at defined concentration gradients, thus providing a stress-free cellular microenvironment for optimum culture conditions. Although liquid handling of nano- and picoliter volumes of fluids dates back to the mid 1980s (Whitesides et al. 2006; Sackmann et al. 2014), the majority of lab-on-a-chip systems for cell analysis still rely on pressure-driven or passively driven flows using external syringe pumps (Stevens et al. 2008; Sun et al. 2010; Chin et al. 2011; X. Li et al. 2012), electrochemical-based (Neagu et al. 1996; Neagu et al. 1997), capillary (Gervais and Delamarche 2009; C. Li et al. 2012) and gravimetrically-driven flow systems (Morier et al. 2004). Among these, syringe pumps are still the most commonly used fluid delivery method to date, because of the low power consumption and stability resulting in constant nutrient supply over long periods of time, which is of great importance

TABLE 6.1
Selected Fabrication Methods and Materials Used for Cell-Based Microfluidics

Fabrication Method	Material	Application	References
Photolithography	SU-8 Photoresist	Chemotaxis	Ayuso et al. 2015
	TMMF Dryfilmresist	Biocompatibility of material tested only	Wangler et al. 2011
	Glass	Osteogenesis study	Jang et al. 2008
Soft lithography	PDMS	Toxicokinetic studies	Nakayama et al. 2008
		Gut model	Kim and Ingber 2013
		Lung model	Dongeun Huh et al. 2010
		Lung disease model	Dongeun Huh et al. 2012
		3D liver cell culture	Leclerc et al. 2003
Hot embossing	Polystyrene COC	Microvascular networks	Borenstein et al. 2010
		3D cell culture	Jeon et al. 2011

TABLE 6.2
Integrated Micropumping Strategies for Cell Chips

Pump Type	Actuation Principle	Flow Rate Q_{max} [µl min^{-1}]	References
Osmotic micropump	passive	30	Xu et al. 2010; Chen et al. 2013
Pneumatic micropump	active	3500	Schomburg et al. 1994; Meng et al. 2000; Unger et al. 2000; Grover et al. 2003b
Rotary micropump	active	5000	Ahn et al. 1995; Du et al. 2009; Du et al. 2013
Piezoelectric micropump	active	16,000	Smits 1990; Forster et al. 1995; Carrozza et al. 1995; Koch et al. 1997

for culturing mammalian cells in microfluidic devices. In turn, the trend toward fully integrated, automated, and miniaturized cell analysis systems has opened new opportunities for the incorporation of microvalves and micropumps (Unger et al. 2000; Grover et al. 2003; Kim et al. 2012) to perform a variety of crucial liquid handling steps such as bidirectional flow control for washing and cell loading procedures, stop-flow regimes for optimum cell culture conditions, defined shear-force application for cell stimulation, and administration of bioactive substances at controlled concentrations and time-points (Table 6.2).

6.2.3 Integrated Sensor Systems

Recent lab-on-a-chip developments show a clear trend toward more complex system architectures that include micropumps and valves, mixers, actuators, degassers, and biosensors, as well as multiple cell cultivation chambers for multiplexed cell analysis (Reichen et al. 2013). Only the heterogeneous integration of liquid handling systems, biointerface solution and biosensing strategies including optical, electrical, magnetic and acoustical sensors (Wu et al. 2011) will allow for standardization and automation of cell-based assays, which is a key requirement for industrial applications and regulatory approvals. Due to the availability of a broad range of fluorescent probes (e.g., fluorescent dyes, reporter genes, etc.) (Reyes et al. 2002; Hata et al. 2003) for selectively staining cellular structures as well as their familiarity to cell biologists, immuno-fluorescence end-point detection remains an important cell analysis method for microfluidic cell cultures. The increasing demand for continuous monitoring cellular responses resulted, however, in the adoption of a variety of optical monitoring techniques, (Charwat et al. 2013; Charwat et al. 2014) including light scattering, (Charwat et al. 2013; Schaefer et al. 1979; Wilson and Foster 2005; Wilson et al. 2005) absorption/transmission (Zhu et al. 2006; Malic et al. 2007) and fluorescence spectroscopy (Reyes et al. 2002; Hata et al. 2003). Additionally, a number of integrated optical sensors have been reported to monitor oxygen consumptions and light scattering of microfluidic cell cultures. (Sud et al. 2006; Nock et al. 2008a, 2008b; Lam et al. 2009; Schapper et al. 2009; Raghavan, et al. 2011; Ungerboeck et al. 2013). While spatially-resolved cellular oxygen consumption was detected using a two-wavelength ratiometric oxygen sensing strategy by laser-induced fluorescence (LIF) (Auroux et al. 2002; Gao et al. 2004) imaging (Ungerboeck et al. 2013) fully spray-coated organic photodiodes (OPDs) were used to detect cell numbers and morphology changes (Hofmann et al. 2005a, 2005b; Tedde et al. 2009; Wang et al. 2009; Ryu et al. 2011). In an attempt to replace standard laser systems that are bulky and costly organic light-emitting diodes (OLEDs) have been successfully integrated and combined with microfluidic systems. (Cai et al. 2010; Liu et al. 2011).

As an alternative to optical labels and complex fluorescence imaging set-up, the application of magnetic sensors in combination with biofunctionalized magnetic-particles have been proposed to manipulate, capture, separate, and analyze microfluidic cell cultures in real time (Miltenyi et al. 1990; Pankhurst et al. 2003; Laurent et al. 2008; Gijs et al. 2010; Kokkinis et al. 2013). As an example, one study integrated magneto-resistive sensors (GMR) to monitor nanoparticle phagocytosis in living cells (Shoshi et al. 2012). Since latest developments of nanodrug delivery systems and *in vivo* imaging systems relay on the efficiency of magnetic particles, this methodology can be used to analyze the interaction of nanodrug carries with e.g., cancerous tissue (Shoshi et al. 2013).

In addition to optical and magnetic detection methods, a variety of electroanalytical techniques such as voltammetry (Kafi et al. 2013; Yea et al. 2013), potentiometry, and impedance spectroscopy have been applied to provide information on cell viability, proliferation, and morphology as well as motility changes. Among these electroanalytical methods, impedance spectroscopy is most often used due to its noninvasive and label-free measurement conditions (Ertl et al. 2009; Sun et al. 2010). Consequently, a large number of applications have been reported over the years including drug and nanomaterial cytotoxicity (Xiao et al. 2003; Yeon et al. 2005; Richter et al. 2011), cell spreading and migration (Wegener et al. 2000), cell-cell junction formation (Wegener et al. 1999), and stem cell differentiation (Cho et al. 2009; Hildebrandt et al. 2010). Furthermore, impedance spectroscopy and light scattering measurements have been combined and integrated into a lab-on-a-chip to monitor cell-to-cell as well as cell-to-surface interactions of adherent and non-adherent mono- as well as co-culture systems. Other promising electrical biosensors for cell analysis include organic field-effect transistors to detect ionic changes caused be cellular metabolic activities and trans-epithelial resistance measurements to assess cell barrier stability and function (Torsi et al. 2002; Sekitani et al. 2009; Someya et al. 2010; Booth et al. 2012; Lin et al. 2012; Koppenhofer et al. 2013; Odijk et al. 2014). A comprehensive list of the various biosensors available can be found in Table 6.3.

6.3 MICROFLUIDIC 2D AND 3D LIFE-CELL MICROARRAYS

The demand for miniaturized high-throughput platforms for pharmaceutical screening applications has led to the development of next generation live-cell microarrays

TABLE 6.3
Biosensors for Lab-on-a-Chip Integration

Method		Application	Label-Free	Measured Response	Readout	References
Optical	LIF	Fluorescent probes (e.g., fluorescent dyes, reporter genes, etc.) for selectively staining of cellular structures and cell biology, (immuno-) fluorescence end-point detection	No	Intensity of fluorescent signal	High specificity (depending of method), single DNA level	Auroux et al. 2002; Reyes et al. 2002; Kazuki Hata et al. 2003; Gao, Yin, and Fang 2004
	Light scattering	Cell morphological changes, structural complexity, internal granularity and size determination	Yes	Intensity of scattered light at specific angles	High (ensemble average) to very high (single cell level)	Schafer et al. 1979; Wilson et al. 2005; Wilson and Foster 2005; Charwat et al. 2013b
	SPR	Sandwich-immunoassays and biomolecular interaction studies	Yes (certain cases)	Shift in refractive index	Very high (nano-molar), single molecules	Adam et al. 2006; Baac et al. 2006; Huang et al. 2006; Kim et al. 2006 Wang et al. 2011
	Optical cavity resonator	Biomolecule detection and immunoassay interactions	Yes	Proportional shift of the resonance wavelength to the change in mass	Very high due to single cell studies	Hu et al. 2009; White et al. 2009

(Continued)

TABLE 6.3 *(Continued)*
Biosensors for Lab-on-a-Chip Integration

Method		Application	Label-Free	Measured Response	Readout	References
	Oxygen sensing	Cell culture and LoC live oxygen monitoring	No	Change in the ratiometric signal of wavelengths or fluorescent signal from analytes	High sensitivity depending on the fluorophore and method	Sud et al. 2006; Nock et al. 2008; Larsen et al. 2011; Sagmeister et al. 2013; Ungerböck et al. 2013
	Optical tweezers	Cell morphological changes, sheer stress application, molecular interaction studies and fluid actuation/manipulation	No	Change in the applied holding force	Very high, single cell and molecular level	Perroud et al. 2008; Zhang et al. 2008; Padgett et al. 2011
	UV/VIS/IR spectroscopy	Dynamic Interaction of biomolecules, kinetics and conformational changes	Yes (certain cases)	Change in the absorbance of molecules	High to very high	Zhu et al. 2006; Malic et al. 2007
	Mass Spectroscopy	Composite analysis	Yes	Change in the size and charge of the (bio)molecules	Very high and distinct analysis capabilities	Baker et al. 2010; Bakstad et al. 2012; Wu et al. 2012; Yin et al. 2012; Wang et al. 2014
Electrical	ECIS	Cell morphological changes, migration, stress responses and differentiation	Yes	Changes in impedance, conductance or capacitance depending on frequency	High sensitivity but mostly ensemble average, but single cell analysis possible	Xiao et al. 2003; Yeon et al. 2005; Ertl et al. 2009; Sun et al. 2010; Richter et al. 2011

(Continued)

TABLE 6.3 *(Continued)*
Biosensors for Lab-on-a-Chip Integration

Method		Application	Label-Free	Measured Response	Readout	References
	FET/OFET	Detection of ionic changes caused by cellular metabolic activities	Yes	Changes of the inherent biomolecular charge	High to very high	Torsi et al. 2002; Sekitani et al. 2009; Someya et al. 2010; Booth et al. 2012; Lin et al. 2012; Koppenhofer et al. 2013; Odijk et al. 2014
	Nanowires	Localized cell-electroporation, biomolecule detection and multiplexing microarrays	Yes	Changes in resistance, current, conductance, or capacitance	Very high	Jokilaakso et al. 2013; Suzuki et al. 2013
	Voltametry	Viability, proliferation, and morphology as well as motility changes	Yes	Change of current produced by the analyte as the potential is varied	High to very high (depending on method)	Kafi et al. 2013; Yea et al. 2013
	Potentiometry	pH sensors, viability, proliferation, and morphology as well as motility changes	Yes	Change in electrical potential	High to very high (MEMS for pH detection)	Brischwein et al. 2003; Yang et al. 2004; Jang et al. 2010
Mechanical	Canitlevers	AFM, structural analysis, shear stress	Yes	Change in the resonant frequency or bending of the cantilever	Very high	Alvarez et al. 2003; Hwang et al. 2009; Waggoner et al. 2010

(Continued)

TABLE 6.3 *(Continued)*
Biosensors for Lab-on-a-Chip Integration

Method		Application	Label-Free	Measured Response	Readout	References
	Nanowires	Surface charge analysis	Yes	Protonation/ deprotonation leads to change in the surface charge	Very high	Cui et al. 2001; Patolsky et al. 2005
Magnetic	Magneto-resistive sensors	Manipulate, capture, separate and analyze microfluidic cell cultures in real-time	No	change in the magnetic field or changed conductance	High to very high (method dependend)	Miltenyi et al. 1990; Pankhurst et al. 2003; Laurent et al. 2008; Gijs et al. 2010; Kokkinis, et al. 2013

by integrating cell micropatterning and capture approaches in lab-on-a-chip devices (Chiu et al. 2000; Whitesides et al. 2001; Khademhosseini et al. 2004a, 2004b; Situma et al. 2006). In the following three sections, the various advancements of microfluidic live-cell microarrays for single-cell, multicell as well as 3D assays are described in more detail. An overview of the discussed microarray technologies can be seen in Figure 6.1.

6.3.1 Microfluidic Single-Cell Microarrays

Microfluidic single-cell microarrays are ideally suited to assess the heterogeneity within a cell population by analyzing the responses of a large number of individual cells with the aim of providing information on subpopulation distribution, cellular activities, and the ratio of responding and nonresponding cells. Practical applications of microfluidic single cell arrays include tumor biology, stem cell biology, antibiotic resistance screening, and single cell immune-typing. As an example, determining the intrinsic cellular heterogeneity of single circulating tumor cells (CTCs) is key in understanding the metastatic potential of CTCs, thus shedding light on the formation and growth of primary tumors, local tumor cell invasion, migration, and extravasation and metastasis. It has been shown that CTC expression profiles diverged distinctly from well-established cancer cell lines, thus questioning the suitability of conventional *in vitro* models for drug discovery and cancer therapy research (Powell et al. 2012). Microfluidic single-cell microarrays can facilitate the study of CTCs by providing diagnostic tools capable of isolating and analyzing CTCs using surface marker-based and marker-free methods. While surface-marker based methods

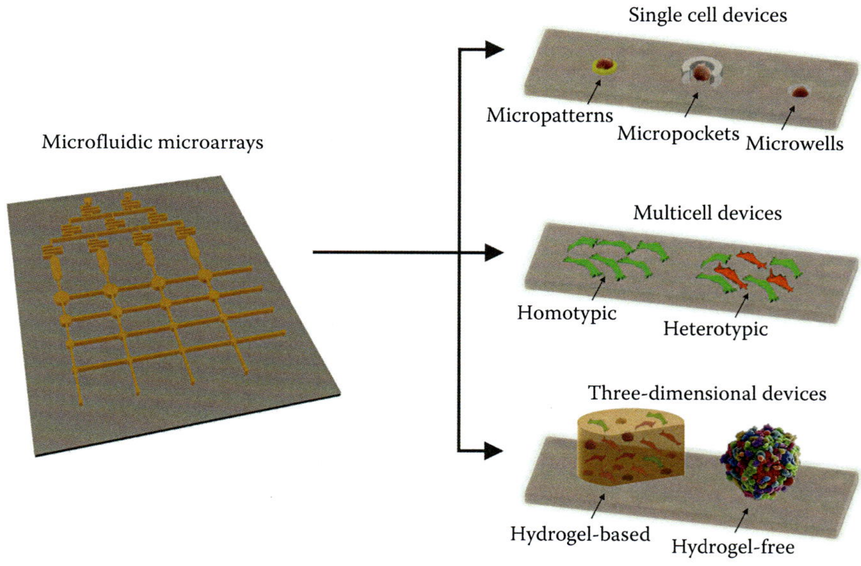

FIGURE 6.1 A schematic overview of microarray technologies in lab-on-chip systems.

predominantly employ magnetic beads for cell capture (Nagrath et al. 2007; Kang et al. 2012), marker-free microfluidic isolation methods use pillars and flow focusing approaches (Hur et al. 2011; Mach et al. 2011; Karabacak et al. 2014). A prominent example of single cell analysis using microfluidic single cell microarrays is the application of geno- and mechanotyping, also called "deformability cytometry," which has been established for identification of malignant and benign cells (Dalerba et al. 2011; Gossett et al. 2012; Tse et al. 2013). Another example of a microfluidic single-cell microarray integrates an array of PDMS-based cell-capture pockets that can be used to detect tumor proliferation and apoptosis following the administration of anticancer agents (Wlodkowic et al. 2009). Similarly, microfluidic single-cell microfluidic microarrays with integrated cell-capture pockets have been applied for analysis of signaling dynamics of hematopoietic stem cells including cell division, time-resolved viability, as well as cell migration and motility analysis (Faley et al. 2009). Another approach uses microarrayed 4.1nl nano-pockets for investigation of rare hematopoietic stem cells, where proliferation studies of single cells have been conducted (Lecault et al. 2011).

Alternative approaches for single-cell analysis are based on single-cell nanowell arrays where, for example, T cells are captured by gravity sedimentation within the nanowells and subsequently stimulated (Han et al. 2012). Cell analysis was accomplished using ELISA and immunofluorescence staining to provide biological information with single-cell resolution (Han et al. 2012; Yamanaka et al. 2012). Similarly, micro-arrayed nanowells can be applied for on-chip secretome analysis using $CD4^+$ positive T cells (Jin et al. 2009).

6.3.2 Microfluidic Multicell Microarrays

While single-cell assays provide information on heterogeneity within a cell population, microfluidic multicell microarrays are used to investigate responses of small-cell populations. An advantage of cultivating small-cell populations is that they allow for cell-to-cell interaction and communication, which is important to maintain appropriate cellular phenotypes. Microfluidic multicell microarrays have predominantly been used for screening applications (Hung et al. 2005a, 2005b; Lee et al. 2006; King et al. 2007; Wada et al. 2008; Kim et al. 2012;) and evaluation of cytotoxic agents (Wada et al. 2008; Song et al. 2010). As an example, Hung et al. presented a 10×10 microchamber array containing circular microbioreactors that are addressable by surrounding perfusion channels and connected to a concentration gradient generator to facilitate creation of dose-response curves (Hung et al. 2005b).

An alternative microfluidic cellular microarray enabled drug screening using 156 spots/cm^2 density in a 3-layered device consisting of a bottom layer with drug-filled microfluidic channels which are separated by a micropatterned nanoporous membrane from a top gel layer supporting the adherent cells (Upadhyaya et al. 2010). More recent work presented a PDMS-based microfluidic multicell microarray that allows large-scale screening of chemotherapeutic efficacy against tumor cells (Song et al. 2010). Results from this study showed that microfluidic parallel testing is comparable to data obtained from conventional cell culture experiments. Another comparison between microfluidic multicell microarrays and traditional 96-well plate cultivation investigated proliferation rates, glucose consumption, growth factor signaling, protein expression, stress markers, and DNA damage revealed that microfluidic multicell microarrays provide biologically meaningful results (Paguirigan et al. 2009).

The application of microfluidic multicell microarray containing pneumatic valves demonstrated reproducible loading of different cell types into individually addressable cultivation chambers as well as delivery of different reagents to specific chambers (Wang et al. 2008). This new-found versatility in cell stimulation and analysis allows for long-term cell cultivation under defined parameters including cell seeding density, feeding routine, medium components, surface compositions (Hattori et al. 2011), micropatterns (Witters et al. 2011), and cell stimulation (Gómez-Sjöberg et al. 2007). The ability to generate stable reagent gradients can also be combined with transfected cell lines such as GFP-reporter genes, which showed dose-dependent fluorescence increase in response to cytotoxic agents (Wada et al. 2008). Furthermore, the integration of microfluidic gas and concentration gradient generators provide the opportunity of gene expression profiling in presence of varying oxygen concentrations (Peng et al. 2013) and drug concentrations (Thompson et al. 2004). In another approach, different reporter cell populations were seeded in microchannels while a variety of compounds believed to be involved in hepatocyte inflammatory processes were injected into each microchannel (King et al. 2007). Live cell imaging of the fluorescent reporters allowed the acquisition of time-resolved responses in 256 nanoliter-scale bioreactors.

A large number of studies that implemented multicell microarray technology investigated the effects of fluid mechanical forces on cell culture physiology (Hung et al. 2005; Lee et al. 2006). Situations where elevated shear stress conditions and increased fluid mechanical forces need to be controlled include endothelial cell activation and stem cell research where mechanical stimulation can determine the differentiation fate of stem cells (Adamo et al. 2011; Toh et al. 2011). In turn, reduced shear stress conditions can readily be established using multilayer designs where outer perfused microchannels that mimic "bloodstreams" are separated by "C"-shaped structures from the inner "interstitial space." This and similar approaches can be used to limit overall shear force exposures by creating temporally stable gradients without the need for continuous or high flow (Sip et al. 2011; Kim et al. 2012) and generate stable gradients with minimal cellular shear exposure using stacked flows (Sip et al. 2011). Although most microfluidic multicell microarrays aim at screening small cell populations, they can also serve a platform for identifying, sorting, and capturing individual cells from a larger population (Chen et al. 2013). As an example, the combination of capturing spots specific to antigens on leukocyte cell-surfaces and detection spots coated with anti-cytokine antibodies allowed the analysis of blood samples. In a similar approach immobilized biofunctionalized magnetic beads were used to selectively capture and analyze cells from clinical samples (Saliba et al. 2010).

6.3.3 Microfluidic 3D Cell Microarrays

Despite recent achievements of microfluidic 2D cell culture systems, they still do not address the fact that *in vivo* cells coexist in 3D communities that are influenced by spatial orientation of cells and cell-to-cell contact within the extracellular matrix (Pampaloni et al. 2007). While multicell microarrays facilitate monitoring of bulk cell responses, many studies have demonstrated the need for more physiological relevant *in vitro* cell culture models. Consequently, 3D microfluidic cell culture microarrays have been developed in recent years to help cells retain their native tissue-specific functions in microfluidic devices (Sorger and Jensen 2006). The combination of 3D-cell cultures with microfluidics offers several advantages including (1) appropriate microscale dimensions that are comparable to *in vivo* microstructures; (2) establishment of chemical gradients to create dynamic 3D microenvironments; and (3) creation of reproducible medium-matrix biointerfaces. In other words, cells cultured in 3D are surrounded by ECM and in direct contact with each other (either homotypic or heterotypic way) and still subjected to controlled nutrient supply and waste removal using a microfluidic channel network. A variety of studies have further shown that 3D culture techniques based on aggregates, spheroids, and hydrogels are comparable with lab-on-a-chip technology. Additionally, the incorporation of microstructures in microfabricated devices further allows control over spheroid geometries including stripes, triangles, and star-shapes (Khademhosseini et al. 2007; Rivron et al. 2012). It is generally accepted that microfluidic 3D cell microarrays represent a valuable tool for improved for high-throughput screening applications of drug targets. In the following two sections, advances on microfluidic systems either employing hydrogel-based and hydrogel-free strategies will be reviewed.

6.3.3.1 Hydrogel-Based Microfluidic Systems

Hydrogels are 3D networks composed of various natural and synthetic polymers that retain water by swelling up to a percentage of 90%, thus mimicking the naturally surrounding of the extracellular matrix (Burdick et al. 2012). Hydrogels can be classified in two basic categories based on the origin of their composing polymer such as natural and synthetic monomers (Cushing et al. 2007). Natural or biological hydrogels used for microfluidic cell culture applications include agarose (Zamora-Mora et al. 2014), chitosan (He et al. 2013), alginate (Meli et al. 2014), hyaluronic acid (HA) (Bian et al. 2013), collagen (Shimizu et al. 2014), dextran (Oh et al. 2014), fibrin (Park et al. 2014), Matrigel (Jin et al. 2013), laminin (Jin et al. 2013), and silk fibroin (He et al. 2013). In turn, synthetic hydrogels have been used in combination with microfluidics and are based on polyethylene glycol (PEG) (Guarnieri et al. 2010), poly(ethylene glycol) diacrylate (PEG-DA) (Sivashankar et al. 2013), 2-hydroxyethyl methacrylate (2-HEMA) (Schwerdt et al. 2014), poly-2-hydroxyethyl methacrylate (PHEMA) (Johnson et al. 2013), poly-L-lactic acid (PLLA) (He et al. 2011), poly-lactic-co-glycolic acid (PLGA) (Heslinga et al. 2014), poly-glycerol sebacate (PGS) (Wu et al. 2014), and PuraMatrix™ (Dereli-Korkut et al. 2014), which is a fully synthetic peptide-based polymer. Among these, the most extensively employed hydrogel for microfluidic cell culture applications is PEG and derivatives thereof.

Important functions of hydrogels in microfluidic devices are the establishment of cellular barriers, the encapsulation of cells and/or drugs and their distribution, as well as the production of scaffolds and wound healing matrices. The main advantage of hydrogels for microfluidic cell cultures, however, is their mimicry of extracellular matrix structures including adequate porosity for cellular organization, biocompatibility, representative stiffness, and influence on cellular fate (Gobaa et al. 2011; Tse et al. 2011), all key parameters that promote native-like tissue function. Moreover, the development of so-called "smart hydrogels" has allowed for time-dependent release of bioactive compounds (Sundararaghavan et al. 2011; Lienemann et al. 2012) and establishment of chemical gradients to trigger cell responses (Kothapalli et al. 2011).

Overall microfluidic 3D cell culture systems have been used to study cell-matrix interactions as well as paracrine signaling in co-cultures of stem cells (Hamilton et al. 2013). A recent example using a microfluidic channel network containing several interconnected chambers investigated the interaction between different cell types and diverse tissues and organ structures such as blood vessels (Sung et al. 2013). Additionally, micropatterned cells have been used in cancer research to assess cell migration and invasive capacity of co-cultures in different hydrogels including collagen type I, Matrigel, and fibrin (Huang et al. 2009). Results of a similar study showed that tissue function was significantly enhanced when hepatocytes were mixed with nonparenchymal cells in varying hydrogel layers with differing stiffness (Kobayashi et al. 2013). Additionally, micropatterning has been applied for neural cells within hydrogel for researching neuronal network formation (Kitagawa et al. 2014). Overall, hydrogels used in 3D cell culture settings mimic the extracellular matrix including chemo- and mechanotransduction events, thus allowing the investigation of cell-cell interaction as well as cell-matrix interactions. Although natural hydrogels are inherently biocompatible and usually biodegradable, synthetic hydrogels offer ease of use and decreased background noise when employing proteomic analyses and other biologic assays (Geckil et al. 2010).

Despite their many advantages, a number of drawbacks for microfluidic cell culture applications still exist and are associated with biodegradability, limited reproducibility, and lack of standardization. For instance, in order to inhibit rapid degradation the addition of supplements, such as Aprotinin, throughout culture life may be required to maintain biodegradable hydrogel structures as ECM (Shikanov et al. 2011). Additional technical limitations include bubble formation and inherent difficulties with introducing cell-laden hydrogels in microfluidic channels prior to polymerization. Finally, the optimum length of time for 3D culture has to be experimentally established for microfluidic devices (Harink et al. 2013).

6.3.3.2 Hydrogel-Free 3D Cell-Based Microfluidic Systems

Similar to single-cell microfluidic systems containing integrated pockets, devices with integrated U-shaped microstructure arrays have been demonstrated as an effective method for generation of multicellular spheroids (MCS) (Fu et al. 2014). The authors demonstrated that *in situ* fabrication of the PEG-based microstructures (pockets) within microchannels can replace an expensive cleanroom setup. The response of epithelial HepG2 tumor cell spheroids to doxorubicin outlines the differences between a 3D liver tissue construct and conventional 2D cultures. Similarly, an increased chemotherapeutic resistance has been shown for terminal epithelial ovarian carcinoma, which was related to an enhanced expression of kallikrein-related peptidases in the presence of the spheroid cell culture (Dong et al. 2010).

PDMS-silicon hybrid devices containing integrated of pyramid-like microcavity arrays were also used for short-term MCF-7 breast cancer and long-term HepG2 liver spheroid culture analysis (Torisawa et al. 2007). Using microfluidic 3D cell microarray technology cell viability, albumin secretion and respiratory activity can be recorded in a high-throughput manner. Another study reported a three-layer PDMS/PC membrane microfluidic system capable of forming prostate cancer co-culture spheroids to recapitulate the growth behavior of PC-3 cancer cells within a bone metastatic prostate cancer microenvironment (Hsiao et al. 2009). Results of this study showed that spheroid culture of $CD133^+$ positive PC-3 cells remained a quiescent and undifferentiated phenotype, thus preserving the relevant surface markers of cancer stem cells (CSCs).

The latest trend in combining microarrays, microfluidics, and 3D cell culture technology includes the reliable establishment of multi-organs-on-a-chip and human-on-a-chip systems that mimic the complex interplay of multiple organs in a single device (Wagner et al. 2013). One prominent multi-organ-chip system for long-term cultivation of liver and skin organoids used a multilayered microfluidic device capable of long-term monitoring of cellular metabolic activity such as glucose consumption, LDH, and lactate production in the absence and presence of troglitazole over a 6-day exposure period (Wagner et al. 2013).

6.4 APPLICATION OF LAB-ON-A-CHIP TECHNOLOGY—SELECTED TOPICS

6.4.1 Quality Control for Cell-Based Therapy Applications

Assuring the quality of biological products has become an important aspect for the cell-manufacturing industry and automation is the most straightforward strategy for assuring a maximum of reproducibility, which is also a core request of regulators. It is also important to highlight that quality control (QC) of cell-based therapies such as cancer vaccines and stem cell-personalized medicine is by far the more labor-intensive procedure in comparison to cell manufacturing, which has essentially become an engineering task (Hinz et al. 2006) Although robotic cell culture systems have been available for about 20 years (Sharma et al. 2011), to date no technological solutions exist that allow for automation and miniaturization of QC measures to ensure product safety by simultaneously reducing manual labor steps and material costs as well as sample and media requirement. In this context, lab-on-a-chip technology has the potential to offer next-generation cell analysis tools capable of inexpensively testing large numbers of single cells or small numbers of cell populations under controlled and reproducible measurement conditions (Whitesides et al. 2006). The application of lab-on-a-chip technology for automated quality control measures has therefore the potential to close the existing product gap by providing fully automated and miniaturized analysis systems with improved reproducibility for (a) assuring compliance with specifications, (b) reduction of hands-on work and corresponding human error, and (c) reduced usage of expensive clinical grade biological reagents.

6.4.1.1 On-Chip Flow Cytometry, Single Cell Manipulation, and Isolation

To date, fluorescence-activated cell sorting (FACS), is considered the gold standard for assessment of viability, purity, and potency of cell-based vaccines in cancer therapy and point-of-care diagnostics. Despite the many advantages, FACS comes with a number of drawbacks including instrument size, cost of equipment, maintenance, and expensive operating liquids as well as its limited throughput capability. The need for increased miniaturization and parallelization has therefore led to the development of a range of modern cytometry techniques (Cho et al. 2010), so-called μFACS devices, that exhibit improved portability and analysis time for various medical applications (Cvetković et al. 2013). Recent developments describe μFACS devices employing electroosmotic (Fu et al. 2004), dielectrophoretic (Lapizco-Encinas et al. 2004), magnetic (Pamme et al. 2006) and hydrodynamic cell sorting and analysis methods (Bang et al. 2006). While electroosmotic cell sorting allows precise flow switching (Fu et al. 2002; Chen et al. 2009) in the presence of low flow rates to analyze tens of particles per second, dielectrophoretic cell sorting enables manipulation and sorting at the single cell level (Chen et al. 2009). In recent years magnetic sorting has also become a popular strategy due to its high selectivity as well as piezoelectric controlled actuators (Chen et al. 2009) and optical tweezers (Perroud et al. 2008) for high precision workflow in the μFACS environment. Hydrodynamically-focused cells are identified based on their fluorescence signal, and selected cells are moved by optical tweezers or laterally deflected using infrared laser into a collection channel (Perroud et al. 2008). On-chip integration of piezoelectric thin-film actuators (Chen et al. 2009) in μFACS devices further enables hydrodynamic focusing of single cells in the sub-nanoliter regime (Cho et al. 2010). As soon as a targeted cell enters the sorting junction, the actuator is activated by a voltage pulse to deflect the cell-containing fluid from the center position toward a lateral collection channel (Chen et al. 2011), thus enabling an operating limit of >1000 cells/s. A commercially available system containing piezoelectric actuators is the NanoCellect μFACS (San Diego, CA). Although current developments in μFACS devices show good reliability and reasonable throughput, the main challenges remain miniaturization and integration of fluidic, optic, and electronic components into compact benchtop-sized instrumentation.

6.4.1.2 Immunoassay-on-Chip for Quality Control Applications

In addition to phenotyping cell cultures, the analysis of secreted biomolecules provides information on the activity and biological function of the e.g., cancer vaccine. The state-of-the-art immunoassay in use today is the enzyme-linked immunosorbent assay, ELISA (Engvall et al. 1971; Van Weemen et al. 1971), which was developed in the 1960s and since then has become a fundamental tool in biological research and the pharmaceutical industry (Fossceco et al. 1996; Lequin et al. 2005). A major drawback of employing ELISA for routine quality control measurements for cell therapy products is its large sample and reagent volumes when using clinical grade materials, its susceptibility to contaminations, and lack of providing information on dynamic changing biological systems. Consequently, a number of immunoassay-on-a-chip systems with satisfactory performance have been developed in the last decade (Honda et al. 2005; Sista et al. 2008; Lee et al. 2009; Sun et al. 2010; Kim et al. 2011;

Miller et al. 2011). As an example, total assay time of ~30 min with very low sample volume can be achieved by capillary fluid delivery of multiple reagents in sequential and parallel manner (Kim et al. 2013).

Alternatives to the widely used ELISA immunoassay format are potency assays where, for instance, the activation of single CD_4^+ T-cells is followed by fluorescent reporting of the regulatory T-cell transcription factor Foxp3 and surface staining of CD_{69} using microwell array technology (Zaretsky et al. 2012). Additionally, a microfluidic single-cell barcode chip (SCBC) developed by Ma et al. (2011) enabled the detection of secreted proteins in 1-nl volume microchambers, each loaded with single cells and small defined numbers of cells. The microfluidic single-cell barcode chip permits on-chip, highly multiplexed detection of less than 1000 copies of proteins and requires only $\sim 1 \times 10^4$ cells for the assay (Ma et al. 2011). Protein concentrations are measured with immunosandwich assays using a spatially encoded antibody barcode. Furthermore, a micromotor-based lab-on-chip uses an "on-the-fly" double-antibody sandwich assay (DASA) to selectively capture target protein in the presence of excess of nontarget proteins (García et al. 2013). This nanomotor-based microchip immunoassay offers many potential applications in clinical diagnostics, environmental and security monitoring fields, as well as further applications in the cell therapy and POC diagnostics sector.

6.4.2 Lab-on-a-Chip for Stem Cell Biology Applications

In 2010 the global stem cell market was estimated at $21.5 billion, and it is projected to double every 5 years (HTStec 2012), thus outlining the importance of stem cell technology for medical therapeutics, drug development, and a variety of health-care applications including toxicological studies, disease modeling, and cell replacement therapies. An important aspect of stem cell research is the availability of well-characterized and validated pluripotent stem cells comparable to those of renowned cell banks. As a consequence, the major challenges associated with culturing stem cells *in vitro* are (a) controlled expansion while maintaining a homogeneous culture of undifferentiated cells and (b) the ability to reliably control and direct stem cell differentiation. To assess the generation of fully functional and specific cell types derived from stem cells, a variety of cell-based assays are routinely used in stem cell cultivations. Lab-on-a-chip technology is expected to provide the next generation of cell analysis tools for the stem cell market (Whitesides et al. 2006), because it is the only technology capable of providing spatial and temporal control over cell growth and stimuli.

6.4.2.1 On-Chip Cultivation of Stem Cells

The application of well-defined chemical and physical stimuli, including spatial and temporal gradients as well as repeated and long-term exposure to soluble factors, is crucial when investigating fundamental aspects of stem cell biology. In this regard microfluidics is ideally suited to monitor stem cell responses to varying reagents, surface properties, and mechanical forces. Micro- and nanofabrication technologies provide the means to (a) create an *in vitro* microenvironment of well-defined structural features including geometry, surface bound chemical factors, and 3D scaffolds, and

(b) precisely define cell cultivation and cell-free areas. The reproducible application of gradients of biomolecules including chemicals and soluble factors within microfluidic devices benefit stem cell maintenance and cultivation. For instance, microfluidic gradient generators have been used in stem cell research to investigate growth factor–dependent differentiation and chemotaxis within a single device (Xu et al. 2013). More sophisticated realizations include the integration of additional mixing meanders or the separation of gradient generation from cell culture chambers (Cimetta et al. 2010; Kim et al. 2012; Xu et al. 2013). Additionally, microchannels loaded with a polyethylene glycol hydrogel containing a concentration gradient cell adhesion motif Arg-Gly-Asp (RGD) within the hydrogel were used to provide *in vivo*-like elasticity during stem cell cultivations. Results of this study showed distinct adhesion behavior of the RGD concentration-dependent stem cell culture (Liu et al. 2012). Furthermore, various surface patterning and entrapment methods have been used to study stem cell growth on spatially defined areas. For instance, a microdevice containing three adjacent flow channels separated by a micropillar array can be used to study osteogenic differentiation employing a polyelectrolyte hydrogel containing bone marrow mesenchymal stem cell culture (Toh et al. 2007). A similar approach using a two-layer microfluidic device containing cell-repellent and cell-attractive areas were employed to cultivate and differentiate human mesenchymal stem-like cells into adipogenic and osteogenic lineages (Tenstad et al. 2010). An alternative stem cell patterning approach involves the creation of chemically tunable alginate hydrogel beads that employ laser direct-write technology to precisely locate single mouse embryonic stem cells within a lab-on-a-chip device (Phamduy et al. 2012).

Besides biochemical and structural cues, stem cell differentiation can also be guided by mechanical and electrical actuators to apply mechanical strain, electromagnetic forces, and ultrasound stimulation. The controlled physical stimulation of mesenchymal stem cells has been demonstrated to promote osteogenic lineage commitment (Chen et al. 2013), thus allowing the investigation of cell behavior in an environment that mimics mechanical forces of a living tissue. Alternative ways to mechanically activate cells include stretching and compression, which are important physiological parameters for cells of the vascular and musculoskeletal system. Mechanical strain is an important factor (Park et al. 2009) that can foster stem cell differentiation *in vitro* into osteogenic (Simmons et al. 2003), chondrogenic (McMahon et al. 2008), smooth muscle (Park et al. 2004) and endothelial (Shojaei et al. 2013) lineages. As examples, pneumatic actuation of PDMS-membranes within biochips allows defined cyclic stretching of a cell-covered membrane via two vacuum lines (Huh et al. 2010), to mechanically stimulate bovine embryos (Bae et al. 2011), and to mimic the hemodynamic microenvironment by using a microfluidic flow-stretch chip (Zheng et al. 2012), while electrostatic actuation of a PDMS capillary valve was used to mechanically stretch cells (Hausherr et al. 2013). Although many studies have evaluated the effects of mechanical strain on stem cell fate, in-depth analysis of the complex interplay between mechanical stimulation and chemical and topographical cues remains largely unexplored. However, the few existing studies performed using microfabricated devices (Hui et al. 2007; Park et al. 2009) and other perfusion systems (Datta et al. 2006; Jeon et al. 2013; Shojaei et al. 2013) indicate that a complex interplay between stimuli and cell responses takes place.

Recent technological advancements have further enabled the multiplexed cultivation and analysis of stem cell cultures (Reichen et al. 2013), including selective pre-screening of human embryonic stem cell clusters (Kamei et al. 2009), and parallelized cell cultivation and analysis using an ECM array (Hattori et al. 2011). Microfluidic systems used for stem cell cultivation and analyses also allow for targeted cell manipulation including trapping, selection, and sorting of stem cells. Among others, geometry-based trapping of mESCs (Khademhosseini et al. 2004b) using integrated microstructures and pairing of two cell types in close proximity for the fusion of embryonic stem cells has been successfully demonstrated (Skelley et al. 2009). Other microfluidic stem cell trapping devices contain up to 2048 single-cell traps (Kobel et al. 2012) and are combined with various sorting approaches including hydrodynamic sorting, deterministic lateral displacement, field-flow fractionation, microstructures, sedimentation, aqueous two-phase systems, inertial separation, and filters (Gossett et al. 2010), and fluorescence and immunomagnetic affinity-based labels (Autebert et al. 2012). Examples of microfluidic cell sorting in stem cell applications include gravitational field-flow fractionation devices where stem cells are transported along a capillary leading to distribution according to size, density, and surface properties (Roda et al. 2009). Additionally, stem cell separation based on cellular stiffness allowed the sorting of deformable metastatic cancer stem cells which escaped the integrated microbarriers. Furthermore, microfluidic dielectrophoretic stem cell trapping (Flanagan et al. 2008) and the application of optical tweezers for microfluidic stem cell isolation was recently demonstrated (Wang et al. 2011). A different approach in stem cell research involves encapsulation (Agarwal et al. 2013) and formation of emulsion droplets to entrap a defined number of stem cells for genetic analysis at the single-cell level (Zeng et al. 2010).

6.4.2.2 3D Stem Cell Cultivation and Application

To obtain a deeper understanding of stem cell-to-tissue cell interactions, a number of 3D co-culture systems have been developed in recent years. A novel method to generate co-cultures is based on droplet microfluidics (Tumarkin et al. 2011) where cell signaling in droplets containing different ratios of the two cell types can be compared. Another approach uses microfluidics for co-culture patterning where hydrodynamic cell focusing is performed in the presence of a perfused membrane (Torisawa et al. 2009). Results from studies employing microfluidic 3D co-culture systems showed (a) improved morphology and functionality of *in vitro* cultivated renal epithelial cells (Huang et al. 2013), and (b) that embryonic stem cells cultivated in close proximity to a feeder layer of mouse embryonic fibroblasts maintain an undifferentiated phenotype (Reichen et al. 2013). A different strategy used the generation of core-shell microcapsules loaded with embryonic aggregate cultures to demonstrate improved cardiac differentiation over conventional hanging drop models (Agarwal et al. 2013). Overall, the majority of microfluidic 3D stem cell cultures employ hydrogel-filled microchannels to study proliferation and differentiation capacities for tissue engineering applications using mesenchymal cells (hMSCs), as they can be easily obtained from patients. For instance, migration studies of hMSCs in fibrin gels of increasing stiffness showed that actin and microtubules are both responsible for migration (Vincent et al. 2013). Additional vascularization studies using hMSCs harvested from three different anatomic locations showed that bone marrow-derived MSCs fostered tubule formation of HUVEC. Moreover, cell characterization after 2 weeks of culture showed that cells were dedifferentiating into pericytes (Trkov et al. 2010). In addition to MSC, human embryonic stem cells (hESC) and murine stem cells (MSC), known to be extremely pluripotent, have been used in combination with microfluidics (Khoury et al. 2010; Park et al. 2009; Moledina et al. 2011).

6.5 FUTURE PERSPECTIVES AND CURRENT CHALLENGES

While the benefits of microfluidic cell cultures have been recognized early on, limited advances in system integration and device operation including tubing, pumps, and actuators have delayed their widespread use to date. To overcome these technological limitations, simple and easy-to-use devices need to be developed (Yu et al. 2007; Meyvantsson et al. 2008; Peng et al. 2013) to be compatible with existing labware and infrastructure such as standard 96-well plate readers, microscopes, and pipetting stations (Yu et al. 2007), thus eliminating the need for microfluidic interface and external tubing and valves (Meyvantsson et al. 2008). Alternatively, fully integrated, miniaturized, and automated microfluidic systems are expected to significantly improve selectivity, efficiency, and sensitivity of lab-on-a-chip devices, thus making them an attractive tool for medicine and biotechnology as well as the pharmaceutical market. Consequently, lab-on-chip technologies for cell analysis are accepted to play a prominent role in addressing the increasing demand for drug screening, disease modeling, pharmaceutical compound optimization, and cytotoxicity testing.

Future improvements of lab-on-a-chip technologies need to address the trend for more automation, parallelization, and integration of sensing and fluid handling components, which are key requirements for industrial-scale applications. While parallelization is vital for developing high-throughput screening tools for pharmaceutical compound testing, lead optimization, and quality control measures for personalized cell therapies, automation is vital for reducing hands-on work and corresponding operating errors. Once these obstacles are overcome and future generations of microfluidic cell analysis systems gain the approval of regulatory agencies, they have the potential to reduce or even replace animal models for pharmaceutical product development as well as revolutionize precision medicine.

REFERENCES

Adam, Pavel, Jakub Dostálek, Jiří Homola. 2006. Multiple Surface Plasmon Spectroscopy for Study of Biomolecular Systems. *Sensors and Actuators B: Chemical* 113 (2): 774–781.

Adamo, Luigi, Guillermo García-Cardeña. 2011. Directed Stem Cell Differentiation by Fluid Mechanical Forces. *Antioxidants & Redox Signaling* 15 (5): 1463–1473.

Agarwal, Pranay, Shuting Zhao, Peter Bielecki, Wei Rao, Jung Kyu Choi, Yi Zhao, Jianhua Yu, Wujie Zhang, Xiaoming He. 2013. One-Step Microfluidic Generation of Pre-Hatching Embryo-like Core-Shell Microcapsules for Miniaturized 3D Culture of Pluripotent Stem Cells. *Lab on a Chip* 3 (23): 4525–4533.

Ahn, Chong H, Mark G Allen. 1995. Fluid Micropumps Based on Rotary Magnetic Actuators. In *Proceedings IEEE Micro Electro Mechanical Systems*, 1995, p. 408.

Alvarez, Mar, Ana Calle, Javier Tamayo, Laura M Lechuga, Antonio Abad, Angel Montoya. 2003. Development of Nanomechanical Biosensors for Detection of the Pesticide DDT. *Biosensors and Bioelectronics* 18 (5–6): 649–653.

Attia, Usama M, Silvia Marson, Jerey R Alcock. 2009. Micro-Injection Moulding of Polymer Microfluidic Devices. *Microfluidics and Nanofluidics* 7 (1): 1–28.

Autebert, Julien, Benoit Coudert, François-Clément Bidard, Jean-Yves Pierga, Stéphanie Descroix, Laurent Malaquin, Jean-Louis Viovy. 2012. Microfluidic: An Innovative Tool for Efficient Cell Sorting. *Methods* 57 (3): 297–307.

Ayuso, Jose Maria, Rosa Monge, Guillermo Llamazares, Marco Moreno, Maria Agirregabiria, Javier Berganzo, Manuel Doblaré, Iñaki Ochoa, Luis J Fernandez. 2015. SU-8 Based Microdevices to Study Self-Induced Chemotaxis in 3D Microenvironments. *Frontiers in Materials* 2 (37). doi: 10.3389/fmats.2015.00037

Baac, Hyoungwon, József P Hajós, Jennifer Lee, Donghyun Kim, Sung June Kim, Michael L Shuler. 2006. Antibody-Based Surface Plasmon Resonance Detection of Intact Viral Pathogen. *Biotechnology and Bioengineering* 94 (4): 815–819.

Bae, Chae Yun, Minseok S Kim, Je-Kyun Park. 2011. Mechanical Stimulation of Bovine Embryos in a Microfluidic Culture Platform. *BioChip Journal* 5 (2): 106–113.

Baker, Monya. 2010. Mass Spectrometry for Biologists. *Nature Methods* 7 (2): 157–161.

Bakstad, Denise, Antony Adamson, David G Spiller, Michael RH White. 2012. Quantitative Measurement of Single Cell Dynamics. *Current Opinion in Biotechnology* 23 (1): 103–109.

Bang, Hyunwoo, Chanil Chung, Jung Kyung Kim, Seong Hwan Kim, Seok Chung, Junha Park, Won Gu Lee, et al. 2006. Microfabricated Fluorescence-Activated Cell Sorter through Hydrodynamic Flow Manipulation. *Microsystem Technologies* 12 (8): 746–753.

Becker, Holger, Laurie E Locascio. 2002. Polymer Microfluidic Devices. *Talanta* 56 (2): 267–287.

Berthier, Erwin, Edmond W K Young, David Beebe. 2012. Engineers Are from PDMS-Land, Biologists Are from Polystyrenia. *Lab on a Chip* 12 (7): 1224–1237.

Bian, Liming, Chieh Hou, Elena Tous, Reena Rai, Robert L Mauck, Jason A Burdick. 2013. The Influence of Hyaluronic Acid Hydrogel Crosslinking Density and Macromolecular Diffusivity on Human MSC Chondrogenesis and Hypertrophy. *Biomaterials* 34 (2): 413–421.

Booth, Ross, Hanseup Kim. 2012. Characterization of a Microfluidic in Vitro Model of the Blood-Brain Barrier. *Lab on a Chip* 12: 1784–1792.

Borenstein, Jeffrey T, Malinda M Tupper, Peter J MacK, Eli J Weinberg, Ahmad S Khalil, James Hsiao, Guillermo García-Cardeña. 2010. Functional Endothelialized Microvascular Networks with Circular Cross-Sections in a Tissue Culture Substrate. *Biomedical Microdevices* 12 (1): 71–79.

Brischwein, Martin, ER Motrescu, E Cabala, Angela M Otto, Helmut Grothe, Bernhard Wolf. 2003. Functional Cellular Assays with Multiparametric Silicon Sensor Chips. *Lab on a Chip—Miniaturisation for Chemistry and Biology* 3 (4): 234–240.

Burdick, Jason A, William L Murphy. 2012. Moving from Static to Dynamic Complexity in Hydrogel Design. *Nature Communications* 3: 1269.

Cai, Yuankun, Alex Smith, Joseph Shinar, Ruth Shinar. 2010. Data Analysis and Aging in Phosphorescent Oxygen-Based Sensors. *Sensors and Actuators B-Chemical* 146 (1): 14–22.

Carlborg, Carl Fredrik, Tommy Haraldsson, Kim Oberg, Michael Malkoch, Wouter van der Wijngaart, Kim Öberg. 2011. Beyond PDMS: Off-Stoichiometry Thiol-Ene (OSTE) Based Soft Lithography for Rapid Prototyping of Microfluidic Devices. *Lab on a Chip* 11 (18): 3136–3147.

Carrozza, MC, N Croce, B Magnani, P Dario. 1995. A Piezoelectric-Driven Stereolithography-Fabricated Micropump. *Journal of Micromechanics and Microengineering* 5 (2): 177.

Charwat, Verena, Martin Joksch, Drago Sticker, Michaela Purtscher, Mario Rothbauer, Peter Ertl. 2014. Monitoring Cellular Stress Responses Using Integrated High-Frequency Impedance Spectroscopy and Time-Resolved ELISA. The Analyst 139 (20). *Royal Society of Chemistry* 5271–5282.

Charwat, Verena, Michaela Purtscher, Sandro F Tedde, Oliver Hayden, Peter Ertl. 2013. Standardization of Microfluidic Cell Cultures Using Integrated Organic Photodiodes and Electrode Arrays. *Lab Chip* 13 (5): 785–797.

Charwat, Verena, Mario Rothbauer, Sandro F Tedde, Oliver Hayden, Jacobus J Bosch, Paul Muellner, Rainer Hainberger, Peter Ertl. 2013. Monitoring Dynamic Interactions of Tumor Cells with Tissue and Immune Cells in a Lab-on-a-Chip. *Analytical Chemistry* 85 (23): 11471–11478.

Chen, Arnold, Tam Vu, Gulnaz Stybayeva, Tingrui Pan, Alexander Revzin. 2013. Reconfigurable Microfluidics Combined with Antibody Microarrays for Enhanced Detection of T-Cell Secreted Cytokines. *Biomicrofluidics* 7 (2): 24105.

Chen, Chun H, Sung H Cho, Hsin-I Chiang, Frank Tsai, Kun Zhang, Yu-Hwa Lo. 2011. Specific Sorting of Single Bacterial Cells with Microfabricated Fluorescence-Activated Cell Sorting and Tyramide Signal Amplification Fluorescence in Situ Hybridization. *Analytical Chemistry* 83 (19): 7269–7275.

Chen, Chun H, Sung Hwan Cho, Frank Tsai, Ahmet Erten, Yu-Hwa Lo. 2009. Microfluidic Cell Sorter with Integrated Piezoelectric Actuator. *Biomedical Microdevices* 11 (6): 1223–1231.

Chen, Julia C, Christopher R Jacobs. 2013. Mechanically Induced Osteogenic Lineage Commitment of Stem Cells. *Stem Cell Research & Therapy* 4 (5): 107.

Chen, Yu-Chih, Patrick Ingram, Xia Lou, Euisik Yoon. 2013. Osmotic Actuation for Microfluidic Components in Point-of-Care Applications. *Micro Electro Mechanical Systems (MEMS), 2013 IEEE 26th International Conference* on January 20–24, 2013.

Chin, Curtis D, Tassaneewan Laksanasopin, Yuk Kee Cheung, David Steinmiller, Vincent Linder, Hesam Parsa, Jennifer Wang, et al. 2011. Microfluidics-Based Diagnostics of Infectious Diseases in the Developing World. *Nature Medicine* 17 (8): 1015–1019.

Chiu, Daniel T, Noo Li Jeon, Sui Huang, Ravi S Kane, Christopher J Wargo, Insung S Choi, Donald E Ingber, George M Whitesides. 2000. Patterned Deposition of Cells and Proteins onto Surfaces by Using Three-Dimensional Microfluidic Systems. *Proceedings of the National Academy of Sciences of the United States of America* 97 (6): 2408–2413.

Cho, Sungbo, Erwin Gorjup, Hagen Thielecke. 2009. Chip-Based Time-Continuous Monitoring of Toxic Effects on Stem Cell Differentiation. *Annals of Anatomy* 191 (1): 145–152.

Cho, Sung Hwan, Chun H Chen, Frank S Tsai, Jessica M Godin, Yu-Hwa Lo. 2010. Human Mammalian Cell Sorting Using a Highly Integrated Micro-Fabricated Fluorescence-Activated Cell Sorter (μFACS). *Lab on a Chip* 10 (12): 1567–1573.

Cimetta, Elisa, Christopher Cannizzaro, Richard James, Travis Biechele, Randall T Moon, Nicola Elvassore, Gordana Vunjak-Novakovic. 2010. Microfluidic Device Generating Stable Concentration Gradients for Long Term Cell Culture: Application to Wnt3a Regulation of B-Catenin Signaling. *Lab on a Chip—Miniaturisation for Chemistry and Biology* 10 (23): 3277–3283.

Coltro, Wendell Karlos Tomazelli, Dosil Pereira de Jesus, José Alberto Fracassi da Silva, Claudimir Lucio do Lago, Emanuel Carrilho. 2010. Toner and Paper-Based Fabrication Techniques for Microfluidic Applications. *Electrophoresis* 31 (15): 2487–2498.

Cui, Yi, Qingqiao Wei, Hongkun Park, Charles M Lieber. 2001. Nanowire Nanosensors for Highly Sensitive and Selective Detection of Biological and Chemical Species. *Science* 293 (5533): 1289–1292.

Cushing, Melinda C, Kristi S Anseth. 2007. Materials Science. Hydrogel Cell Cultures. *Science (New York, N.Y.)* 316 (5828): 1133–1134.

Cvetković, Benjamin Z, Petra S Dittrich. 2013. A Microfluidic Device for Open Loop Stripping of Volatile Organic Compounds. *Analytical and Bioanalytical Chemistry* 405 (8): 2417–2423.

Dalerba, Piero, Tomer Kalisky, Debashis Sahoo, Pradeep S Rajendran, Michael E Rothenberg, Anne A Leyrat, Sopheak Sim, et al. 2011. Single-Cell Dissection of Transcriptional Heterogeneity in Human Colon Tumors. *Nature Biotechnology* 29 (12): 1120–1127.

Datta, Néha, Quynh P Pham, Upma Sharma, Vassilios I Sikavitsas, John A Jansen, Antonios G Mikos. 2006. In Vitro Generated Extracellular Matrix and Fluid Shear Stress Synergistically Enhance 3D Osteoblastic Differentiation. *Proceedings of the National Academy of Sciences of the United States of America* 103 (8): 2488–2493.

Dereli-Korkut, Zeynep, H Dogus Akaydin, AH Rezwanuddin Ahmed, Xuejun Jiang, Sihong Wang. 2014. Three Dimensional Microfluidic Cell Arrays for Ex Vivo Drug Screening with Mimicked Vascular Flow. *Analytical Chemistry* 86 (6): 2997–3004.

Dong, Ying, Olivia L Tan, Daniela Loessner, Carson Stephens, Carina Walpole, Glen M Boyle, Peter G Parsons, Judith A Clements. 2010. Kallikrein-Related Peptidase 7 Promotes Multicellular Aggregation via the $\alpha 5\beta 1$ Integrin Pathway and Paclitaxel Chemoresistance in Serous Epithelial Ovarian Carcinoma. *Cancer Research* 70 (7): 2624–2633.

Dragone, Vincenza, Victor Sans, Mali H Rosnes, Philip J Kitson, Leroy Cronin. 2013. 3D-Printed Devices for Continuous-Flow Organic Chemistry. *Beilstein Journal of Organic Chemistry* 9: 951–959.

Du, Min, Zengshuai Ma, Xiongying Ye, Zhaoying Zhou. 2013. On-Chip Fast Mixing by a Rotary Peristaltic Micropump with a Single Structural Layer. *Science China Technological Sciences* 56 (4): 1047–1054.

Du, Min, Xiongying Ye, Kang Wu, Zhaoying Zhou. 2009. A Peristaltic Micro Pump Driven by a Rotating Motor with Magnetically Attracted Steel Balls. *Sensors* 9 (4): 2611–2620.

Duffy, David C, J Cooper McDonald, Olivier J A Schueller, George M Whitesides. 1998. Rapid Prototyping of Microfluidic Systems in Poly(dimethylsiloxane). *Analytical Chemistry* 70 (23): 4974–4984.

Engvall, Eva, Peter Perlmann. 1971. Enzyme-Linked Immunosorbent Assay (ELISA) Quantitative Assay of Immunoglobulin G. *Immunochemistry* 8 (9): 871–874.

Ertl, Peter, Rudolf Heer. 2009. Interdigitated Impedance Sensors for Analysis of Biological Cells in Microfluidic Biochips. *Elektrotechnik Und Informationstechnik* 126 (1–2): 47–50.

Faley, Shannon L, Mhairi Copland, Donald Wlodkowic, Walter Kolch, Kevin T Seale, John P Wikswo, Jonathan M Cooper. 2009. Microfluidic Single Cell Arrays to Interrogate Signalling Dynamics of Individual, Patient-Derived Hematopoietic Stem Cells. *Lab on a Chip* 9 (18): 2659–2664.

Fiorini, Gina S, Daniel T Chiu. 2005. Disposable Microfluidic Devices: Fabrication, Function, and Application. *Biotechniques* 38 (3): 429–446.

Fiorini, Gina S, Gavin D Jeffries, David S Lim, Christopher L Kuyper, Daniel T Chiu. 2003. Fabrication of Thermoset Polyester Microfluidic Devices and Embossing Masters Using Rapid Prototyped Polydimethylsiloxane Molds. *Lab Chip* 3 (3): 158–163.

Fiorini, Gina S, Robert M Lorenz, Jason Kuo, Daniel T Chiu. 2004. Rapid Prototyping of Thermoset Polyester Microfluidic Devices. *Analytical Chemistry* 76 (16): 4697–4704.

Flanagan, Lisa A, Jente Lu, Lisen Wang, Steve A Marchenko, Noo Li Jeon, Abraham P Lee, Edwin S Monuki. 2008. Unique Dielectric Properties Distinguish Stem Cells and Their Differentiated Progeny. *Stem Cells* 26 (3): 656–665.

Forster, Fred K, Ronald L Bardell, Martin A Afromowitz, Nigel R Sharma, Alan Blanchard. 1995. Design, Fabrication and Testing of Fixed-Valve Micro-Pumps. In *Proceedings of the ASME Fluids Engineering Division* 234: 39–44.

Fossceco, Stewart L, Knoll Pharmaceutical Company, Nathan A Curtis. 1996. Exploring Enzyme-Linked Immunosorbent Assay (ELISA) Data with the SAS® Analyst Application Introduction. *The Journal of Pediatrics* 146: 62–65.

Fu, Anne Yen-Chen. 2002. Microfabricated Fluorescence-Activated Cell Sorters (FACS) for Screening Bacterial Cells. *Etd. Caltech. Edu* 2002.

Fu, Chien-Yu, Sheng-Yang Tseng, Shih-Mo Yang, Long Hsu, Cheng-Hsien Liu, Hwan-You Chang. 2014. A Microfluidic Chip with a U-Shaped Microstructure Array for Multicellular Spheroid Formation, Culturing and Analysis. *Biofabrication* 6 (1): 015009.

Fu, Lung Ming, Ruey Jen Yang, Che-hsin Lin, Yu Jen Pan, Gwo Bin Lee. 2004. Electrokinetically Driven Micro Flow Cytometers with Integrated Fiber Optics for on-Line Cell/particle Detection. *Analytica Chimica Acta* 507: 163–169.

Gao, Jian, Xue-Feng Yin, Zhao-Lun Fang. 2004. Integration of Single Cell Injection, Cell Lysis, Separation and Detection of Intracellular Constituents on a Microfluidic Chip. *Lab on a Chip* 4 (1): 47–52.

García, Miguel, Jahir Orozco, Maria Guix. 2013. Micromotor-Based Lab-on-Chip Immunoassays. *Nanoscale*: 5 (4): 1325–1331.

Geckil, Hikmet, Feng Xu, Xiaohui Zhang, SangJun Moon, Utkan Demirci. 2010. Engineering Hydrogels as Extracellular Matrix Mimics. *Nanomedicine (London, England)* 5 (3): 469–484.

Gervais, Luc, Emmanuel Delamarche. 2009. Toward One-Step Point-of-Care Immunodiagnostics Using Capillary-Driven Microfluidics and PDMS Substrates. *Lab on a Chip* 9 (23): 3330–3337.

Gijs, Martin A M, Frederic Lacharme, Ulrike Lehmann. 2010. Microfluidic Applications of Magnetic Particles for Biological Analysis and Catalysis. *Chemical Reviews* 110 (3): 1518–1563.

Gobaa, Samy, Sylke Hoehnel, Marta Roccio, Andrea Negro, Stefan Kobel, Matthias P Lutolf. 2011. Artificial Niche Microarrays for Probing Single Stem Cell Fate in High Throughput. *Nature Methods* 8 (11): 949–955.

Golden, Andrew P, Joe Tien. 2007. Fabrication of Microfluidic Hydrogels Using Molded Gelatin as a Sacrificial Element. *Lab Chip* 7 (6): 720–725.

Gómez-Sjöberg, Rafael, Anne A Leyrat, Dana M Pirone, Christopher S Chen, Stephen R Quake. 2007. Versatile, Fully Automated, Microfluidic Cell Culture System. *Anal Chem* 79 (22): 8557–8563.

Gossett, Daniel R, Henry T K Tse, Serena A Lee, Yong Ying, Anne G Lindgren, Otto O Yang, Jianyu Rao, Amander T Clark, Dino Di Carlo. 2012. Hydrodynamic Stretching of Single Cells for Large Population Mechanical Phenotyping. *Proceedings of the National Academy of Sciences of the United States of America* 109 (20): 7630–7635.

Gossett, Daniel R, Westbrook M Weaver, Albert J Mach, Soojung Claire Hur, Henry Tat Kwong Tse, Wonhee Lee, Hamed Amini, Dino Di Carlo. 2010. Label-Free Cell Separation and Sorting in Microfluidic Systems. *Anal Bioanal Chem* 397 (8): 3249–3267.

Grover, William H, Alison M Skelley, Chung N Liu, Eric T Lagally, Richard A Mathies. 2003a. Monolithic Membrane Valves and Diaphragm Pumps for Practical Large-Scale Integration into Glass Microfluidic Devices. *Sensors and Actuators B: Chemical* 89 (3): 315–323.

Guarnieri, D, A De Capua, M Ventre, A Borzacchiello, C Pedone, D Marasco, M Ruvo, PA Netti. 2010. Covalently Immobilized RGD Gradient on PEG Hydrogel Scaffold Influences Cell Migration Parameters. *Acta Biomaterialia* 6 (7): 2532–2539.

Han, Qing, Neda Bagheri, Elizabeth M Bradshaw, David A Hafler, Douglas A Lauffenburger, J Christopher Love. 2012. Polyfunctional Responses by Human T Cells Result from Sequential Release of Cytokines. *Proceedings of the National Academy of Sciences of the United States of America* 109 (5): 1607–1612.

Harink, Björn, Séverine Le Gac, Roman Truckenmüller, Clemens van Blitterswijk, Pamela Habibovic. 2013. Regeneration-on-a-Chip? The Perspectives on Use of Microfluidics in Regenerative Medicine. *Lab on a Chip* 13 (18): 3512–3528.

Harrison, D Jed, Andreas Manz, Zhonghui Fan, Hans Ludi, H Michael Widmer. 1992. Capillary Electrophoresis and Sample Injection Systems Integrated on a Planar Glass Chip. *Analytical Chemistry* 64 (17): 1926–1932.

Hattori, Koji, Shinji Sugiura, Toshiyuki Kanamori. 2011. Microenvironment Array Chip for Cell Culture Environment Screening. *Lab on a Chip* 11 (2): 212–214.

Hausherr, Tanja C, Hicham Majd, Damien Joss, Arnaud Müller, Dominique P Pioletti, Martin A M Gijs, Christophe Yamahata. 2013. Capillary-Valve-Based Platform towards Cell-on-Chip Mechanotransduction Assays. *Sensors and Actuators, B: Chemical* 188: 1019–1025.

He, Jiankang, Mao Mao, Yaxiong Liu, Lin Zhu, Dichen Li. 2013. Bottom-up Fabrication of 3D Cell-Laden Microfluidic Constructs. *Materials Letters* 90: 93–96.

He, Tianxi, Qionglin Liang, Kai Zhang, Xuan Mu, Tingting Luo, Yiming Wang, Guoan Luo. 2011. A Modified Microfluidic Chip for Fabrication of Paclitaxel-Loaded Poly(l-Lactic Acid) Microspheres. *Microfluidics and Nanofluidics* 10 (6): 1289–1298.

Heslinga, Michael J, Gabriella M Willis, Daniel J Sobczynski, Alex J Thompson, Omolola Eniola-Adefeso. 2014. One-Step Fabrication of Agent-Loaded Biodegradable Microspheroids for Drug Delivery and Imaging Applications. *Colloids and Surfaces B: Biointerfaces* 116: 55–62.

Hildebrandt, Cornelia, Heiko Büth, Sungbo Cho, Impidjati, Hagen Thielecke. 2010. Detection of the Osteogenic Differentiation of Mesenchymal Stem Cells in 2D and 3D Cultures by Electrochemical Impedance Spectroscopy. *Journal of Biotechnology* 148 (1): 83–90.

Hinz, Thomas, Christian J Buchholz, Ton van der Stappen, Klaus Cichutek, Ulrich Kalinke. 2006. Manufacturing and Quality Control of Cell-Based Tumor Vaccines: A Scientific and a Regulatory Perspective. *Journal of Immunotherapy (Hagerstown, Md.: 1997)* 29 (5): 472–476.

Hofmann, Oliver, Paul Miller, John C deMello, Donal DC Bradley, Andrew J deMello. 2005a. Integrated Optical Detection for Microfluidic Systems Using Thin-Film Polymer Light Emitting Diodes and Organic Photodiodes. *Micro Total Analysis Systems 2004, 2* (297): 506–508.

Hofmann, Oliver, Paul Miller, Paul Sullivan, Timothy S Jones, John C Demello, Donal DC Bradley, Andrew J Demello. 2005b. Thin-Film Organic Photodiodes as Integrated Detectors for Microscale Chemiluminescence Assays. *Sensors and Actuators, B: Chemical* 106 (2): 878–884.

Honda, Nobuo, Ulrika Lindberg. 2005. Simultaneous Multiple Immunoassays in a Compact Disc–shaped Microfluidic Device Based on Centrifugal Force. *Clinical Chemistry* 1961: 1955–1961.

Hsiao, Amy Y, Yu suke Torisawa, Yi Chung Tung, Sudha Sud, Russell S Taichman, Kenneth J Pienta, Shuichi Takayama. 2009. Microfluidic System for Formation of PC-3 Prostate Cancer Co-Culture Spheroids. *Biomaterials* 30 (16): 3020–3027.

HTStec. 2012. *Stem Cells in Research & Drug Discovery Trends 2011.* Cambridge, UK: HTStec Limited.

Hu, Juejun, Xiaochen Sun, Anu Agarwal, Lionel C Kimerling. 2009. Design Guidelines for Optical Resonator Biochemical Sensors. *Journal of the Optical Society of America B* 26 (5): 1032–1041.

Huang, Carlos P, Jente Lu, Hyeryung Seon, Abraham P Lee, Lisa A Flanagan, Ho-Young Kim, Andrew J Putnam, Noo Li Jeon. 2009. Engineering Microscale Cellular Niches for Three-Dimensional Multicellular Co-Cultures. *Lab on a Chip* 9 (12): 1740–1748.

Huang, Hui-Chun, Ya-Ju Chang, Wan-Chun Chen, Hans I-Chen Harn, Ming-Jer Tang, Chia-Ching Wu. 2013. Enhancement of Renal Epithelial Cell Functions through Microfluidic-Based Coculture with Adipose-Derived Stem Cells. *Tissue Engineering. Part A* 19 (17–18): 2024–2034.

Huang, Shih-Chiang, Gwo-Bin Lee, Fan-Ching Chien, Shean-Jen Chen, Wen-Janq Chen, Ming-Chang Yang. 2006. A Microfluidic System with Integrated Molecular Imprinting Polymer Films for Surface Plasmon Resonance Detection. *Journal of Micromechanics and Microengineering* 16 (7): 1251–1257.

Huh, Dongeun, Daniel C Leslie, Benjamin D Matthews, Jacob P Fraser, Samuel Jurek, Geraldine A Hamilton, Kevin S Thorneloe, Michael Allen McAlexander, Donald E Ingber. 2012. A Human Disease Model of Drug Toxicity–Induced Pulmonary Edema in a Lung-on-a-Chip Microdevice. *Science Translational Medicine* 4 (159): 159ra147.

Hui, Elliot E, Sangeeta N Bhatia. 2007. Micromechanical Control of Cell-Cell Interactions. *Proceedings of the National Academy of Sciences of the United States of America* 104 (14): 5722–5726.

Hung, Paul J, Philip J Lee, Poorya Sabounchi, Nima Aghdam, Robert Lin, Luke P Lee. 2005a. A Novel High Aspect Ratio Microfluidic Design to Provide a Stable and Uniform Microenvironment for Cell Growth in a High Throughput Mammalian Cell Culture Array. *Lab on a Chip* 5 (1): 44–48.

Hung, Paul J, Philip J Lee, Poorya Sabounchi, Robert Lin, Luke P Lee. 2005b. Continuous Perfusion Microfluidic Cell Culture Array for High-throughput Cell-based Assays. *Biotechnol Bioeng* 89 (1): 1–8.

Hur, Soojung Claire, Albert J Mach, Dino Di Carlo. 2011. High-Throughput Size-Based Rare Cell Enrichment Using Microscale Vortices. *Biomicrofluidics* 5 (2): 22206.

Hwang, Kyo Seon, Sang-Myung Lee, Sang Kyung Kim, Jeong Hoon Lee, Tae Song Kim. 2009. Micro- and Nanocantilever Devices and Systems for Biomolecule Detection. *Annual Review of Analytical Chemistry* 2 (1): 77–98.

Jang, Am, Zhiwei Zou, Kang Kug Lee, Chong H. Ahn, Paul L. Bishop. 2010. Potentiometric and Voltammetric Polymer Lab Chip Sensors for Determination of Nitrate, pH and Cd(II) in Water. *Talanta* 83 (1): 1–8.

Jang, Kihoon, Kae Sato, Kazuyo Igawa, Ung Il Chung, Takehiko Kitamori. 2008. Development of an Osteoblast-Based 3D Continuous-Perfusion Microfluidic System for Drug Screening. *Analytical and Bioanalytical Chemistry* 390 (3): 825–832.

Jeon, Jessie S, Seok Chung, Roger D Kamm, Joseph L Charest. 2011. Hot Embossing for Fabrication of a Microfluidic 3D Cell Culture Platform. *Biomedical Microdevices* 13 (2): 325–333.

Jeon, Kang Jin, So Hee Park, J W Shin, Yun Gyeong Kang, J S Hyun, Min Jae Oh, Seon Yeon Kim, J W Shin. 2014. Combined Effects of Flow-Induced Shear Stress and Micropatterned Surface Morphology on Neuronal Differentiation of Human Mesenchymal Stem Cells. *Journal of Bioscience and Bioengineering* 117 (2): 242–247.

Jin, Aishun, Tatsuhiko Ozawa, Kazuto Tajiri, Tsutomu Obata, Sachiko Kondo, Koshi Kinoshita, Shinichi Kadowaki, et al. 2009. A Rapid and Efficient Single-Cell Manipulation Method for Screening Antigen-Specific Antibody-Secreting Cells from Human Peripheral Blood. *Nature Medicine* 15 (9): 1088–1092.

Jin, Liang, Tao Feng, Hung Ping Shih, Ricardo Zerda, Angela Luo, Jasper Hsu, Alborz Mahdavi, et al. 2013. Colony-Forming Cells in the Adult Mouse Pancreas Are Expandable in Matrigel and Form Endocrine/acinar Colonies in Laminin Hydrogel. *Proceedings of the National Academy of Sciences of the United States of America* 110 (10): 3907–3912.

Johnson, Renjith P, Eunji Choi, Kyeong Mi Lee, SeongJae Yu, HyukChul Chang, Hongsuk Suh, Il Kim. 2013. Microfluidics Assisted Fabrication of Microspheres by poly(2–hydroxy-ethyl Methacrylate)-Block-Poly(l-Histidine) Hybrid Materials and Their Utilization as Potential Drug Encapsulants. *Microfluidics and Nanofluidics* 14 (1–2): 257–263.

Jokilaakso, Nima, Eric Salm, Aaron Chen, Larry Millet, Carlos Duarte Guevara, Brian Dorvel, Bobby Reddy, et al. 2013. Ultra-Localized Single Cell Electroporation Using Silicon Nanowires. *Lab on a Chip* 13 (3): 336–339.

Kafi, Md Abdul, Cheol Heon Yea, Tae Hyung Kim, Ajay Kumar Yagati, Jeong Woo Choi. 2013. Electrochemical Cell Chip to Detect Environmental Toxicants Based on Cell Cycle Arrest Technique. *Biosensors & Bioelectronics* 41: 192–198.

Kamei, Ken-Ichiro, Shuling Guo, Zeta Tak For Yu, Hiroko Takahashi, Eric Gschweng, Carol Suh, Xiaopu Wang, et al. 2009. An Integrated Microfluidic Culture Device for Quantitative Analysis of Human Embryonic Stem Cells. *Lab on a Chip—Miniaturisation for Chemistry and Biology* 9 (4): 555–563.

Kang, Joo H, Silva Krause, Heather Tobin, Akiko Mammoto, Mathumai Kanapathipillai, Donald E Ingber. 2012. A Combined Micromagnetic-Microfluidic Device for Rapid Capture and Culture of Rare Circulating Tumor Cells. *Lab on a Chip* 12 (12): 2175–2181.

Karabacak, Nezihi Murat, Philipp S Spuhler, Fabio Fachin, Eugene J Lim, Vincent Pai, Emre Ozkumur, Joseph M Martel, et al. 2014. Microfluidic, Marker-Free Isolation of Circulating Tumor Cells from Blood Samples. *Nature Protocols* 9 (3): 694–710.

Khademhosseini, Ali, George Eng, Judy Yeh, Peter A Kucharczyk, Robert Langer, Gordana Vunjak-Novakovic, Milica Radisic. 2007. Microfluidic Patterning for Fabrication of Contractile Cardiac Organoids. *Biomed Microdevices* 9 (2): 149–157.

Khademhosseini, Ali, Kahp Y Suh, Sangyong Jon, George Eng, Judy Yeh, Guan-Jong Chen, Robert Langer. 2004a. A Soft Lithographic Approach to Fabricate Patterned Microfluidic Channels. *Analytical Chemistry* 76 (13): 3675–3681.

Khademhosseini, Ali, Judy Yeh, Sangyong Jon, George Eng, Kahp Y Suh, Jason A Burdick, Robert Langer. 2004b. Molded Polyethylene Glycol Microstructures for Capturing Cells within Microfluidic Channels. *Lab on a Chip* 4 (5): 425–430.

Kim, Chorong, Kristina Kreppenhofer, Jubin Kashef, Dietmar Gradl, Dirk Herrmann, Marc Schneider, Ralf Ahrens, Andreas Guber, Doris Wedlich. 2012. Diffusion- and Convection-Based Activation of Wnt/β-Catenin Signaling in a Gradient Generating Microfluidic Chip. *Lab on a Chip—Miniaturisation for Chemistry and Biology* 12 (24): 5186–5194.

Kim, Hyun Jung, Donald E Ingber. 2013. Gut-on-a-Chip Microenvironment Induces Human Intestinal Cells to Undergo Villus Differentiation. *Integrative Biology: Quantitative Biosciences from Nano to Macro* 5 (9): 1130–1140.

Kim, Hyun-Soo, Se-Hui Jung, Sang-Hyun Kim, In-Bum Suh, Woo Jin Kim, Jae-Wan Jung, Jong Seol Yuk, Young-Myeong Kim, Kwon-Soo Ha. 2006. High-Throughput Analysis of Mumps Virus and the Virus-Specific Monoclonal Antibody on the Arrays of a Cationic Polyelectrolyte with a Spectral SPR Biosensor. *Proteomics* 6 (24): 6426–6432.

Kim, Jungkyu, Minjee Kang, Erik C Jensen, Richard A Mathies. 2012. Lifting Gate Polydimethylsiloxane Microvalves and Pumps for Microfluidic Control. *Analytical Chemistry* 84 (4): 2067–2071.

Kim, Jeongyun, David Taylor, Nitin Agrawal, Han Wang, Hyunsoo Kim, Arum Han, Kaushal Rege, Arul Jayaraman. 2012. A Programmable Microfluidic Cell Array for Combinatorial Drug Screening. *Lab on a Chip* 12 (10): 1813–1822.

Kim, Jungkyu, Erik C Jensen, Mischa Megens, Bernhard Boser, Richard A Mathies. 2011. Integrated Microfluidic Bioprocessor for Solid Phase Capture Immunoassays. *Lab on a Chip* 11 (18): 3106–3112.

Kim, Pilnam, Keon Woo Kwon, Min Cheol Park, Sung Hoon Lee, Sun Min Kim, Kahp Yang Suh. 2008. Soft Lithography for Microfluidics: A Review. *The Korean BioChip Society* 2 (1): 1–11.

Kim, Sung-Jin, Sophie Paczesny, Shuichi Takayama, Katsuo Kurabayashi. 2013. Preprogrammed, Parallel On-Chip Immunoassay Using System-Level Capillarity Control. *Analytical Chemistry* 85 (14): 6902–6907.

King, Kevin R, Sihong Wang, Daniel Irimia, Arul Jayaraman, Mehmet Toner, Martin L Yarmush. 2007. A High-Throughput Microfluidic Real-Time Gene Expression Living Cell Array. *Lab on a Chip* 7 (1): 77–85.

Kitagawa, Yoichi, Yoji Naganuma, Yuya Yajima, Masumi Yamada, Minoru Seki. 2014. Patterned Hydrogel Microfibers Prepared Using Multilayered Microfluidic Devices for Guiding Network Formation of Neural Cells. *Biofabrication* 6 (3): 035011.

Kobayashi, Aoi, Kenta Yamakoshi, Yuya Yajima, Rie Utoh, Masumi Yamada, Minoru Seki. 2013. Preparation of Stripe-Patterned Heterogeneous Hydrogel Sheets Using Microfluidic Devices for High-Density Coculture of Hepatocytes and Fibroblasts. *Journal of Bioscience and Bioengineering* 116 (6): 761–767.

Kobel, Stefan A, Olivier Burri, Alexandra Griffa, Mukul Girotra, Arne Seitz, Matthias P Lutolf. 2012. Automated Analysis of Single Stem Cells in Microfluidic Traps. *Lab on a Chip* 12 (16): 2843–2849.

Koch, M, N Harris, AGR Evans, NM White, A Brunnschweiler. 1997. A Novel Micromachined Pump Based on Thick-Film Piezoelectric Actuation. *Proceedings of International Solid State Sensors and Actuators Conference (Transducers' 97)* 1. Sendai, Japan

Kokkinis, Georgios, Franz Keplinger, Ioanna Giouroudi. 2013. On-Chip Microfluidic Biosensor Using Superparamagnetic Microparticles. *Biomicrofluidics* 7 (5): 54117.

Koppenhofer, D, A Susloparova, D Docter, RH Stauber, S Ingebrandt. 2013. Monitoring Nanoparticle Induced Cell Death in H441 Cells Using Field-Effect Transistors. *Biosensors & Bioelectronics* 40 (1): 89–95.

Kothapalli, Chandrasekhar R, Ed van Veen, Sarra de Valence, Seok Chung, Ioannis K Zervantonakis, Frank B Gertler, Roger D Kamm. 2011. A High-Throughput Microfluidic Assay to Study Neurite Response to Growth Factor Gradients. *Lab on a Chip* 11 (3): 497–507.

Lam, Raymond H W, Min-cheol Cheol Kim, Todd Thorsen. 2009. Culturing Aerobic and Anaerobic Bacteria and Mammalian Cells with a Microfluidic Differential Oxygenator. *Anal Chem* 81 (14): 5918–5924.

Lapizco-Encinas, Blanca H, Blake A Simmons, Eric B Cummings, Yolanda Fintschenko. 2004. Dielectrophoretic Concentration and Separation of Live and Dead Bacteria in an Array of Insulators. *Analytical Chemistry* 76 (6): 1571–1579.

Larsen, Morten, Sergey M Borisov, Björn Grunwald, Ingo Klimant, Ronnie N. Glud. 2011. A Simple and Inexpensive High Resolution Color Ratiometric Planar Optode Imaging Approach: Application to Oxygen and pH Sensing. *Limnology and Oceanography: Methods* 9 (9): 348–60.

Laurent, Sophie, Delphine Forge, Marc Port, Alain Roch, Caroline Robic, Luce Vander Elst, Robert N. Muller. 2008. Magnetic Iron Oxide Nanoparticles: Synthesis, Stabilization, Vectorization, Physicochemical Characterizations and Biological Applications. *Chemical Reviews* 108 (6): 2064–2110.

Lecault, Veronique, Michael VanInsberghe, Sanja Sekulovic, David J H F Knapp, Stefan Wohrer, William Bowden, Francis Viel, et al. 2011. High-Throughput Analysis of Single Hematopoietic Stem Cell Proliferation in Microfluidic Cell Culture Arrays. *Nature Methods* 8 (7): 581–586.

Leclerc, Eric, Yasuyuki Sakai, Teruo Fujii. 2003. Cell Culture in 3-Dimensional Microfluidic Structure of PDMS (polydimenthylsiloxane). *Biomedical Microdevices* 5 (2): 109–114.

Lee, Beom Seok, Jung-Nam Lee, Jong-Myeon Park, Jeong-Gun Lee, Suhyeon Kim, Yoon-Kyoung Cho, Christopher Ko. 2009. A Fully Automated Immunoassay from Whole Blood on a Disc. *Lab on a Chip* 9 (11): 1548–1555.

Lee, Philip J, Paul J Hung, Vivek M Rao, Luke P Lee. 2006. Nanoliter Scale Microbioreactor Array for Quantitative Cell Biology. *Biotechnol Bioeng* 94 (1): 5–14.

Lequin, Rudolf M. 2005. Enzyme Immunoassay (EIA)/enzyme-Linked Immunosorbent Assay (ELISA). *Clinical Chemistry* 2418: 2415–2418.

Li, Chunyu, Chong Liu, Zheng Xu, Jingmin Li. 2012. A Power-Free Deposited Microbead Plug-Based Microfluidic Chip for Whole-Blood Immunoassay. *Microfluidics and Nanofluidics* 12 (5): 829–834.

Li, Xiujun James, Alejandra V. Valadez, Peng Zuo, Zhihong Nie. 2012. Microfluidic 3D Cell Culture: Potential Application for Tissue-Based Bioassays. *Bioanalysis* 5 (12): 1509–1525.

Lienemann, Philipp S, Matthias P Lutolf, Martin Ehrbar. 2012. Biomimetic Hydrogels for Controlled Biomolecule Delivery to Augment Bone Regeneration. *Advanced Drug Delivery Reviews* 64 (12): 1078–1089.

Lin, Peng, Feng Yan. 2012. Organic Thin-Film Transistors for Chemical and Biological Sensing. *Advanced Materials* 24 (1): 34–51.

Liu, Rui, Yuankun Cai, Joong-Mok Park, Kai-Ming Ho, Joseph Shinar, Ruth Shinar. 2011. Organic Light-Emitting Diode Sensing Platform: Challenges and Solutions. *Advanced Functional Materials* 21: 4744–4753.

Liu, Zongbin, Lidan Xiao, Baojian Xu, Yu Zhang, Arthur FT Mak, Yi Li, Wing-yin Man, Mo Yang. 2012. Covalently Immobilized Biomolecule Gradient on Hydrogel Surface Using a Gradient Generating Microfluidic Device for a Quantitative Mesenchymal Stem Cell Study. *Biomicrofluidics* 6 (2): 24111–24112.

Locascio, Laurie E, David J Ross, Peter B Howell, Michael Gaitan. 2006. Fabrication of Polymer Microfluidic Systems by Hot Embossing and Laser Ablation. *Methods in Molecular Biology* 339: 37–46.

Ma, Chao, Rong Fan, Habib Ahmad, Qihui Shi, Begonya Comin-Anduix, Thinle Chodon, Richard C Koya, et al. 2011. A Clinical Microchip for Evaluation of Single Immune Cells Reveals High Functional Heterogeneity in Phenotypically Similar T Cells. *Nature Medicine* 17 (6): 738–743.

Mach, Albert J, Jae Hyun Kim, Armin Arshi, Soojung Claire Hur, Dino Di Carlo. 2011. Automated Cellular Sample Preparation Using a Centrifuge-on-a-Chip. *Lab on a Chip* 11 (17): 2827–2834.

Mair, Dieudonne A, Emil Geiger, Albert P Pisano, Jean M J Frechet, Frantisek Svec. 2006. Injection Molded Microfluidic Chips Featuring Integrated Interconnects. *Lab on a Chip* 6 (10): 1346–1354.

Malic, Lidija, Andrew G Kirk. 2007. Integrated Miniaturized Optical Detection Platform for Fluorescence and Absorption Spectroscopy. *Sensors and Actuators, A: Physical* 135 (2): 515–524.

Manz, Andreas, D Jed Harrison, Elisabeth M J Verpoorte, James C Fettinger, Aran Paulus, Hans Ludi, H Michael Widmer. 1992. Planar Chips Technology for Miniaturization and Integration of Separation Techniques into Monitoring Systems—Capillary Electrophoresis on a Chip. *Journal of Chromatography* 593 (1–2): 253–258.

McDonald, J Cooper, George M Whitesides. 2002. Poly(dimethylsiloxane) as a Material for Fabricating Microfluidic Devices. *Accounts of Chemical Research* 35 (7): 491–499.

McMahon, Louise A, Alan J Reid, Veronica A Campbell, Patrick J Prendergast. 2008. Regulatory Effects of Mechanical Strain on the Chondrogenic Differentiation of MSCs in a Collagen-GAG Scaffold: Experimental and Computational Analysis. *Annals of Biomedical Engineering* 36 (2): 185–194.

Meli, Luciana, Hélder SC Barbosa, Anne Marie Hickey, Leyla Gasimli, Gregory Nierode, Maria Margarida Diogo, Robert J. Linhardt, Joaquim MS Cabral, Jonathan S. Dordick. 2014. Three Dimensional Cellular Microarray Platform for Human Neural Stem Cell Differentiation and Toxicology. *Stem Cell Research* 13 (1): 36–47.

Meng, E, Xuan-Qi Wang, H Mak, Yu-Chong Tai. 2000. A Check-Valved Silicone Diaphragm Pump. *Proceedings IEEE Thirteenth Annual International Conference on Micro Electro Mechanical Systems (Cat. No.00CH36308).*

Meyvantsson, Ivar, Jay W Warrick, Steven Hayes, Allyson Skoien, David J Beebe. 2008. Automated Cell Culture in High Density Tubeless Microfluidic Device Arrays. *Lab on a Chip* 8 (5): 717–724.

Miller, Elizabeth M, Alphonsus H C Ng. 2011. A Digital Microfluidic Approach to Heterogeneous Immunoassays. *Analytical and Bioanalytical Chemistry* 399 (1): 337–345.

Miltenyi, Stefan, W Muller, Walter Weichel, Andreas Radbruch. 1990. High Gradient Magnetic Cell Separation with MACS. *Cytometry* 11 (2): 231–238.

Morier, Patrick, Christine Vollet, Philippe E Michel. 2004. Gravity-induced Convective Flow in Microfluidic Systems: Electrochemical Characterization and Application to Enzyme-linked Immunosorbent Assay Tests. *Electrophoresis* 25 (21–22): 3761–3768.

Nagrath, Sunitha, Lecia V Sequist, Shyamala Maheswaran, Daphne W Bell, Daniel Irimia, Lindsey Ulkus, Matthew R Smith, et al. 2007. Isolation of Rare Circulating Tumour Cells in Cancer Patients by Microchip Technology. *Nature* 450 (7173): 1235–1239.

Nakayama, Hidenari, Hiroshi Kimura, Kikuo Komori, Teruo Fujii, Yasuyuki Sakai. 2008. Development of a Disposable Three-Compartment Micro-Cell Culture Device for Toxicokinetic Study in Humans and Its Preliminary Evaluation. *Aatex* 14: 619–622.

Neagu, Cristina R, Johannes G E Gardeniers, Miko Elwenspoek, John J Kelly. 1996. An Electrochemical Microactuator: Principle and First Results. *Journal of Microelectromechanical Systems* 5 (1): 2–9.

Neagu, Cristina R, Johannes G E Gardeniers, Miko Elwenspoek, John J Kelly. 1997. An Electrochemical Active Valve. *Journal Article Electrochimica Acta* 42 (20–22): 3367–3373.

Nock, Volker, Richard J Blaikie, Tim David. 2008a. Micro-Patterning of Polymer-Based Optical Oxygen Sensors for Lab-on-Chip Applications. *Biomems and Nanotechnology Iii* 6799: Y7990–Y7990.

Nock, Volker, Richard J Blaikie, Tim David. 2008b. Patterning, Integration and Characterisation of Polymer Optical Oxygen Sensors for Microfluidic Devices. *Lab on a Chip* 8 (8): 1300–1307.

Novak, Richard, Navpreet Ranu, Richard A Mathies. 2013. Rapid Fabrication of Nickel Molds for Prototyping Embossed Plastic Microfluidic Devices. *Lab on a Chip* 13 (8): 1468–1471.

Odijk, Mathieu, Andries D van der Meer, Daniel Levner. 2014. Measuring Direct Current Trans-Epithelial Electrical Resistance in Organ-on-a-Chip Microsystems. *Lab on a Chip* 15 (3): 745–752.

Oh, J, K Kim, S Choi, J Jung. 2014. Diffusion-Assisted Spherical Microgel Fabrication Using in Situ Gelable Chitosan and Dextran. *Digest Journal of Nanomaterials and Biostructures* 9 (2): 739–744.

Padgett, Miles, Roberto Di Leonardo. 2011. Holographic Optical Tweezers and Their Relevance to Lab on Chip Devices. *Lab on a Chip* 11 (7): 1196–1205.

Paguirigan, Amy L, David J Beebe. 2009. From the Cellular Perspective: Exploring Differences in the Cellular Baseline in Macroscale and Microfluidic Cultures. *Integrative Biology* 1 (2): 182–195.

Pamme, Nicole, Claire Wilhelm. 2006. Continuous Sorting of Magnetic Cells via on-Chip Free-Flow Magnetophoresis. *Lab on a Chip* 6 (8): 974–980.

Pankhurst, QA, J Connolly, SK Jones, J Dobson. 2003. Applications of Magnetic Nanoparticles in Biomedicine. *Journal of Physics D: Applied Physics* 36 (13): R167–R181.

Park, Jennifer S, Julia SF Chu, Catherine Cheng, Fanqing Chen, David Chen, Song Li. 2004. Differential Effects of Equiaxial and Uniaxial Strain on Mesenchymal Stem Cells. *Biotechnology and Bioengineering* 88 (3): 359–368.

Park, Joong Yull, Sung Ju Yoo, Chang Mo Hwang, Sang-Hoon Lee. 2009. Simultaneous Generation of Chemical Concentration and Mechanical Shear Stress Gradients Using Microfluidic Osmotic Flow Comparable to Interstitial Flow. *Lab on a Chip* 9 (15): 2194–2202.

Park, Young K, Ting Yuan Tu, Sei Hien Lim, Ivan JM Clement, Se Y Yang, Roger D Kamm. 2014. In Vitro Microvessel Growth and Remodeling within a Three-Dimensional Microfluidic Environment. *Cellular and Molecular Bioengineering* 7 (1): 15–25.

Patolsky, Fernando, Charles M Lieber. 2005. Nanowire Nanosensors. *Materials Today* 8 (4): 20–28.

Peng, Chien-Chung, Wei-Hao Liao, Ying-Hua Chen, Chueh-Yu Wu, Yi-Chung Tung. 2013. A Microfluidic Cell Culture Array with Various Oxygen Tensions. *Lab on a Chip* 13 (16): 3239–3245.

Perroud, Thomas D, Julia N Kaiser, Jay C Sy, Todd W Lane, Catherine S Branda, Anup K Singh, Kamlesh D Patel. 2008. Microfluidic-Based Cell Sorting of Francisella Tularensis Infected Macrophages Using Optical Forces Microfluidic-Based Cell Sorting of Francisella Tularensis Infected Macrophages Using Optical Forces. *Analytical Chemistry* 80 (16): 6365–6372.

Phamduy, Theresa B, Nurazhani Abdul Raof, Nathan R Schiele, Zijie Yan, David T Corr, Yong Huang, Yubing Xie, Douglas B Chrisey. 2012. Laser Direct-Write of Single Microbeads into Spatially-Ordered Patterns. *Biofabrication* 4 (2): 25006.

Powell, Ashley A, Amirali H Talasaz, Haiyu Zhang, Marc A Coram, Anupama Reddy, Glenn Deng, Melinda L Telli, et al. 2012. Single Cell Profiling of Circulating Tumor Cells: Transcriptional Heterogeneity and Diversity from Breast Cancer Cell Lines. *PLoS One* 7 (5): e33788.

Reichen, Marcel, Farlan Singh Veraitch, Nicolas Szita. 2013. Development of a Multiplexed Microfluidic Platform for the Automated Cultivation of Embryonic Stem Cells. *Journal of Laboratory Automation* 18: 519–529.

Ren, Kangning, Jianhua Zhou, Hongkai Wu. 2013. Materials for Microfluidic Chip Fabrication. *Accounts of Chemical Research* 46 (11): 2396–2406.

Reyes, Darwin R, Dimitri Iossifidis, Pierre-Alain Auroux, Andreas Manz. 2002. Micro Total Analysis Systems. 1. Introduction, Theory, and Technology. *Analytical Chemistry* 74 (12): 2623–2636.

Rivron, Nicolas C, Erik J Vrij, Jeroen Rouwkema, S Le Gac, A van den Berg, RK Truckenmuller, CA van Blitterswijk. 2012. Tissue Deformation Spatially Modulates VEGF Signaling and Angiogenesis. *Proc Natl Acad Sci U S A* 109 (18): 6886–6891.

Roda, Barbara, Pierluigi Reschiglian, Francesco Alviano, Giacomo Lanzoni, Gian Paolo Bagnara, Francesca Ricci, Marina Buzzi, Pier Luigi Tazzari, Pasqualepaolo Pagliaro, Elisa Michelini. 2009. Gravitational Field-Flow Fractionation of Human Hemopoietic Stem Cells. *Journal of Chromatography A* 1216 (52): 9081–9087.

Rudd, CD, CM Duffy, MS Johnson, PJ Blanchard. 1998. Low Cost, Rapid Processing of Thermoset Composites. *Broader Meaning to Thermosets*: 177–178.

Ryu, Gihan, Jingsong Huang, Oliver Hofmann, Claire A Walshe, Jasmine YY Sze, Gareth D McClean, Alan Mosley, et al. 2011. Highly Sensitive Fluorescence Detection System for Microfluidic Lab-on-a-Chip. *Lab on a Chip* 11: 1664–1670.

Sackmann, Eric K, Anna L Fulton, David JDJ. Beebe. 2014. The Present and Future Role of Microfluidics in Biomedical Research. *Nature* 507 (7491): 181–189.

Sagmeister, Martin, Andreas Tschepp, Elke Kraker, Tobias Abel, Bernhard Lamprecht, Torsten Mayr, Stefan Köstler. 2013. Enabling Luminescence Decay Time-Based Sensing Using Integrated Organic Photodiodes. *In Analytical and Bioanalytical Chemistry*, 405: 5975–5982.

Saliba, Antoine-Emmanuel, Laure Saias, Eleni Psychari, Nicolas Minc, Damien Simon, François-Clément Bidard, Claire Mathiot, Jean-Yves Pierga, Vincent Fraisier, Jean Salamero. 2010. Microfluidic Sorting and Multimodal Typing of Cancer Cells in Self-Assembled Magnetic Arrays. *Proceedings of the National Academy of Sciences* 107 (33): 14524–14529.

Schafer, IA, AM Jamieson, M Petrelli, BJ Price, GC Salzman. 1979. Multiangle Light Scattering Flow Photometry of Cultured Human Fibroblasts: Comparison of Normal Cells with a Mutant Line Containing Cytoplasmic Inclusions. *The Journal of Histochemistry and Cytochemistry* 27 (1): 359–365.

Schapper, Daniel, Muhd Nazrul Alam, Nicolas Szita, Anna Eliasson Lantz, Krist V Gernaey. 2009. Application of Microbioreactors in Fermentation Process Development: A Review. *Analytical and Bioanalytical Chemistry* 395 (3): 679–695.

Schomburg, WK, J Vollmer, B Bustgens, J Fahrenberg, H Hein, W Menz. 1994. Microfluidic Components in LIGA Technique. *Journal of Micromechanics and Microengineering* 4 (4): 186.

Schwerdt, Helen N, Ruth E Bristol, Junseok Chae. 2014. Miniaturized Passive Hydrogel Check Valve for Hydrocephalus Treatment. *IEEE Transactions on Bio-Medical Engineering* 61 (3): 814–820.

Sekitani, Tsuyoshi, Tomoyuki Yokota, Ute Zschieschang, Hagen Klauk, Siegfried Bauer, Ken Takeuchi, Makoto Takamiya, Takayasu Sakurai, Takao Someya. 2009. Organic Nonvolatile Memory Transistors for Flexible Sensor Arrays. *Science* 326 (5959): 1516–1519.

Sharma, Shikha, Ravali Raju, Siguang Sui, Wei-Shou Hu. 2011. Stem Cell Culture Engineering—Process Scale up and Beyond. *Biotechnology Journal* 6: 1317–1329.

Shikanov, Ariella, Min Xu, Teresa K Woodruff, Lonnie D Shea. 2011. A Method for Ovarian Follicle Encapsulation and Culture in a Proteolytically Degradable 3 Dimensional System. *Journal of Visualized Experiments: JoVE* 49: 1–9.

Shimizu, Kazunori, Hiroyuki Araki, Kohei Sakata, Wataru Tonomura, Mitsuru Hashida, Satoshi Konishi. 2015. Microfluidic Devices for Construction of Contractile Skeletal Muscle Microtissues. *Journal of Bioscience and Bioengineering* 119 (2): 212–216.

Shojaei, Shahrokh, Mohamad Tafazzoli-Shahdpour, Mohamad Ali Shokrgozar, Nooshin Haghighipour. 2013. Effects of Mechanical and Chemical Stimuli on Differentiation of Human Adipose-Derived Stem Cells into Endothelial Cells. *The International Journal of Artificial Organs* 36 (9): 663–673.

Shoshi, A, J Schotter, P Schroeder, M Milnera, P Ertl, V Charwat, M Purtscher, et al. 2012. Magnetoresistive-Based Real-Time Cell Phagocytosis Monitoring. *Biosensors & Bioelectronics* 36 (1): 116–122.

Shoshi, A, J Schotter, P Schroeder, M Milnera, P Ertl, R Heer, G Reiss, H Brueckl. 2013. Contemporaneous Cell Spreading and Phagocytosis: Magneto-Resistive Real-Time Monitoring of Membrane Competing Processes. *Biosensors & Bioelectronics* 40 (1): 82–88.

Simmons, Craig A, Sean Matlis, Amanda J Thornton, Shaoqiong Chen, Cun-Yu Wang, David J Mooney. 2003. Cyclic Strain Enhances Matrix Mineralization by Adult Human Mesenchymal Stem Cells via the Extracellular Signal-Regulated Kinase (ERK1/2) Signaling Pathway. *Journal of Biomechanics* 36 (8): 1087–1096.

Sip, Christopher G, Nirveek Bhattacharjee, Albert Folch. 2011. A Modular Cell Culture Device for Generating Arrays of Gradients Using Stacked Microfluidic Flows. *Biomicrofluidics* 5 (2): 22210.

Sista, Ramakrishna S, Allen E Eckhardt, Vijay Srinivasan. 2008. Heterogeneous Immunoassays Using Magnetic Beads on a Digital Microfluidic Platform. *Lab on a Chip* 8 (12): 2188–2196.

Situma, Catherine, Masahiko Hashimoto, Steven A Soper. 2006. Merging Microfluidics with Microarray-Based Bioassays. *Biomolecular Engineering* 23 (5): 213–231.

Sivashankar, Shilpa, Srinivasu Valegerahally Puttaswamy, Ling-Hui Lin, Tz-Shuian Dai, Chau-Ting Yeh, Cheng-Hsien Liu. 2013. Culturing of Transgenic Mice Liver Tissue Slices in Three-Dimensional Microfluidic Structures of PEG-DA (poly(ethylene Glycol) Diacrylate). *Sensors and Actuators B: Chemical* 176: 1081–1089.

Skelley, Alison M, Oktay Kirak, Heikyung Suh, Rudolf Jaenisch, Joel Voldman. 2009. Microfluidic Control of Cell Pairing and Fusion. *Nature Methods* 6 (2): 147–152.

Smits, Jan G. 1990. Piezoelectric Micropump with Three Valves Working Peristaltically. *Sensors and Actuators A: Physical* 21 (1–3): 203–206.

Sollier, Elodie, Coleman Murray, Pietro Maoddi, Dino Di Carlo. 2011. Rapid Prototyping Polymers for Microfluidic Devices and High Pressure Injections. *Lab Chip* 11 (22): 3752–3765.

Someya, Takao, Ananth Dodabalapur, Jia Huang, Kevin C See, Howard E Katz. 2010. Chemical and Physical Sensing by Organic Field-Effect Transistors and Related Devices. *Advanced Materials* 22 (34): 3799–3811.

Song, Huixue, Tan Chen, Baoyue Zhang, Yifan Ma, Zhanhui Wang. 2010. An Integrated Microfluidic Cell Array for Apoptosis and Proliferation Analysis Induction of Breast Cancer Cells. *Biomicrofluidics* 4 (4): 44104.

Sorger, PK, KF Jensen. 2006. Cells on Chip. *Nature* 442: 403–411.

Stevens, Dean Y, Camille R Petri, Jennifer L Osborn. 2008. Enabling a Microfluidic Immunoassay for the Developing World by Integration of on-Card Dry Reagent Storage. *Lab on a Chip* 8 (12): 2038–2045.

Sud, Dhruv, Geeta Mehta, Khamir Mehta, Jennifer Linderman, Shuichi Takayama, Mary-Ann A Mycek. 2006. Optical Imaging in Microfluidic Bioreactors Enables Oxygen Monitoring for Continuous Cell Culture. *Journal of Biomedical Optics* 11 (5): 50504.

Sun, Steven, Minghui Yang, Yordan Kostov, Avraham Rasooly. 2010. ELISA-LOC: Lab-on-a-Chip for Enzyme-Linked Immunodetection. *Lab on a Chip* 10: 2093–2100.

Sun, Tao, Hywel Morgan. 2010. Single-Cell Microfluidic Impedance Cytometry: A Review. *Microfluidics and Nanofluidics* 8 (4): 423–443.

Sundararaghavan, Harini G, Shirley N. Masand, David I Shreiber. 2011. Microfluidic Generation of Haptotactic Gradients through 3D Collagen Gels for Enhanced Neurite Growth. *Journal of Neurotrauma* 28 (11): 2377–2387.

Sung, Jong H, Mandy B Esch, Jean-Matthieu Prot, Christopher J Long, Alec Smith, James J Hickman, Michael L Shuler. 2013. Microfabricated Mammalian Organ Systems and Their Integration into Models of Whole Animals and Humans. *Lab on a Chip* 13 (7): 1201–1212.

Suzuki, Ikuro, Mao Fukuda, Keiichi Shirakawa, Hideyasu Jiko, Masao Gotoh. 2013. Carbon Nanotube Multi-Electrode Array Chips for Noninvasive Real-Time Measurement of Dopamine, Action Potentials, and Postsynaptic Potentials. *Biosensors & Bioelectronics* 49: 270–275.

Tedde, Sandro F, Johannes Kern, Tobias Sterzl, Jens Fürst, Paolo Lugli, Oliver Hayden. 2009. Fully Spray Coated Organic Photodiodes. *Nano Letters* 9 (3): 980–983.

Tenstad, Ellen, Anna Tourovskaia, Albert Folch, Ola Myklebost, Edith Rian. 2010. Extensive Adipogenic and Osteogenic Differentiation of Patterned Human Mesenchymal Stem Cells in a Microfluidic Device. *Lab on a Chip* 10 (11): 1401–1409.

Thomas, Peter C, Srinivasa R Raghavan, Samuel P Forry. 2011. Regulating Oxygen Levels in a Microfluidic Device. *Analytical Chemistry* 83 (22): 8821–8824.

Thompson, Deanna M, Kevin R King, Kenneth J Wieder, Mehmet Toner, Martin L Yarmush, Arul Jayaraman. 2004. Dynamic Gene Expression Profiling Using a Microfabricated Living Cell Array. *Analytical Chemistry* 76 (14): 4098–4103.

Toh, Yi-Chin, Joel Voldman. 2011. Fluid Shear Stress Primes Mouse Embryonic Stem Cells for Differentiation in a Self-Renewing Environment via Heparan Sulfate Proteoglycans Transduction. *The FASEB Journal* 25 (4): 1208–1217.

Toh, Yi-Chin, Chi Zhang, Jing Zhang, Yuet Mei Khong, Shi Chang, Victor D Samper, Danny van Noort, Dietmar W Hutmacher, Hanry Yu. 2007. A Novel 3D Mammalian Cell Perfusion-Culture System in Microfluidic Channels. *Lab Chip* 7 (3): 302–309.

Torisawa, Yu-suke, Bobak Mosadegh, Gary D Luker, Maria Morell, K Sue O'Shea, Shuichi Takayama. 2009. Microfluidic Hydrodynamic Cellular Patterning for Systematic Formation of Co-Culture Spheroids. *Integrative Biology* 1 (11–12): 649–654.

Torisawa, Ys, A Takagi, Y Nashimoto, T Yasukawa, H Shiku, T Matsue. 2007. A Multicellular Spheroid Array to Realize Spheroid Formation, Culture, and Viability Assay on a Chip. *Biomaterials* 28 (3): 559–566.

Torsi, L, AJ Lovinger, B Crone, T Someya, A Dodabalapur, HE Katz, A Gelperin. 2002. Correlation between Oligothiophene Thin Film Transistor Morphology and Vapor Responses. *Journal of Physical Chemistry B* 106 (48): 12563–12568.

Tse, Henry T K, Daniel R Gossett, Yo Sup Moon, Mahdokht Masaeli, Marie Sohsman, Yong Ying, Kimberly Mislick, Ryan P Adams, Jianyu Rao, Dino Di Carlo. 2013. Quantitative Diagnosis of Malignant Pleural Effusions by Single-Cell Mechanophenotyping. *Science Translational Medicine* 5 (212): 212ra163.

Tse, Justin R Adam J Engler. 2011. Stiffness Gradients Mimicking In Vivo Tissue Variation Regulate Mesenchymal Stem Cell Fate (ed. Nic D. Leipzig). *PLoS ONE* 6 (1): e15978.

Tumarkin, Ethan, Lsan Tzadu, Elizabeth Csaszar, Minseok Seo, Hong Zhang, Anna Lee, Raheem Peerani, Kelly Purpura, Peter W Zandstra, Eugenia Kumacheva. 2011. High-Throughput Combinatorial Cell Co-Culture Using Microfluidics. *Integrative Biology* 3 (6): 653–662.

Unger, Marc A. 2000. Monolithic Microfabricated Valves and Pumps by Multilayer Soft Lithography. *Science* 288 (5463): 113–116.

Unger, Marc A, Hou-Pu Chou, Todd Thorsen, Axel Scherer, Stephen R Quake. 2000. Monolithic Microfabricated Valves and Pumps by Multilayer Soft Lithography. *Science (New York)* 288 (5463): 113–116.

Ungerbock, Birgit, Andrej Pohar, Torsten Mayr, Igor Plazl. 2013. Online Oxygen Measurements inside a Microreactor with Modeling of Transport Phenomena. *Microfluidics and Nanofluidics* 14 (3–4): 565–574.

Upadhyaya, Sarvesh, P Ravi Selvaganapathy. 2010. Microfluidic Devices for Cell Based High Throughput Screening. *Lab on a Chip* 10 (3): 341–348.

Van Weemen, B.K, AHWM. Schuurs. 1971. Immunoassay Using Antigen—Enzyme Conjugates. *FEBS Letters* 15 (3): 232–236.

Wada, Ken-Ichi, Akiyoshi Taniguchi, Jun Kobayashi, Masayuki Yamato, Teruo Okano. 2008. Live Cells-based Cytotoxic Sensorchip Fabricated in a Microfluidic System. *Biotechnology and Bioengineering* 99 (6): 1513–1517.

Waggoner, PS, CP Tan, HG Craighead. 2010. Microfluidic Integration of Nanomechanical Resonators for Protein Analysis in Serum. *Sensors and Actuators B: Chemical* 150 (2): 550–555.

Wagner, Ilka, Eva-Maria Materne, Sven Brincker, Ute Süßbier, Caroline Frädrich, Mathias Busek, Frank Sonntag, et al. 2013. A Dynamic Multi-Organ-Chip for Long-Term Cultivation and Substance Testing Proven by 3D Human Liver and Skin Tissue Co-Culture. *Lab on a Chip* 13 (18). The Royal Society of Chemistry: 3538–3547.

Wang, Hsiang-Yu, Ning Bao, Chang Lu. 2008. A Microfluidic Cell Array with Individually Addressable Culture Chambers. *Biosensors and Bioelectronics* 24 (4): 613–617.

Wang, Shujun, Suming Chen, Jianing Wang, Peng Xu, Yuanming Luo, Zongxiu Nie, Wenbin Du. 2014. Interface Solution Isoelectric Focusing with in Situ MALDI-TOF Mass Spectrometry. *Electrophoresis* 35 (17): 2528–2533.

Wang, Xuhua, Maliwan Amatatongchai, Duangjai Nacapricha, Oliver Hofmann, John C de Mello, Donal D C Bradley, Andrew J deMello. 2009. Thin-Film Organic Photodiodes for Integrated on-Chip Chemiluminescence Detection—Application to Antioxidant Capacity Screening. *Sensors and Actuators B-Chemical* 140 (2): 643–648.

Wang, Xiaolin, Shuxun Chen, Marco Kong, Zuankai Wang, Kevin D Costa, Ronald A Li, Dong Sun. 2011. Enhanced Cell Sorting and Manipulation with Combined Optical Tweezer and Microfluidic Chip Technologies. *Lab on a Chip* 11 (21): 3656–3662.

Wang, Yi, Jakub Dostalek, Wolfgang Knoll. 2011. Magnetic Nanoparticle-Enhanced Biosensor Based on Grating-Coupled Surface Plasmon Resonance. *Analytical Chemistry* 83 (16): 6202–6207.

Wangler, N, L Gutzweiler, K Kalkandjiev, C Müller, F Mayenfels, H Reinecke, R Zengerle, N Paust. 2011. High-Resolution Permanent Photoresist Laminate TMMF for Sealed Microfluidic Structures in Biological Applications. *Journal of Micromechanics and Microengineering* 21 (9): 095009.

Wegener, Joachim, CR Keese, Ivar Giaever. 2000. Electric Cell–substrate Impedance Sensing (ECIS) as a Noninvasive Means to Monitor the Kinetics of Cell Spreading to Artificial Surfaces. *Experimental Cell Research* 166: 158–166.

Wegener Joachim, Sigrid Zink, Peter Rösen, H Galla. 1999. Use of Electrochemical Impedance Measurements to Monitor Beta-Adrenergic Stimulation of Bovine Aortic Endothelial Cells. *Pflugers Arch* 437 (6): 925–934.

White, Ian M, Hongying Zhu, Jonathan D Suter, Xudong Fan, Mohammed Zourob. 2009. Label-Free Detection with the Liquid Core Optical Ring Resonator Sensing Platform. In *Biosensors and Biodetection SE—7*, edited by Avraham Rasooly and Keith E. Herold, Vol. 503, pp. 139–165. Methods in Molecular Biology. Springer: Humana Press.

Whitesides, George M. 2006. The Origins and the Future of Microfluidics. *Nature* 442 (7101): 368–373.

Whitesides, George M, E Ostuni, S Takayama, X Jiang, DE Ingber. 2001. Soft Lithography in Biology and Biochemistry. *Annual Review of Biomedical Engineering* 3: 335–373.

Wilson, Jeremy D, Thomas H Foster. 2005. Light Scattering Reports Early Mitochondrial Responses to Photodynamic Therapy. *Optical Methods for Tumor Treatment and Detection: Mechanisms and Techniques in Photodynamic Therapy XIV* 5689: 9–16.

Wilson, Jeremy D, Chad E Bigelow, David J Calkins, Thomas H Foster. 2005. Light Scattering from Intact Cells Reports Oxidative-Stress-Induced Mitochondrial Swelling. *Biophysical Journal* 88 (4): 2929–2938.

Witters, Daan, Nicolas Vergauwe, Steven Vermeir, Frederik Ceyssens, Sandra Liekens, Robert Puers, Jeroen Lammertyn. 2011. Biofunctionalization of Electrowetting-on-Dielectric Digital Microfluidic Chips for Miniaturized Cell-Based Applications. *Lab on a Chip* 11 (16): 2790–2794.

Wlodkowic, Donald, Shannon Faley, Michele Zagnoni, John P Wikswo, Jonathan M Cooper. 2009. Microfluidic Single-Cell Array Cytometry for the Analysis of Tumor Apoptosis. *Anal Chem* 81 (13): 5517–5523.

Wu, Jing, Min Gu. 2011. Microfluidic Sensing: State of the Art Fabrication and Detection Techniques. *Journal of Biomedical Optics* 16 (8): 080901.

Wu, Meiye, Anup K. Singh. 2012. Single-Cell Protein Analysis. *Current Opinion in Biotechnology* 23 (1): 83–88.

Wu, Yaobin, Ling Wang, Baolin Guo, Peter X Ma. 2014. Injectable Biodegradable Hydrogels and Microgels Based on Methacrylated Poly(ethylene Glycol)-Co-Poly(glycerol Sebacate) Multi-Block Copolymers: Synthesis, Characterization, and Cell Encapsulation. *Journal of Materials Chemistry B* 2 (23): 3674–3685.

Xia, Younan, George M Whitesides. 1998. Soft Lithography. *Annual Review of Materials Science* 28: 153–184.

Xiao, Caide C, John H T Luong. 2003. On-Line Monitoring of Cell Growth and Cytotoxicity Using Electric Cell-Substrate Impedance Sensing (ECIS). *Biotechnology Progress* 19 (3): 1000–1005.

Xu, Bi-Yi, Shan-Wen Hu, Guang-Sheng Qian, Jing-Juan Xu, Hong-Yuan Chen. 2013. A Novel Microfluidic Platform with Stable Concentration Gradient for on Chip Cell Culture and Screening Assays. *Lab on a Chip* 13 (18): 3714–3720.

Xu, Hui, Sarah C Heilshorn. 2013. Microfluidic Investigation of BDNF-Enhanced Neural Stem Cell Chemotaxis in CXCL12 Gradients. *Small* 9 (4): 585–595.

Xu, Zhang-Run, Chun-Guang Yang, Cui-Hong Liu, Zhe Zhou, Jin Fang, Jian-Hua Wang. 2010. An Osmotic Micro-Pump Integrated on a Microfluidic Chip for Perfusion Cell Culture. *Talanta* 80 (3): 1088–1093.

Yamanaka, Yvonne J, Gregory L Szeto, Todd M Gierahn, Talitha L Forcier, Kelly F Benedict, Mavis SN Brefo, Douglas A Lauffenburger, Darrell J Irvine, J Christopher Love. 2012. Cellular Barcodes for Efficiently Profiling Single-Cell Secretory Responses by Microengraving. *Anal Chem* 84 (24): 10531–10536.

Yang, L, W Wei, J Xia, H Tao. 2004. Artificial Receptor Layer for Herbicide Detection Based on Electrosynthesized Molecular Imprinting Technique and Capacitive Transduction. *Analytical Letters* 37 (11): 2303–2319.

Yea, Cheol-Heon, Jeung Hee An, Jungho Kim, Jeong-Woo Choi. 2013. In Situ Electrochemical Detection of Embryonic Stem Cell Differentiation. *Journal of Biotechnology* 166: 1–5.

Yeon, Ju Hun, Je Kyun Park. 2005. Cytotoxicity Test Based on Electrochemical Impedance Measurement of HepG2 Cultured in Microfabricated Cell Chip. *Analytical Biochemistry* 341 (2): 308–315.

Yin, Huabing, Damian Marshall. 2012. Microfluidics for Single Cell Analysis. *Current Opinion in Biotechnology* 23 (1): 110–119.

Yu, Hongmei, Caroline M Alexander, David J Beebe. 2007. A Plate Reader-Compatible Microchannel Array for Cell Biology Assays. *Lab on a Chip* 7 (3): 388–391.

Zamora-Mora, Vanessa, Diego Velasco, Rebeca Hernández, Carmen Mijangos, Eugenia Kumacheva. 2014. Chitosan/agarose Hydrogels: Cooperative Properties and Microfluidic Preparation. *Carbohydrate Polymers* 111: 348–355.

Zaretsky, Irina, Michal Polonsky, Eric Shifrut. 2012. Monitoring the Dynamics of Primary T Cell Activation and Differentiation Using Long Term Live Cell Imaging in Microwell Arrays. *Lab on a Chip* 12 (23): 5007–5015.

Zeng, Yong, Richard Novak, Joe Shuga. 2010. High-Performance Single Cell Genetic Analysis Using Microfluidic Emulsion Generator Arrays. *Analytical Chemistry* 82 (8): 3183–3190.

Zhang, Hu, Kuo-Kang Liu. 2008. Optical Tweezers for Single Cells. *Journal of the Royal Society, Interface/the Royal Society* 5 (24): 671–690.

Zhang, Mengying, Jinbo Wu, Limu Wang, Kang Xiao, Weijia Wen. 2010. A Simple Method for Fabricating Multi-Layer PDMS Structures for 3D Microfluidic Chips. *Lab on a Chip* 10 (9): 1199–1203.

Zhang, Zhiyi, Ping Zhao, Gaozhi Xiao. 2009. The Fabrication of Polymer Microfluidic Devices Using a Solid-to-Solid Interfacial Polyaddition. *Polymer* 50 (23): 5358–5361.

Zheng, Wenfu, Bo Jiang, Dong Wang, Wei Zhang, Zhuo Wang, Xingyu Jiang. 2012. A Microfluidic Flow-Stretch Chip for Investigating Blood Vessel Biomechanics. *Lab on a Chip* 12 (18): 3441–3450.

Zhu, L, CS Lee, DL DeVoe. 2006. Integrated Microfluidic UV Absorbance Detector with Attomol-Level Sensitivity for BSA. *Lab on a Chip* 6 (1): 115–120.

Section II

Organs-on-Chips

7 From 2D Culture to 3D Microchip Models

Trachea, Bronchi/Bronchiole and Lung Biomimetic Models for Disease Modeling, Drug Discovery, and Personalized Medicine

Joan E. Nichols, Stephanie P. Vega, Lissenya B. Argueta, Jean A. Niles, Adrienne Eastaway, Michael Smith, David Brown and Joaquin Cortiella

CONTENTS

7.1 INTRODUCTION

We have learned a great deal about respiratory tract and lung physiology or pathophysiology of disease from the study of animal disease models, tissues from patients isolated at autopsy, or growth of monoclonal human cell populations in two-dimensional (2D) cultures. Animal models, mainly mice, have been widely used in research and although animal models can simulate human disease they never fully mirror all aspects of human immune response or pathophysiology of disease.

Because of this many drug treatments and vaccines developed solely using animal models have been ineffective when used in patient care. Use of lung tissue slices to identify events which influence lung physiology has provided some information regarding lung responses and disease development (Smith et al. 1985) but human organ slices can be difficult to obtain and survival in culture is generally limited (Sanderson 2011). Because of this most researchers have utilized *in vitro* cell culture models to answer basic questions related to human cellular responses in the lung. A great deal of what we understand regarding cell-to-cell interactions and cell-extracellular matrix interactions for the trachea, bronchi, bronchioles or alveoli has been garnered from the study of 2D cultures of primary human cells isolated from the respiratory tract, immortalized cells or transformed cell lines.

Recent advances in microfabrication technology, microfluidics, and tissue engineering have provided a new approach to the development of 3D tissue culture models that enable production of robust long-lived human respiratory tract and lung analogs. Use of these models along with more complex 3D human organ culture models, containing multiple cell phenotypes, provides a more reasonable approximation of what occurs in the dynamic *in vivo* microenvironment of the lung. Microfluidic supported 3D respiratory tract and lung models are currently being used as advanced human testing platforms for evaluating drug response or drug toxicity, hopefully reducing the cost of drug development. Human tissue models may also provide a mechanism for development of personalized medical care based on testing of drugs on a patient's own cells or engineered tissues. We hope that this chapter will serve as a guide regarding what has already been accomplished in the development of respiratory tract and lung models as well as for future applications of human respiratory tract and lung organ culture and microchip systems.

7.2 CONSTRUCTION AND SCALING OF RESPIRATORY MODELS

Before a respiratory tract or lung model can be constructed, the specific characteristics of the region of the respiratory tract to be modeled must be carefully considered as well as the model scale. The upper respiratory tract and lung differ significantly in the types of cells found in each region and in the extracellular matrix (ECM) components and structural support characteristics specific to each region. Considerations related to determining model scale include the cellular responses being modeled and the anticipated readout mechanism that will be used to identify or measure that a response has occurred. The first step in scaling is to determine the approximate cell sizes, cell phenotypes, and average number of cells found at the site or region to be modeled. An isometric scale model would contain the same cell types in the same proportions as found in natural human trachea, bronchi, bronchioles, or lung. Allometric scaling refers to any form of scaling that is not isometric. Allometric scaling often provides a starting point for the development of model systems (Wikswo et al. 2013). Most tissue models are allometric in scaling since isometric scaling of tissues and organs is generally difficult to replicate in a model. In our experience, respiratory tract and lung models fall into three categories based on the number of cells used and the complexity in cell type composition: micro-lung (μlung), milli-lung (mlung) or human organ culture (HOC). Microphysiologic systems generally require few cells (1 million or less) and can be maintained in standard Terasaki plates, which hold 11.45 µl, 96-well tissue culture plates, which hold 75–200 µl, or in specially configured microfabricated chambers that vary in chamber volume. Microfluidic technology permits the accommodation and control of micro- to pico-liter amounts on a chip culture device measuring a few square centimeters (or even smaller). Standard culture systems require careful management and manual manipulation to change support media. Microfluidic pumping systems provide oxygen, nutrients, and growth factors and remove metabolic waste products from cultures without the need for manual manipulation of samples. Figure 7.1 shows a microfabricated µlung culture chamber containing 1×10^6 immortalized human lung alveolar epithelial cells maintained in a 100 µl volume.

Peristaltic pumps, syringe pumps, and recirculation chip pumps may be used to provide microfluidic support to microtissue cultures although many pumps do not support movement of small fluid volumes at the speeds or pressures required. µlung or µtrachea/bronchiole models allow for determination of basic changes in cell viability, induction of apoptosis, or simple pathological changes in tissues. Creation of µlung microfluidic system–supported lung-on-a-chip models has allowed examination of cellular events that occur at the alveolar-capillary

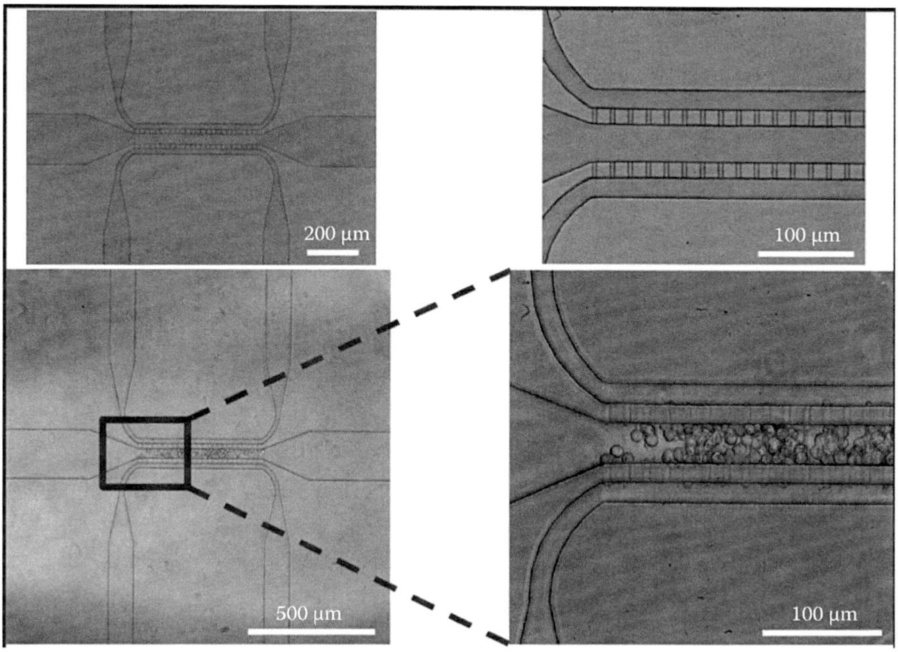

FIGURE 7.1 Microfabricated chamber for support of µlung model containing human immortalized alveolar epithelial cells. This microfluidic chip consisted of a polydimethylsiloxane (PDMS) module bonded to a standard glass microscope slide. This model was maintained using a specialized microprocessor controlled pumping system. (Courtesy of Peter Loskill, collaboration with Kevin Healy, UC Berkeley.)

interface of the human lung or in the trachea, bronchi, or bronchioles of the respiratory tract. Mechanically active "organ-on-a-chip" microdevices have already been used to replicate pathological conditions and responses to drugs or immune therapies (Huh et al. 2010; Huh et al. 2012). New classes of equipment including μ-process automation processing systems, pumps, and microfluidic valves are slowly making their way into the realm of general research use. Many companies offer tutorials for the beginner or advanced researcher on topics ranging from syringe pump use and maintenance, to microchip fabrication using polydimethylsiloxane (PDMS).

Modeling most respiratory diseases often requires more than a million cells in order to consistently replicate specific disease pathologies or allow adequate sampling of cell products. Due to issues related to biologic product "potency," μlung models may not contain enough cells to produce measureable amounts of cell products using current detection methods. mlung systems contain more than a million cells and make up the majority of respiratory tract models. Figure 7.2 shows an example of an mlung model produced using transwell plates as a culture system (Corning Life Sciences, Corning, NY). The transwell system allows for polarization of cells and supports development of an air interface, which is necessary for production of ciliated respiratory cell models. Cell monolayers developed on the membrane (Figure 7.2a–d) can be fixed and stained using standard cytological techniques (Figure 7.2d).

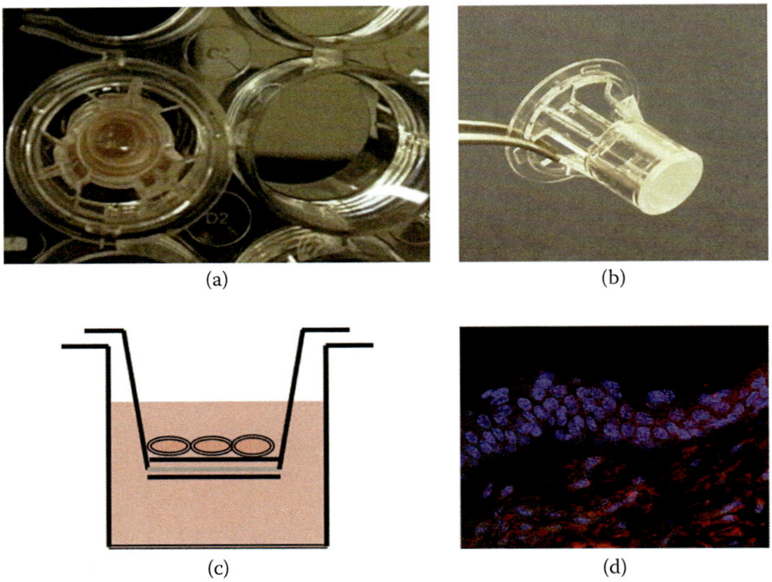

FIGURE 7.2 An example of a mlung transwell plate culture of human bronchial cells is shown in (a–d). (a) Transwell chamber insert in a well. (b) Transwell insert showing porous membrane coated with collagen-1. (c) Diagram showing design of model for bronchial cells in transwell culture. (d) Bronchial cells just after seeding of cells into insert (400× magnification) DAPI nuclear stain, blue.

Portable medium exchange systems are commercially available, which can easily automate the process of media exchange at least for 6-well format or 96-well cultures (Takasago, Westbrough, MA) and, hopefully, other formats in the future.

HOC models or macrophysiologic models are large, complex tissue analogs produced from combinations of cell types in order to recreate tissue interactions. Lung HOC may contain type I and II alveolar epithelial cells (AEC) and endothelial cells as well as fibroblasts, smooth muscle cells and white blood cells (immune cells). Large models such as HOC can be developed and maintained using standard 12- and 24-well plate culture (Figure 7.3a–c) or large microfluidic supported chambers (Synthecon, Houston, TX, special design at request of author) (Figure 7.3d–e) that allow for formation of an air-liquid interface and subsequent maturation of ciliated cells (Figure 7.3f).

The numbers of cells included in an HOC model vary depending on the goal of the model. Pathologic HOC models must contain sufficient numbers of cells to reproduce a desired pathologic response. Many complex HOC models incorporate white

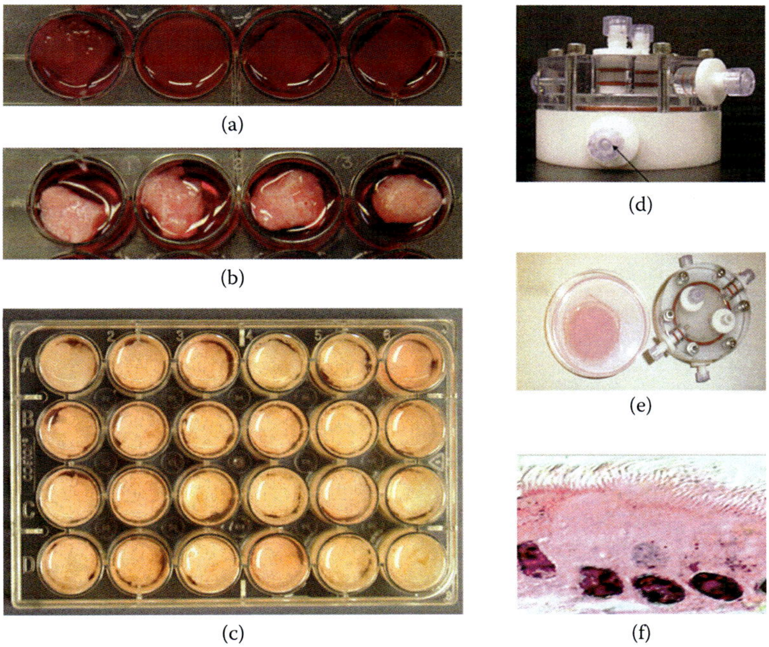

FIGURE 7.3 Example of an *in vitro* human lung HOC culture system. (a) HOC lung cultures following seeding of human AEC onto natural acellular lung scaffolds following culture for 24 hours in a 24-well plate; (b) Human lung HOC following 6 weeks of culture; and (c) replicate 4-week-old HOC lung cultures in a 24-well plate. ([a–c] courtesy of Daniil Weaver in collaboration with Maria Grimaldo, Pennsylvania.) (d) HOC culture chamber (Synthecon, Houston, TX) showing ports for fluid flow and air interface support. (e) Isolated scaffold and (f) maturation of ciliated cells following air interface culture (H&E stain) (630× magnification). (Images [a–c] were in collaboration with Maria Grimaldo and Daniil Weaver, Galveston, Texas.)

blood cells in order to allow for evaluation of specific immune responses. Evaluation of immune responses generally involves the examination of immune cell products such as cytokines or chemokines. Cytokines are protein messengers, produced by immune cells, which mediate cell communication in response to injury or disease. Chemokines are cytokines that influence migration of immune cells and attract cells to the site of injury or microbial invasion. Interleukins are cytokines produced by white blood cells or leukocytes.

The size of *in vitro* HOC models is often limited by oxygen diffusion capacity in the system since models lack vascularization, and therefore systems necessary for support of the natural metabolic processes of cells and tissues. Microfluidic support systems can alleviate some problems encountered due to lack of appropriate vascularization and can allow long-term culture and maturation of tissues. Most commercially available fluidic pumping systems can be used to support HOC cultures.

7.3 CELL SOURCES

Once a model scale has been determined, the next step is to select an appropriate cell source. The design of some models requires that the cells used in the model be isolated directly from human tissues. Cells cultured directly following isolation from a subject are considered to be primary cells. With the exception of primary cells derived from tumors, primary cells are difficult to maintain in culture over time, unlike immortalized cells or transformed cell lines derived from tumors which have the ability to proliferate indefinitely. Although primary cells are useful in production of *in vitro* model systems and remain the "gold standard" for human cell culture, there are obstacles in obtaining primary cells in sufficient quantity to generate replicate models. Human respiratory tract tissues are always difficult to procure, and the primary cell isolation process can be time consuming. Despite the difficulties of isolating and maintaining primary cells, some respiratory tract and lung models require the use of primary cells. A good example of the importance of primary cells in model development utilized primary fibroblasts grown on scaffolds produced form either fibrotic or normal ECM. In this model an increase in transforming growth factor (TGF)-beta-independent differentiation of myofibroblasts was observed by cells grown on the fibrotic but not on normal scaffolds (Booth et al. 2012; Parker et al. 2014). TGF-beta is a cytokine that influences cell proliferation and differentiation and plays a role in asthma, lung fibrosis, and lung disease. Other studies utilizing primary cells include early studies in the development of human lung models focused on the ability of primary AECs to attach and survive on a variety of scaffold materials (Kim et al. 2006; Nichols et al. 2013).

Immortalized and transformed cell lines have both been used in the generation of respiratory tract models. Immortalized 16HBE14o, a bronchial epithelial cell line, have been an excellent choice for modeling cystic fibrosis. 16HBE14o retain mechanisms for chloride secretion, express the cystic fibrosis transmembrane conductance regulator (CFTR) protein and possess conserved transport processes, and trans-epithelial resistance (TEER) properties that are associated with the presence of functional tight junctions (Cozens et al. 1994). This line has also been used to study inflammatory processes. The interaction of 16HBE14o cells in trachea-bronchial models containing neutrophils and granulocytes resulted in secretion of granulocyte macrophage-colony stimulating factor GM-CSF when the 16HBE14o cells were treated with human interleukin-17 (IL-17), a pro-inflammatory cytokine (Laan et al. 2003).

Transformed cell lines, such as the human lung adenocarcinoma cell line, A549, have often been used in lung modeling due to their capability for unrestricted proliferation. A549 are genetically homogenous, derived from a single parent cell (monoclonal), and were classified as type II AECs due to their general morphological characteristics and their ability to secrete surfactant (Giard et al. 1973). Models developed using this cell line include metabolic and macromolecule systems used to examine drug delivery to the pulmonary epithelium (Foster et al. 1998), human lung cancer mimicry *ex vivo* (Mishra et al. 2015), and the observation of cluster of differentiation 4 (CD4)+ and CD8+ T lymphocyte responses proposing type II AEC as accessory antigen-presenting cells (Corbiere et al. 2011). Since transformed cell lines are derived from cancer cells, they may not accurately simulate what occurs naturally in primary cell populations.

Stem cells such as embryonic stem cells (ESCs) and induced pluripotent stem cells (iPSCs) have been used to generate lung epithelium and are an attractive cell source for the development of lung models due to their capacity for unlimited self-renewal and their differentiation plasticity. For the generation of lung models, production of lung tissue from cells derived from a renewable source such as ESC or iPSC is particularly appealing, since it offers the possibility of production of many same-size portions of tissue for replicate culture and exposure under standardized conditions. The first reports of derivation of a lung-specific type II AEC from murine ESC (mESC) were through studies examining the effect of small airway growth medium (SAGM) (Ali et al. 2002) or additives such as retinoic acid in differentiating ESC (Rippon et al. 2006) (Van Vranken et al. 2007). 3D culture of mESC on a scaffold has also been used to induce lung-specific epithelial cell differentiation as has culture in a high-aspect-ratio vessel (HARV) bioreactor containing A549 cell conditioned medium (Siti-Ismail et al. 2012). In 2005, Coraux et al. demonstrated the generation of club cells as well as a fully differentiated airway epithelium from mESC. Human ESC (hESC) have also been successfully differentiated into type II AEC (Samadikuchaksaraei et al. 2006). One important use of ESC cell population is in production of models that allow *in vitro* observation of developmental pathways and cell lineage hierarchy in the lung (Corteilla et al. 2010) (Roszell et al. 2009). Later, use of 3D microgel platforms seeded with mESC have even been used to promote vasculogenesis and tissue formation and provided substantial information regarding extrinsic cues from soluble factors and extracellular matrix in early stages of germ layer formation and embryonic development (Qi et al. 2010).

Human iPSC (hiPSC) have been differentiated into airway epithelium (Firth et al. 2014) and into airway epithelial cells that express the cystic fibrosis transmembrane conductance regulator (CFTR) gene (Wong et al. 2012; Wong et al. 2015). Human iPSCs have also been used to generate type I and II AECs (Ghaedi et al. 2013). IPSC-derived type II AEC were characterized by the expression of pro-surfactant protein C (pro-SPC), pro-surfactant protein B (pro-SPB), surfactant protein A (SPA) and the presence of lamellar bodies (Ghaedi et al. 2013). The production of type I from type II AECs was demonstrated by positive staining for aquaporin 5, podoplanin, and calveolin-1 (type I AEC markers). Derivation of lung progenitors from patient-specific cystic fibrosis iPSCs (Mou et al. 2012) was an important step in the production of differentiated lung epithelium for disease modeling with potential for use in the design of patient-specific therapies for cystic fibrosis or other lung diseases.

Although human ESCs and iPSCs have not been produced in large quantities, they may be the best cell source for use in μlung or lung-on-a-chip systems due to their differentiation potential and, for iPSC, the ability to engineer patient-matched iPSCs. Patient-derived iPSCs could be used for disease modeling, with a particular focus on diseases that burden healthcare and require urgent responses for development of effective therapies, and for use in personalized patient focused medicine.

7.4 SCAFFOLDS USED FOR DEVELOPMENT OF LUNG MODELS

ECM and cell scaffolds play an integral role in influencing the function of cells and the biological properties of the system being modeled. An appropriate scaffold made from either synthetic or natural materials should facilitate cell growth, attachment, and differentiation, which are normal physiologic functions. Other important considerations include scaffold biomechanics, adsorption kinetics, porosity, capacity for remodeling, and ability to establish proper vascularization (Nichols et al. 2008; El-Sherbiny et al. 2013). Recent advances have made possible the development of "smart" hydrogels that can be manipulated to release factors that promote cellular differentiation and migration. These "smart" hydrogels can respond to variations in environmental conditions such as pH, temperature, light, ionic strength, electric field, and protein concentration (Brahim et al. 2002). By manipulating these variables, one can influence the release kinetics of the hydrogel to control the rate of delivery of molecules to the model system (Zhang et al. 2010).

Scaffolds can also be used to provide structural support for cells cultured on microfluidic lung models. Although microfluidic models aim to mimic the *in vivo* microenvironment of the lungs, cells within microchannels often grow as monolayers that lie flat on the micro-channels of the device. Many microfluidic devices contain porous membranes that are coated with ECM proteins such as fibronectin or collagen to improve cell attachment and provide a structure that simulates 3D ECM structure. The architecture of natural lung ECM is complex in its composition and micromechanical properties, which are lung-specific and are difficult to replicate in a microfluidic device (Luque et al. 2013; Nichols et al. 2013; Nichols et al. 2014).

While some success has been achieved with the use of hydrogels in the generation of 3D *in vitro* models, their use has been limited by an inability to recreate the exact complexity, biocompatibility, diffusive profile and mechanical strength of natural extracellular matrix (Geckil et al. 2010; Zhu et al. 2011). Because hydrogels do not adequately mimic natural trachea-bronchial or lung ECM, many groups have developed protocols to produce acellular scaffolds from normal respiratory tissues. Effective tissue decellularization removes cellular debris, DNA, and other immunogenic components from the scaffold while retaining critical scaffold elements that facilitate cell attachment and tissue formation (Balestrini et al. 2015; Hill et al. 2015).

Multiphoton microscopy (MPM) and second harmonic generation (SHG) imaging have been used to examine and quantify the ECM components such as collagen and elastin remaining after decellularization (Nichols et al. 2013), as has mass spectrometry (Wagner et al. 2014). AC lung scaffold has been shown to be superior to Matrigel, Gelfoam, type I collagen, and pluronic F-127 hydrogel matrix at facilitating mESC attachment and survival (Cortiella et al. 2010). AC scaffolds also play an important role in cellular differentiation. Alveolar progenitor (AP) cells seeded on liver-derived AC scaffolds did not display the same degree of type I AEC differentiation as AP cells seeded on lung-derived AC scaffolds, and highly organized alveolar structures only formed when AP cells were seeded onto lung-derived AC scaffolds (Shamis et al. 2011). Advances in proteomics will provide greater characterization of the molecular constituents of AC scaffolds and thus facilitate improvements in tissue modeling by allowing greater correlation of biochemical composition of scaffolds with functional outcomes (Hill et al. 2015). These same advances will contribute to design and production of synthetic materials that more closely match the natural trachea-bronchial or lung ECM in the future for both microfluidic supported microchip and HOC cell culture.

7.5 CURRENT 2D AND 3D LUNG MODELS

Although information provided by 2D models has been helpful in developing a basic understanding of trachea-bronchial or lung responses, 2D models do not recreate the structural and physiological components of the lungs, nor do they reflect many characteristics of normal tissues. In an *in vivo* setting, the 3D structural composition of ECM greatly influences cell behavior and response. 3D trachea-bronchial or lung models can range from simple systems focused on single-cell responses to more complex multicell HOC models containing mixtures of lung cells with or without the incorporation of immune cells. *In vitro* 3D lung models may be constructed of cell aggregates (Birkness et al. 2007), cells cultured on transwell systems (Hermanns et al. 2004; Matrosovich et al. 2004; Hoang et al. 2012; Davis et al. 2015) or cells grown on scaffolds that provide 3D structural support (Cortiella et al. 2010; Kloxin et al. 2012; Melo et al. 2015). Table 7.1 shows a broad overview of trachea-bronchi/bronchiole or lung models that includes scaffold used, cell source, and goal for each model.

Simple models are often capable of being used for a wide range of applications. One relatively simple model system was developed from human bronchial epithelial cells and human fetal lung fibroblasts cultured on transwell plates (Choe et al. 2006). A porous polymeric cylinder placed in each well was filled with a type I collagen gel containing fetal lung fibroblasts (Choe et al. 2006). Human bronchial epithelial cells were cultured on top of the collagen gel containing fibroblasts. The porous polymeric cylinders allowed for formation of an air-liquid interface. Functional features of this model included development of ciliated cells with beating cilia as well as mucus production. The simplicity and practicality in the design of this model allows it to be used for many different studies, including basic airway physiology, pathophysiology of airway inflammation within the mucosa, and transepithelial transport (Choe et al. 2006).

TABLE 7.1
Trachea, Bronchial, Bronchiole, and Lung Models

Reference	Description of Engineered Model	Type of Model	Cell Source	Scaffold	Platform System for Tissue Culture	Goal of the Model
			Trachea, Bronchi/Bronchiole Models			
Chakir et al. 2001	Engineered human bronchial mucosa in an air-liquid interface used to study inflammation and airway repair during asthma	Physiologic Pathologic	Primary bronchial epithelial cells and fibroblasts isolated from lungs of normal and asthmatic patients T lymphocytes isolated from the peripheral blood of asthmatic patients	Collagen gel matrix with embedded human fibroblasts	Petri dishes (35-mm diameter) with an anchorage	Pseudostratified ciliated epithelium with the presence of mucus secretory cells Production of IL-5 by T lymphocytes
Matrosovich et al. 2004	Human airway epithelium used to study differences in cellular tropism for human and avian influenza viruses	Infection Pathologic	Primary human epithelial cells from tracheal/bronchial and nasal tissues	No scaffold used	Transwell (12-mm) culture plate	Sialic acid expression (α2-6– and α2-3–linked) in nonciliated epithelial cells and ciliated cells Secretory cells (identified by Alcian blue-periodic acid Schiff staining)

(Continued)

TABLE 7.1 *(Continued)*
Trachea, Bronchial, Bronchiole, and Lung Models

Reference	Description of Engineered Model	Type of Model	Cell Source	Scaffold	Platform System for Tissue Culture	Goal of the Model
Coraux et al. 2005	Airway epithelium developed from differentiation of murine embryonic stem cells into specific airway epithelial lineages.	Developmental Physiologic	Murine embryonic stem cell line CGR8	Type I collagen-coating, gelatin-coating, type IV collagen-coating, or type VI collagen-coating	Petri dishes—Milicell-HA porous membranes placed on dishes	Basal cells Ciliated cells expressing β-tubulin Active ciliogenesis shown by centriole migration Intermediate cells Club cells expressing CC10 and SPD Cytoplasmic characteristics of club cells → abundant rough endoplasmic reticulum, numerous mitochondria, and secretory granules
Choe et al. 2006	Human bronchial mucosa that mimics structural and functional features of the airway wall	Physiologic	Normal human bronchial epithelial cells Human fetal lung fibroblasts	Type 1 collagen gel	6-well Transwell plates containing porous polyester membranes with 0.4-μm pores	Ciliated cells Ciliogenesis Mucus-secreting cells

(Continued)

TABLE 7.1 *(Continued)*
Trachea, Bronchial, Bronchiole, and Lung Models

Reference	Description of Engineered Model	Type of Model	Cell Source	Scaffold	Platform System for Tissue Culture	Goal of the Model
Hoang et al. 2012	Human airway mucosa composed of an epithelial cell layer and a fibroblast/matrix layer also containing dendritic cells	Physiologic	Human lung fibroblast cell line, MRC-5 Immortalized human bronchial epithelial cell line 16HBE Human monocyte-derived dendritic cells isolated from whole blood	Collagen gel matrix seeded with fibroblasts	6-well transwell plates with 3.0-μm pore inserts	Dendritic cells positive for CD1a and negative for CD14 and expressing DC-SIGN Mucosal junctional proteins claudin I and occludin Expression of E-cadherin and laminin-5 by epithelial cells Production of type IV collagen and tropoelastin by epithelial cells and fibroblasts
Villenave et al. 2012	Pediatric bronchial epithelium used to model the pathogenesis of RSV infection in children	Infection Pathologic	Well-differentiated primary pediatric bronchial epithelial cells	Collagen coating	Transwell plates with 0.4 μm pore membranes	Mucociliary epithelium with beating cilia and mucus production Syncytia formation, apical cell sloughning, and goblet cell hyperplasia after infection with RSV Production of cytokines and chemokines after infection with RSV

(Continued)

TABLE 7.1 *(Continued)*
Trachea, Bronchial, Bronchiole, and Lung Models

Reference	Description of Engineered Model	Type of Model	Cell Source	Scaffold	Platform System for Tissue Culture	Goal of the Model
Cakebread et al. 2014	Pediatric bronchial epithelium monolayer used to examine the role of Th2 cytokines in rhinovirus infection	Infection	Primary bronchial epithelial cells from bronchial brushings of pediatric patients	Collagen coating	12-well plates	Expression of IL-8, IP-10, and GM-CSF after infection Expression of ICAM-1 by bronchial epithelial cells after Th2 cytokine treatment
Davis et al. 2015	Human pseudostratified bronchial epithelium used to study influenza A infections	Physiologic Infection	Primary normal human bronchial/tracheal epithelial cells	Rat tail collagen coating	Transwell plates with 6.5 mm transwell-clear membrane supports	Jacalin expression in goblet cells β-tubulin–positive ciliated cells Cytokeratin-5 positive basal cells
Melo et al. 2015	Trachea-bronchial wall co-culture system	Physiologic	Bronchial epithelial cell line (16HBE14o) Human embryonic lung fibroblast Wi-38 Angiosarcoma endothelial cell line (ISO-HAS-1)	Decellularized porcine trachea	12-well plates with transwell polycarbonate filter membranes (0.4-μm pore size)	Bronchial epithelial cell expression of β-catenin and zona occluden protein 1 Endothelial cell expression of CD 31 Fibroblasts expressing vimentin

(Continued)

TABLE 7.1 ***(Continued)***
Trachea, Bronchial, Bronchiole, and Lung Models

Reference	Description of Engineered Model	Type of Model	Cell Source	Scaffold	Platform System for Tissue Culture	Goal of the Model
Punde et al. 2015	Microfluidic system of a bronchial airway and a blood vessel used to study lung inflammation and airway remodeling	Pathologic	Human bronchial epithelial cell line, Beas-2B Human fibrocytes	Fibronectin	Microfluidic chip composed of two PDMS microchannels and a silicone membrane	Fibrocyte migration induced by ECP Production of pro-inflammatory cytokines and chemokines in response to ECP
			Lung Models			
Hermanns et al. 2004	Human distal lung epithelium and endothelium mimicking the alveolar-capillary barrier	Physiologic Pathologic	HPMEC isolated from normal lung specimens from adult patients undergoing lobectomies for early stage lung cancer Human lung adenocarcinoma cell lines, A549 and NCI H441	Type I collagen gel from calf skin	24-transwell plates (0.4-μm membrane pore size)	Type II AEC expression of TTF-1, SP-A, SP-B, SP-C, SP-D and presence of lamellar bodies A549 cell and NCI H441 cell expression of E-cadherin HPMEC, A549, and NCI H441 expression of zona occluden protein 1 HPMEC expression of CD 31 and VE-cadherin
Birkness et al. 2007	Rounded cell aggregates of lymphocytes and macrophages that mimic human granuloma formation	Pathologic Infection	Human PBMCs	No scaffold used	24-well culture plates	CD 68-positive macrophages CD 3-positive T lymphocytes

(Continued)

TABLE 7.1 *(Continued)*
Trachea, Bronchial, Bronchiole, and Lung Models

Reference	Description of Engineered Model	Type of Model	Cell Source	Scaffold	Platform System for Tissue Culture	Goal of the Model
Cortiella et al. 2010	Tissue-engineered whole-lung rat tissue developed from differentiation of embryonic stem cells into lung specific cell lineages	Developmental Physiologic	Murine embryonic stem cells	Whole acellular rat lung matrix	50 mL rotary bioreactor	Expression of cytokeratin-18 in ciliated epithelial cells Expression of CC10 in club cells Type II AEC expression of pro-SPC Developing epithelium expressing TTF-1 Expression of CD31 and PECAM-1 in endothelial cells Expression of α-SMA in smooth muscle cells
Huh et al. 2010	Microfluidic system of alveolar epithelium and endothelium designed to mimic the alveolar-capillary interface during pulmonary inflammation	Physiologic Pathologic Infection Pharmacological	Human microvascular endothelial cells, HUVEC Human lung adenocarcinoma cell lines, A549 and NCI H441 The immortalized noncancerous lung cell line E10, derived type II AECs Human PBMCs	Collagen gel or fibronectin coating	Microfluidic device composed of two microchannels separated by a flexible PDMS porous (10 μm) membrane that is integrated with computer-controlled vacuum to produce cyclic stretching	Type II AEC surfactant production and presence of lamellar bodies Expression of ICAM-1 and VE-cadherin in endothelial cells Extravasation of neutrophils in response to *Escherichia coli* and nanoparticles

(Continued)

TABLE 7.1 *(Continued)*
Trachea, Bronchial, Bronchiole, and Lung Models

Reference	Description of Engineered Model	Type of Model	Cell Source	Scaffold	Platform System for Tissue Culture	Goal of the Model
Huh et al. 2012	Microfluidic system of alveolar epithelium and endothelium designed to mimic pulmonary edema	Physiologic Pathologic Pharmocological	Human alveolar epithelial cell line NCI-H441 Human pulmonary microvascular endothelial cells	Porous membranes coated with fibronectin	Microfluidic device composed of two microchannels separated by a flexible PDMS porous (10 μm) membrane that is integrated with computer-controlled vacuum to produce cyclic stretching	Expression of microvascular endothelial cell junctional proteins VE-cadherin and occludin Disruption of cell-cell junctions and increased alveolar-capillary barrier permeability in response to IL-2 Inhibition of endothelial permeability with angiopoietin 1
Booth et al. 2012	Tissue-engineered normal lung tissue and fibrotic lung tissue used to study the effect of fibrotic scaffolds on fibroblast phenotype	Physiologic Pathologic	Primary human fibrotic lung fibroblasts isolated from patients undergoing surgical lung biopsy for diagnosis of idiopathic interstitial pneumonia Control normal-lung fibroblasts were obtained from patients undergoing thoracic surgery for nonfibrotic lung diseases, such as lung cancer	Biopsy punch cylinders of decellularized lung matrices from normal healthy lungs and from lung with interstitial pulmonary fibrosis	24-well culture plates	Myofibroblasts expressing α-SMA and fibronectin Expression of TGF-β in normal matrices and fibrotic matrices

(Continued)

TABLE 7.1 *(Continued)*
Trachea, Bronchial, Bronchiole, and Lung Models

Reference	Description of Engineered Model	Type of Model	Cell Source	Scaffold	Platform System for Tissue Culture	Goal of the Model
Nalayanda et al. 2013	Microfluidic system of the alveolar-capillary interface used to study the effects of pressures of varying magnitudes on alveolar cells	Physiologic Pathologic	Human adenocarcinoma cell lines, A549 and NCI H441	No scaffold used	Microfluidic pressure-chip device that sustains the application of static pressure for extended periods	TEER measurements and cell viability after exposure of cells to various pressures as a measure of alveolar monolayer integrity or disruption
Parasa et al. 2014	Generation of human lung tissue containing macrophages used to model granuloma formation after *Mycobacterium tuberculosis* infection	Pathologic Infection	Human lung fibroblasts cell line MRC-5 Immortalized human bronchial epithelial cell line 16HBE14o Monocytes isolated from whole human blood	Type I collagen gel with embedded fibroblasts	6-well transwell plates (3 μm membranes)	CD 68 expression in macrophages Macrophage migration in response to MCP-1

(Continued)

TABLE 7.1 ***(Continued)***
Trachea, Bronchial, Bronchiole, and Lung Models

Reference	Description of Engineered Model	Type of Model	Cell Source	Scaffold	Platform System for Tissue Culture	Goal of the Model
Guirado et al. 2015	Immune cell aggregates that mimic human granuloma formation, used to study the pathogenesis of *M. tuberculosis* and differences in the host response between individuals with and without latent tuberculosis infection	Pathologic Infection	Human PBMCs Autologous serum	No scaffold used	24-well plates with glass cover slips at the bottom of the wells	CD11b-positive mononuclear phagocytes CD3-positive lymphocytes CD4- and CD8-positive T cells Formation of multinucleated giant cells Production of inflammatory cytokines

Note: Models listed in this table were selected based on the design, novelty, purpose, and goal of the model. For each model, the cell type, scaffold, platform, and overall goal are provided. Alveolar epithelial cell (AEC); club cell 10 (CC10) protein; cluster of differentiation (CD); dendritic cell-specific intercellular adhesion molecule-3-grabbing non-integrin (DC-SIGN); epithelial (E) cadherin; eosinophil cationic protein (ECP); granulocyte macrophage-colony stimulating factor (GM-CSF); human primary pulmonary microvascular endothelial cells (HPMEC); human umbilical vein endothelial cell (HUVEC); interleukin-2 (IL-2); interleukin-5 (IL-5); interleukin-8 (IL-8); interferon-gamma-inducible protein 10 (IP-10); intracellular adhesion molecule-1 (ICAM-1); peripheral blood mononuclear cell (PBMC); respiratory syncytial virus (RSV); transepithelial electrical resistance (TEER); transforming growth factor-β (TGF-β); α-smooth muscle actin (α-SMA); pro-surfactant protein C (pro-SPC); macrophage chemoattractant protein-1 (MCP-1); polydimethylsiloxane (PDMS); platelet endothelial cell adhesion molecule (PECAM); thyroid transcription factor-1 (TTF-1); surfactant protein A (SPA); surfactant protein B (SPB); surfactant protein C (SPC); surfactant protein D (SPD); vascular endothelial (VE) cadherin.

Complex models such as HOC models have been developed to allow for examination of numerous pathophysiologic conditions. Asthma is a complex pathologic condition and the link between T lymphocyte-derived IL-5 and eosinophil activation in asthmatic airways is of major importance. Many experimental drugs have been developed to target IL-5 production as a therapeutic approach for control of asthma (Garcia et al. 2013). One example of a well-designed asthma model was created using primary human bronchial epithelial cells and human bronchial fibroblasts isolated from normal and asthmatic patient lung tissues. To examine the mechanistic role of immune cells in airway inflammation during asthma, T lymphocytes isolated from the peripheral blood of asthmatic patients were added (Chakir et al. 2001). Histological examination of the engineered human bronchial mucosa from normal cells exhibited features similar to those in normal bronchial tissues. When T lymphocytes were cultured with asthmatic bronchial epithelial cells, 87% of T lymphocytes in that culture produced IL-5. In contrast, only 2% of T lymphocytes cultured with normal bronchial epithelial cells produced IL-5 (Chakir et al. 2001). The incorporation of T lymphocytes and the ability to recreate an IL-5 mediated asthmatic response in this engineered human bronchial epithelium make this an attractive model for studies focused on the T lymphocyte immune response during asthmatic inflammation. This model may also be used as a preclinical pharmacology model to examine the effect of experimental drugs on IL-5 production in asthmatic human cells and perhaps in development of patient specific therapies in the future.

7.6 MICROFLUIDIC LUNG MODELS

Microfluidic devices are useful tools that can be employed for the development of micro-trachea-bronchi, bronchiole, or lung models focused on small-scale evaluations of single cells or cell-cell interactions in a microenvironment where spatiotemporal gradients can be controlled and maintained during culture (Punde et al. 2015; Ramadan et al. 2015). Microchannels and porous membranes within microfluidic devices can be populated with human lung cells (primary cells or cell lines) or a mixture of cell types. Many of the microfluidic human respiratory system models that are currently available have been designed as alveolar-capillary interface models (Huh et al. 2010; Nalayanda et al. 2010) or models of bronchial epithelium (Punde et al. 2015).

Advancements in the fabrication of microfluidic devices have provided valuable resources for the development of platforms that allow regulated perfusion (delivery and removal of fluids), physiologically relevant fluid flow rates, controlled application of sheer stress and mechanical strain, and air-liquid interface (Ramadan and Gijs 2015). Production of microfluidic-supported respiratory system models will allow evaluation of high-throughput studies of cell-cell interactions, early stages of drug development, and preclinical testing (Nichols et al. 2014). Using simple single-cell phenotype microfluidic trachea-bronchi or lung models, the roles and responses of single cell types can be studied in a controlled microenvironment. By removing many of the variables found *in vivo* and focusing on critical cell-to-scaffold or cell-to-cell interactions, microfluidic lung models can be used to identify cellular and molecular responses directly involved in pathophysiological mechanisms of disease.

Although microfluidic systems have the potential to overcome some of the limitations encountered with *in vitro* cell culture and *in vivo* animal models, there are some disadvantages associated with their use as human lung model systems. Among the major disadvantages is that many do not accurately mimic the 3D architecture of the ECM in human lungs. An important feature in the design of microfluidic lung models is an air-liquid interface that mimics the *in vivo* physiological conditions in the lungs. One of the first microfluidic lung model systems described in the literature was designed to mimic the alveolar-capillary interface and model pulmonary inflammation in response to microbial infection and nanoparticle cytotoxicity (Huh et al. 2010). This microfluidic platform was fabricated using soft lithography and chemical etching of PDMS layers that contained recessed microchannels (Huh et al. 2010). Alignment and bonding of these layers formed two microchannels separated by a flexible porous membrane. Two larger, side vacuum chambers directly adjacent to the microchannels were also formed. The membrane separating the two microchannels was coated with fibronectin or collagen gel to aid in cell attachment. Human alveolar epithelial cells (NCI H441 cell line) were used to seed the upper chamber, representing the epithelium, and human pulmonary microvascular endothelial cells were seeded on the bottom channel to recreate a capillary junction (Huh et al. 2010). Cells formed epithelial and endothelial monolayers and were shown to express the junctional proteins, occludin and vascular endothelial cadherin. When an air-liquid interface was introduced to mimic physiological conditions, there was an increase in surfactant protein production by the epithelium in the upper chamber. To induce mechanical stretching and simulate breathing movements, a computer-controlled vacuum was applied to the side chambers. The effects of stretching were evaluated by measuring changes in the permeability of the epithelial–endothelial barrier (Huh et al. 2010). The ability of this microfluidic alveolar-capillary model to mimic pulmonary inflammation was also examined by introducing human peripheral blood mononuclear cells (PBMCs) to the vascular channel and adding medium containing the pro-inflammatory cytokine tumor necrosis factor-alpha (TNF-α) to the alveolar channel (Huh et al. 2010). Within 5 hours of adding TNF-α, endothelial expression of intracellular adhesion molecule (ICAM)-1 significantly increased and endothelial cells promoted the adhesion of fluorescently labeled neutrophils (Huh et al. 2010). Neutrophils were shown to migrate across endothelial monolayers and the membrane to the alveolar epithelium channel. Addition of *Escherichia coli* expressing green-fluorescent protein (GFP) to the alveolar microchannel resulted in endothelial capture of neutrophils circulating in the vascular channel and transmigration of these neutrophils into the alveolar microchannel. Using time-lapse fluorescence microscopy, red fluorescently-labeled neutrophils could be seen phagocytosing GFP-expressing *E. coli* (Huh et al. 2010). This microfluidic model was also used to examine the toxic effects of nanoparticle delivery on alveolar epithelium. After 5 hours of nanoparticle exposure, the endothelial cells in the vascular channel increased ICAM-1 expression (Huh et al. 2010). This increase in ICAM-1 expression was sufficient to promote neutrophil extravasation into the epithelial channel. This alveolar-capillary microfluidic model was able to mimic some characteristics of the innate immune response during pulmonary inflammation. The ability to recreate physiological functions of the epithelial–endothelial barrier and pathological responses to microbial infections, nanoparticle cytotoxicity, or cyclic mechanical strain enables this system to be used for a wide range of studies. The microfluidic device developed by Huh et al. in 2010 has also been used as an alveolar-capillary model of human drug-induced pulmonary edema (Huh et al. 2012). The alveolar and vascular microchannels for this pulmonary edema model were seeded with the alveolar epithelial cell line NCI-H44I1 and human pulmonary microvascular endothelial cells. To induce pulmonary edema, interleukin-2 (IL-2) was perfused into the microvascular channel at a clinically relevant dose of 1000 U/mL (Huh et al. 2012). By day 4, liquid flooded the air space of the alveolar microchannel (Huh et al. 2012). To mimic the formation of blood clots, the blood plasma proteins prothrombin and fluorescently labeled fibrinogen were perfused into the microvascular channel during IL-2 treatment (Huh et al. 2012). Within a period of 4 days, fluorescent fibrin clots were observed on the surface of alveolar epithelium (Huh et al. 2012). Pathological changes associated with pulmonary edema were characterized by examining alveolar-capillary barrier integrity during IL-2 treatment. Barrier permeability was examined by measuring the transport of fluorescein isothiocyanate (FITC)-labeled insulin from the vascular channel to the alveolar channel in the presence and absence of stretch (Huh et al. 2012). Results showed that IL-2 treatment in the presence of cyclic strain disrupted cell-cell junctions, which resulted in intercellular gaps and significantly increased barrier permeability (Huh et al. 2012). Having shown that this lung-on-a-chip model could mimic specific pathological features of pulmonary edema, Huh and colleagues next examined the use of this model as a pharmacological platform. Angiopoietin 1 (Ang-1), a protein known to inhibit endothelial permeability, was co-administered with IL-2 into the microvascular channel to examine its effects on vascular leakage (Huh et al. 2012). Results showed that Ang-1 was able to prevent IL-2–induced fluid leakage and formation of intercellular gaps in the presence of cyclic strain (Huh et al. 2012).

A microfluidic model of the alveoli-capillary interface with the ability to introduce dynamic forces such as applied pressure and stress can be used to examine ventilator-induced lung injury. One such model was developed using a multiphase microfluidic platform (referred to as a pressure chip) populated with human alveolar epithelial cell lines, A549 and H441, to mimic lung alveoli (Nalayanda et al. 2013). The inner panels of this microfluidic model allowed it to have a basolateral surface where there is a continuous flow of medium and an apical surface, where air was maintained at a steady rate (Nalayanda et al. 2013). In order to allow for examination of ventilator-induced injury, this model was exposed to mechanical disruptions and aerodynamic sheer stress. Results showed that cell responses were dependent on the magnitude and duration of the pressure applied.

Other models have been developed to examine specific protein-cell responses. An effective microfluidic bronchial epithelium model was developed to study the role of eosinophil cationic protein (ECP), a protein released when eosinophils degranulate, on the migration of fibrocytes and the induction of lung inflammation (Punde et al. 2015). This model was constructed using a micropore array silicone chip coated with fibronectin sandwiched between two PDMS channels to create upper and lower layers mimicking a blood vessel (upper layer) and an airway (lower layer) (Punde et al. 2015). This microfluidic human bronchial epithelium model was useful for the examination of pro-inflammatory cytokine and chemokine-induced fibrocyte migration contributing to development of fibrotic conditions (Punde et al. 2015). Stimulation of injured bronchial epithelial cells with inflammatory cytokines induced them to produce chemokines, which recruited fibrocytes expressing the chemokine receptor 4 (CXCR4), to the site of the injured epithelium (Punde et al. 2015). Results from this study highlighted the role of ECP in airway remodeling and development of fibrotic conditions following inflammation. This microfluidic model of human bronchial epithelium could also be useful for studies of other chronic respiratory diseases characterized by airway inflammation and abnormal airway remodeling such as asthma.

7.7 INFECTIOUS DISEASE LUNG MODELS

Current human infectious disease lung models use a variety of human cell types representing the different regions of the trachea/bronchi, bronchioles or lung. In order to develop models of microbial infection or exposure, it is necessary to understand what specific characteristics of human infection the model is meant to reproduce. Progress in antimicrobial vaccine and drug development has often been hampered by the lack of appropriate human *in vitro* models to asses disease pathology or vaccine and drug efficacy. Models developed to examine microbial pathogen infection, replication, and pathophysiology can alleviate this deficiency. One 3D bronchial epithelial model was used to examine responses of epithelial cells to respiratory syncytial virus (RSV) infection. The model was created using primary pediatric bronchial epithelial cells cultured on transwell plates. Transwell membranes were coated with collagen and an air-liquid interface was maintained in the cell culture (Villenave et al. 2012). Infection of the model with RSV produced histopathology and chemokine production similar to that seen in patients with severe and fatal cases of RSV infection (Villenave et al. 2012). The use of primary pediatric bronchial epithelial cells in this system might actually be a good predictive model for the cellular responses seen in pediatric patients.

Complex human bronchial epithelium infectious disease models can be used to examine immune responses to infection. Primary bronchial epithelial cells were isolated from bronchial brushings obtained from non-asthmatic pediatric patients and grown on collagen-coated plates (Cakebread et al. 2014). Characterization of the inflammatory response during rhinovirus infection in this model of human bronchial epithelium allowed for examination of the cellular responses to rhinovirus infection in humans. Identification of signaling pathway inhibitors that blocked inflammatory cytokines and suppressed inflammation in this primary bronchial epithelial cell model may lead to design of improved therapeutic interventions for the treatment of other respiratory infections.

Major research efforts have been directed toward the study of *Mycobacterium tuberculosis*. Granuloma development is a critical part of the immunopathogenesis of *M. tuberculosis* infection. Recent studies suggest that the immunologically dynamic environment within granulomas may contribute to bacterial dissemination and person-to-person transmission, challenging the traditional view of a protective environment where infected cells are secluded to limit dissemination

(Cardona et al 2009; Shaler et al. 2013). This unique immunological response in human granuloma formation is difficult to recreate in animal models. Progress in identifying biomarkers or treatments to limit granuloma formation have been hindered by the lack of a relevant, reproducible infection model that accounts for the complexity of the host immune response as well as pathogen responses that occur following infection (Guirado et al. 2015). Examples of tissue engineered human granuloma models include very basic simple human cell aggregates (Birkness et al. 2007; Guirado et al. 2015) or more complex HOC models (Parasa et al. 2014). A simple model of granuloma formation was developed to examine differences between bacterial determinants and host response occurring in patients with and without latent tuberculosis infection (Guirado et al. 2015). Granuloma-like structures were produced by isolating human peripheral blood mononuclear cells (PBMC) from patients with latent *M. tuberculosis* infection and from uninfected individuals. Multilayered cellular aggregations formed in cultures with *M. tuberculosis*. This model was able to reproduce some features that characterize human tuberculosis granulomas such as formation of multinucleated giant cells, lymphocyte proliferation, inhibited intracellular bacterial growth, and production of a diverse cytokine profile (Guirado et al. 2015). PBMCs isolated from different donors allowed each individual's immune response to *M. tuberculosis* infection and granuloma formation to be studied. Genetic differences that influence immunological response could also be examined (Guirado et al. 2015). One of the limitations of this model was that it was composed only of cell aggregates and lacked the replication of lung structure (Guirado et al. 2015). A more complex HOC model of granuloma formation was engineered in a transwell system, with a human lung fibroblast cell line, human alveolar and bronchial epithelial cells, and human macrophages isolated from peripheral blood (Parasa et al. 2014). A type I collagen gel matrix with embedded human lung fibroblasts was used as a scaffold, on which type II alveolar epithelial cells and macrophages were seeded. The lung tissue engineered in this transwell model displayed a stratified epithelium, mucus secretion, and the ability of cells to actively produce ECM proteins (Parasa et al. 2014). To establish an infection, macrophages were infected with *M. tuberculosis* before being seeded onto the system, and uninfected macrophages were used as a control. Between days 7 and 10 post infection, confocal microscopy revealed formation of macrophage clusters that resembled early granuloma formation at the site of infection (Parasa et al. 2014). These macrophage clusters are similar in appearance to the macrophage aggregates found in *M. tuberculosis*–infected human lymph nodes and human lung tissue from patient biopsies (Parasa et al. 2014). Infection of this model with a virulent strain of *M. tuberculosis* also resulted in necrosis of the epithelial cells, a characteristic of uncontrolled *M. tuberculosis* infection in human lung tissue (Parasa et al. 2014).

7.8 LUNG CANCER MODELS

Lung cancer is the leading cause of cancer death among both men and women, and is accompanied by poor survival rates. New chemotherapies being developed have a high rate of attrition, with only 1 in 10 drugs acquiring FDA approval from phase 1 and only 50% of drugs making it to market from phase 3 (Hay et al. 2014).

The high rate of drug attrition may in part be explained by the reliance on traditional 2D culture platforms that do not adequately account for the effects of human physiology and the *in vivo* tumor microenvironment on gene expression and cancer cell phenotype. Responsiveness of lung carcinoma cell lines HCC827 and A549 to gefitinib, a drug used to treat lung or other types of cancers, was significantly different in 2D and 3D cultures (Stratmann et al. 2014). Other groups have shown that A549 cells grown in 3D expressed 2,954 genes differently than cells grown in a 2D culture plate (Mishra et al. 2014). While 2D models have provided us with invaluable knowledge about cancer biology, they lack the complexity of *in vivo* tumor formation and progression.

3D Lung-on-a-chip models have been developed to study resistance of lung cancer cells to chemotherapy (Siyan et al. 2009). Successful treatment of cancer is impaired by the resistance of cancer cells to chemotherapy. Glucose-regulated proteins (GRPs) have been shown to promote tumor growth, drug resistance (Fu and Lee 2006), and invasive properties of cancer cells in many human cancers (Lee et al. 2014). Although GRPs have been identified as promising targets for development of anticancer therapeutics, a better understanding of their mechanisms in cancer cell responses is needed. A microfluidic gradient chip system developed to study this problem was used to correlate expression of glucose-regulated protein-78 (GRP78) and resistance to the anti-cancer drug VP-16 (Siyan et al. 2009). In this study, the calcium ionophore A23187 was used to induce the expression of GRP78 in cancer cells. The human lung squamous carcinoma cell line, SK-MES-1 in the cell culture chambers was treated with varying concentrations of A23187 through the drug inlet. After 24 hours of treatment with A23187, the expression of GRP78 increased in a dose-dependent manner (Siyan et al. 2009). Results showed that the percentage of A23187-treated cells undergoing apoptosis significantly decreased as the concentration of A23187 increased. The results from this study suggest that VP-16 induced apoptosis of human lung squamous carcinoma cells is suppressed by overexpression of GRP78 indicating the importance of control of GFPs in cancer (Siyan et al. 2009).

A growing body of evidence indicates that the tumor stromal microenvironment greatly influences cancer cell responses through mechano-transduction mediated via cellular integrins (Levental et al. 2009; Goetz et al. 2011). Within the tumor microenvironment, stromal cells such as cancer associated fibroblasts, mesenchymal stem cells, and myeloid-derived tumor suppressor cells produce oncogenic factors and modify ECM components to promote tumor growth and metastasis (El-Nikhely et al. 2012). One commercially available human airway cancer model, OncoCilAirTM, was developed to mimic the tumor-stromal environment. This transwell-supported model was composed of human bronchial epithelial cells, lung fibroblasts, and the non-small cell lung cancer (NSCLC) A549 cell line. The porous membrane within the transwell inserts was seeded with lung fibroblasts (Mas et al. 2015). Human bronchial epithelial cells and A549 cells were then plated on top of the fibroblast-undercoated membrane. After two days in culture, an air-liquid interface was developed, stimulating production of a ciliated pseudostratified epithelium (Mas et al. 2015). A549 cells formed tumor nodules and histology revealed some tumor nodules containing mucin-filled vacuoles invading the normal airway epithelium (Mas et al. 2015). The anti-cancer efficacy of selumetinib, trametinib, docetaxel, and erlotinib, which are drugs that inhibit tumorgenesis, cell proliferation, and inhibition of apoptosis, was examined by monitoring tumor growth after treatment. Tumor growth inhibition rate was highest with trametinib after 20 days of treatment (Mas et al. 2015). The applicability of this human airway cancer model for drug efficacy and toxicity evaluations makes it a valuable tool for preclinical studies and could easily be adapted to a microfluidic system supported format.

Lung cancer models developed with scaffolds composed of natural ECM allow evaluation of cell-matrix interactions involved in invasion and migration of tumor cells. Structural and mechanical cues provided by the ECM influence cancer cell invasiveness and oncogenic gene expression. The rigidity of the ECM has also been shown to have a direct effect on cell behavior. ECM stiffness caused by an increase in collagen cross-linking has been shown to increase the proliferation of cancer cells (Tilghman et al. 2010) and to promote tumor cell invasion and metastasis (Chen et al. 2015). A perfusable lung cancer model designed to mimic *in vivo* tumor formation was developed using decellularized rat lung matrices seeded with human lung cancer cell lines, A549, H460, and H1299 (Mishra et al. 2012). Formation of dense tumor nodules was observed in scaffolds with A549 cells after 11 days of culture and in scaffolds with H460 and H1299 cells after 7 days in culture (Mishra et al. 2012). The tumor nodule formation recreated within this lung model is similar to the tumor growth pattern seen in human lung cancer. The disordered growth of cancer cells along the basement membrane resembled that of metastatic cancer (Mishra et al. 2012).

Metastatic spread of tumor cells results in the spread of cancer cells to secondary sites in the body. The body's vascular system plays a major role in the process of metastasis. Endothelial cells play an active part during both tumor angiogenesis and later movement of tumor cells via the vascular system. Tumor cell-endothelial cell interactions therefore play a significant role in cancer progression and spread to distant tissues such as the lung. 3D microfluidic-based models have been developed to examine the metastatic process, including tumor angiogenesis and cancer cell entry into and out of the bloodstream to examine endothelial barrier function (Zervantonakis et al. 2012). These systems examine basic mechanisms of metastasis but most do not focus specifically on development or treatment of lung cancer.

7.9 PERSONALIZED MEDICINE

Personalized medicine is a form of medical care that proposes the customization of healthcare tailored to the individual patient. High variability in disease patterns between patients often leads to the unsuccessful treatment of disease (Ruppen et al. 2015). Assessment of patient-specific cellular responses to drugs or other therapeutics allows for a more accurate prediction of the individual's response to a particular treatment strategy. Microfluidic devices serve as excellent platforms for the development of personalized models composed of the patient's own cells. Chip-based respiratory models require few cells and could be produced from primary cells obtained from bronchial brush samples or biopsies of tumors from patients. These models could be used to screen drugs or combinations of drugs to design targeted drug-specific therapies.

Microfluidic models of human bronchial epithelium can be used to recreate key pathological cellular responses of human asthma and are better predictive models of drug safety and efficacy. Many current therapies for reducing airway inflammation fail to target the abnormal airway remodeling component of the disease. Airway remodeling in asthma is characterized by goblet cell hyperplasia and hypersecretion of mucus (Parker et al. 2015). Epidermal growth factor (EGF) and its receptor (EGFR) have been shown to be involved in goblet cell hyperplasia (Puddicombe et al. 2000; Burgel et al. 2004) and in induction of mucin MUC4AC, a major mucus-forming protein (Perrais et al. 2002). In asthmatic patients, overexpression of EGF/EGFR has been correlated with disease severity (Amishima et al. 1998; Holgate et al. 1999; Parker et al. 2015). Characterizing each patient's cell response to this EGFR inhibitor serves to identify a potential target strategy for goblet cell hyperplasia and excessive mucus secretion for each patient. In one asthma model, primary cells for this model were obtained from bronchial brushings of five asthmatic and five nonasthmatic children and were cultured on collagen-coated transwell inserts promoting mucociliary differentiation (Parker et al. 2010). After 28 days of culture, cells had fully differentiated into ciliated cells, columnar goblet cells with secretory granules, and basal cells (Parker et al. 2010). The apical surface of this bronchial epithelium was also capable of mucus secretion. Asthmatic and non-asthmatic cell cultures were cultured in medium containing EGF, in medium without EGF, and in medium with EGF and tyrphostin AG1478 (a protein tyrosine kinase inhibitor that inhibits EGFR kinase) (Parker et al. 2015). In asthmatic EGF-positive cultures, the percentage of goblet cells was higher than in asthmatic EGF-negative cultures and non-asthmatic EGF-positive cultures (Parker et al. 2015). When asthmatic EGF-positive cultures were treated with AG1478, there was a reduction in the number of goblet cells present. Treatment with AG1478 also resulted in a significant reduction of mucus secretion from asthmatic EGF-positive cultures (Parker et al. 2015). Although variability between asthmatic cell cultures from different donors exists, the effects of AG1478 followed a consistent trend in reducing goblet cell numbers and mucus secretion (Parker et al. 2015). Results from this study suggest that for these asthmatic pediatric patients, treatment with AG1478 or other EGF/EGFR inhibitors may be a potential therapeutic strategy to target abnormal airway remodeling and promotion of patient-specific disease control.

Microfluidic lung models have already been used to examine anticancer drug sensitivity for individualized treatment of lung cancer (Xu et al. 2013). A drug sensitivity platform was developed using microfluidic chips with gradient concentration generators. This platform consisted of four microfluidic chips connected at the center by a common reservoir. Each microfluidic chip contained a concentration gradient generator with an upstream drug inlet and a downstream medium inlet and three parallel cell culture chambers (Xu et al. 2013). Cell culture chambers were populated with the human NSCLC cell line SPCA-1 and the human lung fibroblast cell line HFL1. Primary lung cells isolated from lung cancer tissue obtained from eight patients undergoing surgical resection were also used to populate separate chambers (Xu et al. 2013). Patient cell responses were compared to the responses of the cancer cell lines following drug treatment. To create a 3D cell culture, cells were resuspended in a soluble thermosensitive basement membrane extract (BME) that formed a gel at 37°C. Cell–BME mixtures were added to individual culture chambers. Morphological features and drug sensitivities were compared (Xu et al. 2013). Drug sensitivities to the anticancer drugs gefitinib, paclitaxel, and gemcitabime were examined. Results showed that all three drugs were able to inhibit primary lung cancer cell growth in a dose-dependent way (Xu et al. 2013). The apoptosis rate for primary lung cancer cells was highest when cultures were treated with a combination of all three drugs. These results indicate that for these patients, a combination of chemotherapeutic drugs might be the best approach for treatment. Microfluidic platforms may also be used to mimic tumor growth and test chemosensitivity by isolating primary lung cancer cells from patients and examining spheroid formation after treatment with a chemotherapeutic drug (Ruppen et al. 2015).

7.10 CONCLUSION

Lung injury due to trauma, microbial pathogenesis, or adverse drug reactions in the lung generally involve the pulmonary parenchyma, the pleura, the airways, and the pulmonary endothelium but can also involve many other cell types. Direct action of toxins or drugs can cause cell death, induce apoptosis, and modulate biomolecule production by lung cells. Immune cells located in lung tissues or migrating into lung tissues can become activated, and immune cell products can influence viability, biomolecule production, and replication of fibroblasts. Good *in vitro* human models and formation of chip-based microfluidic systems of the lung will help to answer many of the questions we have regarding lung pathology and toxicology. At this time there have been only a few lung-on-a chip microfluidic supported trachea-bronchial or lung models produced. The development of specialized pumping systems and microfluidic supported chambers have the potential to revolutionize the way these models can be managed and may provide for production of more robust and longer-lived cultures. Although most of the early models to evaluate trachea-bronchial or lung physiology, pathophysiology, or infection were not designed as microchip systems, these early model designs could be easily transferred to fluidic-supported micro-, milli- or human organ culture systems in the future.

REFERENCES

Ali, Nadire N., Alasdair J. Edgar, Ali Samadikuchaksaraei, Catherine M. Timson, Hanna M. Romanska, Julia M. Polak, and Anne E. Bishop. Derivation of type II alveolar epithelial cells from murine embryonic stem cells. *Tissue Engineering* 8, 541, 2002.

Amishima, Masaru, Mitsuru Munakata, Yasuyuki Nasuhara, Atsuko Sato, Toru Takahashi, Yukihiko Homma, and Yoshikazu Kawakami. Expression of epidermal growth factor and epidermal growth factor receptor immunoreactivity in the asthmatic human airway. *American Journal of Respiratory and Critical Care Medicine* 157 no. 6 (1998): 1907–1912.

Balestrini, Jenna L., and Laura E. Niklason. Extracellular matrix as a driver for lung regeneration. *Annals of Biomedical Engineering* 43 no. 3 (2015): 568–576.

Birkness, Kristin A., Jeannette Guarner, Suraj B. Sable, Ralph A. Tripp, Kathryn L. Kellar, Jeanine Bartlett, and Frederick D. Quinn. An in vitro model of the leukocyte interactions associated with granuloma formation in Mycobacterium tuberculosis infection. *Immunology and Cell Biology* 85 no. 2 (2007): 160–168.

Booth, Adam J., Ryan Hadley, Ashley M. Cornett, Alyssa A. Dreffs, Stephanie A. Matthes, Jessica L. Tsui, Kevin Weiss, et al. Acellular normal and fibrotic human lung matrices as a culture system for in vitro investigation. *American Journal of Respiratory and Critical Care Medicine* 186 no. 9 (2012): 866–876.

Brahim, Sean, Dyer Narinesingh, and Anthony Guiseppi-Elie. Bio-smart hydrogels: Co-joined molecular recognition and signal transduction in biosensor fabrication and drug delivery. *Biosensors and Bioelectronics* 17 no. 11 (2002): 973–981.

Burgel, Pierre-Régis, and Jay A. Nadel. Roles of epidermal growth factor receptor activation in epithelial cell repair and mucin production in airway epithelium. *Thorax* 59 no. 11 (2004): 992–996.

Cakebread, Julie A., Hans Michael Haitchi, Yunhe Xu, Stephen T. Holgate, Graham Roberts, and Donna E. Davies. Rhinovirus-16 induced release of IP-10 and IL-8 is augmented by Th2 cytokines in a pediatric bronchial epithelial cell model. *PloS one* 9 (2014): e94010.

Cardona, Pere Joan. A dynamic reinfection hypothesis of latent tuberculosis infection. *Infection* 37 no. 2 (2009): 80–86.

Chakir, Jamila, Nathalie Pagé, Qutayba Hamid, Michel Laviolette, Louis Philippe Boulet, and Mahmoud Rouabhia. Bronchial mucosa produced by tissue engineering: a new tool to study cellular interactions in asthma. *Journal of Allergy and Clinical Immunology* 107 no. 1 (2001): 36–40.

Chen, Yulong, Masahiko Terajima, Yanan Yang, Li Sun, Young-Ho Ahn, Daniela Pankova, Daniel S. Puperi et al. Lysyl hydroxylase 2 induces a collagen cross-link switch in tumor stroma. *The Journal of Clinical Investigation* 125 no. 3 (2015): 1147–1162.

Choe, Melanie M., Alice A. Tomei, and Melody A. Swartz. Physiological 3D tissue model of the airway wall and mucosa. *Nature Protocols-Electronic Edition* 1 no. 1 (2006): 357–362.

Coraux, Christelle, Béatrice Nawrocki-Raby, Jocelyne Hinnrasky, Claire Kileztky, Dominique Gaillard, Christian Dani, and Edith Puchelle. Embryonic stem cells generate airway epithelial tissue. *American Journal of Respiratory Cell and Molecular Biology* 32 no. 2 (2005): 87–92.

Corbière, Véronique, Violette Dirix, Sarah Norrenberg, Mattéo Cappello, Myriam Remmelink, and Françoise Mascart. Phenotypic characteristics of human type II alveolar epithelial cells suitable for antigen presentation to T lymphocytes. *Respiratory Research* 12 (2011): 15.

Cortiella, Joaquin, Jean Niles, Andrea Cantu, Andrea Brettler, Anthony Pham, Gracie Vargas, Sean Winston, Jennifer Wang, Shannon Walls, and Joan E. Nichols. Influence of acellular natural lung matrix on murine embryonic stem cell differentiation and tissue formation. *Tissue Engineering Part A* 16 no. 8 (2010): 2565–2580.

Cozens, Alison L., Michael J. Yezzi, Karl Kunzelmann, Takashi Ohrui, Lynda Chin, Kai E. Eng, Walter E. Finkbeiner, Jonathan H. Widdicombe, and Dieter C. Gruenert. CFTR expression and chloride secretion in polarized immortal human bronchial epithelial cells. *American Journal of Respiratory Cell and Molecular Biology* 10 no. 1 (1994): 38–47.

Davis, A. Sally, Daniel S. Chertow, Jenna E. Moyer, Jon Suzich, Aline Sandouk, David W. Dorward, Carolea Logun, James H. Shelhamer, and Jeffery K. Taubenberger. Validation of normal human bronchial epithelial cells as a model for influenza A infections in human distal trachea. *Journal of Histochemistry & Cytochemistry* 63 no. 5 (2015): 312–328.

El-Sherbiny IM, and Yacoub MH. Hydrogel scaffolds for tissue engineering: Progress and challenges. *Global Cardiology Science and Practice* 1 no. 3 (2013): 316–342.

El-Nikhely N, Larzabal L, Seeger W, Calvo A, and Savai R. Tumor-stromal interactions in lung cancer: novel candidate targets for therapeutic intervention. *Expert Opin Investig Drugs*. (2012): 21(8): 1107–1122.

Firth, Amy L., Carl T. Dargitz, Susan J. Qualls, Tushar Menon, Rebecca Wright, Oded Singer, Fred H. Gage, Ajai Khanna, and Inder M. Verma. Generation of multiciliated cells in functional airway epithelia from human induced pluripotent stem cells. *Proceedings of the National Academy of Sciences* 111 no. 17 (2014): E1723–E1730.

Foster, Kimberly A., Christine G. Oster, Mary M. Mayer, Michael L. Avery, and Kenneth L. Audus. Characterization of the A549 cell line as a type II pulmonary epithelial cell model for drug metabolism. *Experimental Cell Research* 243 no. 2 (1998): 359–366.

Fu, Yong, and Amy S. Lee. Glucose regulated proteins in cancer progression, drug resistance and immunotherapy. *Cancer Biology & Therapy* 5 no. 7 (2006): 741–744.

Garcia, Gilles, Camille Taillé, Pierantonio Laveneziana, Arnaud Bourdin, Pascal Chanez, and Marc Humbert. Anti-interleukin-5 therapy in severe asthma. *European Respiratory Review* 22 no. 129 (2013): 251–257.

Geckil, Hikmet, Feng Xu, Xiaohui Zhang, SangJun Moon, and Utkan Demirci. Engineering hydrogels as extracellular matrix mimics. *Nanomedicine* 5 no. 3 (2010): 469–484.

Ghaedi, Mahboobe, Elizabeth A. Calle, Julio J. Mendez, Ashley L. Gard, Jenna Balestrini, Adam Booth, Peter F. Bove, Liqiong Gui, Eric S. White, and Laura E. Niklason. Human iPS cell–derived alveolar epithelium repopulates lung extracellular matrix. *The Journal of Clinical Investigation* 123 no. 11 (2013): 4950–4962.

Giard, Donald J., Stuart A. Aaronson, George J. Todaro, Paul Arnstein, John H. Kersey, Harvey Dosik, and Wade P. Parks. In vitro cultivation of human tumors: Establishment of cell lines derived from a series of solid tumors. *Journal of the National Cancer Institute* 51 no. 5 (1973):1417–1423.

Goetz, Jacky G., Susana Minguet, Inmaculada Navarro-Lérida, Juan José Lazcano, Rafael Samaniego, Enrique Calvo, Marta Tello, et al. Del Pozo. Biomechanical remodeling of the microenvironment by stromal caveolin-1 favors tumor invasion and metastasis. *Cell* 146 no. 1 (2011): 148–163.

Guirado, Evelyn, Uchenna Mbawuike, Tracy L. Keiser, Jesus Arcos, Abul K. Azad, Shu-Hua Wang, and Larry S. Schlesinger. Characterization of host and microbial determinants in individuals with latent tuberculosis infection using a human granuloma model. *mBio* 6 no. 1 (2015): e02537–e025314.

Hay, Michael, David W. Thomas, John L. Craighead, Celia Economides, and Jesse Rosenthal. Clinical development success rates for investigational drugs. *Nature Biotechnology* 32 no. 1 (2014): 40–51.

Hermanns, Maria Iris, Ronald E. Unger, Kai Kehe, Kirsten Peters, and Charles James Kirkpatrick. Lung epithelial cell lines in coculture with human pulmonary microvascular endothelial cells: Development of an alveolo-capillary barrier in vitro. *Laboratory Investigation* 84 no. 6 (2004): 736–752.

Hill, Ryan C., Elizabeth A. Calle, Monika Dzieciatkowska, Laura E. Niklason, and Kirk C. Hansen. Quantification of extracellular matrix proteins from a rat lung scaffold to provide a molecular readout for tissue engineering. *Molecular & Cellular Proteomics* 14 no. 4 (2015): 961–973.

Hoang Nguyen, Anh Thu, Puran Chen, Julius Juarez, Patty Sachamitr, Bo Billing, Lidija Bosnjak, Barbro Dahlén, Mark Coles, and Mattias Svensson. Dendritic cell functional properties in a three-dimensional tissue model of human lung mucosa. *American Journal of Physiology-Lung Cellular and Molecular Physiology* 302 no. 2 (2012): L226–L237.

Holgate, Stephen. T., Peter M. Lackie, Donna E. Davies, William R. Roche, and Andrew F. Walls. The bronchial epithelium as a key regulator of airway inflammation and remodelling in asthma. *Clinical & Experimental Allergy* 29, no. s2 (1999): 90–95.

Huh, Dongeun, Benjamin D. Matthews, Akiko Mammoto, Martín Montoya-Zavala, Hong Yuan Hsin, and Donald E. Ingber. Reconstituting organ-level lung functions on a chip. *Science* 328 no. 5986 (2010): 1662–1668.

Huh, Dongeun, Daniel C. Leslie, Benjamin D. Matthews, Jacob P. Fraser, Samuel Jurek, Geraldine A. Hamilton, Kevin S. Thorneloe, Michael Allen McAlexander, and Donald E. Ingber. A human disease model of drug toxicity–induced pulmonary edema in a lung-on-a-chip microdevice. *Science Translational Medicine* 4 no. 159 (2012): 159ra147–159ra147.

Kim, Kevin K., Matthias C. Kugler, Paul J. Wolters, Liliane Robillard, Michael G. Galvez, Alexis N. Brumwell, Dean Sheppard, and Harold A. Chapman. Alveolar epithelial cell mesenchymal transition develops in vivo during pulmonary fibrosis and is regulated by the extracellular matrix. *Proceedings of the National Academy of Sciences* 103 no. 35 (2006): 13180–13185.

Kloxin, April M., Katherine J.R. Lewis, Cole A. DeForest, Gregory Seedorf, Mark W. Tibbitt, Vivek Balasubramaniam, and Kristi S. Anseth. Responsive culture platform to examine the influence of microenvironmental geometry on cell function in 3D. *Integrative Biology* 4 no. 12 (2012): 1540–1549.

Laan, Martti, Olof Prause, Masahide Miyamoto, Margareta Sjöstrand, Ann-Marie Hytönen, Tatsuhiko Kaneko, Jan Lötvall, and Anders Lindén. A role of GM-CSF in the accumulation of neutrophils in the airways caused by IL-17 and TNF-α. *European Respiratory Journal* 23 no. 3 (2003): 387–393.

Lee, Amy S. Glucose-regulated proteins in cancer: molecular mechanisms and therapeutic potential. *Nature Reviews Cancer* 14 no. 4 (2014): 263–276.

Levental, Kandice R., Hongmei Yu, Laura Kass, Johnathon N. Lakins, Mikala Egeblad, Janine T. Erler, Sheri F.T. Fong et al. Matrix crosslinking forces tumor progression by enhancing integrin signaling. *Cell* 139 no. 5 (2009): 891–906.

Luque, Tomás, Esther Melo, Elena Garreta, Joaquin Cortiella, Joan E. Nichols, Ramon Farré, and Daniel Navajas. Local micromechanical properties of decellularized lung scaffolds measured with atomic force microscopy. *Acta Biomaterialia* 9 no. 6 (2013): 6852–6859.

Mas, Christophe, Bernadett Boda, Mireille CaulFuty, Song Huang, Ludovic Wiszniewski, and Samuel Constant. Antitumour efficacy of the selumetinib and trametinib MEK inhibitors in a combined human airway–tumour–stroma lung cancer model. *Journal of Biotechnology* 205 (2015): 111–119.

Matrosovich, Mikhail N., Tatyana Y. Matrosovich, Thomas Gray, Noel A. Roberts, and Hans-Dieter Klenk. Human and avian influenza viruses target different cell types in cultures of human airway epithelium. *Proceedings of the National Academy of Sciences of the United States of America* 101 no. 13 (2004): 4620–4624.

Melo, Esther, Jennifer Y. Kasper, Ronald E. Unger, Ramon Farré, and Charles James Kirkpatrick. Development of a bronchial wall model: Triple culture on a decellularized porcine trachea. *Tissue Engineering Part C: Methods* 21 no. 9 (2015): 909–921.

Mishra, Dhruva K., Chad J. Creighton, Yiqun Zhang, Don L. Gibbons, Jonathan M. Kurie, and Min P. Kim. Gene expression profile of A549 cells from tissue of 4D model predicts poor prognosis in lung cancer patients. *International Journal of Cancer* 134 no. 4 (2014): 789–798.

Mishra, Dhruva K., Michael J. Thrall, Brandi N. Baird, Harald C. Ott, Shanda H. Blackmon, Jonathan M. Kurie, and Min P. Kim. Human lung cancer cells grown on acellular rat lung matrix create perfusable tumor nodules. *The Annals of Thoracic Surgery* 93 no. 4 (2012): 1075–1081.

Mishra, Dhruva K., Steven D. Compean, Michael J. Thrall, Xin Liu, Erminia Massarelli, Jonathan M. Kurie, and Min P. Kim. Human Lung Fibroblasts Inhibit Non-Small Cell Lung Cancer Metastasis in Ex Vivo 4D Model. *The Annals of Thoracic Surgery* (2015): 100 no. 4 (2015): 1167–1174; discussion 1174.

Mou, Hongmei, Rui Zhao, Richard Sherwood, Tim Ahfeldt, Allen Lapey, John Wain, Leonard Sicilian, et al. Generation of multipotent lung and airway progenitors from mouse ESCs and patient-specific cystic fibrosis iPSCs. *Cell Stem Cell* 10 no. 4 (2012): 385–397.

Nalayanda, Divya D., Qihong Wang, William B. Fulton, Tza-Huei Wang, and Fizan Abdullah. Engineering an artificial alveolar-capillary membrane: A novel continuously perfused model within microchannels. *Journal of Pediatric Surgery* 45 no. 1 (2010): 45–51.

Nalayanda, Divya Devaiah, William Benjamin Fulton, Tza-Huei Wang, and Fizan Abdullah. A multiphase fluidic platform for studying ventilator-induced injury of the pulmonary epithelial barrier. *Integrative Biology* 5 no. 9 (2013): 1141–1148.

Nichols, Joan E., and Joaquin Cortiella. Engineering of a complex organ: progress toward development of a tissue-engineered lung. *Proceedings of the American Thoracic Society* 5, no. 6 (2008): 723–730.

Nichols, Joan E., Jean Niles, Michael Riddle, Gracie Vargas, Tuya Schilagard, Liang Ma, Kert Edward, et al. Production and assessment of decellularized pig and human lung scaffolds. *Tissue Engineering Part A* 19 no. 17–18 (2013): 2045–2062.

Nichols, Joan E., Jean A. Niles, Stephanie P. Vega, Lissenya B. Argueta, Adriene Eastaway, and Joaquin Cortiella. Modeling the lung: Design and development of tissue engineered macro-and micro-physiologic lung models for research use. *Experimental Biology and Medicine* 239 no. 9 (2014): 1135–1169. doi: 10.1177/1535370214536679.

Parasa, Venkata Ramanarao, Muhammad Jubayer Rahman, Anh Thu Ngyuen Hoang, Mattias Svensson, Susanna Brighenti, and Maria Lerm. Modeling Mycobacterium tuberculosis early granuloma formation in experimental human lung tissue. *Disease Models & Mechanisms* 7 no. 2 (2014): 281–288.

Parker, Jeremy C., Isobel Douglas, Jennifer Bell, David Comer, Keith Bailie, Grzegorz Skibinski, Liam G. Heaney, and Michael D. Shields. Epidermal Growth Factor Removal or Tyrphostin AG1478 Treatment Reduces Goblet Cells & Mucus Secretion of Epithelial Cells from Asthmatic Children Using the Air-Liquid Interface Model. *PLoS One* 10 no. 6 (2015): e0129546.

Parker, Jeremy, Severine Sarlang, Surendran Thavagnanam, Grace Williamson, Dara O'Donoghue, Remi Villenave, Ultan Power, Michael Shields, Liam Heaney, and Grzegorz Skibinski. A 3-D well-differentiated model of pediatric bronchial epithelium demonstrates unstimulated morphological differences between asthmatic and nonasthmatic cells. *Pediatric Research* 67 no. 1 (2010): 17–22.

Parker, Matthew W., Daniel Rossi, Mark Peterson, Karen Smith, Kristina Sikström, Eric S. White, John E. Connett, Craig A. Henke, Ola Larsson, and Peter B. Bitterman. Fibrotic extracellular matrix activates a profibrotic positive feedback loop. *The Journal of Clinical Investigation* 124 no. 4 (2014): 1622–1635.

Perrais, Michaël, Pascal Pigny, Marie-Christine Copin, Jean-Pierre Aubert, and Isabelle Van Seuningen. Induction of MUC2 and MUC5AC mucins by factors of the Epidermal Growth Factor (EGF) family is mediated by EGF receptor/Ras/Raf/extracellular signal-regulated kinase cascade and Sp1*. *Journal of Biological Chemistry* 277 no. 35 (2002): 32258–32267.

PDMS: A Review, Introduction to PDMS (n.d.), accessed September 3, 2015. http://www.elveflow.com/microfluidic-tutorials/microfluidic-reviews-and-tutorials/the-poly-di-methyl-siloxane-pdms-and-microfluidics

Puddicombe, Sarah M., Ricardo Polosa, Audrey Richter, Muthu T. Krishna, Peter H. Howarth, Stephen T. Holgate, and Donna E. Davies. Involvement of the epidermal growth factor receptor in epithelial repair in asthma. *The FASEB Journal* 14 no. 10 (2000): 1362–1374.

Punde, Tushar H., Wen-Hao Wu, Pei-Chun Lien, Ya-Ling Chang, Ping-Hsueh Kuo, Margaret Dah-Tsyr Chang, Kang-Yun Lee, et al. A biologically inspired lung-on-a-chip device for the study of protein-induced lung inflammation. *Integrative Biology* 7 no. 2 (2015): 162–169.

Qi, Hao, Yanan Du, Lianyong Wang, Hirokazu Kaji, Hojae Bae, and Ali Khademhosseini. Patterned differentiation of individual embryoid bodies in spatially organized 3D hybrid microgels. *Advanced Materials* 22 no. 46 (2010): 5276–5281.

Ramadan, Qasem, and Martin A.M. Gijs. In vitro micro-physiological models for translational immunology. *Lab on a Chip* 15 no. 3 (2015): 614–636.

Rippon, Helen J., Julia M. Polak, Mingde Qin, and Anne E. Bishop. Derivation of distal lung epithelial progenitors from murine embryonic stem cells using a novel three-step differentiation protocol. *Stem Cells* 24 no. 5 (2006): 1389–1398.

Roszell, Blair, Mark J. Mondrinos, Ariel Seaton, Donald M. Simons, Sirma H. Koutzaki, Guo-Hua Fong, Peter I. Lelkes, and Christine M. Finck. Efficient derivation of alveolar type II cells from embryonic stem cells for in vivo application. *Tissue Engineering Part A* 15 no. 11 (2009): 3351–3365.

Ruppen, Janine, Franziska D. Wildhaber, Christoph Strub, Sean R.R. Hall, Ralph A. Schmid, Thomas Geiser, and Olivier T. Guenat. Towards personalized medicine: Chemosensitivity assays of patient lung cancer cell spheroids in a perfused microfluidic platform. *Lab on a Chip* 15 no. 14 (2015): 3076–3085.

Samadikuchaksaraei, Ali, Shahar Cohen, Kevin Isaac, Helen J. Rippon, Julia M. Polak, Robert C. Bielby, and Anne E. Bishop. Derivation of distal airway epithelium from human embryonic stem cells. *Tissue Engineering* 12 no. 4 (2006): 867–875.

Sanderson, Michael J. Exploring lung physiology in health and disease with lung slices. *Pulmonary Pharmacology & Therapeutics* 24 no. 5 (2011): 452–465.

Shaler, Christopher R., Carly N. Horvath, Mangalakumari Jeyanathan, and Zhou Xing. Within the Enemy's Camp: Contribution of the granuloma to the dissemination, persistence and transmission of Mycobacterium tuberculosis. *Frontiers in Immunology* 4 (2013): 30. doi: 10.3389/fimmu.2013.00030.

Shamis, Yulia, Eilat Hasson, Avigail Soroker, Elad Bassat, Yael Shimoni, Tamar Ziv, Ronit Vogt Sionov, and Eduardo Mitrani. Organ-specific scaffolds for in vitro expansion, differentiation, and organization of primary lung cells. *Tissue Engineering Part C: Methods* 17 no. 8 (2011): 861–870.

Siti-Ismail, Norhayati, Ali Samadikuchaksaraei, Anne E. Bishop, Julia M. Polak, and Athanasios Mantalaris. Development of a novel three-dimensional, automatable and integrated bioprocess for the differentiation of embryonic stem cells into pulmonary alveolar cells in a rotating vessel bioreactor system. *Tissue Engineering Part C: Methods* 18 no. 4 (2012): 263–272.

Siyan, Wang, Yue Feng, Zhang Lichuan, Wang Jiarui, Wang Yingyan, Jiang Li, Lin Bingcheng, and Wang Qi. Application of microfluidic gradient chip in the analysis of lung cancer chemotherapy resistance. *Journal of Pharmaceutical and Biomedical Analysis* 49 no. 3 (2009): 806–810.

Smith, Peter F., A. Jay Gandolfi, Carlos L. Krumdieck, Charles W. Putnam, Charles F. Zukoski, William M. Davis, and Klaus Brendel. Dynamic organ culture of precision liver slices for in vitro toxicology. *Life Sciences* 36 no. 14 (1985): 1367–1375.

Stratmann, Anna T., David Fecher, Gaby Wangorsch, Claudia Göttlich, Thorsten Walles, Heike Walles, Thomas Dandekar, Gudrun Dandekar, and Sarah L. Nietzer. Establishment of a human 3D lung cancer model based on a biological tissue matrix combined with a Boolean in silico model. *Molecular Oncology* 8 no. 2 (2014): 351–365.

Tilghman, Robert W., Catharine R. Cowan, Justin D. Mih, Yulia Koryakina, Daniel Gioeli, Jill K. Slack-Davis, Brett R. Blackman, Daniel J. Tschumperlin, and J. Thomas Parsons. Matrix rigidity regulates cancer cell growth and cellular phenotype. *PLoS One* 5 no. 9 (2010): e12905.

Van Vranken, Benjamin E., Helen J. Rippon, Ali Samadikuchaksaraei, Alan O. Trounson, and Anne E. Bishop. The differentiation of distal lung epithelium from embryonic stem cells. *Current Protocols in Stem Cell Biology* Chapter 1 (2007): 1G-1.

Villenave, Rémi, Surendran Thavagnanam, Severine Sarlang, Jeremy Parker, Isobel Douglas, Grzegorz Skibinski, Liam G. Heaney, et al. In vitro modeling of respiratory syncytial virus infection of pediatric bronchial epithelium, the primary target of infection in vivo. *Proceedings of the National Academy of Sciences* 109 no. 13 (2012): 5040–5045.

Wagner, Darcy E., Nicholas R. Bonenfant, Dino Sokocevic, Michael J. DeSarno, Zachary D. Borg, Charles S. Parsons, Elice M. Brooks, et al. Three-dimensional scaffolds of acellular human and porcine lungs for high throughput studies of lung disease and regeneration. *Biomaterials* 35 no. 9 (2014): 2664–2679.

Wikswo, John P., Erica L. Curtis, Zachary E. Eagleton, Brian C. Evans, Ayeeshik Kole, Lucas H. Hofmeister, and William J. Matloff. Scaling and systems biology for integrating multiple organs-on-a-chip. *Lab on a Chip* 13 no. 18 (2013): 3496–3511.

Wong, Amy P., Christine E. Bear, Stephanie Chin, Peter Pasceri, Tadeo O. Thompson, Ling-Jun Huan, Felix Ratjen, James Ellis, and Janet Rossant. Directed differentiation of human pluripotent stem cells into mature airway epithelia expressing functional CFTR protein. *Nature Biotechnology* 30 no. 9 (2012): 876–882.

Wong, Amy P., Stephanie Chin, Sunny Xia, Jodi Garner, Christine E. Bear, and Janet Rossant. Efficient generation of functional CFTR-expressing airway epithelial cells from human pluripotent stem cells. *Nature Protocols* 10 no. 3 (2015): 363–381.

Xu, Zhiyun, Yanghui Gao, Yuanyuan Hao, Encheng Li, Yan Wang, Jianing Zhang, Wenxin Wang, Zhancheng Gao, and Qi Wang. Application of a microfluidic chip-based 3D co-culture to test drug sensitivity for individualized treatment of lung cancer. *Biomaterials* 34 no. 16 (2013): 4109–4117.

Zervantonakis, Ioannis K., Shannon K. Hughes-Alford, Joseph L. Charest, John S. Condeelis, Frank B. Gertler, and Roger D. Kamm. Three-dimensional microfluidic model for tumor cell intravasation and endothelial barrier function. *Proceedings of the National Academy of Sciences* 109 no. 34 (2012): 13515–13520.

Zhang JT, Petersen S, Thunga M, Leipold E, Weidisch R, Liu X, Fahr A, Jandt KD. Microstructured smart hydrogels with enhanced protein loading and release efficiency. *Acta Biomateriala* Apr 6 no. 4 (2010): 1297–1306.

Zhu, Junmin, and Roger E. Marchant. Design properties of hydrogel tissue-engineering scaffolds. *Expert review of medical devices* (2011): 8, no. 5 (2011): 607–626.

8 Liver and Liver Cancer-on-a-Chip

Aleksander Skardal

CONTENTS

8.1 INTRODUCTION: LIVER MODELS

There is a critical need for better bioengineered tissue models to predict efficacy, pharmacokinetics, and potential toxicity for candidate drugs. *In vivo* animal models have long served as the gold standard for testing prior to clinical trials, but the drawbacks associated with animal models are major contributors to the exorbitant costs and uncertainties in bringing a candidate drug from bench to bedside. *In vitro* systems comprised of actual human-derived cells are preferable from a predictive point of view (Greenhough, Medine et al. 2010). However, for these systems to accurately reflect human physiology, cells must retain their *in vivo* functions and remain viable for extended periods of time in *in vitro* settings. These requirements are key for future use in pharmacokinetic and toxicity testing. This is particularly true in the case of liver models, as the liver is commonly the first tissue to be critically assessed for toxic effects during drug and toxicology screening. *In vitro* cultured primary hepatocytes are increasingly being used for screening in the pharmaceutical industry (Gomez-Lechon, Castell et al. 2010). However, there is still a need for an optimal culture system that improves the long-term maintenance of liver cells with retention of liver function for *in vitro* drug screening.

In addition to normal tissue models for screening applications, the same advances in tissue engineering that support fabrication of tissue constructs such as liver can be employed to fabricate models of cancer. Cancer in the liver is one of the leading causes of cancer-based deaths around the world. Liver cancer can take the form of hepatocarcinoma, where the primary tumor begins in the liver, or metastatic disease, in which the primary tumor begins elsewhere, such as the colon, and metastasizes to the liver through the circulatory system. Unfortunately, tumor growth in the liver, an important functional organ in the body, often causes decreased function, eventually resulting in patient death, thus necessitating development of more effective treatments. However, to develop such treatments, better test platforms for conducting research are needed.

In recent years, advances in biotechnology areas such as tissue engineering, biomaterials, and micro- and biofabrication have allowed derivation of new biological systems with massive potential as test platforms. Researchers have developed a wide variety of human-derived *in vitro* models that can be used as specific normal tissues for testing drugs, toxins, and drug candidates (Prestwich, Liu et al. 2007; Prestwich 2008; Skardal, Sarker et al. 2010). Furthermore, through the use of genetics as well as external environmental manipulations, these systems can be employed as specific disease models (Nickerson et al. 2004; Nickerson, Richter et al. 2007; Barrila, Radtke et al. 2010; Benam, Dauth et al. 2015). Further integration with microfabrication and microfluidic technology has resulted in dynamic systems that support multi-tissue interactions, high-throughput testing, and environmental sampling and biosensing, thereby yielding powerful and versatile organ(s)-on-a-chip platforms for applications such as drug discovery (Polini, Prodanov et al. 2014). These organs-on-chips purport to significantly impact the future of medicine. In this chapter, we describe the potential of and current state of liver- and liver cancer-on-a-chip systems.

8.2 MOVING FROM TRADITIONAL CELL CULTURES TO ENGINEERED TISSUE CONSTRUCTS: CHALLENGES AND CRITERIA

Development of new and effective anticancer drugs, and cancer research as a whole, has been limited due to the inability to accurately model tumor progression and signaling mechanisms in a controlled environment. Animal models allow only limited manipulation and study of these mechanisms, and are not necessarily predictive of results in humans. Traditional *in vitro* 2D cultures fail to recapitulate the 3D microenvironment of *in vivo* tissues (Kunz-Schughart, Freyer et al. 2004). Drug diffusion kinetics vary dramatically, drug doses effective in 2D are often ineffective when scaled to patients, and cell-cell/cell-matrix interactions are inaccurate (Ho, Pham et al. 2010; Drewitz, Helbling et al. 2011). Tissue culture dishes have three major differences from the tissue where the tumor was isolated: surface topography, surface stiffness, and most importantly, a 2D rather than 3D architecture. As a consequence, plastic 2D culture places a selective pressure on cells that could substantially alter their original molecular and phenotypic properties. The resulting functional differences between 2D cultures and 3D constructs have been shown repeatedly in liver and many other types of cancer. In fact, we recently demonstrated that on 2D tissue culture dishes, metastatic colon carcinoma cells appeared epithelial, but when transitioned into a 3D liver organoid host environment they "switched" to a mesenchymal and metastatic phenotype (Skardal, Devarasetty et al. 2015b). Bioengineered tissue platforms have evolved that can better mimic the structure and cellular heterogeneity of *in vivo* tissue, and are suitable for mimicking human physiology. Subsequently, these relatively new technologies are vastly superior to their predecessors for drug and toxicology testing, personalized medicine, and tumor bioengineering research. These model organs can be viable for longer periods of time and are cultured to develop functional properties similar to native tissues. This approach has the potential to recapitulate the dynamic role of cell-cell, cell-ECM (extracellular matrix), and mechanical interactions inside the tumor. Further incorporation of cells representative of the tumor stroma, such as endothelial cells and tumor fibroblasts, and physical matrix components, can mimic the *in vivo* tumor microenvironment. Thus, bioengineered tumors are an important resource for *in vitro* study of cancer in 3D, including tumor biomechanics and the effects of anticancer drugs on 3D tumor tissue.

Fortunately, the general concept of performing experiments in 3D versus 2D has gained significant traction over the last 10 or so years. However, there remain hurdles and challenges to overcome. 2D cell culture will almost certainly remain a widely used tool for many years to come. It is far too easy and inexpensive for a complete shift from 2D to 3D. The same reasons that make 2D culture attractive are what have slowed adaptation to 3D systems. Using 3D systems is more complicated. It requires understanding how to harness and implement the innate characteristics of new biomaterials or technologies. Following successful establishment of 3D cultures, cell harvesting and cell passaging, trivial steps in normal 2D cell culture, are significantly more complex, and sometimes not possible without potentially harming the cells. For example, if cells are cultured in a 3D hydrogel matrix, one must effectively dissolve the matrix away to remove the cells. Some hydrogel systems support this (Zhang, Skardal et al. 2008), but most do not. Furthermore, traditional imaging techniques are geared toward 2D cell cultures in which all of the cells of interest are confined to a single focal plane. High quality 3D tissue construct imaging requires confocal or macroconfocal imaging, expensive tools that not all laboratories have access to. Cost of materials is another challenge. More advanced biomaterial systems that have been engineered to be user friendly are more expensive than tissue culture plastic. When it comes to on-a-chip technologies, unless the devices in question are available commercially, researchers must fabricate these systems. Fabrication techniques, such as soft lithography, micromolding, and machining, require additional skillsets and equipment.

However, in the end, when data are critically assessed and published in scientific journals, there is a common occurrence: outcomes derived in 3D systems or dynamic on-a-chip platforms often vastly surpass those in static 2D environments. Very rarely do 2D systems provide results that are better physiological mimics of the human body. As discussed in this chapter, this is particularly true in the context of liver and tumor engineering.

8.3 3D *IN VITRO* LIVER MODELS

In vitro liver models have been employed extensively in the realm of drug testing by researchers in academia and within the pharmaceutical industry. Traditionally, since primary hepatocytes were difficult to maintain in culture until relatively recently, liver-derived cell lines such as HepG2 cells were often employed. Unfortunately, HepG2 cells are derived from hepatomas, and as such, while being robust and easily cultured, they do not retain all of the functionality of primary human hepatocytes.

HepG2 cells only express a subset of cytochrome p450 isoforms, limiting their use in drug metabolism studies. Furthermore, as a robust cell type, they are far less sensitive to environmental stimuli, limiting their use in realistic toxicology studies. Despite these shortcomings, they remain a useful model cell type, and remain a common choice for proof-of-concept work, particularly in the development of new 3D liver systems. However, primary human hepatocytes have become the industry standard for most liver-based screening studies. Yet, only several years ago, researchers had a difficult time maintaining hepatocytes that were viable and functional within *in vitro* settings, thereby limiting their use in long-term studies.

Fortunately, a variety of 3D cell culture and organoid fabrication strategies have arisen—some of which were developed using cell lines—that have enabled formation of and maintenance of relatively high viability and high functioning hepatocyte-based tissue constructs. For example, hanging drop and rotating wallvessel (RWV) bioreactor cultures (Figure 8.1a and b) have successfully supported formation of hepatocyte spheroids (Chang and Hughes-Fulford 2014). Hanging drop systems and the resulting spheroids are now used widely, and are commercially available. These spheroids have quite thoroughly been demonstrated to have superior lifetimes and metabolic functionality to traditional cultures (Messner, Agarkova et al. 2013; Kim, Fluri et al. 2015). Likewise, RWV-generated hepatocyte spheroids are superior to 2D systems in terms of gene expression, function, and cell-cell morphology (Chang and Hughes-Fulford 2014).

A variety of other approaches have been implemented that employ biomaterials such as hydrogels to encapsulate and support hepatocytes in culture (Figure 8.1c). In particular, materials derived from or containing decellularized liver tissue have

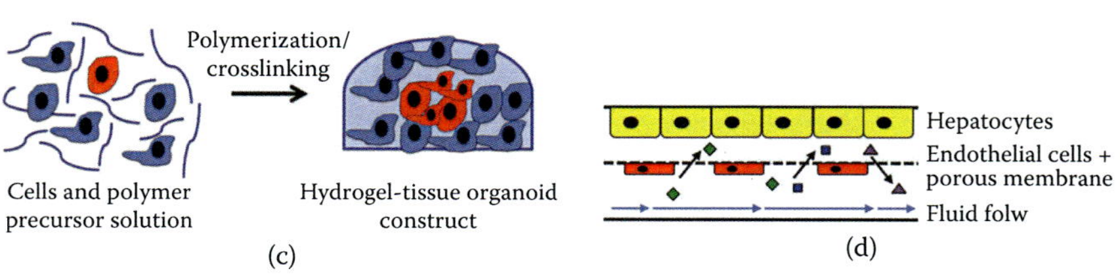

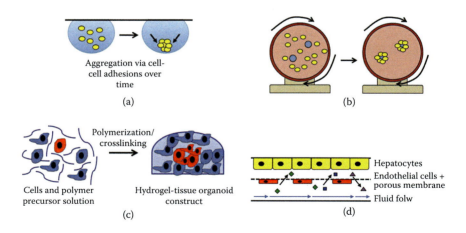

FIGURE 8.1 Approaches for fabricating liver-derived organoids and constructs. (a) Cells aggregate into spheroids in hanging drop cultures. (b) Cells aggregate around cell-adherent microcarrier beads in rotating wall vessel (RWV) bioreactors. (c) Polymer-based hydrogels are crosslinked in the presence of cells, forming 3D cell-hydrogel constructs. (d) Hepatocyte and endothelial cells systems comprised of fluid flow channels and porous membranes form liver sinusoid-like structures.

been employed in several methods to increase lifetime and function of hepatocyte cultures. In one such approach, porcine liver was decellularized through the vasculature using a detergent, after which discs were cut from the ECM. When cultured on the liver ECM discs, which retained key molecular components native to the liver, hepatocytes expressed increased albumin levels compared to cells on tissue culture plastic and in collagen gels. Notably, these cultures could be maintained for 3 weeks, a length of time dramatically longer than traditional 2D cultures could support (Lang, Stern et al. 2011). Building of this use of native liver ECM in *in vitro* cultures, our group further processed decellularized liver by solubilizing it and incorporating it into a heparinized hyaluronic acid hydrogel system. The resulting tissue-specific hydrogel material could be prepared in a fashion more amenable to increased throughput studies. We demonstrated increased albumin and urea production, increased viability, superior morphology, and importantly, increased drug metabolism, in human hepatocytes in these liver hydrogel cultures. Notably, we extended viability and function out to 28 days (Skardal, Smith et al. 2012).

8.4 LIVER-ON-A-CHIP DEVICES

Today, stating that an organoid is "on-a-chip" often conveys an assumption that this is a microfluidic (or mesofluidic) system with fluid channels and housing for the organoid. However, initially the chip component was sometimes a device or element with patterns or wells used to create the organoid (Figure 8.2a and b). An example of this approach is the use of a chip containing microwells of various shapes and sizes that contain regions of cell-adherent collagen versus non-adherent polyethylene glycol. Based on the well conditions, HepG2 cells or rat hepatocytes could be formed into either spheroids or cylindroids in a highly controlled manner that maintained better liver function than 2D controls (Fukuda, Sakai et al. 2006; Mori, Sakai et al. 2008). In another example, HepG2 spheroids were created using an array device of channels and pyramid microfeatures to create functional HepG2 spheroids and multidrug screening (Torisawa, Takagi et al. 2007). Nevertheless, today's liver-on-a-chip systems generally leverage fluid flow for increasing diffusion simulation, delivering drugs or toxins, sampling, or even connecting liver modules to other tissue types forming a multi-organoid system.

One example a fluidic-based liver-on-a-chip used hydrogel matrices to embed HepG2 and NIH-3T3 cells within fluid arrays. These 3D organoids had better function than 2D controls and were demonstrated to respond to acetaminophen in a toxin screening experiment (Au, Chamberlain et al. 2014). Similarly, our group recently took advantage of a versatile photopolymerizable hydrogel system to perform *in situ* device photopatterning to generate HepG2 liver organoids in parallel channel polydimethylsiloxane (PDMS) fluidic devices prepared by soft lithography and molding techniques (Figure 8.2c). We demonstrated toxic agent screening in parallel using multiple alcohol concentrations which, as expected, resulted in a dose-dependent decrease in viability and function with increasing dose (Skardal, Devarasetty et al. 2015c). We are currently modifying this approach—combining *in situ* organoid biofabrication with screening systems—to be substantially

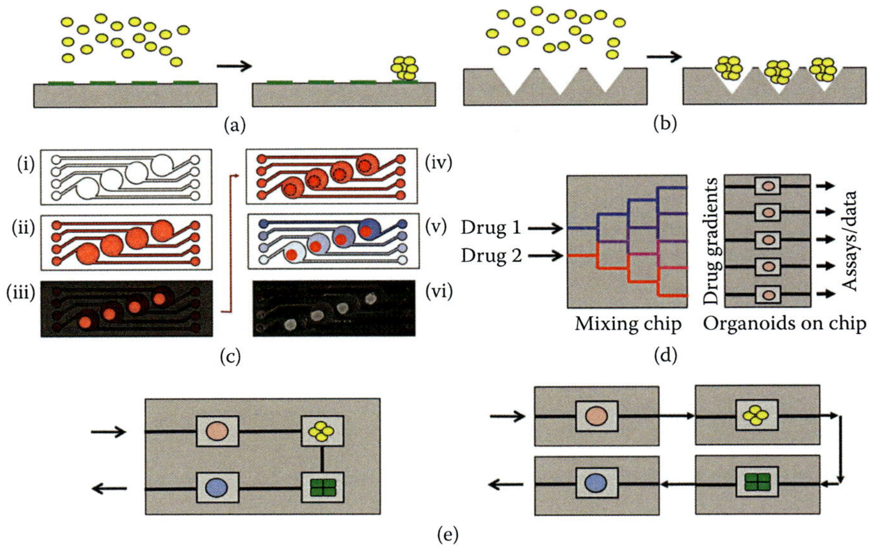

FIGURE 8.2 Liver-on-a-chip systems range in complexity. (a–b) Simple substrate chips can be (a) patterned with cell-adherent proteins or (b) microwells to drive cell aggregation into 3D cell structures. (c) Fluidic channels and chambers can be used to create photopatterned 3D constructs and be used for parallelized drug and toxicology screens. (d) Devices can be designed to facilitate mixing of drugs or other substances to create concentration gradients for increased throughput testing. (e) Linking multiple tissue types together in a platform generates "body-on-a-chip" systems. These systems can be contained within a single device or be comprised of multiple modular components.

miniaturized and parallelized further to increase throughput and the statistical power that can be generated in experiments on single devices the size of traditional microscope slides. Additional devices can provide other non-cell–supporting capabilities, such as facilitating drug or toxin gradient generation over arrays of cell, or organoid cultures by serving as mixing devices (Figure 8.2d).

Using microfabrication approaches, thereby generating more intricate structures, has been explored to create structures such as liver sinusoids (Figure 8.1d). For example, precise layering of rat hepatocytes and endothelial cell co-cultures with fluid flow can generate sinusoid-like models (Kang, Sodunke et al. 2015). In another example, the device containing two distinct chambers separated by a porous membrane with human hepatocytes and endothelial cells was demonstrated to maintain increased albumin and urea secretion under flow conditions compared to static conditions (Prodanov, Jindal et al. 2015). Systems of increased biological complexity have begun to emerge that feature other organoids in addition to the liver (Atac, Wagner et al. 2013; Wagner, Materne et al. 2013; Maschmeyer, Lorenz et al. 2015; Materne, Maschmeyer et al. 2015). These multi-organoid devices, termed "body-on-a-chip" systems (Figure 8.2e), have vast potential in a variety of applications, but to date have been primarily comprised of cell lines, not fully functional hepatocytes, and as such require additional work to demonstrate their ability to accurately mimic human physiology and responses to environmental factors.

8.5 3D *IN VITRO* LIVER CANCER MODELS

Successful development of liver models such as those described to this point allows for a variety of useful implementations. Perhaps the most common use is for drug and toxicology screening. However, with relatively accurate organoid systems, one can expand into other areas of research, including disease modeling. In this setting, cancer is one of the most common pathologies studied. In particular, with respect to liver, several cancers are actively being assessed in microphysiological models, including hepatocellular carcinoma (HCC) and colon carcinoma that has metastasized to the liver.

These models have ranged in complexity and application. In one study, the effect of peroxisome proliferator–activated receptor activation on HCC cell migration and invasion was tested first in a simple 2D wound assay, and subsequently in a 3D Matrigel invasion assay. These assays demonstrated that upregulation by agonists could reduce or prevent invasive behavior (Shen, Chu et al. 2012). A separate HCC Matrigel invasion model was employed to demonstrate cell invasion tracking using quantum dot nanoparticles and several imaging modalities. This platform was able to assess invasive phenotypes, reversal of cell senescence prior to invasion, and quantum dot-highlighted expression of MT1-MMP in filopodia, or invadopodia, of the cells (Fang, Peng et al. 2013).

In addition to assessing migration in invasion models (Figure 8.3a), liver cancer models are being employed to investigate phenomena such as epithelial-to-mesenchymal transition (EMT), which is an integral event in the progression toward metastatic cancers. Co-culture models have been developed that provide the cellular components of the tumor microenvironment that would normally interact with tumor cells. For example, when Bel-7402 HCC cells with normal liver cells or normal vascular endothelial cells, the cells actually underwent an mesenchymal-to-epithelial transition (MET), becoming less invasive. Conversely, when cultured with conditioned media from MRC-5 lung fibroblast cells, they underwent an EMT-like transition, becoming elongated and more invasive (Ding, Zhang et al. 2013). These results demonstrate the importance of the tumor microenvironment in cancer model systems. In addition to stromal cells, the extracellular matrix is an integral component of the tumor microenvironment that should be considered in developing cancer models. For example, when HepG2 cells were formed into 3D heterospheroids together with stromal fibroblasts and embedded in collagen, the cells became significantly more resistant to doxorubicin (Yip and Cho 2013). Our group recently created a platform of liver organoids with colon carcinoma tumor foci using an ECM hydrogel microcarrier and rotating wall vessel bioreactor system. We could demonstrate both the recapitulation of tumor growth over time- and dose-dependent responses to the drug 5-fluorouracil (Figure 8.3b) as well as manipulate the Wnt pathway using small molecule drugs to increase or decrease drug resistance to 5-FU (Skardal, Devarasetty et al. 2015b). Versatile systems such as this will likely provide useful diagnostic and drug screening platforms.

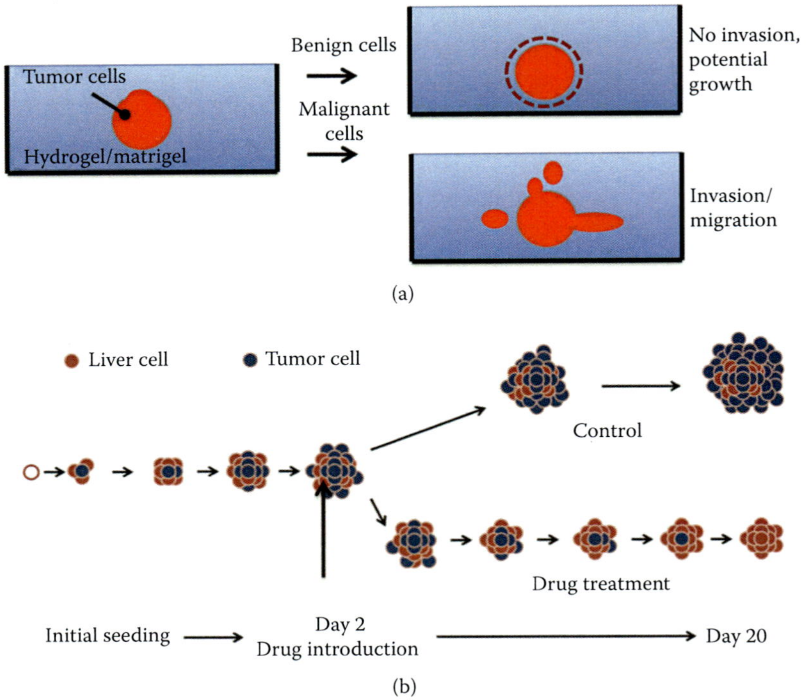

FIGURE 8.3 3D cancer models and the applications. (a) Tumor invasion models often consist of a tumor cell spheroid or cell bolus in or on a hydrogel such as Matrigel. Over time in culture malignant, invasive cells migrate through the matrix while more benign cells show less migratory activity. (b) Liver host organoids containing tumor cells provide a 3D tissue environment in which the total tumor mass can grow unless an intervening drug treatment is administered.

8.6 LIVER CANCER-ON-A-CHIP MODELS

Many cancer model systems such as those described earlier have the potential to be even more powerful and user friendly if integrated appropriately with microfabrication, microfluidic, and sensor technologies, thereby providing on-a-chip platforms for more substantial *in vitro* applications to be performed with.

Early liver cancer-on-a-chip systems in many respects employed the chip component to facilitate measurement taking. Several examples include the development of paper-based microreactors for integration of liver cancer cell cultures with immunoassays (Lei and Huang 2014), quantification of apoptosis of HepG2 tumor cells in microfluidic devices (Ye, Qin et al. 2007a), and electrical impedence measurements of HepG2 cells to determine cell death as an effect of cytotoxic agents (Yeon and Park 2005).

More recent liver cancer-on-a-chip advancements include devices designed for integration with additional high complexity, advanced technologies such as imaging or microarray analyses, allowing more novel investigations to be performed. The small scale of on-a-chip systems was shown to play a significant role in HepG2/C3A metabolism, based on the bioavailability of oxygen. This metabolomics-on-a-chip demonstrated that the microfluidic environment provided more access to oxygen compared to Petri dish cultures, resulting in increase Krebs cycle activity and decreased hypoxia-regulated factor-1 expression (Ouattara, Prot et al. 2012). A high complexity device comprised of multiple drug gradient mixers and parallel cell culture chambers was developed in order to support multiconcentration drug screens paired with cell-labeling and high content imaging data collection on-chip (Ye, Qin et al. 2007b). In another device, HCT-116 colon carcinoma cells and HepG2 cells (used as a liver model), were encapsulated in Matrigel cultures in separate chambers, while myeloblasts (marrow model) were encapsulated in alginate an additional chamber, in order to test the cytotoxic effects of the 5-FU prodrug Tegafur on each cell type, in comparison to 2D control cultures. Interestingly, in 3D the liver was able to metabolize Tegafur to 5-FU, resulting in cell death in the other 3D constructs, while the 2D HepG2 cells could not metabolize the prodrug to its active form (Sung and Shuler 2009). In another example of increased complexity tumor models on a chip, microscale bioreactors were prepared that housed hepatocytes, nonparenchymal cells (NPCs), and breast cancer cells to model the hepatic niche. The device was outfitted with oxygen sensors, micropumps for controlling nutrient distribution, and real-time sampling (Wheeler, Borenstein et al. 2013). This work capitulated in the observation of spontaneous dormancy of the breast cancer cells within the hepatic niche supported by the platform, determined to be due to the NPCs altering microenvironment cytokine profiles. These more complex examples of liver cancer-on-a-chip systems demonstrate the beginnings of the kinds of studies and findings that are projected to be possible in the near future as these systems gain popularity and become widespread within cancer research.

8.7 FUTURE DIRECTIONS AND CONCLUSIONS

The future impact that liver-on-a-chip and cancer-on-a-chip devices may have on drug development, disease modeling, and personalized medicine is highly promising. However, there currently remain limitations that if overcome will improve the overall effectiveness of these devices, including physiological accuracy, response to drugs, and more logistical aspects such as scalability. Implementation of primary cells instead of cell lines, combined with 3D architectures containing extracellular matrix components and supporting cells, appears to be a general framework for improving liver construct function as well as capturing the *in vivo* accuracy of tumor components in such models. Incorporating tumor cells derived from patients will drive development and testing on platforms that are geared toward specific patients, which will almost certainly result in more accurate and effective diagnoses, prognoses, and chemotherapy regimens in the clinic, as well as development of more nuanced and targeted drugs. Combining these high functioning models with platform technologies such as miniaturized devices, automated sensing, and data collection systems will serve to dramatically increase the throughput of the experiments that can be performed. The resulting wealth of data will be integral for more successful and efficient drug candidate development, most likely saving researchers significant amounts of time and reducing the currently massive costs associated with bringing new drugs to patients and the market.

On-a-chip technologies have been gaining momentum in recent years. Organ-on-a-chip and cancer-on-a-chip systems, while relatively new technologies, are already showing promise in the hands of researchers. As these systems improve and become established and more commonplace, we expect that clinicians and industry will begin implementing them in day-to-day operations, resulting in dramatically improved diagnostics, prognostics, and pharmaceutical screening platforms. Ultimately, implementation of these systems will result in improved patient quality of life.

REFERENCES

Atac, B., I. Wagner, R. Horland, R. Lauster, U. Marx, A. G. Tonevitsky, R. P. Azar and G. Lindner (2013). Skin and hair on-a-chip: In vitro skin models versus ex vivo tissue maintenance with dynamic perfusion. *Lab Chip* **13**(18): 3555–3561.

Au, S. H., M. D. Chamberlain, S. Mahesh, M. V. Sefton and A. R. Wheeler (2014). Hepatic organoids for microfluidic drug screening. *Lab Chip* **14**(17): 3290–3299.

Barrila, J., A. Radtke, S. Sarker, A. Crabbé, M. M. Herbst-Kralovetz, C. M. Ott and C. A. Nickerson (2010). Organotypic 3D cell culture models: using the rotating wall vessel to study host-pathogen interactions. *Nat Rev Microbiol.* **8**(11): 791–801.

Benam, K. H., S. Dauth, B. Hassell, A. Herland, A. Jain, K. J. Jang, K. Karalis, et al. (2015). Engineered in vitro disease models. *Annu Rev Pathol* **10**: 195–262.

Chang, T. T. and M. Hughes-Fulford (2014). Molecular mechanisms underlying the enhanced functions of three-dimensional hepatocyte aggregates. *Biomaterials* **35**(7): 2162–2171.

Ding, S., W. Zhang, Z. Xu, C. Xing, H. Xie, H. Guo, K. Chen, et al. (2013). Induction of an EMT-like transformation and MET in vitro. *J Transl Med* **11**: 164.

Drewitz, M., M. Helbling, N. Fried, M. Bieri, W. Moritz, J. Lichtenberg and J. M. Kelm (2011). Towards automated production and drug sensitivity testing using scaffold-free spherical tumor microtissues. *Biotechnol J* **6**(12): 1488–1496.

Fang, M., C. W. Peng, S. P. Liu, J. P. Yuan and Y. Li (2013). In vitro invasive pattern of hepatocellular carcinoma cell line HCCLM9 based on three-dimensional cell culture and quantum dots molecular imaging. *J Huazhong Univ Sci Technolog Med Sci* **33**(4): 520–524.

Fukuda, J., Y. Sakai and K. Nakazawa (2006). Novel hepatocyte culture system developed using microfabrication and collagen/polyethylene glycol microcontact printing. *Biomaterials* **27**(7): 1061–1070.

Gomez-Lechon, M. J., J. V. Castell and M. T. Donato (2010). The use of hepatocytes to investigate drug toxicity. *Methods Mol Biol* **640**: 389–415.

Greenhough, S., C. N. Medine and D. C. Hay (2010). Pluripotent stem cell derived hepatocyte like cells and their potential in toxicity screening. *Toxicology* **278**(3): 250–255.

Ho, W. J., E. A. Pham, J. W. Kim, C. W. Ng, J. H. Kim, D. T. Kamei and B. M. Wu (2010). Incorporation of multicellular spheroids into 3-D polymeric scaffolds provides an improved tumor model for screening anticancer drugs. *Cancer Sci* **101**(12): 2637–2643.

Kang, Y. B., T. R. Sodunke, J. Lamontagne, J. Cirillo, C. Rajiv, M. J. Bouchard and M. Noh (2015). Liver sinusoid on a chip: Long-term layered co-culture of primary rat hepatocytes and endothelial cells in microfluidic platforms. *Biotechnol Bioeng* **112**(12): 2571–2582.

Kim, J. Y., D. A. Fluri, R. Marchan, K. Boonen, S. Mohanty, P. Singh, S. Hammad, et al. (2015). 3D spherical microtissues and microfluidic technology for multi-tissue experiments and analysis. *J Biotechnol* **205**: 24–35.

Kunz-Schughart, L. A., J. P. Freyer, F. Hofstaedter and R. Ebner (2004). The use of 3-D cultures for high-throughput screening: The multicellular spheroid model. *J Biomol Screen* **9**(4): 273–285.
Lang, R., M. M. Stern, L. Smith, Y. Liu, S. Bharadwaj, G. Liu, P. M. Baptista, et al. (2011). Three-dimensional culture of hepatocytes on porcine liver tissue-derived extracellular matrix. *Biomaterials* **32**(29): 7042–7052.
Lei, K. F. and C. H. Huang (2014). Paper-based microreactor integrating cell culture and subsequent immunoassay for the investigation of cellular phosphorylation. *ACS Appl Mater Interfaces* **6**(24): 22423–22429.
Maschmeyer, I., A. K. Lorenz, K. Schimek, T. Hasenberg, A. P. Ramme, J. Hubner, M. Lindner, et al. (2015). A four-organ-chip for interconnected long-term co-culture of human intestine, liver, skin and kidney equivalents. *Lab Chip* **15**(12): 2688–2699.
Materne, E. M., I. Maschmeyer, A. K. Lorenz, R. Horland, K. M. Schimek, M. Busek, F. Sonntag, R. Lauster and U. Marx (2015). The multi-organ chip—A microfluidic platform for long-term multi-tissue coculture. *J Vis Exp* (98): e52526.
Messner, S., I. Agarkova, W. Moritz and J. M. Kelm (2013). Multi-cell type human liver microtissues for hepatotoxicity testing. *Arch Toxicol* **87**(1): 209–213.
Mori, R., Y. Sakai and K. Nakazawa (2008). Micropatterned organoid culture of rat hepatocytes and HepG2 cells. *J Biosci Bioeng* **106**(3): 237–242.
Nickerson, C. A., and C.M. Ott. (2004). A New Dimension in Modeling Infectious Disease (Invited Review). *ASM News* **70**(4): 169–175.
Nickerson, C. A., E. G. Richter and C. M. Ott (2007). Studying host-pathogen interactions in 3-D: Organotypic models for infectious disease and drug development. *J Neuroimmune Pharmacol* **2**(1): 26–31.
Ouattara, D. A., J. M. Prot, A. Bunescu, M. E. Dumas, B. Elena-Herrmann, E. Leclerc and C. Brochot (2012). Metabolomics-on-a-chip and metabolic flux analysis for label-free modeling of the internal metabolism of HepG2/C3A cells. *Mol Biosyst* **8**(7): 1908–1920.
Polini, A., L. Prodanov, N. S. Bhise, V. Manoharan, M. R. Dokmeci and A. Khademhosseini (2014). Organs-on-a-chip: A new tool for drug discovery. *Expert Opin Drug Discov* **9**(4): 335–352.
Prestwich, G. D. (2008). Evaluating drug efficacy and toxicology in three dimensions: Using synthetic extracellular matrices in drug discovery. *Acc Chem Res* **41**(1): 139–148.
Prestwich, G. D., Y. Liu, B. Yu, X. Z. Shu and A. Scott (2007). 3-D culture in synthetic extracellular matrices: New tissue models for drug toxicology and cancer drug discovery. *Adv Enzyme Regul* **47**: 196–207.
Prodanov, L., R. Jindal, S. S. Bale, M. Hegde, W. J. McCarty, I. Golberg, A. Bhushan, M. L. Yarmush and O. B. Usta (2015). Long term maintenance of a microfluidic 3-D human liver sinusoid. *Biotechnol Bioeng* **113**(1): 241–246.
Shen, B., E. S. Chu, G. Zhao, K. Man, C. W. Wu, J. T. Cheng, G. Li, et al. (2012). PPARgamma inhibits hepatocellular carcinoma metastases in vitro and in mice. *Br J Cancer* **106**(9): 1486–1494.
Skardal, A., M. Devarasetty, H. W. Kang, I. Mead, C. Bishop, T. Shupe, S. J. Lee, et al. (2015a). A hydrogel bioink toolkit for mimicking native tissue biochemical and mechanical properties in bioprinted tissue constructs. *Acta Biomater* **25**: 24–34.
Skardal, A., M. Devarasetty, C. Rodman, A. Atala and S. Soker (2015b). Liver-tumor hybrid organoids for modeling tumor growth and drug response in vitro. *Ann Biomed Eng* **43**(10): 2361–2373.
Skardal, A., M. Devarasetty, S. Soker and A. R. Hall (2015c). In situ patterned micro 3-D liver constructs for parallel toxicology testing in a fluidic device. *Biofabrication*. **7**(3): 031001.
Skardal, A., S. F. Sarker, A. Crabbe, C. A. Nickerson and G. D. Prestwich (2010). The generation of 3-D tissue models based on hyaluronan hydrogel-coated microcarriers within a rotating wall vessel bioreactor. *Biomaterials* **31**(32): 8426–8435.

Skardal, A., L. Smith, S. Bharadwaj, A. Atala, S. Soker and Y. Zhang (2012). Tissue specific synthetic ECM hydrogels for 3-D in vitro maintenance of hepatocyte function. *Biomaterials* **33**(18): 4565–4575.

Sung, J. H. and M. L. Shuler (2009). A micro cell culture analog (microCCA) with 3-D hydrogel culture of multiple cell lines to assess metabolism-dependent cytotoxicity of anticancer drugs. *Lab Chip* **9**(10): 1385–1394.

Torisawa, Y. S., A. Takagi, Y. Nashimoto, T. Yasukawa, H. Shiku and T. Matsue (2007). A multicellular spheroid array to realize spheroid formation, culture, and viability assay on a chip. *Biomaterials* **28**(3): 559–566.

Wagner, I., E. M. Materne, S. Brincker, U. Sussbier, C. Fradrich, M. Busek, F. Sonntag, et al. (2013). A dynamic multi-organ-chip for long-term cultivation and substance testing proven by 3D human liver and skin tissue co-culture. *Lab Chip* **13**(18): 3538–3547.

Wheeler, S. E., J. T. Borenstein, A. M. Clark, M. R. Ebrahimkhani, I. J. Fox, L. Griffith, W. Inman, et al. (2013). All-human microphysical model of metastasis therapy. *Stem Cell Res Ther* **4** (Suppl 1): S11.

Ye, N., J. Qin, X. Liu, W. Shi and B. Lin (2007a). Characterizing doxorubicin-induced apoptosis in HepG2 cells using an integrated microfluidic device. *Electrophoresis* **28**(7): 1146–1153.

Ye, N., J. Qin, W. Shi, X. Liu and B. Lin (2007b). Cell-based high content screening using an integrated microfluidic device. *Lab Chip* **7**(12): 1696–1704.

Yeon, J. H. and J. K. Park (2005). Cytotoxicity test based on electrochemical impedance measurement of HepG2 cultured in microfabricated cell chip. *Anal Biochem* **341**(2): 308–315.

Yip, D. and C. H. Cho (2013). A multicellular 3D heterospheroid model of liver tumor and stromal cells in collagen gel for anti-cancer drug testing. *Biochem Biophys Res Commun* **433**(3): 327–332.

Zhang, J., A. Skardal and G. D. Prestwich (2008). Engineered extracellular matrices with cleavable crosslinkers for cell expansion and easy cell recovery. *Biomaterials* **29**(34): 4521–4531.

9 Heart-on-a-Chip

Megan L. McCain

CONTENTS

9.1 INTRODUCTION

Cardiovascular diseases are the foremost cause of death in the United States (Murphy, Xu, and Kochanek 2013) and cardiotoxicity is a leading reason for market withdrawal of pharmaceuticals (Ferri et al. 2013). One reason for both of these statistics is that biomedical research and drug screening has historically been limited to model systems that lack relevance to the human heart, such as rodents or overly-simplified cell culture platforms. In this chapter, we describe human-relevant, biomimetic "heart-on-a-chip" platforms that are under development to improve our ability to study and predict the function of human heart tissue.

9.2 STRUCTURE AND FUNCTION OF THE HUMAN HEART

The primary function of the heart is to pump blood, which is essential for delivering nutrients to and removing waste products from all the cells in our bodies. The heart achieves this function due to the synchronous, pulsatile contraction of aligned, excitable muscle tissue, which is described in detail below.

9.2.1 Regulation of Cardiac Output by Myocardial Architecture

The heart is a pulsatile, electromechanical pump that consists of four chambers: two atria and two ventricles (Figure 9.1). The walls of the ventricles consist of layers of muscular tissue that contract in synchrony to pump deoxygenated blood into the lungs (right ventricle) or oxygenated blood into the aorta (left ventricle). Each layer of ventricular muscle tissue, known as the myocardium, is powered by contractile, striated muscle cells known as cardiac myocytes. Each cardiac myocyte has an elongated, cylindrical shape and is aligned with neighboring myocytes.

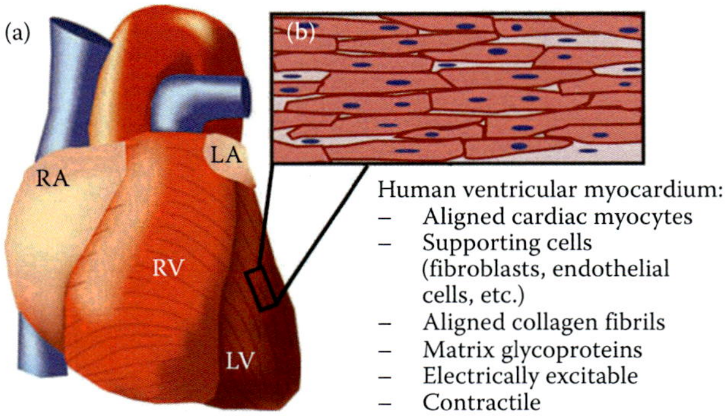

FIGURE 9.1 Structure of the myocardium. The heart consists of four chambers (a): right atrium (RA), left atrium (LA), right ventricle (RV), and left ventricle (LV). Ventricular myocardium (b) consists of aligned cardiac myocytes and supporting cells embedded in an extracellular matrix that consists primarily of collagen fibrils and glycoproteins. Ventricular myocardium is excited to contract by electrical signals. (Modified from McCain, M.L., et al., *Proc. Natl. Acad. Sci. U. S. A.*, 110, 9770–9775, 2013.)

Cardiac myocytes are contractile because they are densely packed with specialized cytoskeletal fibers known as myofibrils (Sarantitis et al. 2012). Each myofibril is a linear array of sarcomeres, and each sarcomere consists of interlocking actin and myosin filaments that slide past each other to shorten the sarcomere in an ATP-dependent process. Because of the hierarchical architecture of the ventricle, sarcomeres within and between cardiac myocytes are oriented in the same direction in each layer of myocardium. Thus, when the ventricle is activated to contract, sarcomeres in each layer of myocardium shorten in the same direction, maximizing the uniaxial force generated by the tissue and the pumping function of the heart.

9.2.2 Electrical Communication in the Ventricle

Cardiac myocytes are activated to contract by action potentials that depolarize the plasma membrane, a phenomenon known as excitation-contraction coupling (Bers 2002). After the plasma membrane depolarizes, calcium enters the cytoplasm from the extracellular space via voltage-sensitive calcium channels. This causes a larger release of calcium ions from the sarcoplasmic reticulum, which then bind to troponin on actin filaments, causing the tropomyosin complex to shift and expose binding sites for myosin heads on actin filaments. Myosin then binds to actin and pulls the filament in an ATP-dependent process, ultimately shortening the sarcomere.

For the ventricle to function as a single, unified pump, electrical current originating from the conduction system must propagate rapidly in the ventricular myocardium. Myocyte-to-myocyte propagation of electrical signals is achieved by specialized cell-to-cell junctions known as intercalated discs (Noorman et al. 2009), which are located primarily at the longitudinal ends of myocytes in mature myocardium. Intercalated discs consist of three protein complexes: adherens junctions and desmosomes, which couple to actin and intermediate filaments, respectively, and gap junctions, which form low-resistance channels that allow ions to easily move from one cardiac myocyte to the next. Thus, due to both the elongated shape of cardiac myocytes and the positioning of low-resistance gap junction channels, electrical signals propagate rapidly across ventricular myocardium in the longitudinal direction, ensuring that the entire ventricle is activated to contract in near synchrony.

9.2.3 Extracellular Matrix

The extracellular matrix (ECM) is a network of proteins that provides mechanical support, resistance, and scaffolding for cells in tissues. In the ventricle, the ECM consists primarily of Type I collagen fibrils aligned parallel to cardiac myocytes (Bowers, Banerjee, and Baudino 2010). The ECM also contains fibronectin and laminin glycoproteins that couple other ECM components, such as collagen, to cardiac myocytes. Cells attach to ECM proteins by specific integrin receptors (Ross and Borg 2001) that link directly to the actin cytoskeleton and activate unique signaling pathways within the cell. Thus, the ECM is not merely a passive support structure, but an important source of structural, mechanical, and chemical cues.

9.2.4 Non-Myocyte Cell Populations

The ventricle contains many cell populations in addition to cardiac myocytes. Cardiac fibroblasts are proliferative, supporting cells that deposit and remodel ECM proteins in the heart throughout development, health, and disease (Sullivan and Black 2013). The ventricle also contains a dense capillary network that consists of vessels lined with endothelial cells. Endothelial cells interface with blood and release signaling factors, such as nitric oxide, that have direct effects on cardiac myocytes (Lim et al. 2015). The ventricle also contains smooth muscle cells, neurons, and immune cells, each of which can affect the function of cardiac myocytes. Thus, although cardiac myocytes are the essential cells for cardiac output, their function can be modulated by the other cell populations in the ventricle.

9.3 THE NEED FOR A "HEART-ON-A-CHIP"

Heart-on-a-chip platforms that mimic the essential structural and functional features of human heart tissue are needed as new tools for disease modeling and cardiotoxicity screening. Current challenges in these fields are described below.

9.3.1 Heart Disease

Cardiovascular diseases are the leading cause of death in the United States, accounting for one out of every four deaths (Murphy, Xu, and Kochanek 2013). Many factors contribute to this statistic, including the following.

1. The heart has limited ability to repair itself after injury because cardiac myocytes are terminally differentiated and do not proliferate to an appreciable extent (Koudstaal et al. 2013). Thus, after an injury such as a myocardial infarction, necrotic muscle tissue is replaced with noncontractile scar tissue that compromises cardiac output.
2. Cardiac myocytes do not regenerate, so acquiring and culturing human cardiac myocytes to study their biology is exceptionally challenging. As a result, researchers have been forced to rely primarily on animal models for cardiac research, which are not necessarily predictive of humans.
3. Several genetic mutations can lead to cardiomyopathies (Cahill, Ashrafian, and Watkins 2013). However, links between genotype and phenotype in human cardiac myocytes are not clearly understood and human genetic cardiomyopathies are often not accurately recapitulated in animal models.
4. Many cardiac diseases are associated with remodeling of the ECM. For example, after an infarction, the scar tissue that forms is rigid and increases the mechanical load on surviving myocytes (Berry et al. 2006). However, understanding how the ECM contributes to cardiac disease progression is difficult because *in vivo* models are too heterogeneous. Furthermore, it has historically been challenging to reproduce the ECM of the heart *in vitro*. Thus, the role of the ECM in cardiac pathogenesis is not well understood.

9.3.2 Preclinical Cardiotoxicity Screening

Cardiotoxicity is a leading cause for market withdrawal of pharmaceuticals (Ferri et al. 2013). One reason for this is that pharmaceutical companies establish cardiotoxicity preclinically using model systems that are not always predictive of the human heart. For example, to determine if compounds are pro-arrhythmic, a common technique is to test if compounds block a single potassium channel in an immortalized cell line that artificially overexpresses the channel (Netzer et al. 2001). However, these cells are missing essential characteristics of cardiac myocytes, such as myofibrils, other ion channels, gap junctions, etc., as well as the native extracellular microenvironment. These intracellular and extracellular factors can have a dramatic effect on overall cell physiology (McCain and Parker 2011). Pharmaceutical companies also test compounds on whole animal models, such as dogs, but these models are not necessarily predictive of humans due to species-dependent differences. Thus, there is a need for more human-relevant, biomimetic platforms for cardiotoxicity screening.

9.4 DESIGN PARAMETERS FOR HEART-ON-A-CHIP

Due to the limitations of current model systems, there is a need for heart-on-a-chip platforms that can better recapitulate human heart tissue *in vitro* for studying human diseases and screening drugs. A fundamental consideration for designing any organ-on-a-chip platform is: How can we capture the essential physiology of the organ with a platform that it is feasible to fabricate and monitor? For heart-on-a-chip, initial efforts have focused on engineering ventricular myocardium because the ventricles are the primary pumping chambers of the heart and the location of most pathologies. To mimic ventricular myocardium *in vitro*, a single cardiac myocyte is too small because it is lacking cell-cell interactions that are essential to the structure and function of the ventricle. Conversely, a 3-dimensional (3D) ventricular pump is too complex to robustly and consistently engineer with current technologies. Thus, most heart-on-a-chip platforms to date consist of micro-scale pieces of ventricular myocardium, engineered to ultimately meet these ideal design parameters:

- Cell source: human-relevant cardiac myocytes
- Biomaterials and scaffolds: platforms that mimic the structural, mechanical, and biochemical aspects of native cardiac ECM
- Metrics: accessible readouts of electrical and contractile function

Efforts toward achieving each of these design parameters are discussed in the following sections.

9.5 CELL SOURCE FOR HEART-ON-A-CHIP

In this section, we describe advantages and disadvantages of currently available cell sources for heart-on-a-chip platforms, which are also summarized in Table 9.1.

TABLE 9.1
Current Cell Sources for Heart on a Chip

Cell Source	Advantages	Disadvantages
Primary human adult cardiac myocytes	• Human-relevant • Mature	• Extremely difficult to acquire • Cannot be expanded in culture
Primary human fetal cardiac myocytes	• Human-relevant • Moderately difficult to acquire	• Not mature • Ethical concerns
Primary neonatal rat cardiac myocytes	• Easy to acquire	• Not human-relevant • Not mature
hESC-derived cardiac myocytes	• Human-relevant • hESCs can be expanded in culture	• Not mature • Ethical concerns
hiPSC-derived cardiac myocytes	• Human-relevant • hiPSCs can be expanded in culture • Can be patient-specific	• Not mature

9.5.1 Human and Rodent Primary Cardiac Myocytes

The ideal cell source for heart-on-a-chip would be primary human adult cardiac myocytes, as adult humans are the primary users of most pharmaceuticals. However, useable quantitates of primary human adult cardiac myocytes cannot be routinely obtained from patients in a practical and safe manner because this would require collecting biopsies of heart tissue from patients on a regular basis. Furthermore, because adult cardiac myocytes are terminally differentiated, they cannot be expanded in culture, further increasing the difficulty of utilizing this cell type. One source of primary human cardiac myocytes that is more readily available to researchers is human fetal cardiac tissue. However, acquiring human fetal tissue raises ethical concerns and still results in a relatively low cell yield. Furthermore, fetal cardiac myocytes are at a very early developmental stage (Goldman and Wurzel 1992) and thus not an ideal surrogate for mature human cardiac myocytes.

Due to the obstacles in acquiring and culturing primary human cardiac myocytes, the gold standard for *in vitro* cardiac experiments for many decades has been primary cardiac myocytes isolated from neonatal rat hearts (Parameswaran et al. 2013). This procedure involves sacrificing neonatal rats and digesting explanted ventricles with enzymes such as collagenase that degrade extracellular matrix proteins. Cardiac myocytes are purified from supporting cell populations, such as fibroblasts, by plating cell suspensions in a culture flask for short periods of time. The supernatant is then collected, which is enriched in cardiac myocytes because supporting cell populations have a higher rate of adhesion to the culture flask. This cell isolation procedure results in a relatively high yield of primary cardiac myocytes that can be used for *in vitro* experiments. Although these cells are not human and also not adult, they are useful for developing and troubleshooting heart-on-a-chip platforms.

9.5.2 Human Embryonic Stem Cell–Derived Cardiac Myocytes

In the late 1990s, Jamie Thomson and colleagues developed protocols to isolate the pluripotent cells from the inner cell mass of human blastocysts (Thomson et al. 1998). These cells, termed human embryonic stem cells (hESCs), could self-renew and be cultured, passaged, and differentiated into cells of all three germ layers. This breakthrough in developmental biology provided researchers with the opportunity to differentiate hESCs into cardiac myocytes for *in vitro* studies. One common protocol for differentiating hESCs into cardiac myocytes entails culturing hESCs in suspension to form them into spherical embryoid bodies. Subpopulations of cells within embryoid bodies spontaneously differentiate into beating cardiac myocytes (Kehat et al. 2001). However, spontaneous differentiation from embryoid bodies is uncontrolled, heterogeneous, and typically requires flow cytometry to separate cardiac myocytes from other cell populations. Thus, to better direct hESC differentiation into cardiac myocytes, researchers began supplementing hESC cultures with growth factors and small molecules that mimic cardiac development, such as activin A and Wnt agonists and antagonists (Mummery et al. 2012). While these protocols have been relatively successful, improving and optimizing the differentiation of hESCs into cardiac myocytes is an ongoing process.

9.5.3 Human Induced Pluripotent Stem Cell–Derived Cardiac Myocytes

Although hESCs hold exciting potential for human disease modeling and regenerative medicine, one significant disadvantage of these cells is that researchers must destroy human embryos to acquire them. Thus, utilizing hESCs for research purposes has raised many ethical concerns, leading to government restrictions on hESC research in the United States (Walters 2004). In response, researchers started investigating alternative approaches for generating human pluripotent stem cells.

In 2006, Shinya Yamanka and colleagues reported a technique for inducing pluripotency in mouse adult fibroblasts (Takahashi and Yamanaka 2006). These reprogrammed cells were termed induced pluripotent stem cells (iPSCs), and reports for generating human iPSCs (hiPSCs) from human adult fibroblasts soon followed (Takahashi et al. 2007). hiPSCs are generated by isolating somatic cells (commonly skin fibroblasts) and introducing a set of four transcription factors into the cells that were initially selected because they are highly expressed in embryonic stem cells. Similar to hESCs, hiPSCs self-renew and can be differentiated into cells of all three germ layers. Although similarities and differences between hESCs and hiPSCs are yet to be fully understood, studies to date have generated cardiac myocytes from hiPSCs using the same or similar protocols that were originally established for differentiating hESCs into cardiac myocytes (Mummery et al. 2012). Thus, many researchers now use hiPSC-derived cardiac myocytes as an alternative to hESC-derived cardiac myocytes.

Because hiPSCs are reprogrammed from skin fibroblasts, they can be generated from individual patients, including those with genetic diseases. Thus, researchers can acquire hiPSC-derived cardiac myocytes with genotypes and phenotypes that match those of cardiac myocytes in patients with inherited diseases, such as long QT syndrome (Moretti et al. 2010). These cells can be used to identify disease mechanisms and/or screen the efficacy of potential therapeutic drugs in human-relevant and disease-specific cardiac myocytes.

Another method for generating hiPSC-derived cardiac myocytes with disease-related genetic mutations is to introduce mutations at the stem cell stage, using gene-editing approaches such as the CRISPR/Cas9 system. This technology entails introducing the Cas9 nuclease into cells together with a guide RNA sequence that targets Cas9 to a specific location within genomic DNA. Cas9 then cleaves the DNA, which is repaired by the cell's machinery in a highly error-prone process. Thus, the gene located at the Cas9 cleavage site is usually mutated into a nonfunctional sequence (Mali, Esvelt, and Church 2013). hiPSCs with the newly introduced genetic mutation can then be differentiated into cardiac myocytes. Thus, hiPSC-derived cardiac myocytes that harbor disease-related genetic mutations can be derived directly from patients with the disease, or be engineered using gene editing. Both methods successfully generate hiPSC-derived cardiac myocytes that can be used for human-relevant disease modeling.

9.5.4 Direct Reprogramming of Cardiac Myocytes

Another approach for generating human cardiac myocytes is direct reprogramming. This process "skips" the hiPSC stage and reprograms somatic cells directly into cardiac myocytes. Similar to the procedure for reprogramming adult somatic cells into hiPSCs, direct cardiac reprogramming is achieved by introducing a defined set of transcription factors associated with cardiac development into adult somatic cells (Ieda et al. 2010; Fu et al. 2013). The ideal factors and conditions for directly reprogramming human somatic cells into cardiac myocytes are still under investigation.

9.6 BIOMATERIALS FOR HEART-ON-A-CHIP

Ideally, biomaterials that are selected as scaffolding for heart-on-a-chip constructs should mimic the mechanical and biochemical properties of the native ECM. Natural and synthetic biomaterials that have potential utility for heart-on-a-chip are described below.

9.6.1 Natural Biomaterials

9.6.1.1 Extracellular Matrix Proteins

Because the ECM in the ventricle is primarily Type I collagen, a well-established strategy for culturing cardiac myocytes is to seed them on Petri dishes or glass coverslips that are coated with Type I collagen protein (Borg et al. 1984). Fibronectin is also present in the heart, especially during cardiac development, and thus has also been used to coat dishes and other surfaces to facilitate cardiac myocyte adhesion (Bursac et al. 2002; Geisse, Sheehy, and Parker 2009). Other ECM proteins, such as fibrin, collagen, and gelatin, have been formed and cross-linked into 3D hydrogels that can be used for culturing cardiac myocytes. One advantage of hydrogels made from ECM proteins is that cells can be directly embedded in the gel (Yuan Ye, Sullivan, and Black 2011) or grown on top of the gel (McCain et al. 2014). Thus, ECM proteins are widely used as substrates and surface coatings for heart-on-a-chip platforms because they are naturally adhesive to cardiac myocytes.

9.6.1.2 Decellularized Cardiac Tissue

In 2008, Doris Taylor and colleagues reported a technique for decellularizing whole rat hearts by perfusing the vasculature with a detergent, leaving behind the intact cardiac ECM (Ott et al. 2008). Decellularized constructs were then used as whole organ scaffolds that could be re-populated with neonatal rat ventricular myocytes. For heart-on-a-chip applications, decellularized heart ECM has been digested, formed into a hydrogel, and used as a scaffold for culturing stem cell-derived cardiac myocytes (Duan et al. 2011). These scaffolds match the composition of ECM proteins in the native heart and thus are highly biomimetic in terms of their biochemical properties. Another approach for utilizing decellularized ECM as a scaffold is to section the heart prior to decellularization. Individual decellularized sections can then be directly used as scaffolds for culturing neonatal rat cardiac myocytes into engineered tissues (Blazeski, Kostecki, and Tung 2015). Thus, decellularized heart tissue has potential applications as a completely natural and biomimetic scaffold for heart-on-a-chip platforms.

9.6.2 Synthetic Biomaterials

9.6.2.1 Polydimethylsiloxane

One of the most common synthetic polymers used for organ-on-chip platforms and microfluidic devices is polydimethylsiloxane (PDMS). PDMS is a synthetic, silicone-based elastomer that is FDA-approved and has been used in biomedical products such as contact lenses and catheters (Abbasi, Mirzadeh, and Katbab 2001). The base of PDMS is a viscoelastic liquid that is mixed with a crosslinking agent and cured to form a flexible solid with a compressive elastic modulus of approximately 1 MPa. To reduce the elastic modulus of PDMS, the ratio of PDMS base to crosslinker can be increased (Brown, Ookawa, and Wong 2005) or PDMS can be mixed with other silicone gels (Palchesko et al. 2012). Cured PDMS is hydrophobic, but can be oxidized using a plasma cleaner or UV ozone cleaner (Abbasi, Mirzadeh, and Katbab 2001). Oxidized PDMS will bind to ECM proteins, such as fibronectin, to make the surface adhesive to cells. Oxidized PDMS will also bond to other PDMS constructs, which is useful for fabricating multicomponent microfluidic devices.

As a cell culture substrate for heart on a chip, PDMS has many desirable qualities. It is inert, nontoxic, and optically clear. Thus, it is easy to monitor cells growing on PDMS surfaces using standard microscopy. Because solid PDMS is fabricated by mixing together two liquid components, it can easily be coated onto surfaces or molded into microfluidic chambers and channels (Huh et al. 2010; Mathur et al. 2015). However, one disadvantage of PDMS is that its surface is hydrophobic and thus resistant to cell adhesion. Even with fibronectin coating, cells often delaminate when cultured on PDMS long-term (Hald et al. 2014; McCain et al. 2014). Therefore, although PDMS has many practical advantages for heart-on-a-chip systems, it is not an ideal surface for cell culture.

9.6.2.2 Polyacrylamide Hydrogels

Another synthetic biomaterial that has potential utility for heart-on-a-chip platforms is polyacrylamide hydrogel. Polyacrylamide hydrogels are synthesized by combining acrylamide and bis-acrylamide in defined ratios, which dictates the elastic modulus of the resulting gel (Pelham and Wang 1997; Aratyn-Schaus et al. 2010; McCain et al. 2012). One of the advantages of polyacrylamide hydrogels is that their elastic moduli are tunable within the range of biological tissues, ranging from brain to muscle to bone (Engler et al. 2006). Thus, for the purposes of heart-on-a-chip, polyacrylamide gels can be fabricated to mimic the elasticity of healthy myocardium, which is approximately 10-15 kPa (Berry et al. 2006), and fibrotic/infarcted myocardium, which is approximately 50 kPa and above (Engler et al. 2008; McCain et al. 2012).

Polyacrylamide hydrogels are nonadhesive to proteins. Thus, to culture cells on polyacrylamide hydrogels, their surface must be modified to facilitate adhesion of ECM proteins (Pelham and Wang 1997; Aratyn-Schaus et al. 2010). One method for attaching ECM proteins to polyacrylamide gels is to dope the hydrogel with streptavidin acrylamide and conjugate biotin residues onto purified ECM proteins. When biotinylated ECM proteins contact the hydrogel surface, they become tethered to the hydrogel via streptavidin-biotin linkages (McCain et al. 2012). With this method, the user chooses which ECM proteins or combination of proteins to attach to the polyacrylamide gel. Thus, polyacrylamide gels are versatile biomaterials because the user has independent control over both the biochemical and mechanical properties of the substrate.

9.7 TECHNIQUES FOR ENGINEERING HEART-ON-A-CHIP TISSUES

To fabricate scaffolds with defined structural features, tissue engineers have leveraged techniques such as photolithography. In this section, we describe methods for regulating cardiac tissue architecture in two dimensions (2D) and 3D for heart-on-a-chip platforms.

9.7.1 Engineering Cardiac Tissues in Two Dimensions: Microcontact Printing and Micromolding

To match the structure and function of native myocardium, cardiac myocytes in heart-on-a-chip constructs should be aligned into confluent, anisotropic tissues. One technique for aligning cardiac myocytes *in vitro* is microcontact printing (Ruiz and Chen 2007; Qin, Xia, and Whitesides 2010). For this process, a user first designs a pattern for cell growth using computer-aided design software and transfers the pattern onto a photomask. In a cleanroom, the photomask is then aligned onto a silicon wafer (typically 4" in diameter) spin-coated with negative photoresist (typically 2–5 μm thick). The entire construct is then exposed to ultraviolet light, which will only expose the unmasked regions of the wafer. The wafer is then immersed in developer solution to dissolve the unexposed regions. The resulting silicon wafer has the features from the photomask raised in photoresist on its surface.

PDMS base mixed with its curing agent is then poured over the wafer in a Petri dish. After the PDMS cures, it is carefully peeled off the wafer and cut into usable "stamps" that are approximately 1 square inch in size. These stamps have the pattern from the wafer molded onto their surface (Qin, Xia, and Whitesides 2010). PDMS stamps are then coated with extracellular matrix protein, such as fibronectin, for approximately 1 hour. The stamp is then dried with compressed air and inverted onto a substrate, such as PDMS-coated coverslips that have been oxidized to facilitate transfer of the protein onto the coverslip. The stamp is then removed and the substrate can be seeded with cells, which only adhere to regions that are patterned with the matrix protein (Geisse, Sheehy, and Parker 2009; Feinberg et al. 2012).

To align cardiac myocytes into confluent, anisotropic tissues, researchers have used microcontact printing to transfer 15 μm-wide lines of fibronectin separated by 2 μm-wide gaps onto PDMS-coated coverslips. Neonatal rat ventricular myocytes (Agarwal, Goss, et al. 2013) as well as mouse (Sheehy et al. 2014) and human (Wang et al. 2014) stem cell–derived cardiac myocytes form confluent tissues with aligned sarcomeres on these or similar patterns. PDMS stamps can also be used as templates for micromolding the surfaces of hydrogels, such as gelatin (McCain et al. 2014) or alginate (Agarwal, Farouz, et al. 2013), prior to cell seeding. Cardiac myocytes cultured on micromolded surfaces also form confluent, aligned tissues, and thus micromolding is another useful approach for engineering aligned cardiac tissues in 2D.

9.7.2 Engineering Cardiac Tissues in Three Dimensions: Bundles and Patches

Because the ventricle is a multilayered structure, 3D cardiac tissue constructs are also under development as potential heart-on-a-chip platforms. One approach for generating miniature, aligned, 3D cardiac tissues is to embed cardiac myocytes into an extracellular matrix hydrogel mixture before the gel has cured. This mixture is then injected into a PDMS microwell that consists of two pillars separated by approximately 500 μm. The mixture solidifies into a hydrogel and, over the course of several days, the cells compact the hydrogel around the pillars (Boudou et al. 2012; Ramade et al. 2014). The final construct is a 3D bundle of cardiac tissue embedded in the gel and suspended between the two PDMS pillars. Due to the geometry of the system, cardiac myocytes naturally self-align between the pillars, forming a 3D, aligned, cardiac tissue bundle.

A similar "self-assembly" approach has been used to engineer larger 3D cardiac patches (Figure 9.2a). Instead of fabricating microwells with two pillars, researchers fabricated a larger PDMS mold with multiple hexagon pillars (Figure 9.2b) and similarly seeded the mold with cardiac myocytes embedded in an extracellular matrix hydrogel solution (Bian, Jackman, and Bursac 2014; Liau et al. 2011; Zhang et al. 2013). The hydrogel solidified over time to form a 3D patch of tissue with pores located in the spaces originally occupied by the PDMS pillars. The pores in the hydrogel induced the cardiac myocytes to align (Figure 9.2c). The final result was a 3D cardiac patch with approximate dimensions of 5 mm × 5 mm × 60 μm (Bian et al. 2009). With this method, neonatal rat, mouse embryonic stem cell–derived, and hESC-derived 3D cardiac tissue constructs have been engineered.

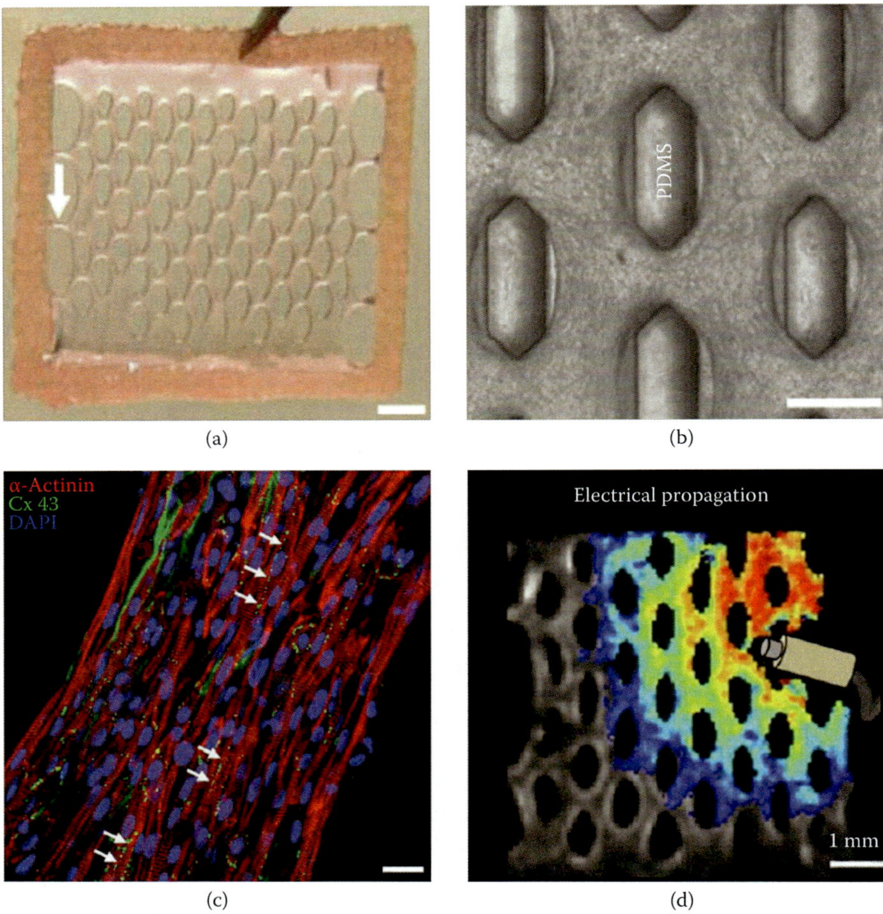

FIGURE 9.2 Engineering 3D cardiac tissues. (a) To engineer a 3D cardiac tissue construct, cardiac myocytes were embedded in an extracellular matrix hydrogel mixture and injected into a mold consisting of an array of PDMS posts (b). (c) Cardiac myocytes compacted the gel around the posts, forming a dense, aligned, 3D tissue. (d) Constructs were loaded with voltage-sensitive dyes to capture action potential propagation. (Modified from Bian, W., et al., *Nat. Prot.*, 4, 1522–1534, 2009; Liau, B., et al., *Biomaterials,* 32, 9180–9187, 2011. With permission.)

9.8 FUNCTIONAL TESTING WITH A HEART-ON-A-CHIP

The essential function of the heart is to pump blood. Thus, heart-on-a-chip systems should ideally incorporate straightforward methods for measuring tissue contractility. Because contractility is dependent on electrical signals, and because many drugs interfere with ion channel function, measuring electrophysiological parameters with heart-on-a-chip constructs is also important. In the following sections, we describe current methods for quantifying both electrophysiology and contractility with current heart-on-a-chip platforms.

9.8.1 Electrophysiology

9.8.1.1 Microelectrode Arrays

To measure electrical activity, cardiac myocytes can be cultured on microelectrode arrays (MEAs). MEAs consist of a glass substrate embedded with stimulating and recording electrodes, each with a diameter in the range of 10 μm. Depending on the configuration, MEAs typically contain up to 60 electrodes separated by distances of tens to hundreds of micrometers and arranged in a square grid. Thus, when cardiac myocytes are cultured on the glass substrate, MEAs can stimulate and/or record extracellular potentials at several locations throughout the tissue (Stett et al. 2003). MEA signals can be used to calculate beat frequency, action potential duration, and propagation velocity from cardiac myocytes cultured on MEAs, as well as changes in these parameters in response to different drugs (Caspi et al. 2009; Navarrete et al. 2013). MEAs have also been integrated into fluidic channels and wells to improve throughput and minimize cell usage for heart-on-a-chip studies (Ma et al. 2012).

One of the advantages of MEA systems is that they are noninvasive to cells. Unlike other techniques, such as patch clamping, MEA signals are recorded from cells without altering the cells. MEA systems can also be placed into incubators so that signals can be recorded from cells and tissues while they are in culture for extended periods of time. However, one disadvantage of MEAs is that they have relatively low spatial resolution and thus can primarily be used only for tissue-level measurements. Furthermore, because cardiac myocytes must be in contact with the electrodes, the user is limited to culturing cells in a 2D monolayer on the glass MEA substrate. Thus, users cannot customize the substrate or make recordings in 3D constructs. Finally, because MEAs record extracellular field potentials, they cannot be used to determine intracellular properties, such as action potential morphology.

9.8.1.2 Voltage- and Calcium-Sensitive Dyes

Another method for measuring the electrophysiological properties of engineered cardiac tissues is to incubate the cells with a fluorescent dye that is sensitive to either membrane voltage or intracellular calcium concentration. With each excitation, the fluorescence signal of the dye increases, which can be detected and recorded with a fluorescent microscope and a sensitive, high-speed camera (McCain et al. 2013) or a fiberoptic bundle (Bursac et al. 2002; Feinberg et al. 2012). One advantage of detecting signals with a fluorescent microscope and high-speed camera is that signals can be recorded from 2D and 3D tissues (Figure 9.2d) cultured on any type of surface, as long as the microscope can focus and acquire signals from the tissue. Furthermore, by using higher power objectives, the user can record signals within individual cells to measure features such as intracellular calcium transients (McCain et al. 2013). However, one disadvantage of this technique is that most dyes are toxic to cells at certain concentrations and exposure times.

As a result, experiments using voltage- and calcium-sensitive dyes are endpoint experiments, so it is not possible to monitor properties of tissues over long-term culture. To combat these disadvantages, myocytes can be transfected with a genetically-encoded fluorescent reporter for intracellular calcium concentration known as GCaMP (Mathur et al. 2015).

9.8.2 Contractility

9.8.2.1 Two-Dimensional Muscular Thin Films

One technique for measuring contractile stresses generated by 2D cardiac tissues is based on muscular thin film (MTF) technology. MTFs are bilayered structures that consist of a monolayer of engineered cardiac tissue adhered to a supportive yet flexible polymer layer (Feinberg et al. 2007). MTFs were first fabricated by spin-coating glass coverslips with the temperature-sensitive polymer poly(N-isopropylacrylamide) (PNIPAm). Above 35°C, PNIPAm is hydrophobic and solidifies in aqueous media. Below 35°C, PNIPAm is hydrophilic and liquefies in aqueous media. PDMS was then spin-coated on top of the PNIPAm layer, microcontacted printed with lines of fibronectin, and seeded with neonatal rat ventricular myocytes, which attached to the fibronectin pattern and formed a confluent, aligned tissue. After several days in culture, constructs were moved to the stage of a stereomicroscope and cooled to room temperature in an aqueous buffer solution. Tissue-PDMS constructs with dimensions of several square millimeters were carefully cut with a scalpel. During cutting, the PNIPAm layer was exposed to the buffer solution and became soluble. The tissue-PDMS constructs, termed MTFs, were then carefully peeled from the underlying glass coverslip. A stimulation electrode placed in the dish activated the tissue to contract and bend the PDMS.

Since their initial development, MTFs have been modified into a higher-throughput heart-on-a-chip platform. First, the fabrication process was altered so that MTF cantilevers remain tethered to the coverslip at one of their longitudinal ends. This reduces the amount of handling done by the user and allows multiple MTFs on one substrate to be imaged from above and analyzed as a group (Grosberg et al. 2011). Stresses generated by tissues can be calculated using a mathematical model based on the curvature and material properties of the MTFs (Alford et al. 2010; Feinberg et al. 2012). To streamline the fabrication process, MTF cantilevers were pre-cut with a laser engraver prior to microcontact printing to eliminate manual cutting with a scalpel. Laser engraving helped to standardize cantilever size and reduce user intervention, further improving throughput and reducing experimental variability (Agarwal, Goss, et al. 2013). MTF constructs have also been miniaturized and integrated into a fluidic chamber to perform controlled drug dosing experiments with wash-in, wash-out capabilities. The PDMS in heart-on-a-chip MTFs has also been replaced with hydrogels, such as micropatterned alginate (Agarwal, Farouz, et al. 2013) and micromolded gelatin (McCain et al. 2014), to better mimic the mechanical properties of native ECM (Figure 9.3). Together, these innovations have improved the ease of use and throughput of MTFs such that they can be utilized as a heart-on-a-chip platform.

The MTF heart-on-a-chip has been used as a platform for disease modeling. Many cardiac diseases are associated with excessive mechanical loading and stretching of

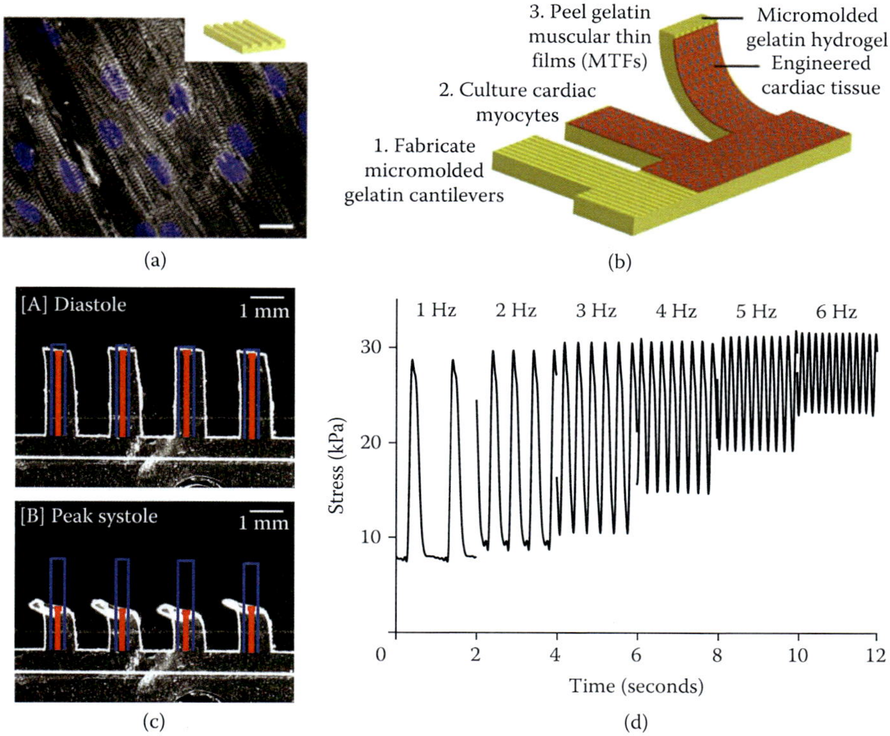

FIGURE 9.3 Gelatin muscular thin film (MTF) heart-on-a-chip constructs. (a) Gelatin hydrogels micromolded with PDMS stamps induced cardiac myocytes to self-assemble into aligned 2D tissues. (b) MTF cantilevers were laser-engraved into gelatin hydrogels prior to cell seeding. After cells formed a confluent tissue, tissue-gelatin cantilevers were peeled from the underlying glass coverslip. (c) As the tissue contracted from diastole [A] to peak systole [B], the MTFs bent away from the glass coverslip. Blue boxes indicate the initial lengths of the MTFs and red bars indicate the x-projection of the film. (d) The x-projection of the film was used to calculate stresses generated by the tissue. (Modified from McCain, M.L., et al., *Biomaterials,* 35, 5462–5471, 2014. With permission.)

the ventricle. To mimic this *in vitro*, MTFs were fabricated onto stretchable silicon membranes such that engineered neonatal rat cardiac tissues could be cyclically stretched (McCain et al. 2013). Cardiac tissues exposed to cyclic stretch for 4 days exhibited pathological remodeling in many different parameters. For example, the ratio of α- to β-myosin heavy chain decreased with stretch, similar to failing myocardium. The magnitude of calcium transients and contractile stresses also decreased with stretch, matching results seen *in vivo* and *ex vivo*. This "failing myocardium-on-a-chip" platform has potential utility as a disease model for assessing drug efficacy with multiple structural and functional outputs.

The MTF heart-on-a-chip platform has also been used for human disease modeling by integrating hiPSC-derived cardiac myocytes. Barth syndrome is caused by a mutation in an enzyme that processes lipids for the mitochondrial membrane of striated muscle cells. As a consequence, cardiac myocytes generate lower levels of ATP and have contractile deficiencies that cause cardiomyopathy. However, the underlying mechanisms that lead to contractile dysfunction are not well understood. To study this disease, researchers seeded MTF heart-on-a-chip constructs with hiPSC-derived cardiac myocytes sourced from wild-type and Barth syndrome patients (Wang et al. 2014). Constructs seeded with hiPSC-derived cardiac myocytes from wild-type patients rhythmically contracted in response to electrical pacing. However, constructs seeded with hiPSC-derived cardiac myocytes from Barth syndrome patients barely moved. These constructs were used to identify disease mechanisms and potential therapies by treating engineered tissues with compounds such as scavengers for reactive oxygen species prior to contractility experiments. This example illustrates how heart-on-a-chip combined with hiPSC-derived cardiac myocytes sourced from patients have exciting potential utility as human-relevant, disease-specific platforms that provide quantitative readouts of tissue function, both at baseline and in response to compounds that could be developed as therapies.

9.8.2.2 Three-Dimensional Tissue Tension Gauges

As described earlier, one method for generating 3D cardiac tissues is to allow cardiac myocytes embedded in an extracellular matrix hydrogel to self-assemble into a muscle bundle suspended between two PDMS pillars. Because PDMS is elastomeric, if the pillars are fabricated with the appropriate geometry, the tissue can bend the pillars toward each other as the cells contract. This movement can be recorded from above with a camera. Based on the dimensions and material properties of the pillars, the amount of force generated by the tissue can be calculated (Boudou et al. 2012). This technology has been used to measure forces from hESC- and hiPSC-derived cardiac myocyte 3D tissue constructs (Chen et al. 2015). Thus, microfabricated 3D cardiac tissue bundles suspended between two PDMS pillars also have applications as functional heart-on-a-chip platforms with readouts of contractile stress.

9.9 REMAINING CHALLENGES AND FUTURE DIRECTIONS

While exciting progress has been made toward developing a human-relevant, biomimetic, functional, heart-on-a-chip platform, several challenges remain:

1. Acquiring mature, human cardiac myocytes. Although researchers can now differentiate cardiac myocytes from human stem cells, one major limitation is that these cells are at a relatively early developmental stage, both genetically and functionally (Sheehy et al. 2014). Thus, human stem cell–derived cardiac myocytes are not equivalent to mature human cardiac myocytes. One of the ongoing challenges for the field is to identify methods to not only differentiate human stem cells into cardiac myocytes but also promote the maturation of human stem cell–derived cardiac myocytes such that they are a more relevant cell type for heart-on-a-chip platforms.
2. Integrating other cell populations. As described earlier, cardiac myocytes are not the only cell population in the ventricle, although they are typically the

only cell type included in most heart-on-a-chip platforms developed to date. Thus, integrating other cell populations, such as cardiac fibroblasts, and other tissue structures, such as blood vessels, are important future directions that can improve the physiological relevance of these model systems.

3. Developing 3D constructs with controlled tissue architecture. Although 3D cardiac tissues are arguably more physiologically relevant than 2D cardiac tissues, it is significantly more challenging to engineer tissues with controlled architecture in 3D compared to 2D. 3D printing of biological structures is an emerging technology that could enable more precise engineering of 3D tissues (Kolesky et al. 2014), but printing living cells remains a significant challenge.
4. Validating and scaling heart-on-a-chip responses. One challenge universal to all organ-on-a-chip platforms is validating that data generated with these microscale platforms match human responses. Another related challenge is understanding how parameters such as drug concentration scale from organ-on-a-chip platforms to the human body.
5. Coupling heart-on-a-chip to other organ-on-a-chip systems. In order to mimic an entire human body, multiple organ-on-a-chip systems will need to be linked together. This introduces new biological challenges, such as identifying media components that support multiple diverse cell types. Coupling organ-on-a-chip systems also requires engineering networks of fluidic channels, pumps, and/or valves that mimic the human vasculature.

In summary, heart-on-a-chip and organs on chips as a whole still face many challenges. However, these platforms are rapidly progressing as researchers continue to integrate breakthroughs in biology with cutting-edge engineering techniques. Because these platforms will likely be more human-relevant than current model systems, especially for the heart, they have potential to truly revolutionize biomedical research and drug screening and have a significant impact on human health.

REFERENCES

Abbasi, F., H. Mirzadeh, and A. A. Katbab. 2001. Modification of polysiloxane polymers for biomedical applications: A review. *Polymer Int* 50 (12):1279–87.

Agarwal, A., Y. Farouz, A. P. Nesmith, L. F. Deravi, M. L. McCain, and K. K. Parker. 2013. Micropatterning alginate substrates for in vitro cardiovascular muscle on a chip. *Adv Funct Mater* 23 (30):3738–46.

Agarwal, A., J. A. Goss, A. Cho, M. L. McCain, and K. K. Parker. 2013. Microfluidic heart on a chip for higher throughput pharmacological studies. *Lab Chip* 13 (18):3599–608.

Alford, P. W., A. W. Feinberg, S. P. Sheehy, and K. K. Parker. 2010. Biohybrid thin films for measuring contractility in engineered cardiovascular muscle. *Biomaterials* 31 (13):3613–21.

Aratyn-Schaus, Y., P. W. Oakes, J. Stricker, S. P. Winter, and M. L. Gardel. 2010. Preparation of complaint matrices for quantifying cellular contraction. *J Vis Exp* (46):pii:2173.

Berry, M. F., A. J. Engler, Y. J. Woo, T. J. Pirolli, L. T. Bish, V. Jayasankar, K. J. Morine, T. J. Gardner, D. E. Discher, and H. L. Sweeney. 2006. Mesenchymal stem cell injection after myocardial infarction improves myocardial compliance. *Am J Physiol Heart Circ Physiol* 290 (6):H2196–203.

Bers, D. M. 2002. Cardiac excitation-contraction coupling. *Nature* 415 (6868):198–205.

Bian, W., C. P. Jackman, and N. Bursac. 2014. Controlling the structural and functional anisotropy of engineered cardiac tissues. *Biofabrication* 6 (2):024109.

Bian, W., B. Liau, N. Badie, and N. Bursac. 2009. Mesoscopic hydrogel molding to control the 3D geometry of bioartificial muscle tissues. *Nat Protoc* 4 (10):1522–34.

Blazeski, A., G. M. Kostecki, and L. Tung. 2015. Engineered heart slices for electrophysiological and contractile studies. *Biomaterials* 55:119–28.

Borg, T. K., K. Rubin, E. Lundgren, K. Borg, and B. Obrink. 1984. Recognition of extracellular matrix components by neonatal and adult cardiac myocytes. *Dev Biol* 104 (1):86–96.

Boudou, T., W. R. Legant, A. Mu, M. A. Borochin, N. Thavandiran, M. Radisic, P. W. Zandstra, J. A. Epstein, K. B. Margulies, and C. S. Chen. 2012. A microfabricated platform to measure and manipulate the mechanics of engineered cardiac microtissues. *Tissue Eng Part A* 18 (9–10):910–19.

Bowers, S. L., I. Banerjee, and T. A. Baudino. 2010. The extracellular matrix: At the center of it all. *J Mol Cell Cardiol* 48 (3):474–82.

Brown, X. Q., K. Ookawa, and J. Y. Wong. 2005. Evaluation of polydimethylsiloxane scaffolds with physiologically-relevant elastic moduli: Interplay of substrate mechanics and surface chemistry effects on vascular smooth muscle cell response. *Biomaterials* 26 (16):3123–9.

Bursac, N., K. K. Parker, S. Iravanian, and L. Tung. 2002. Cardiomyocyte cultures with controlled macroscopic anisotropy: A model for functional electrophysiological studies of cardiac muscle. *Circ Res* 91 (12):e45–54.

Cahill, T. J., H. Ashrafian, and H. Watkins. 2013. Genetic cardiomyopathies causing heart failure. *Circ Res* 113 (6):660–75.

Caspi, O., I. Itzhaki, I. Kehat, A. Gepstein, G. Arbel, I. Huber, J. Satin, and L. Gepstein. 2009. In vitro electrophysiological drug testing using human embryonic stem cell derived cardiomyocytes. *Stem Cells Dev* 18 (1):161–72.

Chen, G., S. Li, I. Karakikes, L. Ren, M. Z. Chow, A. Chopra, W. Keung, et al. 2015. Phospholamban as a crucial determinant of the inotropic response of human pluripotent stem cell-derived ventricular cardiomyocytes and engineered 3-dimensional tissue constructs. *Circ Arrhythm Electrophysiol* 8 (1):193–202.

Duan, Y., Z. Liu, J. O'Neill, L. Q. Wan, D. O. Freytes, and G. Vunjak-Novakovic. 2011. Hybrid gel composed of native heart matrix and collagen induces cardiac differentiation of human embryonic stem cells without supplemental growth factors. *J Cardiovasc Transl Res* 4 (5):605–15.

Engler, A. J., C. Carag-Krieger, C. P. Johnson, M. Raab, H. Y. Tang, D. W. Speicher, J. W. Sanger, J. M. Sanger, and D. E. Discher. 2008. Embryonic cardiomyocytes beat best on a matrix with heart-like elasticity: Scar-like rigidity inhibits beating. *J Cell Sci* 121 (Pt 22):3794–802.

Engler, A. J., S. Sen, H. L. Sweeney, and D. E. Discher. 2006. Matrix elasticity directs stem cell lineage specification. *Cell* 126 (4):677–89.

Feinberg, A. W., P. W. Alford, H. Jin, C. M. Ripplinger, A. A. Werdich, S. P. Sheehy, A. Grosberg, and K. K. Parker. 2012. Controlling the contractile strength of engineered cardiac muscle by hierarchal tissue architecture. *Biomaterials* 33 (23):5732–41.

Feinberg, A. W., A. Feigel, S. S. Shevkoplyas, S. Sheehy, G. M. Whitesides, and K. K. Parker. 2007. Muscular thin films for building actuators and powering devices. *Science* 317 (5843):1366–70.

Ferri, N., P. Siegl, A. Corsini, J. Herrmann, A. Lerman, and R. Benghozi. 2013. Drug attrition during pre-clinical and clinical development: understanding and managing drug-induced cardiotoxicity. *Pharmacol Ther* 138 (3):470–84.

Fu, J. D., N. R. Stone, L. Liu, C. I. Spencer, L. Qian, Y. Hayashi, P. Delgado-Olguin, S. Ding, B. G. Bruneau, and D. Srivastava. 2013. Direct reprogramming of human fibroblasts toward a cardiomyocyte-like state. *Stem Cell Reports* 1 (3):235–47.

Geisse, N. A., S. P. Sheehy, and K. K. Parker. 2009. Control of myocyte remodeling in vitro with engineered substrates. *In Vitro Cell Dev Biol Anim* 45 (7):343–50.

Goldman, B. I., and J. Wurzel. 1992. Effects of subcultivation and culture medium on differentiation of human fetal cardiac myocytes. *In Vitro Cell Dev Biol* 28A (2):109–19.

Grosberg, A., P. W. Alford, M. L. McCain, and K. K. Parker. 2011. Ensembles of engineered cardiac tissues for physiological and pharmacological study: Heart on a chip. *Lab Chip* 11 (24):4165–73.

Hald, E. S., K. E. Steucke, J. A. Reeves, Z. Win, and P. W. Alford. 2014. Long-term vascular contractility assay using genipin-modified muscular thin films. *Biofabrication* 6 (4):045005.

Huh, D., B. D. Matthews, A. Mammoto, M. Montoya-Zavala, H. Y. Hsin, and D. E. Ingber. 2010. Reconstituting organ-level lung functions on a chip. *Science* 328 (5986):1662–8.

Ieda, M., J. D. Fu, P. Delgado-Olguin, V. Vedantham, Y. Hayashi, B. G. Bruneau, and D. Srivastava. 2010. Direct reprogramming of fibroblasts into functional cardiomyocytes by defined factors. *Cell* 142 (3):375–86.

Kehat, I., D. Kenyagin-Karsenti, M. Snir, H. Segev, M. Amit, A. Gepstein, E. Livne, O. Binah, J. Itskovitz-Eldor, and L. Gepstein. 2001. Human embryonic stem cells can differentiate into myocytes with structural and functional properties of cardiomyocytes. *J Clin Invest* 108 (3):407–14.

Kolesky, D. B., R. L. Truby, A. S. Gladman, T. A. Busbee, K. A. Homan, and J. A. Lewis. 2014. 3D Bioprinting of vascularized, heterogeneous cell-laden tissue constructs. *Advanced Materials* 26 (19):3124–30.

Koudstaal, S., S. J. Jansen Of Lorkeers, R. Gaetani, J. M. Gho, F. J. van Slochteren, J. P. Sluijter, P. A. Doevendans, G. M. Ellison, and S. A. Chamuleau. 2013. Concise review: heart regeneration and the role of cardiac stem cells. *Stem Cells Transl Med* 2 (6):434–43.

Liau, B., N. Christoforou, K. W. Leong, and N. Bursac. 2011. Pluripotent stem cell-derived cardiac tissue patch with advanced structure and function. *Biomaterials* 32 (35):9180–7.

Lim, S. L., C. S. Lam, V. F. Segers, D. L. Brutsaert, and G. W. De Keulenaer. 2015. Cardiac endothelium-myocyte interaction: Clinical opportunities for new heart failure therapies regardless of ejection fraction. *Eur Heart J* 36 (31):2050–60.

Ma, Z., Q. Liu, H. Liu, H. Yang, J. X. Yun, C. Eisenberg, T. K. Borg, M. Xu, and B. Z. Gao. 2012. Laser-patterned stem-cell bridges in a cardiac muscle model for on-chip electrical conductivity analyses. *Lab Chip* 12 (3):566–73.

Mali, P., K. M. Esvelt, and G. M. Church. 2013. Cas9 as a versatile tool for engineering biology. *Nat Methods* 10 (10):957–63.

Mathur, A., P. Loskill, K. Shao, N. Huebsch, S. Hong, S. G. Marcus, N. Marks, M. Mandegar, B. R. Conklin, L. P. Lee, and K. E. Healy. 2015. Human iPSC-based cardiac microphysiological system for drug screening applications. *Sci Rep* 5:8883.

McCain, M. L., A. Agarwal, H. W. Nesmith, A. P. Nesmith, and K. K. Parker. 2014. Micromolded gelatin hydrogels for extended culture of engineered cardiac tissues. *Biomaterials* 35 (21):5462–71.

McCain, M. L., H. Lee, Y. Aratyn-Schaus, A. G. Kleber, and K. K. Parker. 2012. Cooperative coupling of cell-matrix and cell-cell adhesions in cardiac muscle. *Proc Natl Acad Sci U S A* 109 (25):9881–6.

McCain, M. L., and K. K. Parker. 2011. Mechanotransduction: The role of mechanical stress, myocyte shape, and cytoskeletal architecture on cardiac function. *Pflugers Arch* 462 (1):89–104.

McCain, M. L., S. P. Sheehy, A. Grosberg, J. A. Goss, and K. K. Parker. 2013. Recapitulating maladaptive, multiscale remodeling of failing myocardium on a chip. *Proc Natl Acad Sci U S A* 110 (24):9770–5.

Moretti, A., M. Bellin, A. Welling, C. B. Jung, J. T. Lam, L. Bott-Flugel, T. Dorn, et al. 2010. Patient-specific induced pluripotent stem-cell models for long-QT syndrome. *N Engl J Med* 363 (15):1397–409.

Mummery, C. L., J. Zhang, E. S. Ng, D. A. Elliott, A. G. Elefanty, and T. J. Kamp. 2012. Differentiation of human embryonic stem cells and induced pluripotent stem cells to cardiomyocytes: A methods overview. *Circ Res* 111 (3):344–58.

Murphy, S. L., J. Xu, and K. D. Kochanek. 2013. Deaths: Final data for 2010. *Natl Vital Stat Rep* 61 (4):1–117.

Navarrete, E. G., P. Liang, F. Lan, V. Sanchez-Freire, C. Simmons, T. Gong, A. Sharma, et al. 2013. Screening drug-induced arrhythmia [corrected] using human induced pluripotent stem cell-derived cardiomyocytes and low-impedance microelectrode arrays. *Circulation* 128 (11 Suppl 1):S3–13.

Netzer, R., A. Ebneth, U. Bischoff, and O. Pongs. 2001. Screening lead compounds for QT interval prolongation. *Drug Discov Today* 6 (2):78–84.

Noorman, M., M. A. van der Heyden, T. A. van Veen, M. G. Cox, R. N. Hauer, J. M. de Bakker, and H. V. van Rijen. 2009. Cardiac cell-cell junctions in health and disease: Electrical versus mechanical coupling. *J Mol Cell Cardiol* 47 (1):23–31.

Ott, H. C., T. S. Matthiesen, S. K. Goh, L. D. Black, S. M. Kren, T. I. Netoff, and D. A. Taylor. 2008. Perfusion-decellularized matrix: Using nature's platform to engineer a bioartificial heart. *Nat Med* 14 (2):213–21.

Palchesko, R. N., L. Zhang, Y. Sun, and A. W. Feinberg. 2012. Development of polydimethylsiloxane substrates with tunable elastic modulus to study cell mechanobiology in muscle and nerve. *PLoS One* 7 (12):e51499.

Parameswaran, S., S. Kumar, R. S. Verma, and R. K. Sharma. 2013. Cardiomyocyte culture—An update on the in vitro cardiovascular model and future challenges. *Can J Physiol Pharmacol* 91 (12):985–98.

Pelham, R. J., Jr., and Yl Wang. 1997. Cell locomotion and focal adhesions are regulated by substrate flexibility. *Proc Natl Acad Sci U S A* 94 (25):13661–5.

Qin, D., Y. Xia, and G. M. Whitesides. 2010. Soft lithography for micro- and nanoscale patterning. *Nat Protoc* 5 (3):491–502.

Ramade, A., W. R. Legant, C. Picart, C. S. Chen, and T. Boudou. 2014. Microfabrication of a platform to measure and manipulate the mechanics of engineered microtissues. *Methods Cell Biol* 121:191–211.

Ross, R. S., and T. K. Borg. 2001. Integrins and the myocardium. *Circ Res* 88 (11):1112–19.

Ruiz, S. A., and C. S. Chen. 2007. Microcontact printing: A tool to pattern. *Soft Matter* 3 (2):168–77.

Sarantitis, I., P. Papanastasopoulos, M. Manousi, N. G. Baikoussis, and E. Apostolakis. 2012. The cytoskeleton of the cardiac muscle cell. *Hellenic J Cardiol* 53 (5):367–79.

Sheehy, S. P., F. Pasqualini, A. Grosberg, S. J. Park, Y. Aratyn-Schaus, and K. K. Parker. 2014. Quality metrics for stem cell-derived cardiac myocytes. *Stem Cell Reports* 2 (3):282–94.

Stett, A., U. Egert, E. Guenther, F. Hofmann, T. Meyer, W. Nisch, and H. Haemmerle. 2003. Biological application of microelectrode arrays in drug discovery and basic research. *Anal Bioanal Chem* 377 (3):486–95.

Sullivan, K. E., and L. D. Black. 2013. The role of cardiac fibroblasts in extracellular matrix-mediated signaling during normal and pathological cardiac development. *J Biomech Eng* 135 (7):71001.

Takahashi, K., K. Tanabe, M. Ohnuki, M. Narita, T. Ichisaka, K. Tomoda, and S. Yamanaka. 2007. Induction of pluripotent stem cells from adult human fibroblasts by defined factors. *Cell* 131 (5):861–72.

Takahashi, K., and S. Yamanaka. 2006. Induction of pluripotent stem cells from mouse embryonic and adult fibroblast cultures by defined factors. *Cell* 126 (4):663–76.

Thomson, J. A., J. Itskovitz-Eldor, S. S. Shapiro, M. A. Waknitz, J. J. Swiergiel, V. S. Marshall, and J. M. Jones. 1998. Embryonic stem cell lines derived from human blastocysts. *Science* 282 (5391):1145–7.

Walters, L. 2004. Human embryonic stem cell research: An intercultural perspective. *Kennedy Inst Ethics J* 14 (1):3–38.

Wang, G., M. L. McCain, L. Yang, A. He, F. S. Pasqualini, A. Agarwal, H. Yuan, et al. 2014. Modeling the mitochondrial cardiomyopathy of Barth syndrome with induced pluripotent stem cell and heart-on-chip technologies. *Nat Med* 20 (6):616–23.

Yuan Ye, K., K. E. Sullivan, and L. D. Black. 2011. Encapsulation of cardiomyocytes in a fibrin hydrogel for cardiac tissue engineering. *J Vis Exp* (55):e3251.

Zhang, D., I. Y. Shadrin, J. Lam, H. Q. Xian, H. R. Snodgrass, and N. Bursac. 2013. Tissue-engineered cardiac patch for advanced functional maturation of human ESC-derived cardiomyocytes. *Biomaterials* 34 (23):5813–20.

10 Skin-on-a-Chip

Claire G. Jeong

CONTENTS

10.1 INTRODUCTION: BACKGROUND

The skin is the most complex and largest organ of the human body, accounting for about 15% of the total adult body weight and serving mainly as a primary protective physical barrier against external environmental signals. It helps to maintain body homeostasis by preventing dehydration, maintaining thermoregulation, and limiting the direct penetration of potentially harmful agents to internal organs (Kanitakis 2002, Groeber, Holeiter et al. 2012; Pereira, Barrias et al. 2013; Mathes, Ruffner et al. 2014). Furthermore, other immunologic, endocrine, metabolic, neurosensory, and psychosocial functions of skin are essential and these multiple roles are closely related to and harmoniously coordinated by three layer structures of skin (composed of epidermis, dermis, and hypodermis) and other skin components such as hair, sensory nerves, the immune system, and various glands (Chuong, Nickoloff et al. 2002; Kanitakis 2002; Brohem, Cardeal et al. 2011; Eungdamrong, Higgins et al. 2014). The epidermis, the outermost stratified epithelium layer of the skin, is made of various cell types to perform such multiple functions as a whole; namely, keratinocytes (90–95%), specialized dendritic Langerhans cells (3–6%), pigment-producing melanocytes, neuroendocrine and epithelial Merkel cells, lymphocytes (<1.3%), and Toker cells. Keratinocytes in the epidermal layer undergo the continuous process of proliferation, differentiation, and cornification (ultimate cell death and shedding), which leads to compartmentalization into a number of complex layers with different stages of keratinocyte maturation. This complex compartmentalization and continuous alignment of epidermal layers (the basal layer [single], the stratum spinous layer [5–15 layers], the granular layer [1–3 ayers], and the cornified layer [5–10 layers]) (Figure 10.1) (Kanitakis 2002; Stark, Szabowski et al. 2004; Brohem, Cardeal et al. 2011; Mathes,

Ruffner et al. 2014). The underlying dermis, separated from avascular epidermis by the basement membrane or basal extracellular matrix (ECM), is composed of two main structural parts: the papillary or superficial dermis and the reticular or deep dermis, altogether protecting the epidermis and epidermal appendages with the vascular and nervous system running through it. The dermis is formed of strong connective tissue and the dermal matrix is rich in collagen (>90%) and elastin conferring flexibility and elastic properties of the skin. The cellular components of the dermis include fibroblasts, endothelial cells, smooth muscle cells, and mast cells, and the important functional units of skin such as nerves, sweat, and sebaceous glands as well as hair follicles and hair shafts are embedded in the dermis (Supp and Boyce 2005; Metcalfe and Ferguson 2007; Brohem, Cardeal et al. 2011). The dermal fibroblasts are known to perform numerous functions in the synthesis and deposition of ECM components including collagen and several cross-talk between neighboring cells for proliferation and migration, and autocrine and paracrine interactions (Wong, McGrath et al. 2007). The deepest part of the skin, the hypodermis, mainly populated by adipocytes, also plays crucial roles in heat regulation and conservation, as a shock absorber, a nutritional energy reservoir, and even as a cytokine depot regulating innate immunity and cell growth (Klein, Permana et al. 2007) (Figure 10.1).

Undoubtedly, loss of small to large parts of this barrier due to illness, injury, and burns render us susceptible to disability or death, and currently available treatment

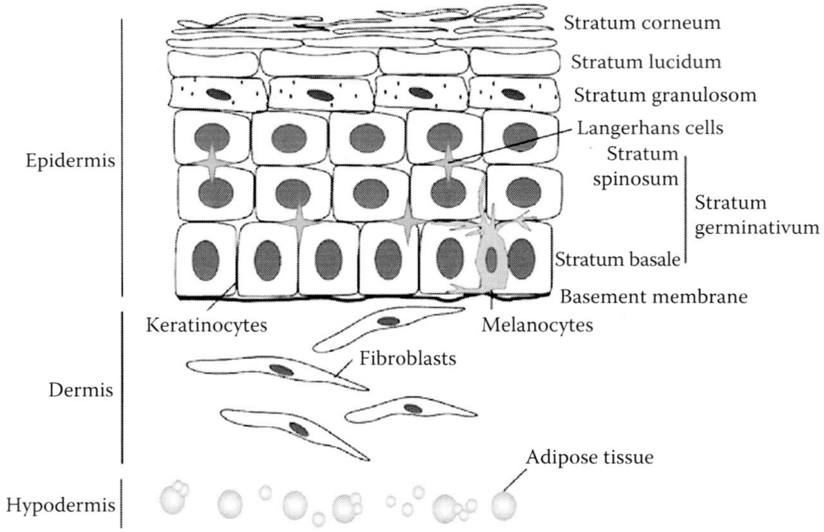

FIGURE 10.1 Schematic rendering of native skin that is subdivided into three main layers: epidermis, dermis, and hypodermis. The epidermis is a stratified squamous epithelium populated by keratinocytes, Langerhans cells, and melanocytes. The dermis is composed of fibroblasts, dermal dendritic cells, mast cells within collagen matrix, and epidermal appendages including hair, sebaceous glands, sweat glands, nerves, and blood vessels are embedded in the dermis layer of skin. The hypodermis is composed mainly of adipocytes. (Adapted from Brohem, C.A., et al., *Pigment. Cell Melanoma Res.*, 24, 35–50, 2011.)

options still do not achieve both functional and cosmetic satisfaction with social and financial burden (Yildirimer, Thanh et al. 2012; Blais, Parenteau-Bareil et al. 2013). Also, there is a significant and increasing demand in modeling and biofabricating three-dimensional (3D) *in vitro* skin tissues for skin disorder therapeutic treatments and drug development (Stark, Szabowski et al. 2004; Yamada and Cukierman 2007; Jean, Lapointe et al. 2009; Rimann and Graf-Hausner 2012; Eungdamrong, Higgins et al. 2014; Mathes, Ruffner et al. 2014). The increasing demand for skin grafts with advances in biofabrication technology of 3D skin architecture has enhanced the development of more complex tissue engineered skin equivalents such as 3D bioprinted skin constructs or "bio-sprayed" skin cells printed directly on wounds in a spatially defined manner (Koch, Kuhn et al. 2010; Koch, Deiwick et al. 2012; Skardal, Mack et al. 2012). These studies not only offer novel and significant advances in wound healing methods and reconfirm the advantages of having cells or tissues cultured in 3D microenvironments compared to two-dimensional (2D) monolayer culture conditions (Griffith and Swartz 2006; Yamada and Cukierman 2007; Mazzoleni, Di Lorenzo et al. 2009) but also underline the need for continuous further developments *in vitro* 3D skin cellular models (Pampaloni, Reynaud et al. 2007). Undoubtedly, cell-based assays play a key role in drug discovery and provide essential information on efficacy and toxicity of potential drug candidates already at an early developmental stage. However, several *in vitro* studies have demonstrated that only in 3D *in vitro* systems keratinocytes proliferate and develop well-ordered epithelia as dermal fibroblasts influence and promote keratinocytes properly to induce the adequate formation of basal membrane proteins and soluble factors to be secreted by epidermis (el-Ghalbzouri, Gibbs et al. 2002; El Ghalbzouri, Lamme et al. 2002; Stark, Szabowski et al. 2004; Wong, McGrath et al. 2007). Moreover, numerous studies have reported that heterogeneous cell-cell interactions are crucial to make artificial *in vitro* skin constructs to recapitulate and mimic the functions of living skin tissues, whereby such skin models could be used to predict more accurate cellular responses in the context of drug screening and toxicity assays. For instance, not only do the mutual interactions between keratinocytes and melanocytes influence and lead to site-specific pigmentation (Duval, Chagnoleau et al. 2012; Bottcher-Haberzeth, Klar et al. 2013; Cichorek, Wachulska et al. 2013), but also blood and lymph vascularization patterns (created by host endothelial cells and lymphocyte recruitments and arrangements) are significantly different for at least early skin maturation and pigmentation processes (Klar, Bottcher-Haberzeth et al. 2014). Moreover, keratinocytes, melanocytes, and fibroblasts contribute to regulation of cell proliferation and innervation (Bottcher-Haberzeth, Klar et al. 2013; Biedermann, Bottcher-Haberzeth et al. 2015) and dermal adipocytes are revealed to affect epidermal morphogenesis and homeostasis during hair follicle regeneration and wound healing while they mediate fibroblast recruitment during skin wound healing (Lu, Yu et al. 2012; Schmidt and Horsley 2013; Driskell, Jahoda et al. 2014). Due to these heterogeneous cell populations and their multicellular cell-cell interactions in skin, the incorporation of various cell types other than single or two-cell types into reconstructed skin models is a new niche of active research. The development of an integrated skin-on-a-chip system is one of these and it is driven to make novel *in vitro* assays or detection methods of compound responses and clinically relevant skin diseased models (i.e., acne, psoriasis, wound/wound healing, and melanoma). Prior to *in vivo* and clinical trials, the functional *in vitro* skin models may reduce and potentially replace animal studies that are often costly, unreliable, and difficult to translate to humans. If the skin models could recapitulate the basic architecture of the human skin and provide more clinically relevant studies for multicellular cell-cell interactions and effects of 3D microenvironments for melanogenesis, proliferation and differentiation of keratinocytes, innervation/vascularization patterns, as well as re-epithelialization after wounds, the models could be used as an intermediate screening/testing tool between monolayer cellular studies and animal/clinical trials to reduce the cost of clinical trials with much more success. Of course, it is not simple to build such *in vitro* skin models many steps are required with positive and negative feedback loops for *in vitro* skin models to be universally accepted as a testing assay form of drug candidates. Astashkina, Mann et al. (2012) illustrated the complex steps and the strategy for use of *in vitro* systems for prediction or replacement of *in vivo* toxicity studies in preclinical trials in a flow chart (Figure 10.2). Regardless of this complexity, the existing skin-equivalent models have been shown to have excellent *in vivo* profile correlation in exposures to UV, corrosive and toxic substances, drugs, chemical agents, and nanoparticles (Netzlaff, Lehr et al. 2005; Astashkina and Grainger 2014; Chapman, Thomas et al. 2014) and can be readily used in the investigation of potential drug targets and related signaling pathways for diseases (i.e., melanoma [Van Kilsdonk, Bergers et al. 2010]). With further advanced technology and manufacture standardization, the bioengineered skin models and the skin-on-a-chip system would not only be a useful platform for cosmetic and pharmacological analyses overcoming dissimilar skin response and functionality between human and animals but also a time- and cost-effective alternative to animal testing prior to clinical trials (Auxenfans, Fradette et al. 2009; Jean, Lapointe et al. 2009; Groeber, Holeiter et al. 2012; Eungdamrong, Higgins et al. 2014; de Vries, Leenaars et al. 2015).

10.2 STATE OF THE ART FOR 3D SKIN TISSUE MODELS

Currently, a broad spectrum of 3D skin models with fabricating conditions and methods exist to mimic different levels of biological complexity and demands. Models are different in terms of cell type (i.e., single-cell to multicell types), architecture (i.e., split vs. full thickness, or flat vs. spheroidal), specific purpose (i.e., explants/wound healing vs. diseases), and bio-fabricating technologies (i.e., bioprinting, bioreactors, or "hanging drop" technology). In general, the higher the complexity of the models, the more reliable and better reflection of the *in vivo* situation than simpler model systems. However, highly complex models still can present low usability for pharmaceutical drug development due to concerns in reproducibility, maintenance, absence of standards for readout and analysis, validation, and costs compared to their effectiveness. For the purposes of drug discovery and development, the most sophisticated *in vitro* skin assay model among existing models or concepts would be probably the skin-on-a-chip approach, as any forms of bioengineered skin constructs could be applied and integrated into such system. Here, with more emphasis on current advancements in tissue-engineered skin constructs and skin-on-a-chip or skin as one part of a multiorgans-on-a-chip (MOC) approach, a selection of

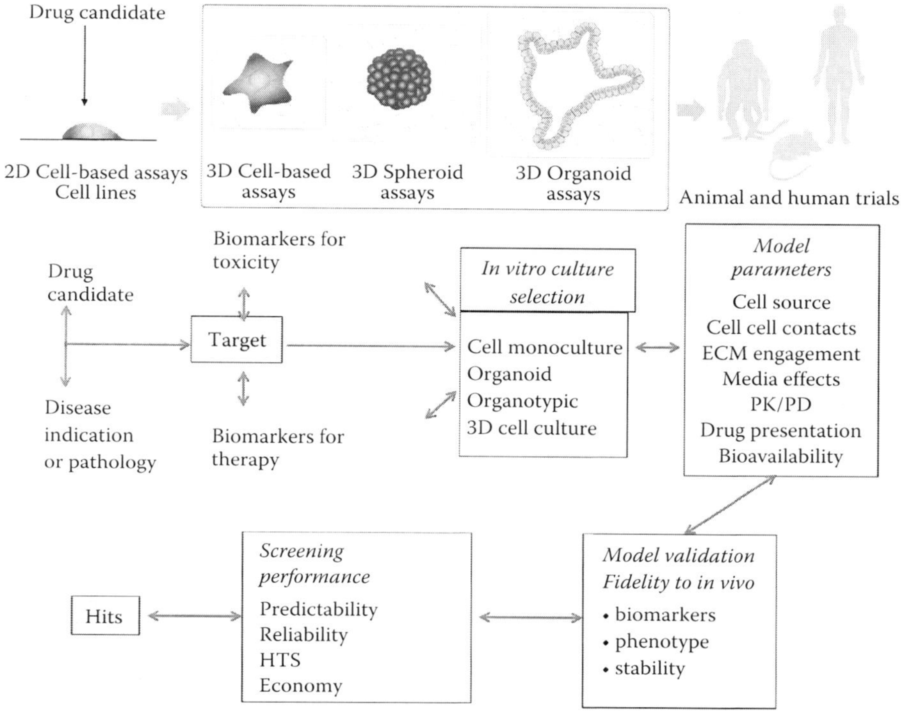

FIGURE 10.2 A flow chart showing the complex steps of a drug candidate being validated to become one of "hits" and illustrating how various different *in vitro* systems can be used efficiently for prediction of *in vivo* and human toxicities before clinical trials. (Adapted from Astashkina, A., et al., *Pharmacol. Ther.*, 134, 82–106, 2012; Ranga, A., et al., *Adv. Drug. Deliv. Rev.* 69–70, 19–28, 2014.)

in vitro skin models are discussed here in detail with increasing complexity from the simplest model, such as reconstructed human epidermis (RHE), to most complex bioengineered or tissue-engineered model, such as skin spheroids or bioprinted skin integrated into a chip microfluidic system (MacNeil 2007; Elliott and Yuan 2011; Groeber, Holeiter et al. 2012; Atac, Wagner et al. 2013; Kimlin, Kassis et al. 2013; Peck and Wang 2013; Wagner, Materne et al. 2013; Lee, Singh et al. 2014; Mathes, Ruffner et al. 2014; Vellonen, Malinen et al. 2014; Maschmeyer, Lorenz et al. 2015).

10.2.1 Reconstructed Human Epidermis

The first *in vitro* skin models were developed to distinguish between corrosive and noncorrosive substances next to the transcutaneous electrical resistance (TER) method (Perkins, Osborne et al. 1996), and after several validation studies and recognition by the European Center for the Validation of Alternative Methods (ECVAM), the first generation of RHE—EpiDerm™, SkinEthic™, EST-1000 method, modified

EpiDerm™ SIT, and even recent LabCyte EPI-MODEL 24—were all evaluated and validated for prediction of skin irritancy (Liebsch, Traue et al. 2000; Katoh, Hamajima et al. 2009). Numerous groups have characterized and compared different commercially available RHE models with their own models to investigate the prediction of human skin toxicity and irritation responses with these *in vitro* models, particularly compared to *in vivo* data (Boelsma, Gibbs et al. 2000; Coquette, Berna et al. 2003; Poumay and Coquette 2007; Tfayli, Piot et al. 2007; Frankart, Malaisse et al. 2012; Guiraud, Hernandez-Pigeon et al. 2014). Due to these combined efforts, the number of necessary animal experiments could be reduced significantly and replaced by these assays for various purposes. RHE with melanocytes (or pigmented RHE) have shown to be able to detect phototoxicity of substances and could be a suitable alternative test for phototoxicity (Augustin, Collombel et al. 1997; Lelievre, Justine et al. 2007). Indeed, the 3D pigmented skin model is one of the most established and physiologically proved models (Duval, Chagnoleau et al. 2012), and these models may serve as valuable tools in the drug assessment and development for patients suffering from vitiligo. However, these simple RHE models are still limited to be a complete replacement of animal models (Schmook, Meingassner et al. 2001; Schreiber, Mahmoud et al. 2005) as they are all lacking in epidermal-dermal crosstalk and cell-cell interactions and skin appendages such as hair follicles and glands, which affect and change skin permeability and penetration of substances to the greater extent (Brohem, Cardeal et al. 2011). Thus, the development of *in vitro* skin models which can recapitulate or simulate the penetration of substances and absorption of drugs is of great interest to both the cosmetic and pharmaceutical industries, and more complex tissue-engineered skin models with heterogeneous cells and features such as hair and glands have been always on demand.

10.2.2 Full Thickness Tissue-Engineered Skin Models with Multiple Cell Types

After numerous inventions and uses of RHE in pharmacological research and industry, full-thickness *in vitro* skin models with dermal and hypodermal layers containing fibroblasts or/and other cells have recently started receiving more attention. As mentioned earlier, heterogeneous multicell populations and interactions seemed to be crucial to recapitulate the main barrier functions of skin, whereby more complex dermatological questions where molecular crosstalk and balanced epidermal and dermal responses involved can be investigated. Epidermal and dermal cells are known to regulate the growth of keratinocytes and fibroblasts via double-paracrine mechanism (Maas-Szabowski, Shimotoyodome et al. 1999; El Ghalbzouri, Lamme et al. 2002; Boehnke, Mirancea et al. 2007; Wong, McGrath et al. 2007) and their interactions play an essential role in wound healing, skin contraction, and remodeling in the contact of toxic chemicals (Falanga, Isaacs et al. 2002; Sun, Jackson et al. 2006; Brohem, Cardeal et al. 2011; Schmidt and Horsley 2013). Based on these facts, there is no doubt that the development and use of full-thickness skin models with multiple cell types are essential to gain more clinically relevant data from toxicological *in vitro* studies and to hope for an alternative validated *in vitro* penetration test system. With TESTSKIN™ as the first commercially available *in vitro* full-thickness skin test system, various different techniques and biomaterials were employed for dermal layer reconstruction. Fibroblasts are seeded into hydrogels or temporary carriers such as collagen, fibrin, hyaluronic material, composite-natural scaffolds, or sponges (Stark, Willhauck et al. 2004; Stark, Boehnke et al. 2006; Boehnke, Mirancea et al. 2007; Auxenfans, Fradette et al. 2009; Kuroyanagi, Yamamoto et al. 2014), and even collagen-hybrid synthetic polymers (Ng, Khor et al. 2004; Chen, Sato et al. 2005; Venugopal, Zhang et al. 2006) for better mechanical stability. With additional dermal layer, the barrier properties of full thickness models are much closer to those of *in vivo* skin models and significantly better than simple RHE models. As shown in Figure 10.3a, physical and morphological structures look a lot more similar to those of native human skin (Asbill, Kim et al. 2000; Tai, Goto et al. 2004; Schafer-Korting, Mahmoud et al. 2008; Shibayama, Hisama et al. 2008). This advancement due to addition of fibroblasts and dermal components led many researchers to speculate on the effects of additional cell types and features which have not yet been integrated into existing skin models, as fibroblasts are not the only other cell type that is crucial to skin reconstruction and regeneration. Furthermore, dermal modulation of human epidermal pigmentation and the influence of fibroblasts on melanocyte proliferation and melanin distribution/degradation were reported (Hedley, Layton et al. 2002; Cario-Andre, Pain et al. 2006). Since it was not too difficult to add melanocytes to the epidermal compartment (Figure 10.3b), the 3D pigmented skin model has been developed and fully used as one of the most established and physiologically

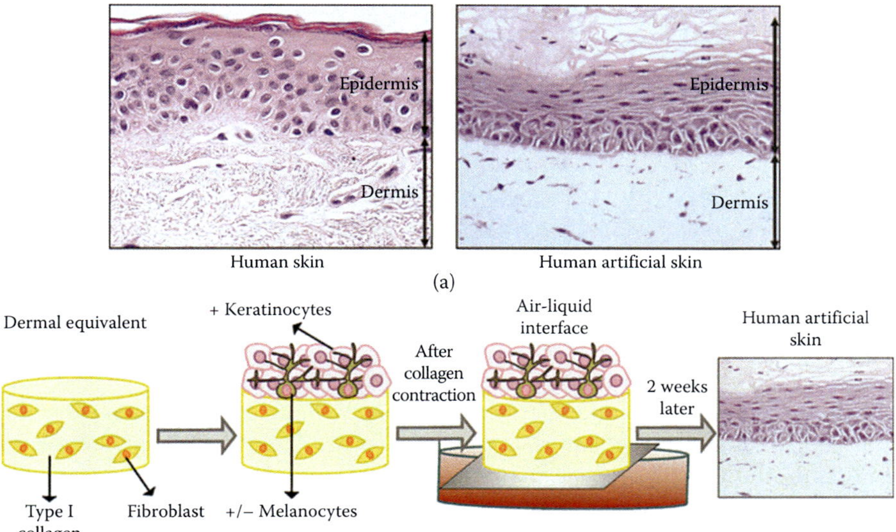

FIGURE 10.3 Histological analysis (HE) of human facial skin and human artificial skin (a) and a schematic diagram of preparation for artificial bilayer skin model (b). HE demonstrates that the bilayered reconstructed skin equivalent well resembles human skin in terms of the thickness of the epidermis and dermis. (Based on Brohem, C.A., et al., *Pigment. Cell Melanoma Res.*, 24, 35–50, 2011.)

proved skin models (Duval, Chagnoleau et al. 2012). Another important cell type to be implemented in the *in vitro* skin reconstruct are Langerhans cells (LCs), which are epithelial dendritic cells responsible for epidermal immunological reactions (Figure 10.1) (Kanitakis 2002; Groeber, Holeiter et al. 2012). Various groups have tried to implement this cell type (Regnier, Staquet et al. 1997; Regnier, Patwardhan et al. 1998; Sivard, Dezutter-Dambuyant et al. 2003; Facy, Flouret et al. 2004; Facy, Flouret et al. 2005; Uchino, Takezawa et al. 2009), and some recent studies indicate the success of Langerhans cells integration and the maturation and migration of LCs in an *in vitro* skin model observed when stimulated by skin sensitizers (Ouwehand, Spiekstra et al. 2011; Ouwehand, Spiekstra et al. 2012), however the integration of LC into the *in vitro* skin models has remained a challenge due to LCs' limited or lack of subculture and expansion methods *in vitro*.

Another important feature of *in vitro* skin equivalents useful for drug development would be a capillary network in the dermis formed by endothelial cells (ECs) and hair follicles formed by primary dermal papilla and outer root sheath cells. Several groups integrated ECs into the dermis of their full-thickness skin models in an attempt to mimic the *in vivo* models and synergistic interactions between ECs and other surrounding cells (Black, Berthod et al. 1998; Ponec, El Ghalbzouri et al. 2004; Tonello, Vindigni et al. 2005; Bellas, Seiberg et al. 2012), however so far there has been no skin model developed or reported with a fully reconstructed capillary network *in vitro*. Also, integration of hair follicles into the *in vitro* skin models for drug development is at an even less mature stage, with only recent findings in de novo formation of hair follicles with sebaceous glands by co-injection of primary dermal papilla and outer root sheath cells into human skin explants (Krugluger, Rohrbacher et al. 2005; Higgins, Chen et al. 2013; Wu, Scott et al. 2014). Other stem cells derived from a variety of tissues such as mesenchymal stem cells (MSCs), adipose-derived stem cells, amniotic fluid–derived stem cells, and pluripotent stem cells have been introduced to reconstructed *in vitro* skin models and indicate their influence on other surrounding cells *in vitro*, therapeutic effects in wound healing, and potential cell sources as a replacement of other cell types (Lorenz, Rupf et al. 2007; Koch, Kuhn et al. 2010; Keck, Haluza et al. 2011; Lu, Yu et al. 2012; Skardal, Mack et al. 2012; van de Kamp, Kramann et al. 2012), yet much elucidation and clarification of how integration of stem cells develop and evolve to be in the *in vitro* system are still required in order to be used in the drug development.

Interestingly and surprisingly, most previous *in vitro* studies related to skin have ignored the possible influence of the hypodermis, the layer located beneath the dermis and various cell types mentioned earlier were integrated either into epidermis or dermis, forming bilayered structures instead of trilayered structures. When adipose-derived stem cells or adipocytes (or pre-adipocytes) were used, they were used as an alternative cell source for other cell types (Lorenz, Rupf et al. 2007; Keck, Haluza et al. 2011; Lu, Yu et al. 2012)—they were not included and induced to create hypodermis. Only recently, a few groups have attempted to create trilayer tissue-engineered *in vitro* skin models (Sugihara, Toda et al. 2001; Trottier, Marceau-Fortier et al. 2008; Bellas, Seiberg et al. 2012; Monfort, Soriano-Navarro et al. 2013), and Bellas et al. exhibited physiological morphologies of each layer with key markers like keratin 10, collagens I and IV, and also glycerol and leptin production due to adipose metabolism. This trend of creating trilayered full-thickness skin models is likely to continue with advancements in *in vitro* tissue biofabrication.

10.3 NOVEL BIOFABRICATION TECHNOLOGY-DRIVEN SKIN MODELS

There is no doubt that recent advancements in the field of tissue engineering and regenerative medicine have been driven and accelerated by advances and novel applications of biotechnology and biofabrication methods such as use of 3D printing or bioprinting, bioreactors for scale-up cell production, electrospinning, "hanging drop" technology for tissue spheroid formation, complex *in vitro* microfluidic systems, and "tissue/organ/lab/body-on-a-chip" systems manufactured with a combination of technologies. Even though there have been myriad novel 3D tissue culture studies and test systems reported, these 3D cell screening models have been presented in the literature with little validation (i.e., simple immunohistochemistry of a few features, cell adhesion, viability, and proliferation activities). These systems are often created by manual, inconsistent, and unreproducible fabrication methods, without any options for scale-up production or experimental flexibility, such that no universal acceptance of any 3D cultured tissue models has yet been achieved (Astashkina, Mann et al. 2012). There is no exception for the tissue-engineered *in vitro* skin models for drug development, and only recently a few research groups have attempted to use these innovative fabrication technologies to develop high-throughput *in vitro* tissue-engineered skin assays, which have greater potential to be used for drug development. In this section, recent reports and advancements in 3D *in vitro* skin models manufactured by a selection of biofabrication technologies, including a complex microfluidic skin-on-a-chip system, are reviewed and discussed.

10.3.1 3D Bioprinting

3D bioprinting, a flexible, automated on-demand platform for the free-form fabrication of complex living architectures, offers significant advantages compared to conventional skin tissue engineering. This approach facilitates the precise control of multiple viable cell compositions and positioning in a layer-by-layer assembly mode, appropriate micro and macro architectures in a wide range of different sizes, and provides the capability to achieve this in a high-throughput, time/cost-effective, and highly reproducible manner (Murphy and Atala 2014). The flat, yet multilayered and highly stratified structures of skin make it a perfect candidate for organ printing or printed tissue assay models compared to other tissues and organs, and several groups have recently presented proof-of-concept studies with this bioprinting technology applied to reconstruct an *in vitro* skin equivalent using keratinocytes or/and fibroblasts as representative cell types (Lee, Debasitis et al. 2009; Koch, Kuhn et al. 2010; Koch, Deiwick et al. 2012; Michael, Sorg et al. 2013; Lee, Singh et al. 2014; Algzlan and Varada 2015; Rimann, Bono et al. 2015) and to prove the feasibility and therapeutic effects of printed cells for skin wound healing (Skardal, Mack et al. 2012). The current commercially available skin equivalent models (i.e., RHE) are still not available in the 96- or 384-well format; however, bioprinting technology will make this high-throughput compatible and feasible for bioprinted skin models with single or multicell types. Although the concept if 3D printing was first introduced by Charles W. Hull decades ago (Murphy and Atala 2014), the application of this technology in the field of tissue engineering and regenerative medicine for personalized medicine and drug assay development is at a very early stage. Originally, 3D printing was developed for nonbiological applications, and proven printable materials vary from metals to polymers. Thus, many of the current challenges present in this field of tissue engineering and drug development are due to or related to technical, material, and cellular aspects of the bioprinting process such as need for optimized resolution, speed, pressure, and limited options for bioinks (materials to deliver/print cells with) that are suitable to print, mechanically stable after printing, and biologically friendly for specific cell types. Materials currently used in the field of soft tissue engineering and regenerative medicine like skin are predominantly limited to naturally derived polymers (i.e., gelatin, collagen, fibrin, and hyaluronic acid) or/and synthetic polymers (i.e., polyethylene glycol [PEG]) (Billiet, Vandenhaute et al. 2012). Due to these limitations, almost all studies related to bioprinted *in vitro* skin models are focusing more on optimizing printing parameters and only limited to simple biochemical analysis of their printed skin constructs and characterization; for instance, printability of skin cells with desired viability, printed cell localization, proliferation, layer-by-layer arrangements, and morphological assessments in 3D (histology and immunofluorescence, or immunohistochemistry). In the most recent published work, Rimann, Bono et al. (2015) performed a fair comparison between commercially available RHE (CellnTech) and their bioprinted skin constructs (bioprinted dermal layer + hand-seeded keratinocytes) for various culture time points (7–42 days [dermal layer: immersion culture] + 5 and 14 days [bi-layers: air-liquid interphase {ALI} culture] and different culture modes [immersion and ALI]). This is the only report on bioprinted *in vitro* skin constructs cultured for such a long time (up to 8 weeks) and included a comparison to commercially available skin products. Koch et al. demonstrated one of the most thorough biological characterizations of the printed bi-layered skin constructs: the 3D arrangement of printed skin cells, post-printing cell proliferation, and basement membrane development (confirmed by the presence of Ki67 and laminin, respectively), and how those printed skin cells evolve intercellular adhesion and communication via adherens and gap junction formation, which are essential for tissue formation and maturation (Figure 10.4) (Koch, Kuhn et al. 2010; Koch, Deiwick et al. 2012). All these studies corroborate and strongly support the positive enhancement of *in vitro* skin fabrication with a combination of advanced technology like 3D printing. Yet much more cellular and molecular characterization over longer culture periods are still needed before bioprinted skin can become widely used for drug screening and development.

Additionally, there are still unresolved issues that we need to improve. First, most printed tissue models including skin are produced with in-house–built or customized bioprinters, resulting in limited reproducibility and transferability of these achievements to other research groups and industries (Lee, Debasitis et al. 2009; Koch, Kuhn et al. 2010; Koch, Deiwick et al. 2012; Michael, Sorg et al. 2013; Lee, Singh et al. 2014; Algzlan and Varada 2015; Rimann, Bono et al. 2015). Second, there are no uniform or standardized bioinks (hydrogels) that have been optimized

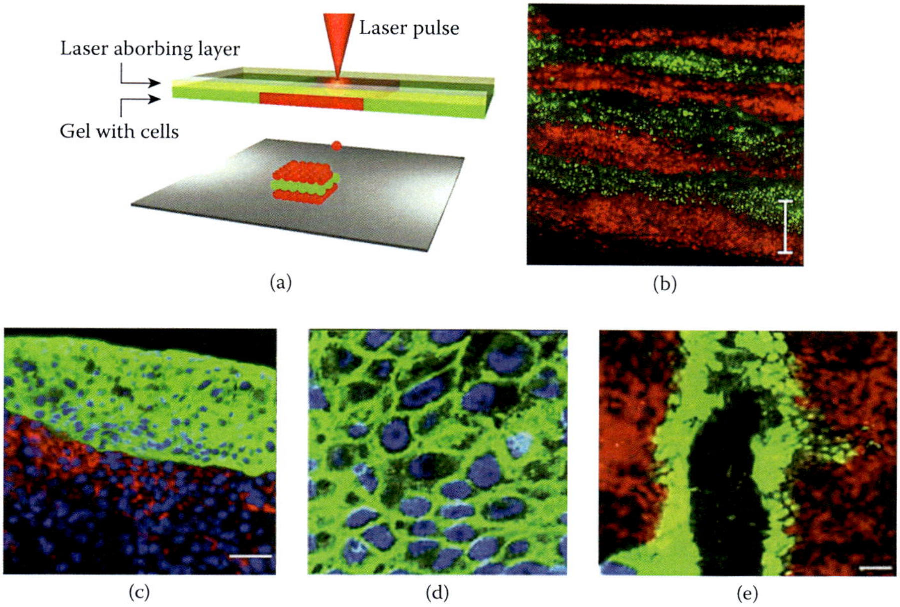

FIGURE 10.4 The delicate layer-by-layer 3D bioprinting of skin constructs is feasible ([a–b]: scale bars = 500 μm) and skin-mimicking bilayered constructs composed of keratinocytes and fibroblasts embedded in collagen were bioprinted and imaged (up to 10 days of culture after printing) (c–e). (c) The keratinocytes stained in green (cytokeratin 14) formed a compact cell-cell organization with fibroblasts stained in red (panreticular fibroblast) and cell nuclei stained in blue (Hoechst 33342). (d–e) Qualitative analysis of adherens and gap junction formation of bioprinted skin constructs. (d) Pan-cadherin–staining (green) and cell nuclei staining with Hoechst 33342 (blue). (e) Gap junction coupling visualized in green with lucifer yellow; the keratinocytes are depicted in red (as transfected with mCherry) (scale bars = 50 μm). (Adapted and modified from Koch, L., et al., *Biotechnol. Bioeng.*, 109, 1855–1863, 2012.)

for printing applications for specific cell types or tissues. So far, the outcomes are promising and there is a great potential for bioprinted *in vitro* skin constructs to be used in drug development, but further optimization of this system in terms of technical and biological aspects is necessary for the complete standardization of 3D skin tissue model generation for drug development. Furthermore, both short term (<3 weeks) and long-term culture studies of *in vitro* skin tissue models and their careful characterization compared to existing models are of great interest, and are crucial next steps to monitor the short- and long-term effects of multiple drug administration (Rimann, Bono et al. 2015). In addition, in order to fully utilize the advantages of bioprinting, bioprinted *in vitro* skin constructs based on multiheterogeneous cell populations more than single- or two-cell types, with carefully arranged stratified structures need to be created and thoroughly characterized and analyzed, whereby superiority of such bioprinted 3D skin constructs would be validated.

10.3.2 3D Multicellular Spheroids

Spheroid 3D culture or microtissue models originally stem from studies of spontaneous formation of cell aggregates after co-culturing heterogeneous cells with stationary and rotational techniques in as early as 1952 (Moscona and Moscona 1952; Moscona 1961), and they have been quickly adapted and evolved for generating *in vitro* tissue culture multicellular spheroids using both normal and cancer cell lines. Several assays have been developed for the use of spheroids for efficacy and penetration of drugs and cellular interactions (Landry and Freyer 1984; Durand 1990; Kunz-Schughart 1999; Smalley, Lioni et al. 2006; Ma, Jiang et al. 2012), and recently the tumor spheroids model, including skin spheroids, has become a platform to screen anticancer compounds (Ho, Pham et al. 2010; Patel, Aryasomayajula et al. 2015). Traditionally, spinner flasks (Franko, Freedman et al. 1984; Mueller-Klieser 1984), microgravity with rotating-wall vessels (Schrader, Kremling et al. 2007; Tanaka, Tanaka et al. 2013), and spontaneous aggregation (Kelm, Timmins et al. 2003) methods were employed for formation of cell aggregates. Most updated methods of 3D spheroid fabrication expand these traditional approaches to include spontaneous aggregate formation from single cell to multicellular levels (co-cultures) with rocking/rotary cell cultures and automated hanging drop technology (Timmins, Dietmair et al. 2004; Timmins and Nielsen 2007; Foty 2011). Hanging drop technologies are developed to cultivate spheroids *in vitro* by placing drops of suspended cells on a surface incubated upside down, and this laborious process has been greatly simplified and improved in reproducibility up to the level of high-throughput screening (HTS) system by combined technologies of modified plate and fluidic handling (commercially available as GravityPlus™ technology [InSphero] or PERFECTA3D™ hanging drop plates [3D Biomatrix]) (Pampaloni, Reynaud et al. 2007; Mazzoleni, Di Lorenzo et al. 2009; Astashkina, Mann et al. 2012; Astashkina and Grainger 2014). In particular, the skin spheroid model has been extensively employed for the study of melanoma (Rofstad, Wahl et al. 1986; Ahammer, DeVaney et al. 2001; Rieber and Rieber 2006; Marrero, Messina et al. 2009; Na, Seok et al. 2009; Schatton and Frank 2010; Kuch, Schreiber et al. 2013; Mo, Sun et al. 2013; Vorsmann, Groeber et al. 2013; Haass and Schumacher 2014; Larson, Lee et al. 2014), as it has proven to be a versatile, relatively simple, and easily augmentable model system for preclinical drug development and toxicity testing. The model does not require external scaffolds to aggregate or much time to generate, could be used with other tissue engineering and regenerative medicine inspired technologies (i.e., use of bioreactors, bioprinters, with automated manufacture capability), and most importantly, unlike bioprinting, this method is much more convenient and standardized as it is already commercially available through several companies (i.e., InSphero, 3D Biomatrix, Microtissues) for technical support. The 3D melanoma spheroids model implanted into a collagen gel more closely represents the *in vivo* tumor architecture and microenvironment than adherent cell culture, and it even mimics the tumor heterogeneity observed *in vivo* with new oxygen/nutrient gradient created in a hypoxic zone allowing melanoma cells to be interacting with their stroma (Smalley, Haass et al. 2006; Smalley, Lioni et al. 2008; Smalley 2013). Moreover, the 3D spheroid model recapitulates the behavior of melanomas *in vivo* such that cell lines from different origins show different growth and invasion characteristics reflecting the original state of aggressiveness (Smalley, Haass et al. 2006). In addition to this close resemblance to the nature of *in vivo*, the 3D skin spheroids model seems to be significantly more resistant to certain antitumor drugs and hence provides a more realistic prediction of *in vivo* drug responses (Nicholson, Bibby et al. 1997; Smalley, Haass et al. 2006; Elliott and Yuan 2011). However, while the heterogeneity of multicellular skin spheroids provides multiple advantages, it also involves challenges to control their size. As large cellular aggregates usually require a careful control of the diffusion of nutrients and chemical agents as well as the gas exchange, reliability and reproducibility of this model for compound diffusion/penetration could not be guaranteed for each trial (Pampaloni, Reynaud et al. 2007). Also, the spheroid model is limited in normal cell-cell interactions and maintains its shape and functions for a shorter culture time period such that long-term toxicology analysis is not yet feasible or reliable. Perhaps due to these reasons, there have not been many studies of drug responses reported directly correlating or comparing to *in vivo* models even with such superior advantages and similarity to *in vivo* models (Elliott and Yuan 2011). Nevertheless, the skin spheroid model is one of the more significant, successful, and well-established 3D *in vitro* skin models currently in use and is in the process of further development, especially for melanoma drug metabolism and toxicity in a more physiological setting (Vorsmann, Groeber et al. 2013). This can be relatively easily integrated into a bioprinting system or microfluidic system as an organoid (a representative of a tissue or organ). Thus there is no doubt of its great potential in future drug development.

10.3.3 Skin-on-a-Chip

The field of microfluidics or lab-on-a-chip technology aims to improve and extend the possibilities of bioassays with the idea of miniaturization and automation. Recent advances in stem cell research, regenerative medicine, biomaterials, tissue engineering, and microfluidics accelerated the development of new 3D *in vitro* models closely mimicking human tissues and organs, and the lab-on-a-chip technology has led to a paradigm shift in drug discovery and delivery (LaVan, Lynn et al. 2002; Weigl, Bardell et al. 2003; Ghaemmaghami, Hancock et al. 2012; Neuzi, Giselbrecht et al. 2012; Vladisavljevic, Khalid et al. 2013; Polini, Prodanov et al. 2014; Ramadan and Gijs 2015). The lab-on-a-chip system could be built to be a single organ on a chip to MOC (Figure 10.5, top), and the ultimate goal of the MOC is to provide analytical platforms mimicking human normal and pathological physiology *in vitro,* with enabled continuous monitoring and quantification of the cellular responses to various controlled external stimuli as one organ to multi-organs' response. The combined biofabrication technology of engineered tissues (i.e., 3D bioprinting or hanging drop) and microfluidics technologies enable precise control of cell/tissue positions and interactions and of static or dynamic fluid flows, creating better in vitro cell culture microenvironments with spatiotemporal chemical gradients and mechanical cues. In turn, this resembles in vivo microenvironments for cells and tissues more closely, therefore makes the model more physiologically relevant (Ghaemmaghami, Hancock et al. 2012; Polini, Prodanov et al. 2014). Other advantages of the lab-on-a-chip technology include reliable reproducibility

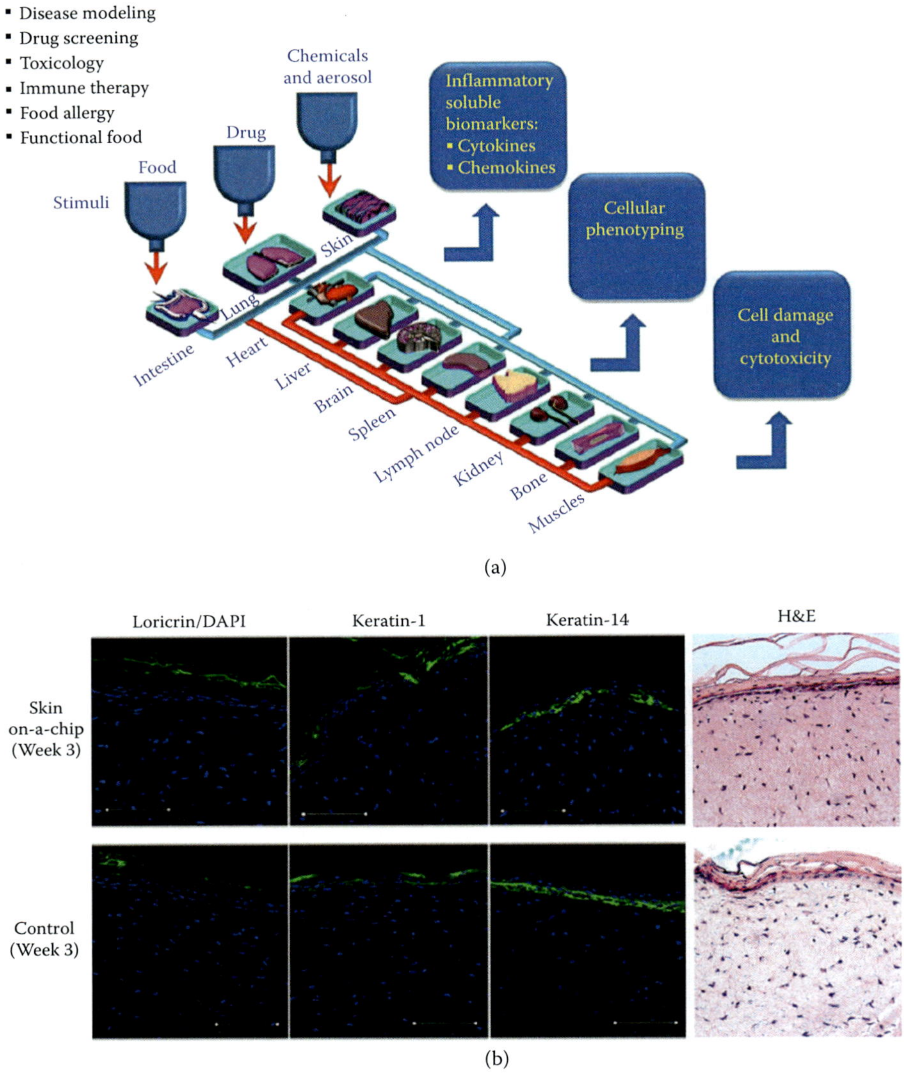

FIGURE 10.5 A schematic diagram of multi-organs-on-a-chip (MOC) (a) and histological and imunohistofluorscent image analysis of skin constructs integrated in on a chip device (b). Advancements in tissue engineering, regenerative medicine, and microfluidics allow screening of the whole body immunity for many different applications with a device like MOC, such as for disease modeling, drug screening, toxicology, food allergy analysis, immune therapy development, etc. (Reproduced with permission from Ramadan, Q. and Gijs, M.A., *Lab. Chip.*, 15, 614–636, 2015.) Abaci et al. (2015) demonstrated that their skin equivalent cultured on a chip (bottom) is capable of recirculating the media at desired flow rates and the system maintains the barrier function of skin equivalents well for up to 3 weeks, with proliferating keratinocytes and localized and differentiated keratinocytes starting at 1 week *in vitro*. (Adapted from Ramadan, Q. and Gijs, M.A., *Lab. Chip.*, 15, 614–636, 2015; Abaci, H.E., et al., *Lab. Chip.*, 15, 882–888, 2015.)

and precise implementation of new assay design principles in a cost-effective manner (Esch, Smith et al. 2014). Due to these advantages, various tissue models on a chip have been developed in academia and industry, and excellent comprehensive reviews exist for bone, breast, cardiac, corneal, intestine, liver, lung, muscle, and tumor tissue models (Elliott and Yuan 2011; Ghaemmaghami, Hancock et al. 2012; Huh, Torisawa et al. 2012; Tsui, Lee et al. 2013; Young 2013; Lancaster and Knoblich 2014; Ranga, Gjorevski et al. 2014; Ramadan and Gijs 2015). Unfortunately, to date there are only a few groups who have reported or explored specifically for the skin cell or tissue-related model on a chip (An, Lee et al. 2012; Atac, Wagner et al. 2013; Brauchle, Johannsen et al. 2013; Hou, Hagen et al. 2013; Wagner, Materne et al. 2013; Abaci, Gledhill et al. 2015; Maschmeyer, Hasenberg et al. 2015; Maschmeyer, Lorenz et al. 2015), and a relatively small number of companies such as Hurel (New Brunswick, NJ), MatTek Corp. (Ashland, MA), and Phenion (Henkel AG & Co, Dusseldorf, Germany) are publicly known to develop skin platforms to drug screening. Also, skin-on-a-chip–related study is still in its infancy. None of the studies currently published have described experiments involving compound/drug screening and testing capabilities through skin-on-a-chip as a single organ on a chip. Our most advanced tissue-engineered 3D skin models with complexity have never been integrated and tested as the *in vitro* skin model used in microfluidics systems. Most studies are at the level of showing how the microfluidics system helps cell culture conditions for integrated simple *in vitro* skin models (i.e., RHE or bilayer skin equivalents) by providing dynamic perfusion and nutrient gradients whereby morphological and physiological characteristics of *in vitro* skin equivalents are well developed and maintained (Figure 10.5, bottom) (Atac, Wagner et al. 2013; Abaci, Gledhill et al. 2015; Maschmeyer, Lorenz et al. 2015). The human full-thickness skin models were integrated into the MOC, and after 9 days of dynamic culture conditions Atac et al. observed that rearrangement and compression of dermal layer and MOC-aided culture systems significantly improved spatiotemporal control of cellular microenvironments compared to traditional *in vitro* assays (Atac, Wagner et al. 2013). The small number of scientific reports directly related to skin-on-a-chip is somewhat surprising, as the barrier tissue analogs like skin must be incorporated into multi-organs-on-a-chip systems to take into account the significantly reduced bioavailability of drugs absorbed through barrier tissue like skin and to better resemble the body response of drug candidates, as shown in Figure 10.5 and as suggested by researchers (Ghaemmaghami, Hancock et al. 2012; Huh, Torisawa et al. 2012; Guo, Higgins et al. 2013). Since this lab-on-a-chip technology is relatively new and is rapidly developing, further integration of advanced tissue-engineered *in vitro* skin constructs (i.e., 3D bio-printed skin or 3D skin spheroids) is anticipated. As Frey, Misun et al. (2014) and Ramadan and Gijs (2015) comprehensibly reviewed and discussed the potential and challenges of multiple 3D spheroids into microfluidic network integration, there is no doubt that the convergence of multiple advanced biotechnologies (i.e., bioprinting + hanging drop + microfluidics: 3D bioprinted tissue spheroids/organs on chips) will make greater advances in biomimetic *in vitro* human skin models for drug testing and development, and many academia and industrial sectors could benefit from this technology and translational research outcomes.

One can also envision creation of diseased skin models on a chip (Figure 10.5). As mentioned earlier, the interaction of heterogeneous multicell populations (i.e., keratinocytes, fibroblasts, melanocytes, and other immune cells) is crucial and tightly controlled in skin, and the disruption of such interactions is usually the cause or trigger of an uncontrolled proliferation of keratinocytes (i.e., psoriasis) or melanocytes (i.e., melanoma) and diseased skin (Jean, Lapointe et al. 2009; Vorsmann, Groeber et al. 2013; Haass and Schumacher 2014). The *in vitro* investigation of these processes in skin equivalents aided with a dynamic perfusion system (i.e., diseased skin-on-a-chip) will offer better understanding of the complex processes that underlie these diseases and thus enhance novel therapies and drug development. Several researchers successfully developed and characterized new psoriatic skin equivalents *in vitro* (Barker, McHale et al. 2004; Harrison, Layton et al. 2007; Tjabringa, Bergers et al. 2008; Jean, Lapointe et al. 2009; Groeber, Holeiter et al. 2012) through use of transglutaminase inhibitors and cytokines (i.e., TNF-α, IL-1α, IL-2 and IL-6). Also, as reviewed earlier in Section 10.3.2, the various 3D forms of the melanoma skin equivalents have already been a great help in closing the gap between the 2D cell culture and *in vivo* studies and in elucidating its aggressive and invasive behavior. However, in order to fully investigate hyper proliferation and tumor angiogenesis/metastasis development *in vitro*, the microfluidic system allowing vascularization, dynamic fluid perfusion, and the circulation and contribution of the immune cells will be necessary. Another interesting skin model on a chip would be wound and wound-healing skin models on a chip. Wound and tissue repair after wounding is the outcome of a series of overlapping events: hemostasis, inflammation, granulated tissue formation, and scar formation. In order to study the wound healing process in detail, it is paramount to simulate this cascade of events in the *in vitro* wound model systems, and the wound/wound healing skin model on a chip could offer extra information on protein synthesis, re-epithelialization, and proliferation. The existing *in vitro* models (i.e., a monolayer cell culture assay and *ex vivo* skin cultures) can only give useful information on migratory and proliferative responses of cells and cytokine releases upon creation of wounds but not on remodeling and repair processes. Some groups already have attempted to simulate wound-related conditions using multiple types of cells on a chip and the influence of chemical/mechanical cues to wound healing and the performance of different wound dressings observed on a chip and its therapeutic potential were reported (Nie, Yamada et al. 2007; Doran, Mills et al. 2009; Xie, Zhang et al. 2011; Sun, Peng et al. 2012; Zhao, Wang et al. 2012; Li, Wang et al. 2015). One group found that the results acquired by microchip model corroborate well with animal experiments for fibrous wound dressings (Zhao, Wang et al. 2012). Despite the lack of commercially available chip-based drug discovery tools, the maturation of lab-on-a-chip technology-inspired devices and platforms has an increasing impact on many aspects of the drug discovery process and continues to give benefits for miniaturization, automatization, higher reliability, and better control of various *in vitro* conditions compared to traditional *in vitro* assays. In this regard, the diseased or damaged *in vitro* skin tissue models or even patient-specific *in vitro* skin models on a chip would offer more accurate investigation of the molecular signals, cell-cell interactions, and biological responses toward compounds and potential drug candidates for various skin diseases (Ghaemmaghami, Hancock et al. 2012; Esch, Smith et al. 2014; Esch, Bahinski et al. 2015).

10.4 CONCLUSIONS

Despite the tremendous advancements in *in vitro* skin models, there is still much more work to be done before these models approach the complexity and function of native *in vivo* skin. The convergence of these various emerging 3D biofabrication and microfluidics technologies in skin tissue engineering and regenerative medicine will not immediately revolutionize drug discovery and development. However, there is no doubt that it will significantly contribute to bridge the gaps between simple monolayer *in vitro* assays and *in vivo* models and eventually to the development of novel drug screening methods and tools at cellular or tissue levels. As highlighted in this chapter, skin-on-a-chip and skin models integrated into the MOCs are probably the most complex and advanced skin-based platform that could be developed and used for drug discovery and development. There are still significant challenges and concerns that must be tackled and overcome—for instance, standardization and validation for *in vitro* skin equivalents manufacture, storage, and packaging, optimization of skin-on-a-chip device design and development, and development of optimized common culture medium for multicellular skin equivalents. Given that existing skin models already serve well as predictive tools to investigate new therapeutically active agents for specific purposes and much biofabrication and microfluidics technologies are emerging and evolving rapidly, the 3D *in vitro* skin models recapitulating critical features and responses of normal or diseased human skin will represent valuable drug screening and developing tools with great potential.

REFERENCES

Abaci, H. E., K. Gledhill, Z. Guo, A. M. Christiano and M. L. Shuler (2015). Pumpless microfluidic platform for drug testing on human skin equivalents. *Lab Chip* **15**(3): 882–888.

Ahammer, H., T. T. DeVaney and H. A. Tritthart (2001). Fractal dimension of K1735 mouse melanoma clones and spheroid invasion in vitro. *Eur Biophys J* **30**(7): 494–499.

Algzlan, H. and S. Varada (2015). Three-dimensional printing of the skin. *JAMA Dermatol* **151**(2): 207.

An, J. H., J. S. Lee, J. R. Chun, B. K. Oh, M. D. Kafi and J. W. Choi (2012). Cell chip-based monitoring of toxic effects of cosmetic compounds on skin fibroblast cells. *J Nanosci Nanotechnol* **12**(7): 5143–5148.

Asbill, C., N. Kim, A. El-Kattan, K. Creek, P. Wertz and B. Michniak (2000). Evaluation of a human bio-engineered skin equivalent for drug permeation studies. *Pharm Res* **17**(9): 1092–1097.

Astashkina, A. and D. W. Grainger (2014). Critical analysis of 3-D organoid in vitro cell culture models for high–throughput drug candidate toxicity assessments. *Adv Drug Deliv Rev* **69–70**: 1–18.

Astashkina, A., B. Mann and D. W. Grainger (2012). A critical evaluation of in vitro cell culture models for high-throughput drug screening and toxicity. *Pharmacol Ther* **134**(1): 82–106.

Atac, B., I. Wagner, R. Horland, R. Lauster, U. Marx, A. G. Tonevitsky, R. P. Azar and G. Lindner (2013). Skin and hair on-a-chip: In vitro skin models versus ex vivo tissue maintenance with dynamic perfusion. *Lab Chip* **13**(18): 3555–3561.

Augustin, C., C. Collombel and O. Damour (1997). Use of dermal equivalent and skin equivalent models for identifying phototoxic compounds in vitro. *Photodermatol Photoimmunol Photomed* **13**(1–2): 27–36.

Auxenfans, C., J. Fradette, C. Lequeux, L. Germain, B. Kinikoglu, N. Bechetoille, F. Braye, F. A. Auger and O. Damour (2009). Evolution of three dimensional skin equivalent models reconstructed in vitro by tissue engineering. *Eur J Dermatol* **19**(2): 107–113.

Barker, C. L., M. T. McHale, A. K. Gillies, J. Waller, D. M. Pearce, J. Osborne, P. E. Hutchinson, G. M. Smith and J. H. Pringle (2004). The development and characterization of an in vitro model of psoriasis. *J Invest Dermatol* **123**(5): 892–901.

Bellas, E., M. Seiberg, J. Garlick and D. L. Kaplan (2012). In vitro 3D full-thickness skin-equivalent tissue model using silk and collagen biomaterials. *Macromol Biosci* **12**(12): 1627–1636.

Biedermann, T., S. Bottcher-Haberzeth, A. S. Klar, D. S. Widmer, L. Pontiggia, A. D. Weber, D. M. Weber, C. Schiestl, M. Meuli and E. Reichmann (2015). The influence of stromal cells on the pigmentation of tissue-engineered dermo-epidermal skin grafts. *Tissue Eng Part A* **21**(5–6): 960–969.

Billiet, T., M. Vandenhaute, J. Schelfhout, S. Van Vlierberghe and P. Dubruel (2012). A review of trends and limitations in hydrogel-rapid prototyping for tissue engineering. *Biomaterials* **33**(26): 6020–6041.

Black, A. F., F. Berthod, N. L'Heureux, L. Germain and F. A. Auger (1998). In vitro reconstruction of a human capillary-like network in a tissue-engineered skin equivalent. *FASEB J* **12**(13): 1331–1340.

Blais, M., R. Parenteau-Bareil, S. Cadau and F. Berthod (2013). Concise review: Tissue-engineered skin and nerve regeneration in burn treatment. *Stem Cells Transl Med* **2**(7): 545–551.

Boehnke, K., N. Mirancea, A. Pavesio, N. E. Fusenig, P. Boukamp and H. J. Stark (2007). Effects of fibroblasts and microenvironment on epidermal regeneration and tissue function in long-term skin equivalents. *Eur J Cell Biol* **86**(11–12): 731–746.

Boelsma, E., S. Gibbs, C. Faller and M. Ponec (2000). Characterization and comparison of reconstructed skin models: Morphological and immunohistochemical evaluation. *Acta Derm Venereol* **80**(2): 82–88.

Bottcher-Haberzeth, S., A. S. Klar, T. Biedermann, C. Schiestl, C. Meuli-Simmen, E. Reichmann and M. Meuli (2013). "Trooping the color": Restoring the original donor skin color by addition of melanocytes to bioengineered skin analogs. *Pediatr Surg Int* **29**(3): 239–247.

Brauchle, E., H. Johannsen, S. Nolan, S. Thude and K. Schenke-Layland (2013). Design and analysis of a squamous cell carcinoma in vitro model system. *Biomaterials* **34**(30): 7401–7407.

Brohem, C. A., L. B. Cardeal, M. Tiago, M. S. Soengas, S. B. Barros and S. S. Maria-Engler (2011). Artificial skin in perspective: Concepts and applications. *Pigment Cell Melanoma Res* **24**(1): 35–50.

Cario-Andre, M., C. Pain, Y. Gauthier, V. Casoli and A. Taieb (2006). In vivo and in vitro evidence of dermal fibroblasts influence on human epidermal pigmentation. *Pigment Cell Res* **19**(5): 434–442.

Chapman, K. E., A. D. Thomas, J. W. Wills, S. Pfuhler, S. H. Doak and G. J. Jenkins (2014). Automation and validation of micronucleus detection in the 3D EpiDerm human reconstructed skin assay and correlation with 2D dose responses. *Mutagenesis* **29**(3): 165–175.

Chen, G., T. Sato, H. Ohgushi, T. Ushida, T. Tateishi and J. Tanaka (2005). Culturing of skin fibroblasts in a thin PLGA-collagen hybrid mesh. *Biomaterials* **26**(15): 2559–2566.

Chuong, C. M., B. J. Nickoloff, P. M. Elias, L. A. Goldsmith, E. Macher, P. A. Maderson, J. P. Sundberg, et al. (2002). What is the 'true' function of skin? *Exp Dermatol* **11**(2): 159–187.

Cichorek, M., M. Wachulska, A. Stasiewicz and A. Tyminska (2013). Skin melanocytes: Biology and development. *Postepy Dermatol Alergol* **30**(1): 30–41.

Coquette, A., N. Berna, A. Vandenbosch, M. Rosdy, B. De Wever and Y. Poumay (2003). Analysis of interleukin-1alpha (IL-1alpha) and interleukin-8 (IL-8) expression and release in in vitro reconstructed human epidermis for the prediction of in vivo skin irritation and/or sensitization. *Toxicol In vitro* **17**(3): 311–321.

de Vries, R. B., M. Leenaars, J. Tra, R. Huijbregtse, E. Bongers, J. A. Jansen, B. Gordijn and M. Ritskes-Hoitinga (2015). The potential of tissue engineering for developing alternatives to animal experiments: A systematic review. *J Tissue Eng Regen Med* **9**(7): 771–8.

Doran, M. R., R. J. Mills, A. J. Parker, K. A. Landman and J. J. Cooper-White (2009). A cell migration device that maintains a defined surface with no cellular damage during wound edge generation. *Lab Chip* **9**(16): 2364–2369.

Driskell, R. R., C. A. Jahoda, C. M. Chuong, F. M. Watt and V. Horsley (2014). Defining dermal adipose tissue. *Exp Dermatol* **23**(9): 629–631.

Durand, R. E. (1990). Multicell spheroids as a model for cell kinetic studies. *Cell Tissue Kinet* **23**(3): 141–159.

Duval, C., C. Chagnoleau, F. Pouradier, P. Sextius, E. Condom and F. Bernerd (2012). Human skin model containing melanocytes: Essential role of keratinocyte growth factor for constitutive pigmentation-functional response to alpha-melanocyte stimulating hormone and forskolin. *Tissue Eng Part C Methods* **18**(12): 947–957.

El-Ghalbzouri, A., S. Gibbs, E. Lamme, C. A. Van Blitterswijk and M. Ponec (2002). Effect of fibroblasts on epidermal regeneration. *Br J Dermatol* **147**(2): 230–243.

El Ghalbzouri, A., E. Lamme and M. Ponec (2002). Crucial role of fibroblasts in regulating epidermal morphogenesis. *Cell Tissue Res* **310**(2): 189–199.

Elliott, N. T. and F. Yuan (2011). A review of three-dimensional in vitro tissue models for drug discovery and transport studies. *J Pharm Sci* **100**(1): 59–74.

Esch, E. W., A. Bahinski and D. Huh (2015). Organs–on-chips at the frontiers of drug discovery. *Nat Rev Drug Discov* **14**(4): 248–260.

Esch, M. B., A. S. Smith, J. M. Prot, C. Oleaga, J. J. Hickman and M. L. Shuler (2014). How multi-organ microdevices can help foster drug development. *Adv Drug Deliv Rev* **69–70**: 158–169.

Eungdamrong, N. J., C. Higgins, Z. Guo, W. H. Lee, B. Gillette, S. Sia and A. M. Christiano (2014). Challenges and promises in modeling dermatologic disorders with bioengineered skin. *Exp Biol Med (Maywood)* **239**(9): 1215–1224.

Facy, V., V. Flouret, M. Regnier and R. Schmidt (2004). Langerhans cells integrated into human reconstructed epidermis respond to known sensitizers and ultraviolet exposure. *J Invest Dermatol* **122**(2): 552–553.

Facy, V., V. Flouret, M. Regnier and R. Schmidt (2005). Reactivity of Langerhans cells in human reconstructed epidermis to known allergens and UV radiation. *Toxicol In vitro* **19**(6): 787–795.

Falanga, V., C. Isaacs, D. Paquette, G. Downing, N. Kouttab, J. Butmarc, E. Badiavas and J. Hardin-Young (2002). Wounding of bioengineered skin: Cellular and molecular aspects after injury. *J Invest Dermatol* **119**(3): 653–660.

Foty, R. (2011). A simple hanging drop cell culture protocol for generation of 3D spheroids. *J Vis Exp* (51) e2720.

Frankart, A., J. Malaisse, E. De Vuyst, F. Minner, C. L. de Rouvroit and Y. Poumay (2012). Epidermal morphogenesis during progressive in vitro 3D reconstruction at the air-liquid interface. *Exp Dermatol* **21**(11): 871–875.

Franko, A. J., H. I. Freedman and C. J. Koch (1984). Oxygen supply to spheroids in spinner and liquid-overlay culture. *Recent Results Cancer Res* **95**: 162–167.

Frey, O., P. M. Misun, D. A. Fluri, J. G. Hengstler and A. Hierlemann (2014). Reconfigurable microfluidic hanging drop network for multi-tissue interaction and analysis. *Nat Commun* **5**: 4250.

Ghaemmaghami, A. M., M. J. Hancock, H. Harrington, H. Kaji and A. Khademhosseini (2012). Biomimetic tissues on a chip for drug discovery. *Drug Discov Today* **17**(3–4): 173–181.

Griffith, L. G. and M. A. Swartz (2006). Capturing complex 3D tissue physiology in vitro. *Nat Rev Mol Cell Biol* **7**(3): 211–224.

Groeber, F., M. Holeiter, M. Hampel, S. Hinderer and K. Schenke-Layland (2012). Skin tissue engineering—In vivo and in vitro applications. *Clin Plast Surg* **39**(1): 33–58.

Guiraud, B., H. Hernandez-Pigeon, I. Ceruti, S. Mas, Y. Palvadeau, C. Saint-Martory, N. Castex-Rizzi, H. Duplan and S. Bessou-Touya (2014). Characterization of a human epidermis model reconstructed from hair follicle keratinocytes and comparison with two commercially models and native skin. *Int J Cosmet Sci* **36**(5): 485–493.

Guo, Z., C. A. Higgins, B. M. Gillette, M. Itoh, N. Umegaki, K. Gledhill, S. K. Sia and A. M. Christiano (2013). Building a microphysiological skin model from induced pluripotent stem cells. *Stem Cell Res Ther* **4**(Suppl 1): S2.

Haass, N. K. and U. Schumacher (2014). Melanoma never says die. *Exp Dermatol* **23**(7): 471–472.

Harrison, C. A., C. M. Layton, Z. Hau, A. J. Bullock, T. S. Johnson and S. MacNeil (2007). Transglutaminase inhibitors induce hyperproliferation and parakeratosis in tissue-engineered skin. *Br J Dermatol* **156**(2): 247–257.

Hedley, S. J., C. Layton, M. Heaton, K. H. Chakrabarty, R. A. Dawson, D. J. Gawkrodger and S. MacNeil (2002). Fibroblasts play a regulatory role in the control of pigmentation in reconstructed human skin from skin types I and II. *Pigment Cell Res* **15**(1): 49–56.

Higgins, C. A., J. C. Chen, J. E. Cerise, C. A. Jahoda and A. M. Christiano (2013). Microenvironmental reprogramming by three-dimensional culture enables dermal papilla cells to induce de novo human hair-follicle growth. *Proc Natl Acad Sci U S A* **110**(49): 19679–19688.

Ho, W. J., E. A. Pham, J. W. Kim, C. W. Ng, J. H. Kim, D. T. Kamei and B. M. Wu (2010). Incorporation of multicellular spheroids into 3-D polymeric scaffolds provides an improved tumor model for screening anticancer drugs. *Cancer Sci* **101**(12): 2637–2643.

Hou, L., J. Hagen, X. Wang, I. Papautsky, R. Naik, N. Kelley-Loughnane and J. Heikenfeld (2013). Artificial microfluidic skin for in vitro perspiration simulation and testing. *Lab Chip* **13**(10): 1868–1875.

Huh, D., Y. S. Torisawa, G. A. Hamilton, H. J. Kim and D. E. Ingber (2012). Microengineered physiological biomimicry: Organs-on-chips. *Lab Chip* **12**(12): 2156–2164.

Jean, J., M. Lapointe, J. Soucy and R. Pouliot (2009). Development of an in vitro psoriatic skin model by tissue engineering. *J Dermatol Sci* **53**(1): 19–25.

Kanitakis, J. (2002). Anatomy, histology and immunohistochemistry of normal human skin. *Eur J Dermatol* **12**(4): 390–399; quiz 400–391.

Katoh, M., F. Hamajima, T. Ogasawara and K. Hata (2009). Assessment of human epidermal model LabCyte EPI-MODEL for in vitro skin irritation testing according to European Centre for the Validation of Alternative Methods (ECVAM)-validated protocol. *J Toxicol Sci* **34**(3): 327–334.

Keck, M., D. Haluza, D. B. Lumenta, S. Burjak, B. Eisenbock, L. P. Kamolz and M. Frey (2011). Construction of a multi-layer skin substitute: Simultaneous cultivation of keratinocytes and preadipocytes on a dermal template. *Burns* **37**(4): 626–630.

Kelm, J. M., N. E. Timmins, C. J. Brown, M. Fussenegger and L. K. Nielsen (2003). Method for generation of homogeneous multicellular tumor spheroids applicable to a wide variety of cell types. *Biotechnol Bioeng* **83**(2): 173–180.

Kimlin, L., J. Kassis and V. Virador (2013). 3D in vitro tissue models and their potential for drug screening. *Expert Opin Drug Discov* **8**(12): 1455–1466.

Klar, A. S., S. Bottcher-Haberzeth, T. Biedermann, C. Schiestl, E. Reichmann and M. Meuli (2014). Analysis of blood and lymph vascularization patterns in tissue-engineered human dermo-epidermal skin analogs of different pigmentation. *Pediatr Surg Int* **30**(2): 223–231.

Klein, J., P. A. Permana, M. Owecki, G. N. Chaldakov, M. Bohm, G. Hausman, C. M. Lapiere, et al. (2007). What are subcutaneous adipocytes really good for? *Exp Dermatol* **16**(1): 45–70.

Koch, L., A. Deiwick, S. Schlie, S. Michael, M. Gruene, V. Coger, D. Zychlinski, et al. (2012). Skin tissue generation by laser cell printing. *Biotechnol Bioeng* **109**(7): 1855–1863.

Koch, L., S. Kuhn, H. Sorg, M. Gruene, S. Schlie, R. Gaebel, B. Polchow, et al. (2010). Laser printing of skin cells and human stem cells. *Tissue Eng Part C Methods* **16**(5): 847–854.

Krugluger, W., W. Rohrbacher, K. Laciak, K. Moser, C. Moser and J. Hugeneck (2005). Reorganization of hair follicles in human skin organ culture induced by cultured human follicle-derived cells. *Exp Dermatol* **14**(8): 580–585.

Kuch, V., C. Schreiber, W. Thiele, V. Umansky and J. P. Sleeman (2013). Tumor-initiating properties of breast cancer and melanoma cells in vivo are not invariably reflected by spheroid formation in vitro, but can be increased by long-term culturing as adherent monolayers. *Int J Cancer* **132**(3): E94–105.

Kunz-Schughart, L. A. (1999). Multicellular tumor spheroids: Intermediates between monolayer culture and in vivo tumor. *Cell Biol Int* **23**(3): 157–161.

Kuroyanagi, M., A. Yamamoto, N. Shimizu, E. Ishihara, H. Ohno, A. Takeda and Y. Kuroyanagi (2014). Development of cultured dermal substitute composed of hyaluronic acid and collagen spongy sheet containing fibroblasts and epidermal growth factor. *J Biomater Sci Polym Ed* **25**(11): 1133–1143.

Lancaster, M. A. and J. A. Knoblich (2014). Organogenesis in a dish: Modeling development and disease using organoid technologies. *Science* **345**(6194): 1247125.

Landry, J. and J. P. Freyer (1984). Regulatory mechanisms in spheroidal aggregates of normal and cancerous cells. *Recent Results Cancer Res* **95**: 50–66.

Larson, A. R., C. W. Lee, C. Lezcano, Q. Zhan, J. Huang, A. H. Fischer and G. F. Murphy (2014). Melanoma spheroid formation involves laminin-associated vasculogenic mimicry. *Am J Pathol* **184**(1): 71–78.

LaVan, D. A., D. M. Lynn and R. Langer (2002). Moving smaller in drug discovery and delivery. *Nat Rev Drug Discov* **1**(1): 77–84.

Lee, V., G. Singh, J. P. Trasatti, C. Bjornsson, X. Xu, T. N. Tran, S. S. Yoo, G. Dai and P. Karande (2014). Design and fabrication of human skin by three-dimensional bioprinting. *Tissue Eng Part C Methods* **20**(6): 473–484.

Lee, W., J. C. Debasitis, V. K. Lee, J. H. Lee, K. Fischer, K. Edminster, J. K. Park and S. S. Yoo (2009). Multi-layered culture of human skin fibroblasts and keratinocytes through three-dimensional freeform fabrication. *Biomaterials* **30**(8): 1587–1595.

Lelievre, D., P. Justine, F. Christiaens, N. Bonaventure, J. Coutet, L. Marrot and J. Cotovio (2007). The EpiSkin phototoxicity assay (EPA): Development of an in vitro tiered strategy using 17 reference chemicals to predict phototoxic potency. *Toxicol In vitro* **21**(6): 977–995.

Li, Y., S. Wang, R. Huang, Z. Huang, B. Hu, W. Zheng, G. Yang and X. Jiang (2015). Evaluation of the effect of the structure of bacterial cellulose on full thickness skin wound repair on a microfluidic chip. *Biomacromolecules* **16**(3): 780–789.

Liebsch, M., D. Traue, C. Barrabas, H. Spielmann, P. Uphill, S. Wilkins, J. P. McPherson, et al. (2000). The ECVAM Prevalidation Study on the Use of EpiDerm for Skin Corrosivity Testing. *Altern Lab Anim* **28**(3): 371–401.

Lorenz, K., T. Rupf and A. Bader (2007). In vitro induction of endothelial cell differentiation of adipose tissue derived stem cells (ADSC) and vascularisation in dermal skin equivalents. *J Stem Cells Regen Med* **2**(1): 135.

Lu, W., J. Yu, Y. Zhang, K. Ji, Y. Zhou, Y. Li, Z. Deng and Y. Jin (2012). Mixture of fibroblasts and adipose tissue-derived stem cells can improve epidermal morphogenesis of tissue-engineered skin. *Cells Tissues Organs* **195**(3): 197–206.

Ma, H. L., Q. Jiang, S. Han, Y. Wu, J. Cui Tomshine, D. Wang, Y. Gan, G. Zou and X. J. Liang (2012). Multicellular tumor spheroids as an in vivo-like tumor model for three-dimensional imaging of chemotherapeutic and nano material cellular penetration. *Mol Imaging* **11**(6): 487–498.

Maas-Szabowski, N., A. Shimotoyodome and N. E. Fusenig (1999). Keratinocyte growth regulation in fibroblast cocultures via a double paracrine mechanism. *J Cell Sci* **112** (Pt 12): 1843–1853.

MacNeil, S. (2007). Progress and opportunities for tissue-engineered skin. *Nature* **445**(7130): 874–880.

Marrero, B., J. L. Messina and R. Heller (2009). Generation of a tumor spheroid in a microgravity environment as a 3D model of melanoma. *In vitro Cell Dev Biol Anim* **45**(9): 523–534.

Maschmeyer, I., T. Hasenberg, A. Jaenicke, M. Lindner, A. K. Lorenz, J. Zech, L. A. Garbe, et al. (2015). Chip-based human liver-intestine and liver-skin co-cultures—A first step toward systemic repeated dose substance testing in vitro. *Eur J Pharm Biopharm* **95**(Pt A): 77–87.

Maschmeyer, I., A. K. Lorenz, K. Schimek, T. Hasenberg, A. P. Ramme, J. Hubner, M. Lindner, et al. (2015). A four-organ-chip for interconnected long-term co-culture of human intestine, liver, skin and kidney equivalents. *Lab Chip* **15**(12): 2688–2699.

Mathes, S. H., H. Ruffner and U. Graf-Hausner (2014). The use of skin models in drug development. *Adv Drug Deliv Rev* **69–70**: 81–102.

Mazzoleni, G., D. Di Lorenzo and N. Steimberg (2009). Modelling tissues in 3D: The next future of pharmaco-toxicology and food research? *Genes Nutr* **4**(1): 13–22.

Metcalfe, A. D. and M. W. Ferguson (2007). Bioengineering skin using mechanisms of regeneration and repair. *Biomaterials* **28**(34): 5100–5113.

Michael, S., H. Sorg, C. T. Peck, L. Koch, A. Deiwick, B. Chichkov, P. M. Vogt and K. Reimers (2013). Tissue engineered skin substitutes created by laser-assisted bioprinting form skin-like structures in the dorsal skin fold chamber in mice. *PLoS One* **8**(3): e57741.

Mo, J., B. Sun, X. Zhao, Q. Gu, X. Dong, Z. Liu, Y. Ma, et al. (2013). The in-vitro spheroid culture induces a more highly differentiated but tumorigenic population from melanoma cell lines. *Melanoma Res* **23**(4): 254–263.

Monfort, A., M. Soriano-Navarro, J. M. Garcia-Verdugo and A. Izeta (2013). Production of human tissue-engineered skin trilayer on a plasma-based hypodermis. *J Tissue Eng Regen Med* **7**(6): 479–490.

Moscona, A. (1961). Rotation-mediated histogenetic aggregation of dissociated cells. A quantifiable approach to cell interactions in vitro. *Exp Cell Res* **22**: 455–475.

Moscona, A. and H. Moscona (1952). The dissociation and aggregation of cells from organ rudiments of the early chick embryo. *J Anat* **86**(3): 287–301.

Mueller-Klieser, W. (1984). Microelectrode measurement of oxygen tension distributions in multicellular spheroids cultured in spinner flasks. *Recent Results Cancer Res* **95**: 134–149.

Murphy, S. V. and A. Atala (2014). 3D bioprinting of tissues and organs. *Nat Biotechnol* **32**(8): 773–785.

Na, Y. R., S. H. Seok, D. J. Kim, J. H. Han, T. H. Kim, H. Jung, B. H. Lee and J. H. Park (2009). Isolation and characterization of spheroid cells from human malignant melanoma cell line WM-266-4. *Tumour Biol* **30**(5–6): 300–309.

Netzlaff, F., C. M. Lehr, P. W. Wertz and U. F. Schaefer (2005). The human epidermis models EpiSkin, SkinEthic and EpiDerm: An evaluation of morphology and their suitability for testing phototoxicity, irritancy, corrosivity, and substance transport. *Eur J Pharm Biopharm* **60**(2): 167–178.

Neuzi, P., S. Giselbrecht, K. Lange, T. J. Huang and A. Manz (2012). Revisiting lab-on-a-chip technology for drug discovery. *Nat Rev Drug Discov* **11**(8): 620–632.

Ng, K. W., H. L. Khor and D. W. Hutmacher (2004). In vitro characterization of natural and synthetic dermal matrices cultured with human dermal fibroblasts. *Biomaterials* **25**(14): 2807–2818.

Nicholson, K. M., M. C. Bibby and R. M. Phillips (1997). Influence of drug exposure parameters on the activity of paclitaxel in multicellular spheroids. *Eur J Cancer* **33**(8): 1291–1298.

Nie, F. Q., M. Yamada, J. Kobayashi, M. Yamato, A. Kikuchi and T. Okano (2007). On-chip cell migration assay using microfluidic channels. *Biomaterials* **28**(27): 4017–4022.

Ouwehand, K., S. W. Spiekstra, T. Waaijman, M. Breetveld, R. J. Scheper, T. D. de Gruijl and S. Gibbs (2012). CCL5 and CCL20 mediate immigration of Langerhans cells into the epidermis of full thickness human skin equivalents. *Eur J Cell Biol* **91**(10): 765–773.

Ouwehand, K., S. W. Spiekstra, T. Waaijman, R. J. Scheper, T. D. de Gruijl and S. Gibbs (2011). Technical advance: Langerhans cells derived from a human cell line in a full-thickness skin equivalent undergo allergen-induced maturation and migration. *J Leukoc Biol* **90**(5): 1027–1033.

Pampaloni, F., E. G. Reynaud and E. H. Stelzer (2007). The third dimension bridges the gap between cell culture and live tissue. *Nat Rev Mol Cell Biol* **8**(10): 839–845.

Patel, N. R., B. Aryasomayajula, A. H. Abouzeid and V. P. Torchilin (2015). Cancer cell spheroids for screening of chemotherapeutics and drug-delivery systems. *Ther Deliv* **6**(4): 509–520.

Peck, Y. and D. A. Wang (2013). Three-dimensionally engineered biomimetic tissue models for in vitro drug evaluation: Delivery, efficacy and toxicity. *Expert Opin Drug Deliv* **10**(3): 369–383.

Pereira, R. F., C. C. Barrias, P. L. Granja and P. J. Bartolo (2013). Advanced biofabrication strategies for skin regeneration and repair. *Nanomedicine (Lond)* **8**(4): 603–621.

Perkins, M. A., R. Osborne and G. R. Johnson (1996). Development of an in vitro method for skin corrosion testing. *Fundam Appl Toxicol* **31**(1): 9–18.

Polini, A., L. Prodanov, N. S. Bhise, V. Manoharan, M. R. Dokmeci and A. Khademhosseini (2014). Organs-on-a-chip: A new tool for drug discovery. *Expert Opin Drug Discov* **9**(4): 335–352.

Ponec, M., A. El Ghalbzouri, R. Dijkman, J. Kempenaar, G. van der Pluijm and P. Koolwijk (2004). Endothelial network formed with human dermal microvascular endothelial cells in autologous multicellular skin substitutes. *Angiogenesis* **7**(4): 295–305.

Poumay, Y. and A. Coquette (2007). Modelling the human epidermis in vitro: Tools for basic and applied research. *Arch Dermatol Res* **298**(8): 361–369.

Ramadan, Q. and M. A. Gijs (2015). In vitro micro-physiological models for translational immunology. *Lab Chip* **15**(3): 614–636.

Ranga, A., N. Gjorevski and M. P. Lutolf (2014). Drug discovery through stem cell-based organoid models. *Adv Drug Deliv Rev* **69–70**: 19–28.

Regnier, M., A. Patwardhan, A. Scheynius and R. Schmidt (1998). Reconstructed human epidermis composed of keratinocytes, melanocytes and Langerhans cells. *Med Biol Eng Comput* **36**(6): 821–824.

Regnier, M., M. J. Staquet, D. Schmitt and R. Schmidt (1997). Integration of Langerhans cells into a pigmented reconstructed human epidermis. *J Invest Dermatol* **109**(4): 510–512.

Rieber, M. and M. S. Rieber (2006). Signalling responses linked to betulinic acid-induced apoptosis are antagonized by MEK inhibitor U0126 in adherent or 3D spheroid melanoma irrespective of p53 status. *Int J Cancer* **118**(5): 1135–1143.

Rimann, M., E. Bono, H. Annaheim, M. Bleisch and U. Graf-Hausner (2016). Standardized 3D Bioprinting of Soft Tissue Models with Human Primary Cells. *J Lab Autom* **21**(4): 496–509.

Rimann, M. and U. Graf-Hausner (2012). Synthetic 3D multicellular systems for drug development. *Curr Opin Biotechnol* **23**(5): 803–809.

Rofstad, E. K., A. Wahl and T. Brustad (1986). Radiation response of human melanoma multicellular spheroids measured as single cell survival, growth delay, and spheroid cure: Comparisons with the parent tumor xenograft. *Int J Radiat Oncol Biol Phys* **12**(6): 975–982.

Schafer-Korting, M., A. Mahmoud, S. Lombardi Borgia, B. Bruggener, B. Kleuser, S. Schreiber and W. Mehnert (2008). Reconstructed epidermis and full-thickness skin for absorption testing: Influence of the vehicles used on steroid permeation. *Altern Lab Anim* **36**(4): 441–452.

Schatton, T. and M. H. Frank (2010). The in vitro spheroid melanoma cell culture assay: Cues on tumor initiation? *J Invest Dermatol* **130**(7): 1769–1771.

Schmidt, B. A. and V. Horsley (2013). Intradermal adipocytes mediate fibroblast recruitment during skin wound healing. *Development* **140**(7): 1517–1527.

Schmook, F. P., J. G. Meingassner and A. Billich (2001). Comparison of human skin or epidermis models with human and animal skin in in-vitro percutaneous absorption. *Int J Pharm* **215**(1–2): 51–56.

Schrader, S., C. Kremling, M. Klinger, H. Laqua and G. Geerling (2007). Generation of organized Lacrimal gland cell spheroids by simulated microgravity. *J Stem Cells Regen Med* **2**(1): 158.

Schreiber, S., A. Mahmoud, A. Vuia, M. K. Rubbelke, E. Schmidt, M. Schaller, H. Kandarova, et al. (2005). Reconstructed epidermis versus human and animal skin in skin absorption studies. *Toxicol In vitro* **19**(6): 813–822.

Shibayama, H., M. Hisama, S. Matsuda and M. Ohtsuki (2008). Permeation and metabolism of a novel ascorbic acid derivative, disodium isostearyl 2-O-L-ascorbyl phosphate, in human living skin equivalent models. *Skin Pharmacol Physiol* **21**(4): 235–243.

Sivard, P., C. Dezutter-Dambuyant, J. Kanitakis, J. F. Mosnier, H. Hamzeh, N. Bechetoille, O. Berthier, et al. (2003). In vitro reconstructed mucosa-integrating Langerhans' cells. *Exp Dermatol* **12**(4): 346–355.

Skardal, A., D. Mack, E. Kapetanovic, A. Atala, J. D. Jackson, J. Yoo and S. Soker (2012). Bioprinted amniotic fluid-derived stem cells accelerate healing of large skin wounds. *Stem Cells Transl Med* **1**(11): 792–802.

Smalley, K. S. (2013). Overcoming melanoma drug resistance through metabolic targeting? *Pigment Cell Melanoma Res* **26**(6): 793–795.

Smalley, K. S., N. K. Haass, P. A. Brafford, M. Lioni, K. T. Flaherty and M. Herlyn (2006). Multiple signaling pathways must be targeted to overcome drug resistance in cell lines derived from melanoma metastases. *Mol Cancer Ther* **5**(5): 1136–1144.

Smalley, K. S., M. Lioni and M. Herlyn (2006). Life isn't flat: Taking cancer biology to the next dimension. *In vitro Cell Dev Biol Anim* **42**(8–9): 242–247.

Smalley, K. S., M. Lioni, K. Noma, N. K. Haass and M. Herlyn (2008). In vitro three-dimensional tumor microenvironment models for anticancer drug discovery. *Expert Opin Drug Discov* **3**(1): 1–10.

Stark, H. J., K. Boehnke, N. Mirancea, M. J. Willhauck, A. Pavesio, N. E. Fusenig and P. Boukamp (2006). Epidermal homeostasis in long-term scaffold-enforced skin equivalents. *J Investig Dermatol Symp Proc* **11**(1): 93–105.

Stark, H. J., A. Szabowski, N. E. Fusenig and N. Maas-Szabowski (2004). Organotypic cocultures as skin equivalents: A complex and sophisticated in vitro system. *Biol Proced Online* **6**: 55–60.

Stark, H. J., M. J. Willhauck, N. Mirancea, K. Boehnke, I. Nord, D. Breitkreutz, A. Pavesio, P. Boukamp and N. E. Fusenig (2004). Authentic fibroblast matrix in dermal equivalents normalises epidermal histogenesis and dermoepidermal junction in organotypic co-culture. *Eur J Cell Biol* **83**(11–12): 631–645.

Sugihara, H., S. Toda, N. Yonemitsu and K. Watanabe (2001). Effects of fat cells on keratinocytes and fibroblasts in a reconstructed rat skin model using collagen gel matrix culture. *Br J Dermatol* **144**(2): 244–253.

Sun, T., S. Jackson, J. W. Haycock and S. MacNeil (2006). Culture of skin cells in 3D rather than 2D improves their ability to survive exposure to cytotoxic agents. *J Biotechnol* **122**(3): 372–381.

Sun, Y. S., S. W. Peng and J. Y. Cheng (2012). In vitro electrical-stimulated wound-healing chip for studying electric field-assisted wound-healing process. *Biomicrofluidics* **6**(3): 34117.

Supp, D. M. and S. T. Boyce (2005). Engineered skin substitutes: Practices and potentials. *Clin Dermatol* **23**(4): 403–412.

Tai, A., S. Goto, Y. Ishiguro, K. Suzuki, T. Nitoda and I. Yamamoto (2004). Permeation and metabolism of a series of novel lipophilic ascorbic acid derivatives, 6-O-acyl-2-O-alpha-D-glucopyranosyl-L-ascorbic acids with a branched-acyl chain, in a human living skin equivalent model. *Bioorg Med Chem Lett* **14**(3): 623–627.

Tanaka, H., S. Tanaka, K. Sekine, S. Kita, A. Okamura, T. Takebe, Y. W. Zheng, Y. Ueno, J. Tanaka and H. Taniguchi (2013). The generation of pancreatic beta-cell spheroids in a simulated microgravity culture system. *Biomaterials* **34**(23): 5785–5791.

Tfayli, A., O. Piot, F. Draux, F. Pitre and M. Manfait (2007). Molecular characterization of reconstructed skin model by Raman microspectroscopy: Comparison with excised human skin. *Biopolymers* **87**(4): 261–274.

Timmins, N. E., S. Dietmair and L. K. Nielsen (2004). Hanging-drop multicellular spheroids as a model of tumour angiogenesis. *Angiogenesis* **7**(2): 97–103.

Timmins, N. E. and L. K. Nielsen (2007). Generation of multicellular tumor spheroids by the hanging-drop method. *Methods Mol Med* **140**: 141–151.

Tjabringa, G., M. Bergers, D. van Rens, R. de Boer, E. Lamme and J. Schalkwijk (2008). Development and validation of human psoriatic skin equivalents. *Am J Pathol* **173**(3): 815–823.

Tonello, C., V. Vindigni, B. Zavan, S. Abatangelo, G. Abatangelo, P. Brun and R. Cortivo (2005). In vitro reconstruction of an endothelialized skin substitute provided with a microcapillary network using biopolymer scaffolds. *FASEB J* **19**(11): 1546–1548.

Trottier, V., G. Marceau-Fortier, L. Germain, C. Vincent and J. Fradette (2008). IFATS collection: Using human adipose-derived stem/stromal cells for the production of new skin substitutes. *Stem Cells* **26**(10): 2713–2723.

Tsui, J. H., W. Lee, S. H. Pun, J. Kim and D. H. Kim (2013). Microfluidics-assisted in vitro drug screening and carrier production. *Adv Drug Deliv Rev* **65**(11–12): 1575–1588.

Uchino, T., T. Takezawa and Y. Ikarashi (2009). Reconstruction of three-dimensional human skin model composed of dendritic cells, keratinocytes and fibroblasts utilizing a handy scaffold of collagen vitrigel membrane. *Toxicol In vitro* **23**(2): 333–337.

van de Kamp, J., R. Kramann, J. Anraths, H. R. Scholer, K. Ko, R. Knuchel, M. Zenke, S. Neuss and R. K. Schneider (2012). Epithelial morphogenesis of germline-derived pluripotent stem cells on organotypic skin equivalents in vitro. *Differentiation* **83**(3): 138–147.

Van Kilsdonk, J. W., M. Bergers, L. C. Van Kempen, J. Schalkwijk and G. W. Swart (2010). Keratinocytes drive melanoma invasion in a reconstructed skin model. *Melanoma Res* **20**(5): 372–380.

Vellonen, K. S., M. Malinen, E. Mannermaa, A. Subrizi, E. Toropainen, Y. R. Lou, H. Kidron, M. Yliperttula and A. Urtti (2014). A critical assessment of in vitro tissue models for ADME and drug delivery. *J Control Release* **190**: 94–114.

Venugopal, J. R., Y. Zhang and S. Ramakrishna (2006). In vitro culture of human dermal fibroblasts on electrospun polycaprolactone collagen nanofibrous membrane. *Artif Organs* **30**(6): 440–446.

Vladisavljevic, G. T., N. Khalid, M. A. Neves, T. Kuroiwa, M. Nakajima, K. Uemura, S. Ichikawa and I. Kobayashi (2013). Industrial lab-on-a-chip: Design, applications and scale-up for drug discovery and delivery. *Adv Drug Deliv Rev* **65**(11–12): 1626–1663.

Vorsmann, H., F. Groeber, H. Walles, S. Busch, S. Beissert, H. Walczak and D. Kulms (2013). Development of a human three-dimensional organotypic skin-melanoma spheroid model for in vitro drug testing. *Cell Death Dis* **4**: e719.

Wagner, I., E. M. Materne, S. Brincker, U. Sussbier, C. Fradrich, M. Busek, F. Sonntag, et al. (2013). A dynamic multi-organ-chip for long-term cultivation and substance testing proven by 3D human liver and skin tissue co-culture. *Lab Chip* **13**(18): 3538–3547.

Weigl, B. H., R. L. Bardell and C. R. Cabrera (2003). Lab-on-a-chip for drug development. *Adv Drug Deliv Rev* **55**(3): 349–377.

Wong, T., J. A. McGrath and H. Navsaria (2007). The role of fibroblasts in tissue engineering and regeneration. *Br J Dermatol* **156**(6): 1149–1155.

Wu, X., L. Scott, Jr., K. Washenik and K. Stenn (2014). Full-thickness skin with mature hair follicles generated from tissue culture expanded human cells. *Tissue Eng Part A* **20**(23–24): 3314–3321.

Xie, Y., W. Zhang, L. Wang, K. Sun, Y. Sun and X. Jiang (2011). A microchip-based model wound with multiple types of cells. *Lab Chip* **11**(17): 2819–2822.

Yamada, K. M. and E. Cukierman (2007). Modeling tissue morphogenesis and cancer in 3D. *Cell* **130**(4): 601–610.

Yildirimer, L., N. T. Thanh and A. M. Seifalian (2012). Skin regeneration scaffolds: A multimodal bottom-up approach. *Trends Biotechnol* **30**(12): 638–648.

Young, E. W. (2013). Cells, tissues, and organs on chips: Challenges and opportunities for the cancer tumor microenvironment. *Integr Biol (Camb)* **5**(9): 1096–1109.

Zhao, Q., S. Wang, Y. Xie, W. Zheng, Z. Wang, L. Xiao, W. Zhang and X. Jiang (2012). A rapid screening method for wound dressing by cell-on-a-chip device. *Adv Healthc Mater* **1**(5): 560–566.

11 Tissue-Engineered Kidney Model

Erica P. Kimmerling and David L. Kaplan

CONTENTS

11.1 INTRODUCTION

Through a process of filtration, secretion, and absorption, the kidney regulates the excretion of waste and retention of solutes such as electrolytes and amino acids. In processing the blood the kidney also functions to maintain acid-base homeostasis, plasma osmolality, and blood pressure. These functional roles of the kidney are the basis for the recent interest in developing *in vitro* kidney models that can be applied to drug development, disease studies, and nephrotoxicity testing.

Recapitulating the functions of the human kidney *in vitro* remains challenging due to the anatomical complexity required to mimic renal physiology. The kidney is composed of numerous region-specific epithelial cell types, interstitial cells, and a complex microvasculature (Kriz 1981). Although cell type and matrix composition are the primary considerations for achieving a specific function, the incorporation of mechanical stimulation also needs to be considered due to the influence of fluid flow in maintaining epithelial cell phenotype. Based on the downstream application of the system, three-dimensional (3D) tissue culture and kidney-on-a-chip systems are typically designed to mimic only a narrow set of functions. Accordingly, these simplified approaches preclude important, multicomponent functions such as the renin-angiotension-aldosterone system, which regulates blood pressure and has been shown to drive progression of chronic kidney disease (Remuzzi et al. 2005). In this chapter, we explore the anatomy and physiology of the kidney, the design considerations, common methods for assessing phenotype and function, and the benefits and limitations of state-of-the-art *in vitro* kidney model systems.

11.2 CLINICAL MOTIVATIONS

There is a significant clinical need for the development of *in vitro* kidney tissue models.

Tissue-engineered approaches are required to better understand the causes of renal failure and for the development of new treatment options. Upon renal failure, due to acute or chronic causes, renal replacement therapies such as dialysis or transplantation are required to restore function. Currently, 185,000 Americans have a functioning kidney transplant and 450,000 are on dialysis (United States Renal Data System 2014). Dialysis is able to replace the essential filtration functions of the kidney but long-term dialysis can lead to quality of life issues due to malnutrition and depression. Additionally, patients in chronic renal failure undergoing dialysis are still at risk for anemia as a result of a decreased production of erythropoietin, which stimulates the production of red blood cells. While kidney transplants are able to completely restore function, this treatment option is limited by donor supply, with the majority of people on the organ transplant waiting list requiring a kidney. Tissue-engineered models can be utilized to study the causes of renal disease and assess the efficacy of preventative and restorative treatments, which would ultimately reduce the need for renal replacement therapies.

In addition to drug development and disease modeling, these approaches have considerable utility in assessing drug-induced nephrotoxicity during pharmaceutical development. The kidneys are inherently exposed to a high proportion of pharmaceuticals that enter the body, considering that a quarter of the total cardiac output passes through the kidneys. Renal toxicity is a significant contributor to late-stage failure of drugs undergoing clinical trials and this failure adds to the high cost of drug development. These occurrences can potentially be attributed to the limited capability of two-dimensional cell culture and animal models to accurately predict human toxicity outcomes. Additionally, it is unknown whether drugs that cause toxicity in animals are representative of the human response (Dieterle et al. 2010). As such, tissue-engineered systems that better predict the human response to treatment would be capable of reducing drug development costs and improving patient outcomes.

11.3 OVERVIEW OF KIDNEY ANATOMY

11.3.1 Organization of the Kidney

The human kidney can be distilled down to three primary structural components: the nephron, the microvasculature, and the collecting system, as shown in Figure 11.1. These components are located beneath the renal capsule, a tough, fibrous protective layer, within the outer and inner layers known as the cortex and medulla, respectively. Each of the one million nephrons in the human kidney consists of a renal corpuscle and a tubular portion. The precise localization of components is vital to renal function with filtration occurring in the cortical layer and reabsorption within the medulla.

A basic understanding of renal anatomy is necessary in order to assess the strengths and limitations of *in vitro* systems designed to mimic renal functions. Simplified models typically attempt to mimic nephron function, so more emphasis is placed on this structural component. However, the role of the endothelium, interstitium, and lymphatic system should be considered before designing a kidney tissue model for a specific application.

11.3.2 Renal Corpuscle

A renal corpuscle consisting of a glomerulus and a Bowman capsule is at the start of each nephron. The renal artery that supplies blood to each kidney branches to eventually form the interlobular arteries. These arteries are the source of the afferent

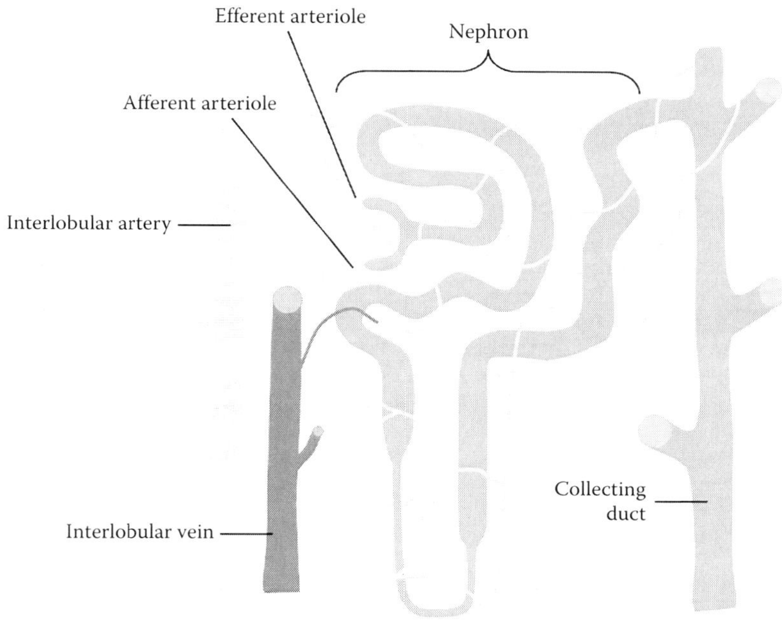

FIGURE 11.1 Structural components of the kidney.

arterioles at the leading end of the glomerulus (see Figure 11.1). Blood is filtered as it passes through the capillary tuft of the glomerulus and exits through the efferent arterioles. The Bowman capsule, consisting of a parietal epithelial layer, collects the filtrate before entering the lumen of the tubular portion of the nephron. Glomerular filtration is regulated by the glomerular basement membrane (GBM) in addition to glomerular endothelial cells, specialized epithelial cells known as podocytes, and specialized smooth muscle cells known as mesangial cells which interact with each other, as shown in Figure 11.2.

Mesangial cells and their anchoring mesangial matrix, known as the mesangium, are the support structure of the glomerular capillaries. In this role, these contractile cells are able to regulate the fluid flow and filtration through the capillaries. These cells are also known to generate their own matrix and regulate matrix turnover and as such are implicated in diseases, including diabetic nephropathy, where glomerulosclerosis occurs (Mason and Wahab 2003). Mesangial cells also play a significant role in glomerular paracrine signaling as exemplified by the secretion of transforming growth factor beta-1 (TGF-β1) and vascular endothelial growth factor (VEGF), in response to changes in capillary tension (Schlondorff and Banas 2009).

Glomerular endothelial cells, which form the capillary tuft, are surrounded by the mesangium on one side and the GBM on the other. These highly fenestrated endothelial cells, with pore sizes ranging 60 to 80 nanometers and a negatively charged glycocalyx, are the initial component of the three-layer glomerular filtration barrier (Satchell and Braet 2009). The GBM is a thick (approximately 350 nanometers), specialized extracellular matrix consisting of specific isoforms of type IV collagen

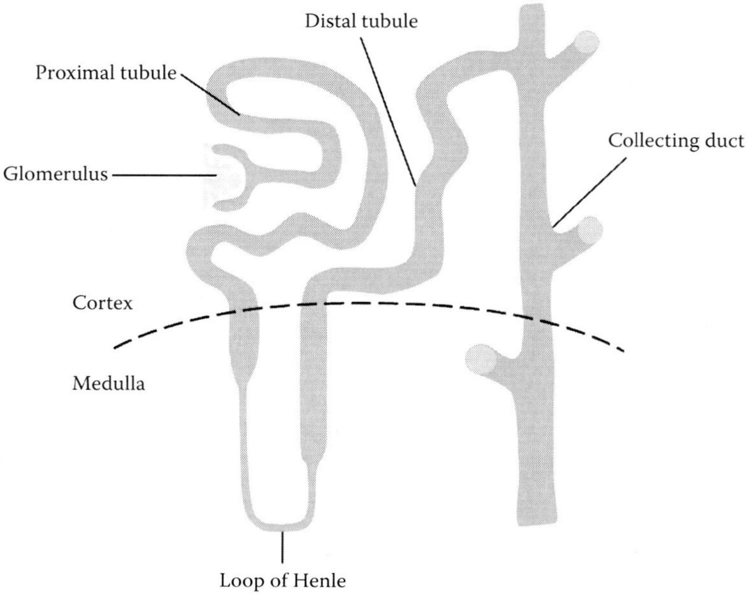

FIGURE 11.2 Components of the nephron.

(α3, α4, and α5), laminin 11 (α5, β2 and γl) nidogen, and heparan sulfate proteoglycan (primarily agrin) (Miner 2012). The negative charge of the heparan sulfate proteoglycans and the endothelial glycocalyx are the basis of the charge-based selectivity of the filtration barrier. The final component of the filtration barrier is the highly differentiated cell-type known as podocytes, which also carry a negative surface charge due to sialoglycoproteins. These cells are fixed to the GBM via interdigitating extensions known as foot processes with transmembrane proteins specific to the GBM. Podocytes are particularly susceptible to paracrine signaling with surface receptors including, but not limited to, transient receptor potential cation channels and the angiotension II type 1 receptor known to influence cytoskeletal organization (Greka and Mundel 2012). Ultimately, the glomerular filtration barrier allows for uncharged macromolecules smaller than 1.8 nanometers to progress through uninhibited, and essentially prohibits molecules larger than 4 nanometers. In between these sizes, fractional clearances are dependent on size and charge. While there are currently few tissue-engineered glomerular models, the incorporation of this component will be critical to mimicking basic renal functions in future *in vitro* systems.

11.3.3 Tubule and Collecting Ducts

The remaining portion of the nephron, the renal tubule system, consists of distinct segments: the proximal tubule, thick and thin segments of the descending loop of Henle, the thin and thick segments of the ascending loop of Henle, and the distal tubule, as shown in Figure 11.3. The connecting tubule links the nephron with the collecting duct system. Transport through the renal epithelium of the renal tubules is accomplished by both paracellular and transcellular transport. Particular epithelial phenotypes and transport mechanisms are specific to the different tubular segments.

Reabsorption of water and solutes begins in the proximal tubule. The luminal surface of the proximal tubule consists of a brush border for increasing surface area and leaky tight junctions. The basolateral sodium potassium adenosine triphosphatase (Na^+K^+ ATPase) pumps work in conjunction with the luminal sodium

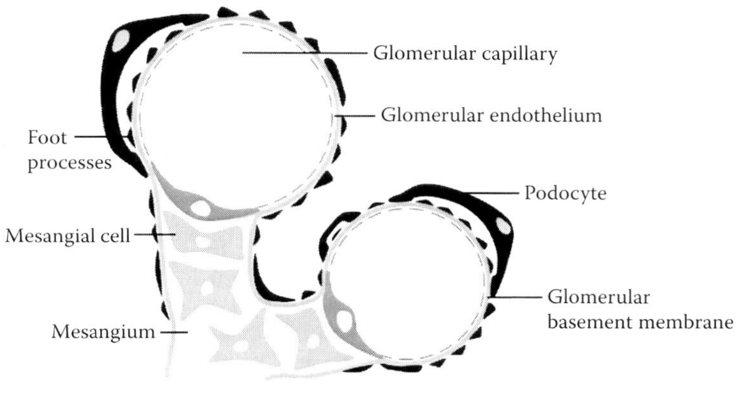

FIGURE 11.3 Cross-section of the glomerular lobule.

hydrogen (Na^+H^+) ion exchanger for the reabsorption of ions. In addition to ion transport, aquaporin 1 (AQP1) channels and lysosomes are used for transporting water and proteins, respectively.

Generally, the loop of Henle is water permeable in the descending loop and water impermeable in the ascending loop. In addition to being water impermeable, the epithelium of the thick ascending loop contains luminal sodium potassium chloride ($Na^+K^+2Cl^-$) cotransporters and robust tight junctions. This differential permeability establishes an osmotic gradient within the renal medulla that drives the countercurrent system known for establishing urine concentration before excretion.

The distal convoluted tubule segment is differentiated from the loop of Henle by its specific luminal sodium chloride (Na^+Cl^-) transporter, known for its sensitivity to thiazide diuretics, basolateral interdigitations, and apical microvilli. Both the collecting tubules and the collecting duct system consist of principle and intercalated cells with the number of intercalated cells decreasing as the collecting duct system progresses. Principle cells possess luminal aquaporin 2 (AQP2) channels sensitive to vasopressin, also known as arginine vasopressin or antidiuretic hormone, which induces water permeability within these segments. Intercalated cells are responsible for active secretion and reabsorption of protons.

The segment distinct sodium transporters discussed in this section exemplify the unique epithelial phenotypes within the renal tubule and collecting duct system but are not an inclusive list of the all the phenotypic differences. The specialization of renal epithelia is an important consideration when designing tissue engineered models, specifically with respect to cell choice and functional outcomes. Accordingly, most systems are limited to a specific set of applications based on the cell phenotypes utilized in the kidney tissue model.

11.4 CELL SOURCES

11.4.1 Animal Cell Sources

Although the differences between animal and human cell biology are not completely understood, the availability and ease of culture of animal cell lines supports their use in tissue engineered models. More importantly, some animal cell lines, such as Madin-Darby canine kidney (MDCK) epithelial cells, have been in use for over 50 years, and as such have been characterized in a plethora of biological contexts and for a range of phenotypes. Another common animal cell line originating from pig kidney, known as LLC-PK_1, has a proximal tubule-like phenotype and divides rapidly in culture (Hull, Cherry, and Weaver 1976).

The MDCK cell line was originally derived from the kidney of a healthy female adult cocker spaniel with the specific kidney segment of origin remaining unknown. MDCK cultures display characteristics such as tight junction formation and epithelial polarization and transport but are considered to be a heterogeneous cell population (Lang and Paulmichl 1995). MDCK subpopulations and clones have been shown to have distinct phenotypes with characteristics indicative of different segments such as the collecting duct (Kriz 1981). Accordingly, MDCK cells can be used to model specific renal functions but the heterogeneity of the specific culture must be adequately understood. While the use of animal cell lines will not provide a direct understanding of human renal cell behavior, the ease of culture and the extent of prior characterization support their use in targeted model systems.

11.4.2 Human Cell Sources

When working with human cells there is often a trade-off between ease of culture and well-characterized cell behavior, as with cell lines, and improved phenotypic characteristics, as seen in primary cells. Immortalized cell lines have a well characterized phenotype, can be continuously passaged, and eliminate the patient-to-patient variability affiliated with primary cell culture. Conversely, considering primary cells can be isolated from normal biopsies, systems which seek to observe patient-specific responses to perturbations would need to work with primary cell sources. While primary human cells are known for improved phenotypic characteristics, these cells are limited to only a few passages before dedifferentiation occurs.

Human kidney-2 (HK-2) cells are one of the more common immortalized normal adult kidney cell lines. This cell line was established from proximal tubule cell culture using the human papilloma virus (HPV 15) E6/E7 genes (Ryan et al. 1994). These cells maintain some characteristics of the proximal tubule phenotype such as a parathyroid hormone–stimulated adenylate cyclase response and irresponsiveness to antidiuretic hormone. RPTEC/TERT1 (renal proximal tubule epithelial cells/telomerase) is an alternative human cell line that possesses a similar phenotype and functional responses to HK-2 cells (Wieser et al. 2008). The immortalization of RPTEC/TERT1 cells is the result of ectopic expression of the catalytic subunit of telemorase as opposed to the use of oncogenes. Most available human renal cells are a mixed cortical epithelial cell population or cells that have been verified to possess proximal tubule-like phenotypes. As a result of the available cell sources, human-based tissue models are significantly limited in the amount of phenotypes that can be studied. To address this problem most current approaches focus on a specific mechanism or function required to model a particular aspect of a disease or toxic response. However, when establishing these tissue models it is important to recognize that cell phenotypes have been shown to change in response to microenvironmental cues such as a 3D matrix or fluid induced shear stresses. Therefore, targeted cell phenotypes should be reassessed in the context of an optimized microenvironment over the course of system development.

11.4.3 Stem Cells

Using insights from renal development there have been recent advancements in the ability to differentiate human stem cells along renal lineages. For instance, embryonic stem cells have previously been consistently differentiated through the posterior primitive streak to an intermediate mesodermal phenotype (Takasato et al. 2014). Similarly, uretic bud and metanephric mesenchyme committed renal progenitor cells have been formed from human-induced pluripotent stem cells (Xia et al. 2013;

Taguchi et al. 2014). The differentiation of renal progenitors from stem cells has the potential to significantly influence the development of future tissue-engineered models. Induced stem cell sources offer the opportunity for personalized tissue models for treatment response. While significant research is still required before these cells can be differentiated into specialized renal cell types, these cells may be used one day to accurately recapitulate kidney cell function.

11.5 GLOMERULAR TISSUE MODELS

A significant scientific and clinical need exists for the development of new methods for understanding of the development of glomerular diseases and the effect of drugs on the glomerular filtration barrier. The nature of patient studies limits the ability to study early mechanisms involved in drug response and disease development. Currently, the primary method for assessing kidney function and the extent of chronic kidney disease is a metric known as glomerular filtration rate. Glomerular filtration rate is indirectly determined using algorithms which associate the level of serum creatinine in a patient's urine to their sex, race, and age (Levey et al. 2009). *In vitro* tissue models would provide direct insight into human cell behavior that cannot be obtained from clinical metrics. Despite the need for these models, limited progress has been made in their development due to limited sources of the highly differentiated glomerular cell phenotypes.

While primary glomerular endothelial and mesangial cells can be obtained commercially, these specialized cells do not readily propagate in culture and easily lose their phenotype. To circumvent this issue, intact glomeruli have been used to investigate drug-induced nephrotoxicity of known glomerular toxicants such as gentamicin and cisplatin (Rodriguez-Barbero et al. 2000). Intact glomerular tufts can be isolated from human tissue through a process of mechanical and enzymatic tissue degradation and size dependent sieving. In this approach the cross-sectional area of the isolated glomeruli was used as a metric of toxicity and confirmed *in vivo* drug response. While this method circumvents issues associated with glomerular cell culture, working with whole glomeruli requires repeated isolation of primary tissue for each experiment.

The use of conditional immortalization has been employed over the last ten years to establish glomerular cell lines. Transfection of cells with the *SV40-T* gene causes cells to proliferate at 33°C and differentiate at 37°C (Jat et al. 1991). Conditionally immortalized glomerular endothelial cells capable of expressing fenestrations and podocytes expressing *in vivo* podocyte markers have been established (Satchell et al. 2006; Saleem et al. 2002). This cell source has the benefit of both continual propagation and a differentiated phenotype, but there is still a limited understanding of their ability to mimic glomerular function.

A functional glomerular tissue model is dependent not only on the cell sources but also on the structural organization and the incorporation of a matrix mimicking the glomerular basement membrane. While the basement membrane isolated from Engelbreth-HolmSwarm sarcoma is commonly used in 3D tissue–engineered models, this matrix does not contain the specific isoforms of collagen and laminin of the glomerular basement membrane. One model used electrospun collagen 1 and polycaprolactone on a nickel mesh support to co-culture conditionally immortalized glomerular endothelial and podocytes cells (Slater et al. 2011). Although this approach does not use a physiologically relevant mimic of the glomerular basement membrane, it does provide a method for studying glomerular cell paracrine signaling. This glomerular tissue model exemplifies how downstream applications need to be considered when developing new systems.

11.6 MICROFLUIDIC MODELS

Microfluidic systems are intrinsically desirable for renal cell culture, considering the channel sizes are typically within one order of magnitude of renal tubule diameters which can range from 20 to 200 microns depending on the segment. The scale of these systems enables mechanical stimulation with physiologically relevant fluid induced shear stresses which are typically estimated to be between 0.2 and 20 dyn/cm^2. Renal microfluidic systems are typically of either a single or dual layer construction with the latter design supporting basolateral stimulation in addition to apical/luminal flow. Most renal microfluidic systems are designed as potential drug-induced nephrotoxicity models.

11.6.1 Animal Cell Systems

MDCK cells were a commonly used cell type in early renal microfluidic systems. A number of studies have cultured MDCK cells for 96 hours on a fibronectin-coated polydimethylsiloxane (PDMS) device (Baudoin et al. 2007; Snouber et al. 2012a, 2012b). The upregulation of drug metabolism enzymes on the microfluidic device compared to static culture emphasizes the benefit of a dynamic environment (Snouber et al. 2012b). This study also highlights the ability to still perform standard transcriptomic and proteomic analysis after perfusion culture and the need to perform additional cell characterization to understand the effect of mechanical stimulation on phenotype. The functionality of the approach was demonstrated using a two-chamber device for liver cell culture upstream of the MDCK cell chamber (Choucha-Snouber et al. 2013). MDCK cells in this device experienced an increased toxicity to ifosfamide treatment compared to MDCK cells alone. This response is suggested to be a result of the toxicity of the liver metabolized byproduct of ifosfamide, chloroacetylaldehyde, which caused the same toxicity response when cells were treated directly. These microfluidic systems demonstrate the benefits of dynamic stimulation and the ability to do physiologically relevant co-cultures. While Snouber et al. were able to recapitulate a specific example of drug-induced nephrotoxicity, further optimization is necessary to develop a model suitable for nephrotoxicity screening applications considering the phenotypical heterogeneity of MDCK cell populations. The use of a homogenous cell source enables specific research questions to be tackled in the context of mechanical stimulation.

A multilayer renal microfluidic device capable of mimicking the luminal and interstitial spaces has been established, as depicted in Figure 11.4 (Jang and Suh 2010). This approach enables cell stimulation through osmotic gradients, localized

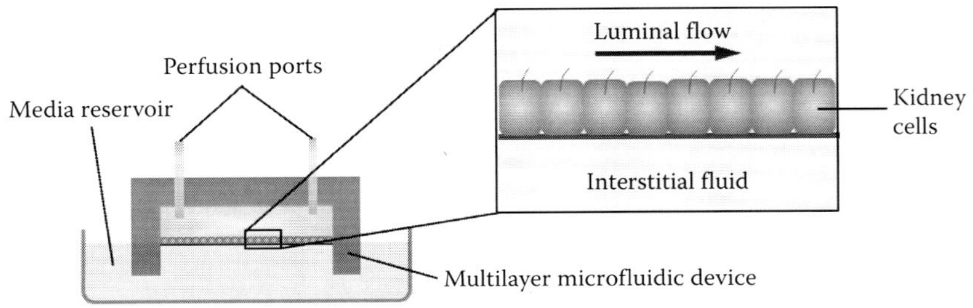

FIGURE 11.4 Schematic of a multilayer renal microfluidic device. This design is representative of the device developed by Jang et al. which enables stimulation by fluid induced shear stresses and hormonal or osmotic gradients.

hormone treatment, and shear stress. Rat inner medullary collecting duct (IMCD) cells incorporated into the device demonstrated a more physiologically relevant phenotype, such as a columnar phenotype and AQP2 basolateral localization, after 5 hours of fluid-induced shear stress (Jang and Suh 2010). The applicability of this two-layer system was further tested by using the interstitial compartment to stimulate with arginine vasopressin and establish an osmotic gradient; both conditions are known to influence AQP2 localization in collecting duct cells (Jang et al. 2011). The observed translocation of AQP2 in this system upon perturbation demonstrates the benefit of choosing cell types and a culture environment specific to the desired outcome.

11.6.2 Human Cell Models

In human cell microfluidic models there has been a recent push to develop more advanced designs to mimic the physiological microenvironment and to explore the applications of established designs.

A dual perfusion system meant to mimic both tubular flow and extracapillary flow was recently developed (Gao et al. 2011). In this approach, serpentine microchannels were etched into a glass substrate and the channels were sealed around a polycarbonate membrane. While cell phenotype within this device was not extensively studied, the approach supported long-term cell viability of RPTEC cells under dynamic conditions for 10 days. Additionally, techniques such as hot embossing are now being used to topographically pattern the culture surface of renal microfluidic devices in order to mimic the complex structural organization of tubular basement membrane. HK-2 and RPTEC cells cultured on porous, 1-micron ridges of polycarbonate substrates demonstrated tight junction staining and alignment in the direction of the topography (Frohlich, Zhang, and Charest 2012). Microfluidic devices have also been designed specifically to measure functional outcomes. Human cortical epithelial cells and MDCK cells were cultured in a bilayer microfluidic system that was able to measure transepithelial electrical resistance (TEER) (Ferrell et al. 2010). This quantitative measurement of tight junction formation is both an indicator of cell phenotype and cell health and as such is a useful metric for disease models and nephrotoxicity.

While additional fabrication techniques are being used to develop mimetic microenvironments, established systems are being tested for their utility for conducting translatable research. For example, in 2013, a human kidney proximal tubule-on-a-chip model was described for use in nephrotoxicity studies (Jang et al. 2013). This study used a similar multilayer device to the previous studies with IMCD but instead incorporated human RPTEC cells on a collagen IV–coated membrane. Under dynamic conditions these cells had upregulated expression of the Na^+K^+ ATPase transporter and AQP2. Additionally, these cells had higher albumin uptake, glucose transporter, and alkaline phosphatase activity compared to static conditions. Dosing in cisplatin, a known proximal tubule toxin, caused cellular injury as would be expected. This toxicity was shown to be through physiologically mimetic mechanisms when the cisplatin transporter, organic cation transporter OCT1, was inhibited and toxicity was prevented.

11.6.3 Microfluidic Models Summary

Microfluidic systems optimized for renal cell culture have great potential as future tools for modeling drug-induced nephrotoxicity and drug screening applications. These approaches inherently mimic physiological forces, and complex microenvironments can be fabricated using common techniques. The current progress with these microfluidic systems reiterates the need for application-targeted designs such as the use of osmotic gradients and hormone stimulation to induce a phenotypic response in collecting duct cells, as Jang et al. demonstrated (2011).

While proper consideration of cellular phenotype and environment stimulation can yield the best outcomes with microfluidic systems, there are also intrinsic limitations to this approach. Most of the established microfluidic culture systems are unable to recapitulate a 3D tissue environment due to the use of solid membrane supports for cell culture. Without a bulk matrix environment these systems are unable to model changes in the interstitium associated with diseases states and diseases with structural phenotypic outcomes such as polycystic kidney disease. Moreover, most of the established systems have only been conducted as short-term culture, less than 4 days, with only one system maintaining culture for over a week, as seen in Table 11.1. Approaches capable of sustaining viability over long-term culture will be necessary for investigating chronic diseases and assessing nephrotoxicity. It is important to keep in mind that while significant progress has been made in optimizing kidney cell culture in microfluidic environments, these systems are not to the point of capturing kidney function as a whole, and new devices should be characterized with respect to the specific renal functions they are designed to recapitulate.

11.7 THREE-DIMENSIONAL MODELS

Although 3D tissue–engineered kidney models lack the fluid-induced stimulation offered by microfluidic approaches, these methods allow for a more accurate recapitulation of structural morphologies. To enable more physiologically relevant outcomes, these systems are typically fabricated using biopolymer-based hydrogels such as type 1 collagen and basement membrane isolates consisting of proteins including

TABLE 11.1
Kidney Cell Microfluidic Culture Systems

Cell Type	Matrix Coating	Culture Duration	Functional Outcomes	Author, Year
MDCK	Fibronectin	96 hrs (24 hrs static, 72 hrs dynamic	Glucose consumption, ammonia production	Baudoin et al. 2007
	Fibronectin	96 hours (24 hrs static, 72 hrs dynamic)	Transcriptomic and proteomic analyses; ifosamide treatment; liver co-culture	Snouber et al. 2012a; Snouber et al. 2012b; Choucha-Snouber et al. 2013; Jang et al. 2013
	Collagen IV	>7 days	TEER	Ferrell et al. 2010
IMCD	None	4 days (5 hrs dynamic)	Arginine vasopressin stimulation, osmotic gradient	Jang and Suh 2010; Jang et al. 2011
RPTEC	Collagen IV	6 days (3 days dynamic)	Albumin uptake, cisplatin toxicity	Jang et al. 2013
	Matrigel	10 days dynamic	Cell viability	Gao et al. 2011
	Collagen IV	N/A	Cell alignment	Frohlich, Zhang, and Charest 2012
HK-2	Cultrex	72 hrs dynamic	Epithelial-to-mesenchymal transition	Zhou et al. 2014
	Collagen IV	N/A	Cell alignment	Frohlich, Zhang, and Charest 2012

laminin and type IV collagen. These hydrogels have the benefit of being susceptible to cellular remodeling, a response particularly relevant to disease phenotypes. In addition to dissociated cell culture, 3D environments have been used to maintain organoid culture *ex vivo*. The utility of these approaches has been demonstrated for study areas including, but not limited to, polycystic kidney disease and drug-induced nephrotoxicity (DesRochers, Palma, and Kaplan 2014).

11.7.1 Polycystic Kidney Disease Models

Polycystic kidney disease is a disorder characterized by the formation of fluid-filled cysts within the kidney, the most common form of which is autosomal-dominant polycystic kidney disease (ADPKD). While animal models and patient studies have been the primary means of studying disease pathogenesis and treatment efficacy, the development of *in vitro* disease models has offered a complementary method of characterizing the structural phenotype associated with this disease. The models have thus far been used to investigate disease development in the context of fluid secretion, extracellular matrix interactions, and cell growth kinetics.

One of the most common approaches for modeling PKD involves culturing MDCK cells in type 1 collagen hydrogels. This method is known to yield physiologically relevant cystic structures with a layer of polarized epithelial cells surrounding a fluid filled cavity. While these cells automatically form cystic structures in 3D, they can be stimulated—through hormonal stimulation with hepatocyte growth factor—to undergo tubulogenesis (Bukanov et al. 2002). The localization of polycystin-1, a key protein implicated in this disease, to the cell membrane of luminal structures in a manner similar to the *in vivo* phenotype highlights the benefits of structural organization for PKD studies. 3D cell culture techniques can also be used to address the inherent heterogeneity of MDCK cell populations discussed earlier. Studies have been conducted on MDCK cysts formed from cells subcloned from a single cyst (Grantham et al. 1989). The cystic phenotype and the highly investigated epithelial characteristics of MDCK support the use of this model to investigate mechanisms behind cyst growth.

3D PKD models have also been adapted for mechanistic, high-throughput, and long-term applications. Characterizing potential disease mechanisms in a 3D environment enables a direct understanding of how a cell process affects structure formation and organization. In recent work, the easily-transfected LLC-PK1 cell line was used to study how the genetic loss of intracellular calcium release channels altered cyst structure development in a 3D tissue model (Kuo et al. 2014). Observations of the cystic structures in this model enabled researchers to associate cyst growth with a loss of the primary cilia, a functional cellular component. A tubule-to-cyst conversion model was developed using immortalized mouse collecting duct cells in a 20 microliter type 1 collagen-Matrigel drop culture (Montesano et al. 2009). The nature of this approach is amenable to high-throughput screening applications for identifying therapeutics capable of stalling early cyst development. At the same time, a method was established for extended *in vitro* culture that would enable long-term studies of targeted therapeutics. To prevent the degradation of a collagen-Matrigel hydrogel over time, matrix and cellular components were infused into a porous silk protein scaffold (Subramanian et al. 2010). The inclusion of a silk scaffold support also allowed for the use of a perfusion bioreactor system that maintained culture for 8 weeks. Over the course of long-term culture, polycystin 1–deficient immortalized mouse collecting duct cells developed cystic phenotypes in conjunction with abnormal extracellular matrix deposition and cell cycle progression (Subramanian et al. 2012). These tissue-engineered approaches highlight the ability to design 3D PKD models for specific disease progression and drug screening applications. However, a significant weakness of the approaches discussed thus far is the use of animal cell types.

Tissue models using diseased human ADPKD cell sources are well established and allow for drug screening on 3D cystic structures formed with the same underlying genetic mechanisms that cause cyst formation in patients (Carone et al. 1995; Wallace, Grantham, and Sullivan 1996). ADPKD epithelial cells isolated from extracted cysts are cultured in type 1 collagen and stimulated with forskolin to induce cyst formation. These physiologically relevant cysts have been tested with treatments suspected of halting cyst progression such as Sorafenib (Yamaguchi et al. 2010). While these 3D models are useful for assessing compounds meant to slow cyst progression, they are not applicable to studies looking to elucidate early disease pathogenesis before cyst formation.

11.7.2 Drug-Induced Nephrotoxicity

3D tissue culture systems specifically designed for testing drug-induced nephrotoxicity are limited. For this purpose, an organoid culture system where isolated mouse proximal tubules were cultured in a modified hyaluronic acid-based hydrogel has been developed (Astashkina et al. 2012). Culture of proximal tubules within this gel allowed for phenotypic stability for up to 6 weeks as demonstrated by maintained AQP1 and megalin expression along with glucose regulation similar to *in vivo* proximal tubules. Treatment of this culture with the nephrotoxic compounds such as cisplatin resulted in the production of a clinical biomarker of kidney injury, Kim-1. This type approach, which uses primary structures from animal cell sources, could potentially be used to better understand the causes of nephrotoxicity in animal models.

Similar to the organoid approach to drug-induced nephrotoxicity, immortalized human proximal tubule–like cells have been stably cultured for 8 weeks (DesRochers et al. 2013). In this model, cells suspended in a type 1 collagen-Matrigel mix formed tubule-like structures without additional stimulation. These structures were positive for organic ion transporters and cytokeratin in addition to displaying sodium-dependent glucose uptake. Cells within this system displayed a greater sensitivity to gentamicin and doxorubicin when compared to standard cell culture techniques. The ability to observe drug effects over the course of long-term culture is beneficial for studies of chronic tissue response to treatment.

11.8 SUMMARY

3D tissue models have the benefit of recapitulating structural phenotypes over prolonged culture periods. The inclusion of an extracellular matrix environment has been shown to support long-term functional viability, a valuable parameter for certain kidney disease and drug response studies. Moreover, a 3D culture environment *in vitro* is necessary to study disease development for diseases such as ADPKD where a shift in structural phenotype *in vivo* is part of the pathogenesis.

Despite their demonstrated utility, current 3D tissue models fall significantly short of mimicking the *in vivo* kidney environment, and as such disease models and toxicity studies will continue to provide an incomplete picture. 3D models are readily adaptable to co-culture experiments, and the inclusion of multiple cell types would enable the exploration of more complex biological mechanisms. A culture method which established a confluent monolayer of human proximal tubule cells atop a type 1 collagen gel seeded with dermal fibroblasts was able to show epithelial-based regulation of fibroblast phenotype in response to injury (Moll et al. 2013). Additionally, in line with recent evidence suggesting macrophages play a role in PKD progression, macrophage inclusion with cystic ADPKD cells yielded a higher total number of microcysts (Swenson-Fields et al. 2013).

11.9 FUTURE DIRECTIONS

Current *in vitro* kidney tissue systems provide a physiologically relevant environment capable of supporting specific kidney cell phenotypes and functions. Microfluidic platforms for renal cell culture enable mechanical stimulation via fluid induced shear stresses. Moreover, 3D tissue models have supported structure formation within long-term cultures. In order to advance these systems for drug development and disease modeling applications, future systems will either need to be tailored for higher throughput or more complex culture environments. Renal microfluidic systems have the potential to be adapted for high-throughput applications for screening drug toxicity or efficacy due to the simplicity of the design. For mechanistic studies, systems which incorporate both fluidic stimulation and a 3D culture environment are desirable. Furthermore, co-culture of renal epithelial cells with endothelial and interstitial cells will be necessary to mimic *in vivo* paracrine signaling between cell types. A significant challenge to establishing complete renal function *in vitro* is the limited availability of segment specific cell types. Future advancements in stem cell differentiation through renal lineages will be important to developing systems capable of recapitulating a broader range of kidney cell phenotypes and functions.

REFERENCES

Astashkina, Anna I., Brenda K. Mann, Glenn D. Prestwich, and David W. Grainger. 2012. A 3-D Organoid Kidney Culture Model Engineered for High-Throughput Nephrotoxicity Assays. *Biomaterials* 33 (18): 4700–4711.

Baudoin, Régis, Laurent Griscom, Matthieu Monge, Cécile Legallais, and Eric Leclerc. 2007. Development of a Renal Microchip for in Vitro Distal Tubule Models. *Biotechnology Progress* 23 (5): 1245–1253.

Bukanov, Nikolay O., Hervé Husson, William R. Dackowski, Brandon D. Lawrence, Patricia A. Clow, Bruce L. Roberts, Katherine W. Klinger, and Oxana Ibraghimov-Beskrovnaya. 2002. Functional Polycystin-1 Expression is Developmentally Regulated during Epithelial Morphogenesis in Vitro: Downregulation and Loss of Membrane Localization during Cystogenesis. *Human Molecular Genetics* 11 (8): 923–936.

Carone, Frank A., Sakie Nakamura, Robert Bacallao, W. James Nelson, Mustafa Khokha, and Yashpal S. Kanwar. 1995. Impaired Tubulogenesis of Cyst-Derived Cells from Autosomal Dominant Polycystic Kidneys. *Kidney International* 47 (3): 861–868.

Choucha-Snouber, Leila, Caroline Aninat, Laurent Grsicom, Geoffrey Madalinski, Céline Brochot, Paul Emile Poleni, Florence Razan, Christiane Guguen Guillouzo, Cécile Legallais, and Anne Corlu. 2013. Investigation of Ifosfamide Nephrotoxicity Induced in a Liver–kidney Co-culture Biochip. *Biotechnology and Bioengineering* 110 (2): 597–608.

DesRochers, Teresa M., Erica Palma, and David L. Kaplan. 2014. Tissue-Engineered Kidney Disease Models. *Advanced Drug Delivery Reviews* 69: 67–80.

DesRochers, Teresa M., Laura Suter, Adrian Roth, and David L. Kaplan. 2013. Bioengineered 3D Human Kidney Tissue, a Platform for the Determination of Nephrotoxicity. *PLoS One* 8 (3): e59219.

Dieterle, Frank, Frank Sistare, Federico Goodsaid, Marisa Papaluca, Josef S. Ozer, Craig P. Webb, William Baer, Anthony Senagore, Matthew J. Schipper, and Jacky Vonderscher. 2010. Renal Biomarker Qualification Submission: A Dialog between the FDA-EMEA and Predictive Safety Testing Consortium. *Nature Biotechnology* 28 (5): 455–462.

Ferrell, Nicholas, Ravi R. Desai, Aaron J. Fleischman, Shuvo Roy, H. David Humes, and William H. Fissell. 2010. A Microfluidic Bioreactor with Integrated Transepithelial Electrical Resistance (TEER) Measurement Electrodes for Evaluation of Renal Epithelial Cells. *Biotechnology and Bioengineering* 107 (4): 707–716.

Frohlich, Else M., Xin Zhang, and Joseph L. Charest. 2012. The use of Controlled Surface Topography and Flow-Induced Shear Stress to Influence Renal Epithelial Cell Function. *Integrative Biology* 4 (1): 75–83.

Gao, Xiaofang, Yo Tanaka, Yasuhiko Sugii, Kazuma Mawatari, and Takehiko Kitamori. 2011. Basic Structure and Cell Culture Condition of a Bioartificial Renal Tubule on Chip Towards a Cell-Based Separation Microdevice. *Analytical Sciences* 27 (9): 907.

Grantham, Jared J., Marie Uchic, EJ Cragoe Jr, James Kornhaus, J. Aaron Grantham, Vicki Donoso, Roberto Mangoo-Karim, Andrew Evan, and James McAteer. 1989. Chemical Modification of Cell Proliferation and Fluid Secretion in Renal Cysts. *Kidney Int* 35 (6): 1379–1389.

Greka, Anna and Peter Mundel. 2012. Cell Biology and Pathology of Podocytes. *Annu Rev Physiol* 74: 299–323.

Hull, Robert N., William R. Cherry, and GW Weaver. 1976. The Origin and Characteristics of a Pig Kidney Cell Strain, LLC-PK1. *In Vitro* 12 (10): 670–677.

Jang, Kyung-Jin, Hye Sung Cho, Won Gyu Bae, Tae-Hwan Kwon, and Kahp-Yang Suh. 2011. Fluid-Shear-Stress-Induced Translocation of Aquaporin-2 and Reorganization of Actin Cytoskeleton in Renal Tubular Epithelial Cells. *Integrative Biology* 3 (2): 134–141.

Jang, Kyung-Jin, Ali Poyan Mehr, Geraldine Hamilton, Lori McPartlin, Seyoon Chung, Kahp-Yang Suh, and Donald Ingber. 2013. Human Kidney Proximal Tubule-on-a-Chip for Drug Transport and Nephrotoxicity Assessment. *Integrative Biology*

Jang, Kyung-Jin and Kahp-Yang Suh. 2010. A Multi-Layer Microfluidic Device for Efficient Culture and Analysis of Renal Tubular Cells. *Lab on a Chip* 10 (1): 36–42.

Jat, Parmjit S., Mark D. Noble, Paris Ataliotis, Yujiro Tanaka, Nikos Yannoutsos, Lena Larsen, and Dimitris Kioussis. 1991. "Direct Derivation of Conditionally Immortal Cell Lines from an H-2K^b–tsA58 Transgenic Mouse."

Kriz, Wilhelm. 1981. Structural Organization of the Renal Medulla: Comparative and Functional Aspects. *The American Journal of Physiology* 241 (1): R3–16.

Kuo, Ivana Y., Teresa M. DesRochers, Erica P. Kimmerling, Lily Nguyen, Barbara E. Ehrlich, and David L. Kaplan. 2014. Cyst Formation Following Disruption of Intracellular Calcium Signaling. *Proceedings of the National Academy of Sciences of the United States of America* 111 (39): 14283–14288.

Lang, Florian and Markus Paulmichl. 1995. Properties and Regulation of Ion Channels in MDCK Cells. *Kidney International* 48 (4): 1200–1205.

Levey, Andrew S., Lesley A. Stevens, Christopher H. Schmid, Yaping Lucy Zhang, Alejandro F. Castro, Harold I. Feldman, John W. Kusek, Paul Eggers, Frederick Van Lente, and Tom Greene. 2009. A New Equation to Estimate Glomerular Filtration Rate. *Annals of Internal Medicine* 150 (9): 604–612.

Mason, Roger M. and Nadia A. Wahab. 2003. Extracellular Matrix Metabolism in Diabetic Nephropathy. *Journal of the American Society of Nephrology: JASN* 14 (5): 1358–1373.

Miner, Jeffrey H. 2012. The Glomerular Basement Membrane. *Experimental Cell Research* 318 (9): 973–978.

Moll, Solange, Martin Ebeling, Franziska Weibel, Annarita Farina, Andrea Araujo Del Rosario, Jean Christophe Hoflack, Silvia Pomposiello, and Marco Prunotto. 2013. Epithelial Cells as Active Player in Fibrosis: Findings from an in Vitro Model. *PLoS One* 8 (2): e56575.

Montesano, Roberto, Hafida Ghzili, Fabio Carrozzino, Bernard C. Rossier, and Eric Féraille. 2009. cAMP-Dependent Chloride Secretion Mediates Tubule Enlargement and Cyst Formation by Cultured Mammalian Collecting Duct Cells. *American Journal of Physiology-Renal Physiology* 296 (2): F446–57.

Remuzzi, Giuseppe, Norberto Perico, Manuel Macia, and Piero Ruggenenti. 2005. The Role of Renin-Angiotensin-Aldosterone System in the Progression of Chronic Kidney Disease. *Kidney International* 68: S57–S65.

Rodriguez-Barbero, Alicia, Beatrice, L'azou, Jean Cambar, and Jose M. Lopez-Novoa. 2000. Potential use of Isolated Glomeruli and Cultured Mesangial Cells as in Vitro Models to Assess Nephrotoxicity. *Cell Biology and Toxicology* 16 (3): 145–153.

Ryan, Michael J., Gretchen Johnson, Judy Kirk, Sally M. Fuerstenberg, Richard A. Zager, and Beverly Torok-Storb. 1994. HK-2: An Immortalized Proximal Tubule Epithelial Cell Line from Normal Adult Human Kidney. *Kidney International* 45 (1): 48–57.

Saleem, Moin A., Michael J. O'Hare, Jochen Reiser, Richard J. Coward, Carol D. Inward, Timothy Farren, Chang Y. Xing, Lan Ni, Peter W. Mathieson, and Peter Mundel. 2002. A Conditionally Immortalized Human Podocyte Cell Line Demonstrating Nephrin and Podocin Expression. *Journal of the American Society of Nephrology: JASN* 13 (3): 630–638.

Satchell, Simon C., Candida H. Tasman, Anurag Singh, Lan Ni, Joyce M. Geelen, Christopher J. Von Ruhland, Michael J. O'Hare, Moin A. Saleem, Van Den Heuvel, Lambertus P, and Peter W. Mathieson. 2006. Conditionally Immortalized Human Glomerular Endothelial Cells Expressing Fenestrations in Response to VEGF. *Kidney International* 69 (9): 1633–1640.

Satchell, Simon C. and Filip Braet. 2009. Glomerular Endothelial Cell Fenestrations: An Integral Component of the Glomerular Filtration Barrier. *American Journal of Physiology. Renal Physiology* 296 (5): F947–56.

Schlondorff, Detlef and Bernhard Banas. 2009. The Mesangial Cell Revisited: No Cell is an Island. *Journal of the American Society of Nephrology: JASN* 20 (6): 1179–1187.

Slater, Sadie C., Vince Beachley, Thomas Hayes, Daming Zhang, Gavin I. Welsh, Moin A. Saleem, Peter W. Mathieson, Xuejun Wen, Bo Su, and Simon C. Satchell. 2011. An in Vitro Model of the Glomerular Capillary Wall using Electrospun Collagen Nanofibres in a Bioartificial Composite Basement Membrane. *PLoS One* 6 (6): e20802.

Snouber, Leila Choucha, Sébastien Jacques, Matthieu Monge, Cécile Legallais, and Eric Leclerc. 2012a. Transcriptomic Analysis of the Effect of Ifosfamide on MDCK Cells Cultivated in Microfluidic Biochips. *Genomics* 100 (1): 27–34.

Snouber, Leila Choucha, Franck Letourneur, Philippe Chafey, Cedric Broussard, Matthieu Monge, Cécile Legallais, and Eric Leclerc. 2012b. Analysis of Transcriptomic and Proteomic Profiles Demonstrates Improved Madin–Darby Canine Kidney Cell Function in a Renal Microfluidic Biochip. *Biotechnology Progress* 28 (2): 474–484.

Subramanian, Balajikarthick, Wei-Che Ko, Vikas Yadav, Teresa M. DesRochers, Ronald D. Perrone, Jing Zhou, and David L. Kaplan. 2012. The Regulation of Cystogenesis in a Tissue Engineered Kidney Disease System by Abnormal Matrix Interactions. *Biomaterials* 33 (33): 8383–8394.

Subramanian, Balajikarthick, Darya Rudym, Chris Cannizzaro, Ronald Perrone, Jing Zhou, and David L. Kaplan. 2010. Tissue-Engineered Three-Dimensional in Vitro Models for Normal and Diseased Kidney. *Tissue Engineering Part A* 16 (9): 2821–2831.

Swenson–Fields, Katherine I., Carolyn J. Vivian, Sally M. Salah, Jacqueline D. Peda, Bradley M. Davis, Nico van Rooijen, Darren P. Wallace, and Timothy A. Fields. 2013. Macrophages Promote Polycystic Kidney Disease Progression. *Kidney International* 83 (5): 855–864.

Taguchi, Atsuhiro, Yusuke Kaku, Tomoko Ohmori, Sazia Sharmin, Minetaro Ogawa, Hiroshi Sasaki, and Ryuichi Nishinakamura. 2014. Redefining the in Vivo Origin of Metanephric Nephron Progenitors Enables Generation of Complex Kidney Structures from Pluripotent Stem Cells. *Cell Stem Cell* 14 (1): 53–67.

Takasato, Minoru, Pei X. Er, Melissa Becroft, Jessica M. Vanslambrouck, Ed G. Stanley, Andrew G. Elefanty, and Melissa H. Little. 2014. Directing Human Embryonic Stem Cell Differentiation Towards a Renal Lineage Generates a Self-Organizing Kidney. *Nature Cell Biology* 16 (1): 118–126.

United States Renal Data System. 2014. *2014 USRDS Annual Data Report: An Overview of the Epidemiology of Kidney Disease in the United States. National Institutes of Health. National Institute of Diabetes and Digestive and Kidney Diseases*, Bethesda, MD.

Wallace, Darren P., Jared J. Grantham, and Lawrence P. Sullivan. 1996. Chloride and Fluid Secretion by Cultured Human Polycystic Kidney Cells. *Kidney International* 50 (4): 1327–1336.

Wieser, Matthias, Guido Stadler, Paul Jennings, Berthold Streubel, Walter Pfaller, Peter Ambros, Claus Riedl, Hermann Katinger, Johannes Grillari, and Regina Grillari-Voglauer. 2008. hTERT Alone Immortalizes Epithelial Cells of Renal Proximal Tubules without Changing their Functional Characteristics. *American Journal of Physiology. Renal Physiology* 295 (5): F1365–75.

Xia, Yun, Emmanuel Nivet, Ignacio Sancho-Martinez, Thomas Gallegos, Keiichiro Suzuki, Daiji Okamura, Min-Zu Wu, Ilir Dubova, Concepcion Rodriguez Esteban, and Nuria Montserrat. 2013. Directed Differentiation of Human Pluripotent Cells to Ureteric Bud Kidney Progenitor-Like Cells. *Nature Cell Biology* 15 (12): 1507–1515.

Yamaguchi, Tamio, Gail A. Reif, James P. Calvet, and Darren P. Wallace. 2010. Sorafenib Inhibits cAMP-Dependent ERK Activation, Cell Proliferation, and in Vitro Cyst Growth of Human ADPKD Cyst Epithelial Cells. *American Journal of Physiology. Renal Physiology* 299 (5): F944–51.

12 Body-on-a-Chip

Mahesh Devarasetty, Steven D. Forsythe, Thomas D. Shupe, Aleksander Skardal and Shay Soker

CONTENTS

12.1 INTRODUCTION

Animals, and especially mice, have been used as "*in vivo* test tubes" in scientific experiments for many years for developmental biology, disease modeling, and many other research applications (Hughes et al. 2011). Mice and rats have been used specifically to test the toxicity of new drugs prior to initiation of human clinical drug trials. However, in recent years it has become clear that animal models have significant limitations due to their genetic differences when compared to humans. This is especially critical for assessment of toxic side effects of drugs that might target the liver, heart, and other organs, due to the differences in enzymatic activity between humans and rodents, which results in different processing of drugs.

Traditional tissue culture models make use of plastic dishes to grow cells, and this method has been used for a variety of applications throughout scientific history. However, these models are poorly representative of the normal environment in which cells are found, the human body. First, tissues are made of a three-dimensional (3D) structure, while tissue culture plates are two-dimensional (2D). Cells behave differently in 2D versus 3D where they are surrounded by other cells, showered with signaling factors, and attached to the complex extracellular matrix (ECM). Second, the plastic used in tissue culture-ware is several magnitudes higher in stiffness compared with normal tissue, which has a drastic effect on cellular function (Engler et al. 2006). Cells found in a native tissue are "embedded" within the ECM, which provides tissue-specific stiffness and elasticity. Finally tissues are made of multiple cell types, while cell cultures are usually composed of a single cell type. Each of the various cell types within a tissue has a different contribution to the tissue's overall function.

Recent advances in tissue engineering and biomaterial research have resulted in successful fabrication of multicellular tissue constructs and micro-organs (organoids) that demonstrate authentic functional properties of the corresponding human tissue/organ. For example, there are liver organoids with metabolic activity, contracting skeletal and cardiac muscle constructs, "breathing" lung organoids, gut, kidney, and brain constructs, and more (Huh, Hamilton, and Ingber 2011; Bhatia and Ingber 2014).

Microfabrication techniques based on a variety of nanotechnologies resulted in the fabrication of miniature microfluidic systems that can control the flow of liquids through small channels and chambers (Smith et al. 2013). Tissues and organoids can be placed inside these chambers using sophisticated biomaterials to immobilize them and provide a proper microenvironment, and then be continuously perfused for long-term maintenance or drug deposition. These systems have been dubbed body-on-a-chip systems since they create a system representing not only multiple tissue and organ types but also the circulatory system that connects them, similar to the vascular system of the human body (Bhatia and Ingber 2014; Verhulsel et al. 2014). These properties allow the body-on-a-chip system to model normal physiology and disease states of whole organs. The microchannels can be used to perfuse the organoids with various pharmacological agents, similar to the way these agents are introduced to the tissues in the body. Furthermore, the organoids may be constructed from abnormal cells, affected by specific mutations that impair their function, to yield a diseased organoid, and drugs aimed at treating the specific disease are perfused through the microfluidic system. One such example is tumor-on-a-chip in which healthy cells make up the normal part of the tissue and the cancerous cells create the tumors within the normal tissue. These tumor organoids may be used for screening and testing of anticancer drugs, measuring the direct effect on the tumor cells and the side effects on the healthy tissue.

The ability of the body-on-a-chip platform to control the flow between chambers allows delivery of a test compound through one organoid type, then to another, different organoid, and so on (Bhatia and Ingber 2014). The compound may be metabolized by each of these tissue-specific organoids, and the downstream organoids will be exposed to the processed compound, similar to drugs being metabolized by the liver prior to circulating through to other organs. As such, the body-on-a-chip platform is ideal for testing newly developed drugs and assessing potential toxic side effects in human tissues and organs. Furthermore, the body-on-a-chip platform may be used for pharmacological studies to determine the specific activities of newly developed drugs and reduce the time of dose escalation studies in clinical trials.

In this chapter we describe techniques to create functional 3D human cell cultures. We then describe the microengineered, microfluidic systems designed to contain these cultures (organoids) under physiological perfusion conditions. Finally, we describe potential uses of the body-on-a-chip systems for drug screening and other applications.

12.2 3D CELL CULTURE SYSTEMS

Body-on-a-chip systems rely on highly functioning cellular models to generate physiologically relevant outputs such as basal metabolism, drug response, or other physiological functions (e.g., beating, or contraction, rate in heart models).

Systems that can produce this level of physiologic function are complex, and researchers have found that 3D cell culture systems are the most reliable, efficient, and cost-effective methods for constructing such systems. Although 2D cell culture has been used historically in the literature, the modality has inherent flaws. 2D systems do not fully recapitulate the *in vivo* environment or the function and phenotype of tissues and organs. In simpler cases, this aspect of 2D culture isn't considered a drawback, but in complex studies which rely on accurate cellular output, such as pharmaceutical discovery or diagnostic study, it represents a major hurdle to overcome. This limitation can be mitigated through the use of animal models, but even animal models do not completely simulate human outputs. Together, animal models and 2D culture systems comprise much of the contemporary research repertoire, and thus there has existed a limitation in the modeling of functional, physiologic systems.

3D cell culture has arisen as a mode for addressing the inherent limitations of 2D systems and animal models. 3D cell culture allows the development of physiologically relevant microarchitecture, chemotactic gradients, and cellular composition that drive cellular reorganization and ultimately produce functional output similar to that of *in vivo* systems. The core tenet is that recreating the cell-cell and cell-matrix interactions present *in vivo* will stimulate cells to act more physiologically. In addition, the use of human-derived cells eliminates the species-to-species variability found in animal models.

3D cell culture can be separated into two major types: scaffold-based technologies and scaffold-free technologies. Scaffold-based technologies rely on a premade substrate material for the attachment of cells: for example, cells can be resuspended in hydrogel solutions of collagen or hyaluronic acid which allow the attachment and reorganization of cells, or cells can be seeded on the ECM of a previously decellularized tissue (Baptista et al. 2013; Ruedinger et al. 2015). Scaffolds can also be constructed from synthetic materials such as polycaprolactone (PCL) or poly(lactic-co-glycolic acid) (PLGA), which are biocompatible and biodegradable polymers. Polymeric scaffolds can be constructed in a range of sizes, shapes, and stiffnesses using 3D extrusion-based printers, electrospinners, or basic molding techniques (Place et al. 2009; Gentile et al. 2014).

Scaffold-based technologies can leverage the scaffold as an instructive device for guiding cells into specific phenotypes through the conjugation of specific growth factors to the scaffold surface, incorporating ECM components, and even the modulation of scaffold mechanical stiffness (Engler et al. 2006; Motamedian et al. 2015). Tissue-engineered bone cultures rely on porous scaffolds to mimic the structure of native bone; for instance, some of the earliest engineered bone uses porous PLGA seeded with osteoblasts which eventually produce calcium deposits and alkaline-phosphatase activity resembling native bone (Crane, Ishaug, and Mikos 1995). Alternatively, scaffold-free technologies forego the use of a scaffold and rely on the cells to naturally develop an ECM or create cell-cell interactions that serve to anchor the cells together, forming a larger, more complex structure. Scaffold-free technologies can rely on low-adherence substrates to eliminate the development of 2D monolayers. For example, nonadherent plates can be used to force cells to attach to one another and aggregate, or hanging drop techniques can be used to eliminate all cell-substrate interaction and facilitate 3D structure formation.

For example, functional, vascularized liver buds were formed using non-adherent substrates and hepatic-destined cells. The nonadherent surface promoted the migration and aggregation of cells where they self-organized into immature liver-like structures (Takebe et al. 2013).

As evidenced before, there are many methods for producing 3D culture models; another example is micropatterning of ECM components, which allows precision placement of adhesion proteins. Specific protein micropatterns can drive differentiation of mesenchymal stem cells into osteogenic, adipogenic, and chondrogenic lineages. Decreasing the area of cellular attachment points forces the cell to contract, driving the cell toward a adipogenic lineage which represents softer tissue; conversely, allowing the cell to spread over a large attachment area shows the opposite effect, and pushes cells toward a stiffer, osteogenic lineage (McBeath et al. 2004; Kilian et al. 2010; Gao, McBeath, and Chen 2010). Rotating wall vessel bioreactors, a suspension culture system, can be used to create microgravity environments where cells attach to microcarrier beads while in a constant state of free-fall, resulting in the formation of cellular aggregates. Epithelial cells (from many organs) grown in this way develop into tissue constructs replete with tight junctions, correct apical/basal polarity, and the ability to produce mucus (Radtke and Herbst-Kralovetz 2012).

Although 3D culture represents a significant improvement over 2D culture in terms of physiologic relevance, there are still some critics. The complexity involved in 3D culture increases the difficulty of scaling up and results in high expense when implemented into high-throughput screening methods. Although 3D cell culture may not see immediate incorporation into clinical drug discovery schemes, it remains a versatile and informative tool for research.

12.3 MICROENGINEERING

The cellular content of a body-on-a-chip model is only one part of the puzzle. Even with the perfect composition of cells, proper function and response cannot be guaranteed. Another aspect that has to be considered is the localization of cells and what proteins or ECM components those cells are allowed to interact with. As mentioned earlier, an immature cell can be pushed toward any number of mature lineages purely by restricting cell spreading; thus, it is imperative to deliberately generate the environment for a cellular model because cells are greatly influenced by the makeup of the surrounding microenvironment. Techniques to customize and tailor the microenvironment fall into the category of microengineering.

Micropatterning is the precise placement of proteins on a cell culture substrate, which is achieved through several different methods: (1) microcontact printing, where a stamp coated in ECM components is pressed against the substrate to create the pattern similar to an ink stamp on paper; (2) photo-patterning, where ultraviolet (UV) light is shined through a photomask to activate the substrate surface and allow protein binding in specific areas; once activated, the substrate can be incubated with ECM solution to facilitate binding; and (3) laser-patterning, where laser light is used to activate the substrate with high precision (Thery 2010).

Many components of cell regulatory elements can be controlled using micropatterning. For instance, using islands of micropatterned ECM adhesion proteins that restrict cell spreading will also induce apoptosis in bovine adrenal capillary cells and terminal differentiation in epidermal keratinocytes. Conversely, using patterns that induce cell spreading allowed cellular proliferation, and the same bovine capillary cells and keratinocytes showed increased levels of DNA synthesis. These experiments hint at a link between ECM adhesion and cell cycle regulation (Chen et al. 1997; Watt, Jordan, and O'Neill 1988). Stem cells can also be differentiated through the use of micropatterns: mesenchymal stem cells grown on small micropatterns that reduce cell spreading differentiate toward adipogenic lineages, whereas micropatterns that induce cell spreading promote differentiation into osteogenic lineages. These mesenchymal stem cell studies also showed that cell shape was enough to modulate the expression of signaling proteins Rac1 and N-cadherin, which have a role in cell lineage specification (McBeath et al. 2004; Gao, McBeath, and Chen 2010). Micropatterning represents a powerful tool for microenvironment replication and simulation *in vitro*. By just changing the shape of a cell, whole expression cascades can be altered—or even more drastically, a cell can be induced to hit the kill-switch and apoptose. Micropatterning has wide applications and can be utilized to generate more physiologic output from *in vitro* systems.

If the idea of micropatterning is to replicate the *in vivo* environment in 2D, bioprinting is the 3D analog. Bioprinting involves the additive, layer-by-layer deposition of matrix, cells, and growth factors in a predefined manner. This method allows the production of multicellular 3D constructs with physiologic ECM composition and growth factors that facilitate function and self-organization of embedded cells.

Bioprinting is modular and customizable based on the desired structure. Applications requiring rigid matrices can be printed using high mechanical stiffness biomaterials or materials that can be crosslinked post-printing. For example, dental implants were bioprinted using polycaprolactone and hydroxyapatite (Kim et al. 2010). On the other hand, soft tissues like vascular grafts can be printed using low mechanical stiffness poly(ethylene glycol) (PEG) hydrogels (Stosich et al. 2009). Due to the consistency and convenience of bioprinters, bioprinted structures are becoming exceedingly complex and proving that recapitulation of physiologic 3D environments is integral to proper tissue level function.

Liver organoids can be made using a microextrusion bioprinter that prints liver-based spheroids suspended in a tailor-made hydrogel. These constructs show high levels of functionality and viability, making them ideal candidates for body-on-a-chip models (Skardal, Devarasetty, Kang, et al. 2015). Skin substitutes have been created using laser-assisted bioprinting which allowed precise placement of each layer of skin. These constructs were then implanted into mouse wound models and demonstrated vascularization, differentiation of keratinocytes, and collagen producing fibroblasts—all hallmarks of native skin (Michael et al. 2013).

3D printing is still in its infancy and logistical obstacles have hindered its application in whole-organ recapitulation. However, the inherent consistency and modularity make it a perfect complement to body-on-a-chip modeling. 3D printing can be used to generate high-throughput scales with low variability while incorporating the complex microengineering required for high functioning organ models.

12.4 MICROENGINEERED MODELS

Historically, *in vitro* models have relied on monolayer, or 2D, cell culture. The pharmaceutical industry uses monolayer disease lines for high-throughput testing of candidate drugs. Diagnostic firms use a more complex system with primary cells isolated from *in vivo* systems, but these cells are still studied in 2D, on traditional cell culture plastic. As previously described in Section 12.2, 2D cultures are not always physiologically accurate, and produce results that may not be relevant to *in vivo* applications.

The growing ubiquity of tissue engineering and microengineering in research has shifted the paradigm of model systems, though. Given the expense and ethical responsibility of animal models, researchers have turned to 3D microengineered models to study biological interactions on a variety of scales, from the interaction of a single cancer cell with the surrounding microenvironment to the interaction between separate, different tissue-engineered organ systems as in body-on-a-chip platforms. Microengineered models promise to expedite the rate of drug discovery while also progressing our understanding of disease dynamics.

Microengineered models are as varied as their applications. Some require the use of microfluidic devices, some rely on scaffold manipulation, while others utilize 3D printing for high throughput and consistency.

12.4.1 Microengineered Organ Models

Otherwise known as "organs-on-chips," microengineered organ models are meant to model healthy organ systems with as much accuracy as possible. These models enable system-level modeling through incorporation into body-on-a-chip platforms where each organ model is linked through microfluidic circulation. Organs-on-chips represent the functional unit of the body-on-a-chip, and they integrate aspects of microengineering, 3D cell culture, and microfluidics. Organ models alone must produce enough output to be measurable, but outputs must also be tuned in relation to other models when integrated into body-on-a-chip systems for balanced organ dynamics.

Cardiac models represent one of the simpler cases of microengineered organ models. The heart's main function in the body is to pump blood, and thus an *in vitro* model is mainly served by replicating this beating action. Monolayer cultures of human cardiomyocytes grown on Matrigel will spontaneously beat in culture (Goldman and Wurzel 1992). To produce 3D constructs, monolayer sheets can be layered, resulting in high-density tissues that spontaneously contract (Shimizu et al. 2002). These monolayer-based techniques replicate the heart's beating action but are not ideal for measuring 3D mechanics such as contractile force. 3D cardiac constructs can be developed by embedding cardiomyocytes in collagen I hydrogels and molded into rings; these constructs self-organize and develop physiologic action potential propagation (Zimmermann et al. 2002).

Lung models prove complex because they introduce an air-liquid interface which can be difficult to maintain *in vitro*. A simple model using two parallel plates (one is seeded with lung epithelial cells) flanking a microfluidic channel can be used to model the effects of airway reopening due to mechanical ventilation (Bilek, Dee, and Gaver 2003). Another model uses lung epithelial cells grown on a semiporous membrane; on either side of the membrane are microfluidic channels. As the cells grow to confluency, the liquid in one channel is evacuated to form an air liquid interface around the lung epithelium (Huh et al. 2007; Tavana et al. 2011). Other models utilize the same format but include pneumatic side-channels to introduce the lung's characteristic stretching to the cellularized membrane (Huh et al. 2010).

Vascular models typically utilize a single layer of endothelial cells. This 2D model can serve as a basic model of the blood-contacting surface of a blood vessel but does little to replicate the dilation and constriction, permeabilization, and clotting phenomenon found in native vasculature. Microfluidic models of vasculature have been constructed using collagen I hydrogels encapsulating human endothelial cells and pericytes within a microfluidic channel. The cells self-organize into a tube like morphology complete with cell-cell junctions, and the model correctly responds to drugs (van der Meer et al. 2013). Another microfluidic model used concentric, cellularized hydrogels molded into rings. The inner portion of the ring contained endothelial cells while the outer portion contained smooth muscle cells, thereby mimicking the cellular composition of vasculature. The rings were then connected to create tubes and then perfused to create the vascular model (Du et al. 2011).

Liver models are typically produced to meet a certain criterion: drug toxicity, metabolism, or the production of bile. The liver houses many vital functions and it is difficult to replicate them all *in vitro*. The simplest system of recapitulating liver function *in vitro* uses a sandwich culture of hepatocytes between collagen or Matrigel layers (LeCluyse 2001). Spheroid models incorporating nonparenchymal cells of the liver can be made to increase model output and function (Chang and Hughes-Fulford 2009); other techniques to increase liver function suspend hepatocytes in hydrogels doped with liver-specific ECM (Skardal et al. 2012). A microfluidic model uses immobilized hepatocyte spheroids inside a microfluidic chamber. The spheroids are then perfused with medium for maintenance and drug deposition (Toh et al. 2009).

The models mentioned here are only a small sample of those found in the literature, and increasingly more complex models are developed every year. The most high profile models are those dubbed body-on-a-chip systems which incorporate several model organs as outlined throughout this text. These systems can assess the system-wide effects of a drug or pathogen: for instance, how drug metabolites produced by the liver affect the function of cardiac tissue. Systems like this, incorporating many major organ systems, have the potential to replace animal models for drug testing and also allow rapid response to dangerous new pathogens.

12.4.2 Microengineered Disease Models

Sometimes called "disease-in-a-dish" models, microengineered disease models seek to replicate physiologic disease characteristics *in vitro*. This allows close scrutiny of disease progression and a high functioning model for drug screening.

Tumor models attempt to model the unique microenvironment found around cancer cells. Rotating wall vessel (RWV) bioreactors can be used to create tumor aggregates centered around a non-degradable microcarrier bead (Hammond and Hammond 2001; Skardal et al. 2010; Becker and Souza 2013). These aggregates accurately model the diffusion characteristics of tumors found in the body (Becker and Souza 2013). Metastatic cells can also be integrated with normal cells to create a tumor model complete with stroma (Skardal, Devarasetty, Rodman, et al. 2015). Microfluidic models can integrate 3D cultured tissues with complex geometry and flow characteristics to model tumor migration. Devices can be developed with discrete areas for stromal and cancer cells, and they have shown that stromal cells, such as fibroblasts, can induce tumor growth and invasion (Domenech et al. 2009; Domenech et al. 2012). Cancer cell intravasation of vasculature is a crucial point in metastasis; devices incorporating endothelial cells allow the observation of cancer cell interaction with a pseudo-vasculature (Song et al. 2009).

Fibrosis changes the mechanical properties of organ systems: muscle fibrosis can change range of motion or cardiac output, while pulmonary fibrosis can reduce lung function. Microengineered fibrosis models tend to incorporate normal tissue with fibroblastic cells to study the mechanistic cascade of fibrosis while also providing a relevant platform for drug testing. Cardiac fibrosis can be modeled using cardiac fibroblasts grown on a micropatterned surface of soft and stiff areas that mimic the varying stiffness found in fibrotic tissue (Zhao et al. 2014). Collagen hydrogel matrices of varying stiffness used to culture lung fibroblasts can be used to model the environment of pulmonary fibrosis (Vicens-Zygmunt et al. 2015).

12.5 MICROFLUIDIC DEVICES

The advent of microelectromechanical systems (MEMS), such as those found in silicone chips, brought with it a slew of advanced manufacturing techniques. An entire repertoire of technologies was developed to produce micron scale components for the rapidly growing electronics field. This included the production of silicone or metal wafers on the order of 1 to 100 micrometers which were then fashioned into arrays of transistors for processors and microchips. These technologies have bled into other fields in a variety of forms. In biomedical sciences, the ability to produce micron scale architecture for cell interaction was particularly intriguing; in the past it was difficult to produce mechanical force and topographical features similar to that found at the cell level, but MEMS technologies promised to change that (Sackmann, Fulton, and Beebe 2014).

Microfluidics is a field borne out of the incorporation of MEMS technology into biomedical science. By producing thin wafers of silicone, polymer, or metal, then machining or etching those wafers into negative molds for lithography, structures with built-in microchannels and microarchitectures can be fashioned. Originally these technologies were used for producing micron scale topographies to study cellular interaction before being applied to more complex systems such as polymerase chain reaction (PCR) chips (Cady et al. 2005) or devices that can fractionate a small blood sample for disease testing (Hou et al. 2011).

Microfluidic devices are produced in several main steps. First, the deposition of molding material or resist, typically polymethylmethacrylate (PMMA) or SU-8, onto a glass substrate. This material should be light-reactive and thus is called a photoresist. Next, a photomask is used to isolate areas of interest (or exclude areas of interest in the case of negative photoresists) before a UV light is shined over the photomask. The UV light crosslinks the material in positive photoresists (PMMA) or degrades the material in negative photoresists (SU-8). Afterward, the unneeded photoresist material is discarded from the substrate and the result should be a thin structure that matches the photomask. This structure is a negative mold of the desired microfluidic device. Finally, a curable silicone can be poured over the negative mold, allowed to cure and removed to produce the final microfluidic channel or system molded in silicone (Qin, Xia, and Whitesides 2010).

Microfluidic device designs are highly varied depending on the application. Devices can be fabricated to separate a blood sample using laminar flow to allow many tests to be achieved using a small volume of blood; such devices have found use in third-world and developing countries where low-cost alternatives are required (Yager et al. 2006; Hou et al. 2011; Emani et al. 2012). Other designs can be used to mix two independent fluids and produce a gradient of concentration ratios between them; this approach can be useful for expediting the assay of drug concentrations on cell cultures (Lee et al. 2011; Wunderlich et al. 2013). Even more complex body-on-a-chip designs involve the integration of 3D cultured cells that form organoids which are then combined with other organoids and biosensors into a single microfluidic system (Huh, Hamilton, and Ingber 2011). Bodies-on-a-chip serve to replicate the system-level dynamics of the human body while retaining efficient analysis methods found in cell culture.

Like many technologies before, microfluidics was derived from work meant for a completely different field. Although MEMS technology was meant for silicone chip manufacturing, it has found a home in biomedical research and development. Microfluidic devices allow the integration of biosensors, organ models, and tunable fluid flow characteristics into a single system that derives much more information than the sum of its parts. This powerful combination of attributes makes microfluidics the perfect platform for body-on-a-chip systems.

12.6 ADVANTAGES, SHORTCOMINGS, AND FUTURE IMPROVEMENTS

As multiple organ types are envisioned in a single body-on-a–chip system, in order to replicate the response of as many as possible target organs to a specific drug, an appropriate sourcing for the different cell types composing these organoids is required. Embryonic and iPS (induced pluripotent stem cell)–derived cells represent an ideal source (Takahashi and Yamanaka 2006; Kim et al. 2009; Takebe et al. 2013). Detailed protocols are now available to differentiate these cells into mature and functional cell types of almost every tissue in the body. For example, iPS-derived cardiac muscle organoids, inside a microfluidic system, were used to (1) test the response to drugs that were eliminated from clinical use due to cardiac toxicity and (2) compare the response of cardiac cells derived from individuals with different genetic backgrounds. The immune system plays an important role in response to tissue damage well as in the response to toxicants. Many of the described body-on-a–chip systems lack this cellular component and thus may not accurately represent the full spectrum of drug response/toxicity. In addition, in the liver, stimulation of the immune system activates the nonparenchymal stellate cells, which leads to tissue fibrosis. Recapitulation of this process in an *in vitro* body-on-a–chip system is still a major challenge (Bhatia and Ingber 2014).

The material used to fabricate the microfluidic system has special requirements. It has to be transparent for better monitoring, biocompatible, and most importantly, has to provide tight sealing. Polydimethylsiloxane (PDMS) is widely used to build microfluidic systems, as it allows gas diffusion and appropriate transparency (Gunther et al. 2010; Grosberg et al. 2011; Nakao et al. 2011). However, it absorbs biomolecules nonspecifically, which may impact the accuracy of toxicity measurements (Wong and Ho 2009; Berthier, Young, and Beebe 2012). Surface modification of PDMS may alleviate some of these shortcomings (Wong and Ho 2009; Zhou, Ellis, and Voelcker 2010; Zhou et al. 2012). Materials used to encapsulate cells and form organoids were initially collagen and fibrin. They are easy to work with, are biocompatible, and allow cell growth and migration. However, they lack tissue-specific signals. We fabricated liver specific biogel, made of hyaluronic acid (HA) and liver extracellular matrix extract (Skardal et al. 2012). Primary human hepatocytes cultured on this biogel showed better growth, morphology, and metabolic activity compared with cells cultured on Type I collagen.

Liver organoids are the most "popular" microphysiological system for drug screening. However, drug metabolites generated in the liver circulate through the body and have multiple tissue targets. As such, effort needs to be made to fabricate multitissue body-on-a-chip systems. The most important tissues for pharmacokinetic and drug toxicity studies are the kidney, brain, heart, lung, and the gastrointestinal system (van Midwoud et al. 2010; Choucha-Snouber et al. 2013). With the generation of multitissue type systems, there will be a need for a universal "blood substitute" media that will be compatible with different cell types. Such specialized media will have to provide the necessary growth factors to each tissue. However, a growth factor may have a positive effect on one tissue type and a negative effect on another (Zhang et al. 2009). To overcome this problem, one can deposit a local source of the growth factor inside a specific tissue chamber, or within the organoid. Alternatively, a parallel microchannel system can be created that will "feed" each organoid with its own specific media.

Other areas for improvement are scaling up fluid flow rates and pressures to perfuse larger organoids in order to match tissue perfusion as in the human body. Scaling up is especially important to provide meaningful data for pharmacokinetic/pharmacodynamic (PK/PD) studies (Wikswo, Block, et al. 2013; Wikswo, Curtis, et al. 2013). In order to perform accurate PK/PD studies the body-on-a–chip system will likely require development of automated multiparametric monitoring approach. Such a system will not only provide real-time quantitative measurements of metabolites and other biomarkers but will also support feedback control and continuous regulation of levels of oxygen, pH, temperature, etc. Finally, the body-on-a–chip system will eventually be designed to represent different human diseases (Sung et al. 2014). One approach is to derive the organoids using cells, such as iPS cells, from people with specific genetic backgrounds prone to particular diseases (Bellin et al. 2012).

Another approach is to expose the "healthy" organoids to an environment that will induce a disease state, such as cytokine-induced fibroblastic cell proliferation to create fibrosis, liver enzyme toxicants to cause hepatocyte death, and improper metabolic function and application of drugs that affect the beating of cardiac organoids.

12.7 CONCLUSIONS

Body-on-a–chip systems are revolutionary model platforms to represent human physiology *in vitro*. They are based on true 3D tissue architecture and a fluid circulation system that delivers reagents from one tissue to the other in a controlled manner (reviewed in Esch, King, and Shuler 2011; Huh, Hamilton, and Ingber 2011; Bhatia and Ingber 2014). These systems have the potential to recapitulate both normal and disease tissue states and response to drugs and toxicants at the organ level. Combined with sophisticated microfluidic and monitoring systems it could become a standard for drug toxicity screening, validation, and prioritization of newly developed drugs. Furthermore, the body-on-a-chip system may identify new physiologically relevant biomarkers for drug efficacy and/or toxicity and serve for PK/PD modeling of drug dosing studies. The use of patient-specific cells, tissues, and organoids will support recent efforts toward establishment of personalized medicine guidelines. The existing body-on-a–chip systems are still in their infancy, and many more technical and biological improvements are needed in order to create a true comprehensive human physiological system that will completely replace *in vivo* experimentation in animals and humans.

REFERENCES

Baptista, P. M., D. Vyas, E. Moran, et al. 2013. Human liver bioengineering using a whole liver decellularized bioscaffold. *Methods Mol Biol* 1001:289–98. doi: 10.1007/978-1-62703-363-3_24.

Becker, Jeanne L., and Glauco R. Souza. 2013. Using space-based investigations to inform cancer research on Earth. *Nat Rev Cancer* 13 (5):315–27. doi: 10.1038/nrc3507.

Bellin, M., M. C. Marchetto, F. H. Gage, et al. 2012. Induced pluripotent stem cells: The new patient? *Nat Rev Mol Cell Biol* 13 (11):713–26. doi: 10.1038/nrm3448.

Berthier, Erwin, Edmond W. K. Young, and David Beebe. 2012. Engineers are from PDMS-land, biologists are from polystyrenia. *Lab Chip* 12 (7):1224–37. doi: 10.1039/C2LC20982A.

Bhatia, Sangeeta N., and Donald E. Ingber. 2014. Microfluidic organs-on-chips. *Nat Biotech* 32 (8):760–72. doi: 10.1038/nbt.2989.

Bilek, Anastacia M., Kay C. Dee, and Donald P. Gaver. 2003. Mechanisms of surface-tension-induced epithelial cell damage in a model of pulmonary airway reopening. *J Appl Physiol* 94 (2):770–83.

Cady, Nathaniel C., Scott Stelick, Madanagopal V. Kunnavakkam, et al. 2005. Real-time PCR detection of Listeria monocytogenes using an integrated microfluidics platform. *Sensors and Actuators B: Chemical* 107 (1):332–41. doi: http://dx.doi.org/10.1016/j.snb.2004.10.022.

Chang, T. T., and M. Hughes-Fulford. 2009. Monolayer and spheroid culture of human liver hepatocellular carcinoma cell line cells demonstrate distinct global gene expression patterns and functional phenotypes. *Tissue Eng Part A* 15 (3):559–67. doi: 10.1089/ten.tea.2007.0434.

Chen, Christopher S., Milan Mrksich, Sui Huang, et al. 1997. Geometric control of cell life and death. *Science* 276 (5317):1425–8. doi: 10.1126/science.276.5317.1425.

Choucha-Snouber, L., C. Aninat, L. Grsicom, et al. 2013. Investigation of ifosfamide nephrotoxicity induced in a liver-kidney co-culture biochip. *Biotechnol Bioeng* 110 (2):597–608. doi: 10.1002/bit.24707.

Crane, Genevieve M., Susan L. Ishaug, and Antonios G. Mikos. 1995. Bone tissue engineering. *Nat Med* 1 (12):1322–4.

Domenech, M., R. Bjerregaard, W. Bushman, et al. 2012. Hedgehog signaling in myofibroblasts directly promotes prostate tumor cell growth. *Integr Biol (Camb)* 4 (2):142–52. doi: 10.1039/c1ib00104c.

Domenech, M., H. Yu, J. Warrick, et al. 2009. Cellular observations enabled by microculture: Paracrine signaling and population demographics. *Integr Biol (Camb)* 1 (3):267–74. doi: 10.1039/b823059e.

Du, Y., M. Ghodousi, H. Qi, et al. 2011. Sequential assembly of cell-laden hydrogel constructs to engineer vascular-like microchannels. *Biotechnol Bioeng* 108 (7):1693–703. doi: 10.1002/bit.23102.

Emani, S., R. Sista, H. Loyola, et al. 2012. Novel microfluidic platform for automated lab-on-chip testing of hypercoagulability panel. *Blood Coagul Fibrinolysis* 23 (8):760–8. doi: 10.1097/MBC.0b013e328358e982.

Engler, A. J., S. Sen, H. L. Sweeney, et al. 2006. Matrix elasticity directs stem cell lineage specification. *Cell* 126 (4):677–89. doi: 10.1016/j.cell.2006.06.044.

Esch, M. B., T. L. King, and M. L. Shuler. 2011. The role of body-on-a-chip devices in drug and toxicity studies. *Annu Rev Biomed Eng* 13:55–72. doi: 10.1146/annurev-bioeng-071910-124629.

Gao, L., R. McBeath, and C. S. Chen. 2010. Stem cell shape regulates a chondrogenic versus myogenic fate through Rac1 and N-cadherin. *Stem Cells* 28 (3):564–72. doi: 10.1002/stem.308.

Gentile, P., V. Chiono, I. Carmagnola, et al. 2014. An overview of poly(lactic-co-glycolic) acid (PLGA)-based biomaterials for bone tissue engineering. *Int J Mol Sci* 15 (3):3640–59. doi: 10.3390/ijms15033640.

Goldman, B. I., and J. Wurzel. 1992. Effects of subcultivation and culture medium on differentiation of human fetal cardiac myocytes. *In Vitro Cell Dev Biol* 28a (2):109–19.

Grosberg, A., P. W. Alford, M. L. McCain, et al. 2011. Ensembles of engineered cardiac tissues for physiological and pharmacological study: Heart on a chip. *Lab Chip* 11 (24):4165–73. doi: 10.1039/c1lc20557a.

Gunther, A., S. Yasotharan, A. Vagaon, et al. 2010. A microfluidic platform for probing small artery structure and function. *Lab Chip* 10 (18):2341–9. doi: 10.1039/c004675b.

Hammond, T. G., and J. M. Hammond. 2001. Optimized suspension culture: The rotating-wall vessel. *Am J Physiol Renal Physiol* 281 (1):F12–25.

Hou, Han Wei, Ali Asgar S. Bhagat, Wong Cheng Lee, et al. 2011. Microfluidic devices for blood fractionation. *Micromachines* 2 (3):319.

Hughes, J. P., S. Rees, S. B. Kalindjian, et al. 2011. Principles of early drug discovery. *Br J Pharmacol* 162 (6):1239–49. doi: 10.1111/j.1476-5381.2010.01127.x.

Huh, Dongeun, Hideki Fujioka, Yi-Chung Tung, et al. 2007. Acoustically detectable cellular-level lung injury induced by fluid mechanical stresses in microfluidic airway systems. *Proc Natl Acad Sci* 104 (48):18886–91. doi: 10.1073/pnas.0610868104.

Huh, Dongeun, Geraldine A. Hamilton, and Donald E. Ingber. 2011. From 3D cell culture to organs-on-chips. *Trends Cell Biol* 21(12):745–54. doi: http://dx.doi.org/10.1016/j.tcb.2011.09.005.

Huh, Dongeun, Benjamin D. Matthews, Akiko Mammoto, et al. 2010. Reconstituting organ-level lung functions on a chip. *Science* 328 (5986):1662–8. doi: 10.1126/science.1188302.

Kilian, K. A., B. Bugarija, B. T. Lahn, and et al. 2010. Geometric cues for directing the differentiation of mesenchymal stem cells. *Proc Natl Acad Sci U S A* 107 (11):4872–7. doi: 10.1073/pnas.0903269107.

Kim, J. B., V. Sebastiano, G. Wu, et al. 2009. Oct4-induced pluripotency in adult neural stem cells. *Cell* 136 (3):411–19. doi: 10.1016/j.cell.2009.01.023.

Kim, K., C. H. Lee, B. K. Kim, et al. 2010. Anatomically shaped tooth and periodontal regeneration by cell homing. *J Dental Res* 89 (8):842–7. doi: 10.1177/0022034510370803.

LeCluyse, Edward L. 2001. Human hepatocyte culture systems for the in vitro evaluation of cytochrome P450 expression and regulation. *Eur Pharmaceut Sci* 13 (4):343–68. doi: http://dx.doi.org/10.1016/S0928-0987(01)00135-X.

Lee, Chia-Yen, Chin-Lung Chang, Yao-Nan Wang, et al. 2011. Microfluidic mixing: A review. *Int J Mol Sci* 12 (5):3263–87. doi: 10.3390/ijms12053263.

McBeath, R., D. M. Pirone, C. M. Nelson, et al. 2004. Cell shape, cytoskeletal tension, and RhoA regulate stem cell lineage commitment. *Dev Cell* 6 (4):483–95.

Michael, S., H. Sorg, C. T. Peck, et al. 2013. Tissue engineered skin substitutes created by laser-assisted bioprinting form skin-like structures in the dorsal skin fold chamber in mice. *PLoS One* 8 (3):e57741. doi: 10.1371/journal.pone.0057741.

Motamedian, S. R., S. Hosseinpour, M. G. Ahsaie, et al. 2015. Smart scaffolds in bone tissue engineering: A systematic review of literature. *World J Stem Cells* 7 (3):657–68. doi: 10.4252/wjsc.v7.i3.657.

Nakao, Y., H. Kimura, Y. Sakai, et al. 2011. Bile canaliculi formation by aligning rat primary hepatocytes in a microfluidic device. *Biomicrofluidics* 5 (2):22212. doi: 10.1063/1.3580753.

Place, Elsie S., Julian H. George, Charlotte K. Williams, et al. 2009. Synthetic polymer scaffolds for tissue engineering. *Chem Soc Rev* 38 (4):1139–51. doi: 10.1039/B811392K.

Qin, Dong, Younan Xia, and George M. Whitesides. 2010. Soft lithography for micro- and nanoscale patterning. *Nat Protocols* 5 (3):491–502.

Radtke, A. L., and M. M. Herbst-Kralovetz. 2012. Culturing and applications of rotating wall vessel bioreactor derived 3D epithelial cell models. *J Vis Exp* (62). doi: 10.3791/3868.

Ruedinger, F., A. Lavrentieva, C. Blume, et al. 2015. Hydrogels for 3D mammalian cell culture: A starting guide for laboratory practice. *Appl Microbiol Biotechnol* 99 (2):623–36. doi: 10.1007/s00253-014-6253-y.

Sackmann, Eric K., Anna L. Fulton, and David J. Beebe. 2014. The present and future role of microfluidics in biomedical research. *Nature* 507 (7491):181–9. doi: 10.1038/nature13118.

Shimizu, T., M. Yamato, Y. Isoi, et al. 2002. Fabrication of pulsatile cardiac tissue grafts using a novel 3-dimensional cell sheet manipulation technique and temperature-responsive cell culture surfaces. *Circ Res* 90 (3):e40.

Skardal, A., M. Devarasetty, C. Rodman, et al. 2015. Liver-tumor hybrid organoids for modeling tumor growth and drug response in vitro. *Ann Biomed Eng* 43 (10):2361–73. doi: 10.1007/s10439–015-1298-3.

Skardal, A., S. F. Sarker, A. Crabbe, et al. 2010. The generation of 3-D tissue models based on hyaluronan hydrogel-coated microcarriers within a rotating wall vessel bioreactor. *Biomaterials* 31 (32):8426–35. doi: 10.1016/j.biomaterials.2010.07.047.

Skardal, A., L. Smith, S. Bharadwaj, et al. 2012. Tissue specific synthetic ECM hydrogels for 3-D in vitro maintenance of hepatocyte function. *Biomaterials* 33 (18):4565–75. doi: 10.1016/j.biomaterials.2012.03.034.

Skardal, Aleksander, Mahesh Devarasetty, Hyun-Wook Kang, et al. 2015. A hydrogel bioink toolkit for mimicking native tissue biochemical and mechanical properties in bioprinted tissue constructs. *Acta Biomaterialia*. doi: http://dx.doi.org/10.1016/j.actbio.2015.07.030.

Smith, Alec S. T., Christopher J. Long, Bonnie J. Berry, et al. 2013. Microphysiological systems and low-cost microfluidic platform with analytics. *Stem Cell Res Ther* 4 (Suppl 1):S9–S9. doi: 10.1186/scrt370.

Song, Jonathan W., Stephen P. Cavnar, Ann C. Walker, et al. 2009. Microfluidic endothelium for studying the intravascular adhesion of metastatic breast cancer cells. *PLoS One* 4 (6):e5756. doi: 10.1371/journal.pone.0005756.

Stosich, Michael S., Eduardo K. Moioli, June K. Wu, et al. 2009. Bioengineering strategies to generate vascularized soft tissue grafts with sustained shape. *Methods* 47 (2):116–21. doi: http://dx.doi.org/10.1016/j.ymeth.2008.10.013.

Sung, J. H., B. Srinivasan, M. B. Esch, et al. 2014. Using physiologically-based pharmacokinetic-guided "body-on-a-chip" systems to predict mammalian response to drug and chemical exposure. *Exp Biol Med (Maywood)* 239 (9):1225–39. doi: 10.1177/1535370214529397.

Takahashi, K., and S. Yamanaka. 2006. Induction of pluripotent stem cells from mouse embryonic and adult fibroblast cultures by defined factors. *Cell* 126 (4):663–76. doi: 10.1016/j.cell.2006.07.024.

Takebe, Takanori, Keisuke Sekine, Masahiro Enomura, et al. 2013. Vascularized and functional human liver from an iPSC-derived organ bud transplant. *Nature* 499 (7459):481–4. doi: 10.1038/nature12271.

Tavana, H., P. Zamankhan, P. J. Christensen, et al. 2011. Epithelium damage and protection during reopening of occluded airways in a physiologic microfluidic pulmonary airway model. *Biomed Microdevices* 13 (4):731–42. doi: 10.1007/s10544-011-9543-5.

Thery, M. 2010. Micropatterning as a tool to decipher cell morphogenesis and functions. *J Cell Sci* 123 (Pt 24):4201–13. doi: 10.1242/jcs.075150.

Toh, Y. C., T. C. Lim, D. Tai, et al. 2009. A microfluidic 3D hepatocyte chip for drug toxicity testing. *Lab Chip* 9 (14):2026–35. doi: 10.1039/b900912d.

van der Meer, A. D., V. V. Orlova, P. ten Dijke, et al. 2013. Three-dimensional co-cultures of human endothelial cells and embryonic stem cell-derived pericytes inside a microfluidic device. *Lab Chip* 13 (18):3562–8. doi: 10.1039/c3lc50435b.

van Midwoud, P. M., M. T. Merema, E. Verpoorte, et al. 2010. A microfluidic approach for in vitro assessment of interorgan interactions in drug metabolism using intestinal and liver slices. *Lab Chip* 10 (20):2778–86. doi: 10.1039/c0lc00043d.

Verhulsel, M., M. Vignes, S. Descroix, et al. 2014. A review of microfabrication and hydrogel engineering for micro-organs on chips. *Biomaterials* 35 (6):1816–32. doi: 10.1016/j.biomaterials.2013.11.021.

Vicens-Zygmunt, V., S. Estany, A. Colom, et al. 2015. Fibroblast viability and phenotypic changes within glycated stiffened three-dimensional collagen matrices. *Respir Res* 16:82. doi: 10.1186/s12931-015-0237-z.

Watt, F. M., P. W. Jordan, and C. H. O'Neill. 1988. Cell shape controls terminal differentiation of human epidermal keratinocytes. *Proceedings of the National Academy of Sciences of the United States of America* 85 (15):5576–80.

Wikswo, J. P., E. L. Curtis, Z. E. Eagleton, et al. 2013. Scaling and systems biology for integrating multiple organs-on-a-chip. *Lab Chip* 13 (18):3496–511. doi: 10.1039/c3lc50243k.

Wikswo, John P., Frank E. Block, David E. Cliffel, et al. 2013. Engineering challenges for instrumenting and controlling integrated organ-on-chip systems. *IEEE Trans Biomed Eng* 60(3):682–90. doi: 10.1109/TBME.2013.2244891.

Wong, I., and C. M. Ho. 2009. Surface molecular property modifications for poly(dimethylsiloxane) (PDMS) based microfluidic devices. *Microfluid Nanofluidics* 7 (3):291–306. doi: 10.1007/s10404-009-0443-4.

Wunderlich, Bengt, Daniel Nettels, Stephan Benke, et al. 2013. Microfluidic mixer designed for performing single-molecule kinetics with confocal detection on timescales from milliseconds to minutes. *Nat Protocols* 8 (8):1459–74. doi: 10.1038/nprot.2013.082.

Yager, Paul, Thayne Edwards, Elain Fu, et al. 2006. Microfluidic diagnostic technologies for global public health. *Nature* 442 (7101):412–18.

Zhang, C., Z. Zhao, N. A. Abdul Rahim, et al. 2009. Towards a human-on-chip: Culturing multiple cell types on a chip with compartmentalized microenvironments. *Lab Chip* 9 (22):3185–92. doi: 10.1039/b915147h.

Zhao, H., X. Li, S. Zhao, Y. Zeng, L. Zhao, et al. 2014. Microengineered in vitro model of cardiac fibrosis through modulating myofibroblast mechanotransduction. *Biofabrication* 6 (4):045009. doi: 10.1088/1758-5082/6/4/045009.

Zhou, J., A. V. Ellis, and N. H. Voelcker. 2010. Recent developments in PDMS surface modification for microfluidic devices. *Electrophoresis* 31 (1):2–16. doi: 10.1002/elps.200900475.

Zhou, Jinwen, Dmitriy A. Khodakov, Amanda V. Ellis, et al. 2012. Surface modification for PDMS-based microfluidic devices. *Electrophoresis* 33 (1):89–104. doi: 10.1002/elps.201100482.

Zimmermann, W. H., K. Schneiderbanger, P. Schubert, et al. 2002. Tissue engineering of a differentiated cardiac muscle construct. *Circ Res* 90 (2):223–30.

Section III

Applications

13 Integrated Multi-Organoid Dynamics

Aleksander Skardal, Mahesh Devarasetty, Sean Murphy and Anthony Atala

CONTENTS

13.1 INTRODUCTION

There is a critical need for improved systems to model the effects of chemical and biological agents on the body.[1,2] Currently, animal models serve as gold standards for testing, but the drawbacks associated with such models are high costs and uncertainties in interpretation of the results. Interspecies differences and variability means that animal models are often poor predictors of human efficacy and toxicology. *In vitro* systems that use human tissues would be preferable; however, for these systems to serve as tools that reflect human biology, key physiological features and toxicology endpoints need to be included in their design for informative and reliable efficacy, pharmocokinetic, and toxicity testing. Traditional *in vitro* 2D cultures, currently the norm for early drug compound screening, fail to recapitulate the 3D microenvironment of *in vivo* tissues.[3,4] Tissue culture dishes have three major differences from native tissue microenvironments: surface topography, surface stiffness, and most importantly, a 2D rather than 3D architecture. As a consequence, 2D culture places a selective pressure on cells, substantially altering their original phenotypic properties. Drug diffusion kinetics are not accurately modeled in 2D tissue culture, drug doses effective in 2D are often ineffective when scaled to patients, and the lack of cell-cell/cell-matrix interactions in 2D often lead to loss of cell function.[3,5,6] Instead, "organ-on-a-chip" devices that can simulate 3D tissue architectures and physiological fluid flow conditions are better options. Such devices are capable of producing rapid, reliable predictions of biological processes, disease states, and drug and toxicology responses. Additionally, these engineering platforms benefit by having reproducible hardware, a simple scale up transition, high throughput assay capability, automation potential, and control over physical environmental conditions such as fluid flow and shear stress. These microengineering and microfluidics technologies have resulted in advanced studies in creating and culturing 3D human tissue on a chip.[7] Currently, many organ-on-chip systems exist,[8,9] as well as several on-chip disease models.[8] By biofabricating the respective cell type or combining several cell types into 3D tissue constructs, these model organs can be viable for longer periods of time and are cultured to develop functional properties similar to the native tissues. The next challenge is to combine several organs on the same microfluidic device to model a simple organism-on-a-chip for drug and therapeutic studies. Ultimately, in the human body, tissues and organs are interconnected and interdependent on one another.

13.2 SINGLE VERSUS MULTIPLE ORGANOID FUNCTION

In vitro models that accurately recapitulate human tissues and model disease are limited, and fewer still exist in which multiple tissues are represented in an integrated fashion. This is an important technological limitation as tissue and organ development and function within the body does not occur in isolation, and it is essential that organs receive vascular, neural, metabolic, and hormonal signals and support for normal tissue function. In drug screening, for instance, toxic effects in secondary tissues can be as important as effects at the target site. If undetected, these effects can lead to an unnecessarily high rate of failure or withdrawal due to side effects. Likewise, in cancer metastasis, in which malignant tumor cells migrate from one location to another, multiple tissue or organ sites and a circulatory system (vascular or lymphatic) are involved. As such, while useful for many applications, single organoid models have limited efficacy for recapitulating the complex physiological interactions between multiple tissues that occur often in the human body. Instead multi-organoid platforms are necessary. Here, we describe several examples of multi-organ interactions that demonstrate the importance of considering deployment of multi-organoid platforms over single organoid systems.

13.2.1 Cancer

As described earlier, one important phenomenon that requires a multi-organoid approach for recapitulating it *in vitro* is modeling of cancer metastasis. In metastasis, cells in a tumor typically under go epithelial-to-mesenchymal (EMT) transition, proliferate rapidly in the primary tumor site, intravasate through endothelial cells into the blood stream or lymphatic system, after which they extravasate and colonize a downstream tissue. Currently, few *in vitro* systems exist that employ a multi-organoid approach. However, they are in development, and our laboratory has demonstrated that it is possible to recapitulate metastasis *in vitro*, albeit in a reductionist manner. Our metastasis-on-a-chip platform was devised to allow tracking of tumor cell metastasis from a colon organoid to a liver organoid within a simple microfluidic device (Figure 13.1a). We demonstrated that metastatic HCT116 cells were able to migrate and disseminate out of the colon organoid into the circulating media system and engraft in the downstream liver organoid. Conversely, nonmetastatic cell type SW480 proliferated at the primary site but never migrated to the liver. Using this platform we performed drug screens with common chemotherapy agents and also showed that we could manipulate the physical tumor microenvironment through hydrogel chemistry, altering environmental elastic modulus, which could stunt or accelerate tumor cell migration from tumor foci into the surrounding organoid space. We are currently advancing this system in several ways. We have recently integrated endothelial barriers in order to support intravasation and extravasation. We have also begun adding additional downstream organoids, with a goal being to explore what is more important in metastasis—proximity and location of the downstream site, or the cellular and extracellular matrix composition of the downstream site.

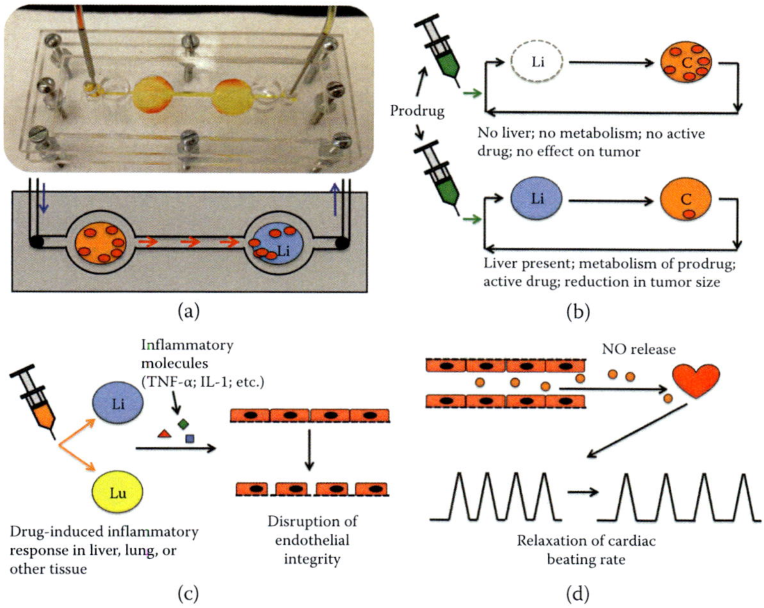

FIGURE 13.1 Multi-organ interactions. (a) Metastasis of tumor cells from one organ/organoid site to another, demonstrated *in vitro* in a metastasis-on-a-chip device in which colorectal carcinoma metastasizes from the colon to the liver. (b) Reliance of a prodrug (e.g., an anticancer 5-fluorouracil prodrug) on liver metabolism to activate the drug to generate a positive effect. (c) Inflammatory molecules secreted from organs such as the liver and lung can affect downstream integrity of vascular endothelium by disrupting the barrier function of endothelial cells. (d) Nitric oxide release from the vasculature in high levels can result in relaxation of cardiac beating rates.

13.2.2 Drug Testing/Toxicology

A broad area of importance is how multiple organs and tissues respond to administration of particular drugs. A variety of examples exist that demonstrate this concept. For example, 5-fluorouracil (5-FU) is one of the many common chemotherapy agents employed in treating colorectal cancer. Unfortunately, 5-FU can induce a variety of detrimental side effects in patients. In an attempt to reduce toxicity, several 5-FU prodrugs have been developed. These prodrugs are inactive in the native form, only becoming active after metabolism, generally in the liver. Consequently, without including a metabolically active liver organoid in *in vitro* 5-FU prodrug drug studies, results would be essentially meaningless. Building a platform with a liver and a tumor would allow metabolism of the prodrug followed by assessment of the activated drug on the tumor (Figure 13.1b).

13.2.3 Disease Modeling

An additional example incorporates a variety of organs and the vasculature. There are many drugs that are known to cause inflammatory responses in different tissues. For example, large doses of the analgesic acetaminophen (i.e., Panadol, Tylenol) cause inflammation and toxicity in the liver. Similarly, chemotherapeutics such as bleomycin cause inflammation, toxicity, and irreversible fibrosis in the lungs. In both cases, toxicity and cell death results in release of inflammatory molecules such as TNF-α and interleukin-1 into the circulation—both of which, in turn, can cause disruption and loss of integrity in the vascular endothelium (Figure 13.1c), similar to a response to histamine. Likewise, insult to the vasculature can lead to release of nitric oxide, which at high enough levels can change the beating profile of the heart (Figure 13.1d). These are just a few examples of significant downstream impacts of upstream drug treatments. Integrated multi-organoid model systems are required to detect these complex and multi-organ drug effects in a predictive and physiologically relevant manner.

13.3 BIOFABRICATING A HIGHLY PHYSIOLOGICALLY ACCURATE MULTI-ORGANOID PLATFORM

Our team's advancement of biofabrication and tissue engineering techniques has enabled formation of artificial constructs that model the organization of diverse tissues and tumors and can be serialized and addressed with fluidic technologies. One of the main problems in the implementation of tissue-engineered 3D model systems, whether for cancer applications, drug development, or toxicology, is that they simply do not sufficiently mimic their *in vivo* human counterparts. To provide a model system that can be employed to predict how a drug will affect a patient, or how malignant a tumor will be, model tissue constructs necessitate a baseline level of physiological accuracy. This includes morphology and multicellular organizations, functionalities such as protein synthesis and secretion, metabolic rates, drug metabolism, response to drugs and their metabolites, and long-term stability and viability. Our team is at the forefront of *in vitro* model development and implementation and has developed a portfolio of 3D tissue models. Crucially, these discrete "organoids" reproduce functionality of *in vivo* systems; for example, they synthesize and secrete typical biomarkers, respond to toxins, and metabolize drugs. Snapshots of these advanced systems are given below.

Our team was fortunate to secure a competitive Department of Defense contract through the Defense Threat Reduction Agency in 2013 to develop an integrated body-on-a-chip system for use in bioweapon and chemical weapon assessment and countermeasure development. Our team, lead by Dr. Anthony Atala at the Wake Forest Institute for Regenerative Medicine, led a multi-institutional collaboration between investigators at Brigham & Women's Hospital, Harvard Medical School, Johns Hopkins University, the University of Michigan, and the Edgewood Chemical and Biological Center in this program, termed the X vivo Capability for Evaluation and Licensure (X.C.E.L) Program. During the 3 years of this program, our team developed and tested a multi-organoid body-on-a-chip platform, ECHO (*ex vivo* console of human organoids). This platform would be comprised of four engineered tissue organoid types—liver, cardiac, vascular, and lung—which would be developed independently and integrated into a single system that could provide more complex physiological responses to toxic agents and pharmaceuticals than more simple 2D cell cultures and single organoid platforms. The purpose of this program was also to supplement and in some cases replace some animal testing methods. To date, our team has a robust four-organoid system incorporating online, real-time biosensors for accelerated assessment of tissue function and toxic responses during toxic agent and pharmaceutical compound screening studies.

13.3.1 Overview of ECHO Platform Organoids

With our multi-organ body-on-a-chip system, we have compiled a comprehensive set of data demonstrating the high functioning characteristics of a variety of tissue types. For both our liver and cardiac organoids, we began by employing the hanging drop method to form spherical tissue organoids. For liver, the organoids were comprised of primary human hepatocytes, stellate cells, and Kupffer cells. For cardiac, human iPS cell-derived cardiomyocytes were utilized. Following spherical organoid formation and individual characterization studies, the organoids were integrated into microfluidic hardware—described in more detail below—for on-a-chip studies. This was performed by using bioprinting to deposit the organoids into tissue-supportive hydrogels to create a 3D extracellular matrix–derived environment for organoid maintenance.[10,11] Liver organoids fabricated using liver ECM-derived hydrogels maintain viability and function *in vitro* for 4 weeks.[11] We have demonstrated the presence of key liver markers (albumin, multiple CYPs, epithelial cell-cell adhesion markers, dipeptidyl peptidase IV, and OST-α). These organoids produce albumin, urea, respond to toxins such as acetaminophen in a dose-dependent manner, and can be rescued from such insults with clinically relevant molecules such as *N*-acetyl-L-cysteine (Figure 13.2a). Cardiac organoids also remain viable beyond 4 weeks, support transport of fluorescent dye molecules throughout the organoids (suggesting high levels of cell-cell communication), beat spontaneously (Figure 13.2b), and respond to a variety of drugs by changing beating rates, which we capture using an onboard camera system[12,13] and custom software for analysis (Figure 13.2c). Additionally, we have incorporated a vascular endothelium module that responds to agents such as histamine by losing membrane integrity (Figure 13.2d), which can allow increased transfer of larger MW molecules. This module was comprised of a membrane-based fluid module in which human vascular endothelial cells resided as a confluent monolayer on a semipermeable membrane while the ECHO platform fluid flow passed over the monolayer. This endothelium organization allowed integration of a trans-endothelium electrical resistance (TEER) sensor, which afforded us the capability to measure membrane integrity in real time.

While still under development, a significant progress has also been made toward incorporation of a lung organoid module. This lung organoid comprises primary lung epithelium and supporting cells layered on an ECM-coated transwell membrane module. Proper differentiation of the airway cells was induced by air-liquid interface culture, as measured by TEER, formation of ciliated epithelial cells, and production of physiologically relevant levels of mucus.

Few research endeavors in the *in vitro* model space employ multiple organ models within one platform—most focus on a single tissue of interest. This can be shortsighted as inter-organ interactions occur in the body constantly and therefore should be recapitulated in these systems. We have integrated liver organoids,

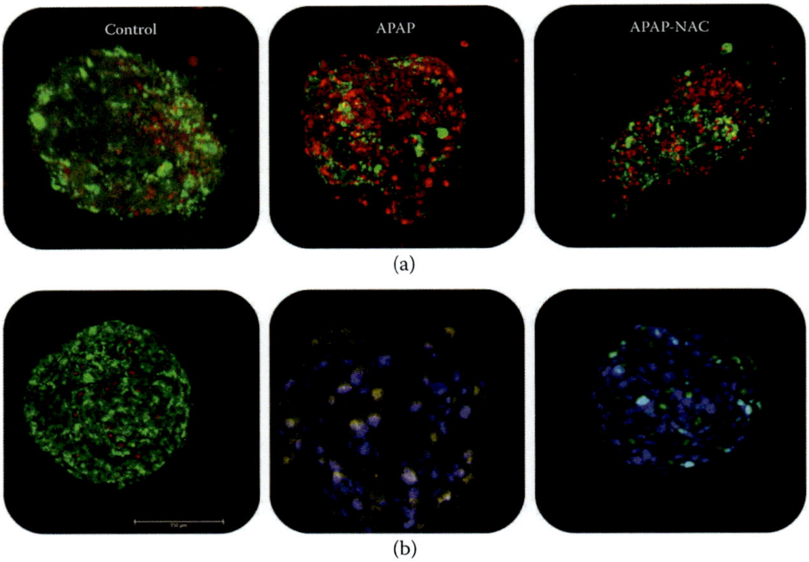

FIGURE 13.2 Highly functional organoids for a multi-organoid body-on-a-chip platform. (a) Acetaminophen (APAP) toxicity in liver organoids and reduction in toxicity by *N*-acetyl-L-cysteine (NAC). (b) Cardiac organoids remain viable long-term and support transport of fluorescent dyes (lucifer yellow [yellow stain] and fluorescein [green stain]) through interconnected ion channels suggesting high levels of cell-cell communication. *(Continued)*

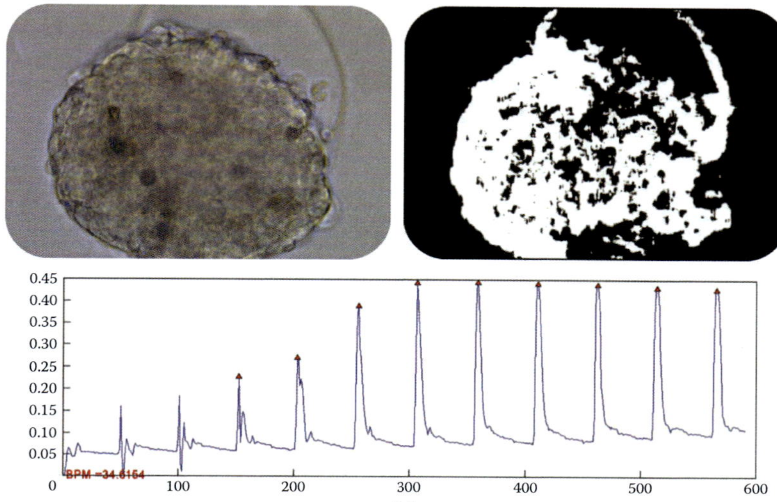

(c)

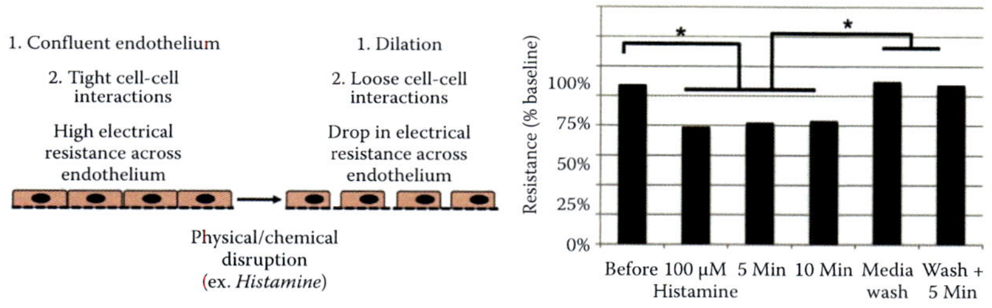

(d)

FIGURE 13.2 (Continued) Highly functional organoids for a multi-organoid body-on-a-chip platform. (c) Beating analysis of cardiac organoids: an onboard camera captures video of beating organoids, after which beating rates are calculated by quantifying pixel movement, generating beat plots. (d) Vascular endothelium devices respond to changes in endothelium integrity as measured by a trans-endothelium electrical resistance sensor.

cardiac organoids, and endothelial modules in microfluidic devices (Figure 13.3a) under common media and have shown integrated multi-organoid responses to drugs very similar to those encountered in humans. For example, Figure 13.3b describes the effects of propranolol and epinephrine on these cardiac organoids, with or without liver organoids. Without liver, propranolol, a beta-blocker, blocks cardiac beating increases by epinephrine. However, with both organoids present, propranolol is metabolized by the liver organoid, resulting in a measurable epinephrine-induced increase in beating rates. To our knowledge, these experiments are some of the first truly interactive multi-organoid systems to be tested successfully.

Finally, we have explored the use of our system in drug screening. Unfortunately, many drugs have made it through preclinical studies and clinical trials, after which they existed on the market for years in some cases, before being recalled

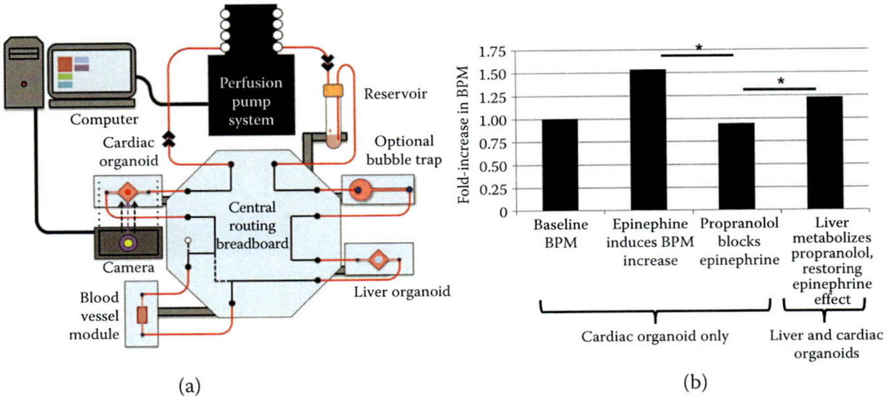

FIGURE 13.3 A multi-organoid body-on-a-chip. (a) Depiction of a liver, cardiac, and vascular organoid-containing body-on-a-chip platform. (b) The effects of propranolol and epinephrine on cardiac organoids, with or without liver organoids. Without liver, propranolol, a beta-blocker, blocks cardiac beating increases by epinephrine. However, with both organoids present, propranolol is metabolized by the liver organoid resulting in a measurable epinephrine-induced increase in beating rates.

by the Food and Drug Administration (FDA) for having toxic effects in humans. Notably, 90% of drugs that are pulled from the market are pulled due to effects on the liver and the heart. To demonstrate the utility of our platform, we screened several of these drugs (Figure 13.4). They include liver toxins troglitazone (Rezulin), an antidiabetic and anti-inflammatory that was recalled for causing liver failure, and Mibefradil, an ion channel blocker that was recalled for having fatal interactions with other drugs, including antibiotics. In our platform, troglitazone and mibefradil both result in liver toxicity. We also screened the cardiac toxin rofecoxib (Vioxx), an NSAID that was recalled due to serious cardiovascular side effects such as heart attack and stroke, as well as skin reactions and gastrointestinal bleeding. In addition, we screened the anticancer drugs 5-FU and isoproterenol, a beta-adrenergic agonist, both of which are known to induce cardiac toxicity. In the cardiac organoids, rofecoxib, 5-FU, and isoproterenol each result in increased levels of dead cells as doses increase (Figure 13.4). More importantly, however, using the onboard camera, beating effects were observed to decrease with dose increases as well. This is notable, as drugs recalled for cardiac toxicity are typically not recalled for inducing cell death, but rather are recalled for causing changes in heartbeat kinetics.

13.3.2 Microfluidic Hardware Integration

The ECHO platform organoids described earlier are integrated into a single system using a collection of modular microfluidic devices. As depicted in Figure 13.3a, this system has several components. First, the organoids themselves reside within polydimethylsiloxane (PDMS) modules formed using conventional soft lithography and replica molding to form chambers and fluid channels in pieces of PDMS.[9,14,15]

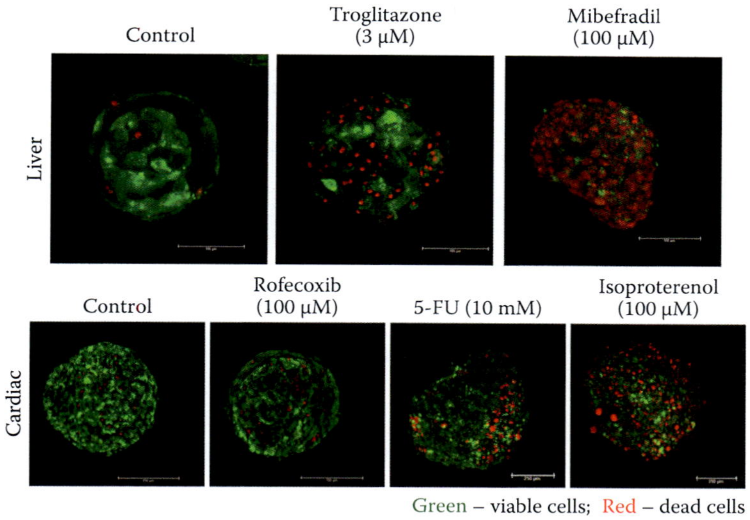

FIGURE 13.4 Screening of FDA-recalled drugs using multi-organoid systems. Liver toxins, troglitazone and mibefradil, and cardiac toxins rofecoxib, 5-fluorouracil, and isoproterenol are shown (green, viable cells; red, dead cells).

These pieces are bonded to glass slides or coverslips through plasma treatment, leaving the glass surface as the substrate of the channels and chambers. For organoid integration, at the beginning of experiments, organoids would be bioprinted into the chambers using tissue-specific hydrogel bio-inks,[10,16] after which chambers were closed with another PDMS piece containing inlet and outlet ports, and clamped. A central fluid routing unit (octagon-shaped component in Figure 13.3a) was also formed using soft lithography and replicate molding. This unit allows modular "plug-and-play" addition of each organoid module to the overall platform. Additionally, our platform included a microperistaltic pump, media reservoir, and onboard camera for organoid monitoring.[12,13] In general, all the components described were linked together in series using microfluidic tubing. During most experiments, flow was perfused through the system by the microperistaltic pump at a rate of 10 uL/min.

13.3.3 Common Media Development

One concern for multi-organoid systems is the need for a medium that will support all cell types in the platform. In previous work, we have quite easily maintained intestinal epithelial cells, liver-derived HEPG2 cells, and HCT116 colon carcinoma cells within a common medium. In our body-on-a-chip work, we have maintained up to four organoid types in culture simultaneously, most of which were biofabricated from primary cells or iPS cells. This common media contains factors important to the different cell types in the system. Development of versatile media formulations has the potential to commercially impact the catalog of cell culture media and supplement products. Reducing the wide variety of cell culture media formulations would help to standardize experimental conditions and reduce overall costs of many laboratories.

13.3.4 Miniaturization and High-Throughput

Currently, the majority of multi-organoid systems, as well as single organoid systems, that do a sufficient job at mimicking *in vivo* physiology, are created and maintained in low-throughput systems. This currently places a significant limit on the number of drug compounds that can be screened with these systems. Fortunately, many laboratories are working on miniaturization of organ-on-a-chip systems in order to place multiple 3D organoids and multiple multi-organoid systems within single devices.[9] This reduction of size, together with improvements in sensing and biomarker monitoring technology,[12,13,17,18] has the potential to dramatically increase the number of replicants and experimental conditions that can be screened at once, eventually realizing the full potential that multi-organoid screening platforms will have in biomedical research and clinical and commercial deployment.

13.4 CURRENT AND FUTURE APPLICATIONS

Multi-organoid body-on-a-chip systems are rapidly advancing and are positioned to be deployed into drug screening in the very near future. These platforms have significant utility in other areas, which have been alluded to above. These include precision medicine, cancer models, common cell culture media development, and improvements in miniaturization and high-throughput implementation of these organoid systems.

13.4.1 Drug Testing

Billions of dollars and years of scientific expertise are spent on drug candidates that eventually turn out to be toxic to humans or ineffective in curing disease.[19] Even approved and marketed drugs must sometimes be withdrawn from the market because subsequent experience shows them to be harmful or ineffective. The massive investments made in these drugs are lost. Because of the billions of dollars invested in failed drugs, capital that could have been directed to drugs that could control or cure a wide range of diseases is never invested. Millions of patients throughout the world are denied cures that might otherwise be available.

The current drug development process is flawed in several ways. First, the use of cell culture systems that do not represent tissue complexity. The process fails to reproduce several important aspects of native human tissue including the 2D nature of a tissue culture dish versus 3D structure of normal tissues, the use of cell lines and not primary cells, and the lack of extracellular matrix—all of which play important roles in tissue function. As a result, drugs selected based on tissue culture assays exhibit very different activity when tested in whole tissue. Second, drugs that have succeeded in animal testing often fail in humans. Animal testing suffers from the most fundamental flaw—metabolism in mice, rats, and other test animals is different than in humans. What is safe for those animals is often toxic to humans;

what cures disease in animals regularly fails to do so in people. On the other hand, some drugs that have failed in animal tests might possibly work in humans. Libraries of these failed drugs sit on pharmaceutical company shelves. But using current technology, we will never know whether the differences in human and animal physiology may have contributed to the failure in animal testing. Drugs that fail in the animal model will never be approved and therefore never be tested in people. Finally, human patients have a large genetic variability compared with inbred lines of mice and rats used for toxicology studies. Clinical data clearly show that some drugs may be toxic for some patients but safe in others, efficacious in some patients but not others. Many drugs approved for testing in humans fail in Phase 1 trials. Others pass Phase 1 but fail in later development. Some drugs fail even after they reach the market for reasons that were not apparent in trials. Because we cannot readily distinguish between those patients whom the drug will help and whom it will hurt, many drugs with great potential have been discarded. Targeting treatments to only those that will benefit is one of the greatest challenges in medicine. It is clear now that the ideal approach is "personalized medicine" rather than "generic medicine." Yet our tools to achieve personalized medicine are very limited. Needless to say, animal models are of little or no use in this process. But testing in humans exacts a high cost in the suffering of those who are harmed or fail to respond and still may not answer critical questions.

The search for a better pharmaceutical testing model is at the forefront of biomedical research. The enormous potential of the body-on-a-chip platform in terms of recapitulating the physiology of human organs is a strong argument for implementation of this platform for drug screening and toxicology evaluation.[20] Below we describe several examples for the use of a body-on-a-chip platform for drug screening of specific interests.

Since the liver is the main target for most drug toxicity testing the first body-on-a chip systems, and most contemporary systems, employed liver organoids inside a microfluidic system, alone or together with other tissues. A microfluidic system with two separate chambers, one with encapsulated human liver microsomes and the second with a hepatic liver cell line HepG2, was used to test the effect of acetaminophen of liver cell viability. The drug that was applied to the liver microsomes showed a significant effect on HepG2 viability in the second chamber.[21]

Platforms to test the toxicity of chemotherapeutic drugs used liver organoids in combination with tumor cells. A microfluidic system containing the hepatic cell line HepG2, colon carcinoma cells (HCT116), and myeloblasts (Kasumi-1) was used to test whether Tegafur, an oral prodrug of the chemotherapeutic drug 5-FU, can be metabolized by the liver organoids.[22] Tegafur has a better bioavailability than 5-FU due to rapid enzymatic degradation of 5-FU.[23] The body-on-a chip system was able to reproduce the metabolism of Tegafur to 5-FU in the liver organoids and it was carried by the microfluidic system to the cancer cells where it induced cytotoxicity. In contrast, when Tegafur was added directly to the cancer cells it didn't have the same effect because it was not converted to 5-FU.

Another example is a study testing the effects of doxorubicin alone or in combination with cyclosporine and nicardipine (chemosensitizers). A body-on-a chip system containing liver HepG2 cells, uterine cancer cells, and bone marrow cells in separate chambers was used to test the effect of these drugs and their combination.[24]

Drug combinations significantly inhibited cancer cell proliferation better than doxorubicin alone. The liver and bone marrow cells were not affected by the drug combinations compared with treatments with a chemosensitizing drug alone. Results from standard cell culture experiments showed an additive rather than synergistic effect of the drug combinations.

Body-on-a chip systems containing cells representing different organs, each in a different microfluidic chamber that are connected by channel, were used to assess the physiological response to drugs.[22,24–27] The toxicant naphthalene added to the liver chamber was metabolically processed and the converted metabolites were circulated through the chambers containing the other cell types. In the lung chamber these metabolites depleted the cellular glutathione levels.[25,26] The adipocyte in the fat chamber attenuated the glutathione depletion by naphthalene and instead accumulated hydrophobic compounds.[25,26]

Altogether, these body-on-a chip systems demonstrated that the organoids inside them are metabolically active and they can, to a certain degree, replicate the physiological inter-organ metabolic relationships of the human body. An additional component of the body-on-a chip system is a measuring/sensing capability. It is important to be able to measure the cell/organoid response in the most direct and real-time manner in order to reliably determine the effect of the tested drugs. Among the common parameters measured are cell viability/death, outcomes of applying mechanical forces on the cells/organoids, electrophysiological measures such as electrical resistance, and a detection element to analyze various chemical metabolites.[2]

13.4.2 Precision Medicine

While numerous *in vitro* organoid systems are being explored for general drug development screening, few exist for deployment in the clinic for optimizing therapies for specific patients. This is a clinical need, since in many cases, prescribing therapies to some cancer patients is almost a trial-and-error process. With personalized cancer models, therapies can be screened using patients' own tumor cells in a physiologically accurate 3D model system. Accurate prediction of a patient's tumor progression and response to therapy is one of the most challenging areas in oncology. Treatment decisions are usually made based on the general success rate of a drug, not on how a drug might affect the specific individual. Recently, the concept of precision, or personalized, medicine (PM) has been employed to address these problems by using the patient's genetic profile to identify "drugable" targets for treatment.[28–30] However, in practice, the utility of PM is less clear;[31] even after identification of key mutations, oncologists are often left with several drug options with little information about potential side effects, thus creating a need to further develop tools to help predict the personalized response to cancer drugs.[32,33] We are currently working to develop a platform consisting of patient-specific tumor organoids in which anticancer drugs will be tested for efficacy. By implementing the 3D tumor-on-a-chip microfluidic platform with a circulatory system and multiple tissue organoid sites as described earlier, we can visualize and track the kinetics of tumor progression and metastasis to a distant site *in vitro*. Realization of this kind of technology will allow screening of these drugs prior to treatment while monitoring all organ systems for side effects, dramatically transforming patient care and improving treatment outcomes. Such powerful and comprehensive technology does not currently exist and would represent a quantum leap in oncology.

13.4.3 Multi-Organoid Cancer Systems

Multi-site metastasis-on-a-chip platforms as described earlier, comprised of tumor foci within multiple host tissue constructs, really have not been explored widely. There has been a recent surge in on-a-chip devices that assess certain aspects of metastasis. For example, the Kamm group has developed a device that includes an endothelium and a bone mimic chamber, allowing modeling of extravasation of circulating breast cancer cells into bone.[34,35] Other recent systems include multichannel devices to analyze the process by which tumor aggregates migrate through a collagen gel and an endothelial layer,[36] devices for assessing the effects of interstitial pressure on cell migration,[37] and a system for screening anti-angiogenic drugs.[38] These systems illustrate the potential of these technologies. However, there is still a lack of platforms integrating both primary and metastatic sites and the zones in between (i.e., circulation and endothelium) onboard one device. The nexus of tissue engineering with microscale devices, paired with real-time imaging and sensing, results in a powerful investigative tool. By providing circulating flow through the system, we can recapitulate the dissemination of tumor cells from primary site organoids into circulation, after which metastatic cells can colonize one or more organoids downstream. This is notable and novel because phenotypes of cells in the originating malignant tumors and metastases can vary significantly—for example, resulting in varying levels of invasiveness due to matrix metalloproteinase secretion and stem cell-like genes[39,40]—making the ability to study both sites and microenvironments extremely important.

REFERENCES

1. Jamieson LE, Harrison DJ, Campbell CJ. Chemical analysis of multicellular tumour spheroids. *Analyst* 2015; **140**(12): 3910–20.
2. Sung JH, Srinivasan B, Esch MB, et al. Using physiologically-based pharmacokinetic-guided "body-on-a-chip" systems to predict mammalian response to drug and chemical exposure. *Exp Biol Med* 2014; **239**(9): 1225–39.
3. Kunz-Schughart LA, Freyer JP, Hofstaedter F, Ebner R. The use of 3-D cultures for high-throughput screening: The multicellular spheroid model. *J Biomol Screen* 2004; **9**(4): 273–85.
4. Pasirayi G, Scott SM, Islam M, O'Hare L, Bateson S, Ali Z. Low cost microfluidic cell culture array using normally closed valves for cytotoxicity assay. *Talanta* 2014; **129**: 491–8.
5. Ho WJ, Pham EA, Kim JW, et al. Incorporation of multicellular spheroids into 3-D polymeric scaffolds provides an improved tumor model for screening anticancer drugs. *Cancer Sci* 2010; **101**(12): 2637–43.
6. Drewitz M, Helbling M, Fried N, et al. Towards automated production and drug sensitivity testing using scaffold-free spherical tumor microtissues. *Biotechnol J* 2011; **6**(12): 1488–96.
7. Polini A, Prodanov L, Bhise NS, Manoharan V, Dokmeci MR, Khademhosseini A. Organs-on-a-chip: A new tool for drug discovery. *Expert Opin Drug Discov* 2014; **9**(4): 335–52.

8. Benam KH, Dauth S, Hassell B, et al. Engineered in vitro disease models. *Ann Rev Pathol* 2015; **10**: 195–262.
9. Skardal A, Devarasetty M, Soker S, Hall AR. In situ patterned micro 3D liver constructs for parallel toxicology testing in a fluidic device. *Biofabrication* 2015; **7**(3): 031001.
10. Skardal A, Devarasetty M, Kang HW, et al. A hydrogel bioink toolkit for mimicking native tissue biochemical and mechanical properties in bioprinted tissue constructs. *Acta Biomater* 2015; **25**: 24–34.
11. Skardal A, Smith L, Bharadwaj S, Atala A, Soker S, Zhang Y. Tissue specific synthetic ECM hydrogels for 3-D in vitro maintenance of hepatocyte function. *Biomaterials* 2012; **33**(18): 4565–75.
12. Kim SB, Koo KI, Bae H, et al. A mini-microscope for in situ monitoring of cells. *Lab Chip* 2012; **12**(20): 3976–82.
13. Zhang YS, Ribas J, Nadhman A, et al. A cost-effective fluorescence mini-microscope for biomedical applications. *Lab Chip* 2015; **15**(18): 3661–9.
14. McDonald JC, Duffy DC, Anderson JR, et al. Fabrication of microfluidic systems in poly(dimethylsiloxane). *Electrophoresis* 2000; **21**(1): 27–40.
15. Xia Y, Whitesides GM. Soft lithography. *Ann Rev Mater Sci* 1998; **28**: 153–84.
16. Skardal A, Devarasetty M, Kang HW, et al. Bioprinting cellularized constructs using a tissue-specific hydrogel bioink. *J Vis Exp* 2016; **110**: e53606.
17. Kim SB, Bae H, Cha JM, et al. A cell-based biosensor for real-time detection of cardiotoxicity using lensfree imaging. *Lab Chip* 2011; **11**(10): 1801–7.
18. Shaegh SAM, Ferrari FD, Zhang YS, et al. A microfluidic optical platform for real-time monitoring of pH and oxygen in microfluidic bioreactors and organ-on-chip devices. *Biosens Bioelectron* 2016; **10**(4): 044111.
19. Hughes JP, Rees S, Kalindjian SB, Philpott KL. Principles of early drug discovery. *Br J Pharmacol* 2011; **162**(6): 1239–49.
20. Chan CY, Huang P-H, Guo F, et al. Accelerating drug discovery via organs-on-chips. *Lab Chip* 2013; **13**(24): 4697–710.
21. Ma B, Zhang G, Qin J, Lin B. Characterization of drug metabolites and cytotoxicity assay simultaneously using an integrated microfluidic device. *Lab Chip* 2009; **9**(2): 232–8.
22. Sung JH, Shuler ML. A micro cell culture analog (microCCA) with 3-D hydrogel culture of multiple cell lines to assess metabolism-dependent cytotoxicity of anti-cancer drugs. *Lab Chip* 2009; **9**(10): 1385–94.
23. Malet-Martino M, Martino R. Clinical studies of three oral prodrugs of 5-fluorouracil (capecitabine, UFT, S-1): A review. *Oncologist* 2002; **7**(4): 288–323.
24. Tatosian DA, Shuler ML. A novel system for evaluation of drug mixtures for potential efficacy in treating multidrug resistant cancers. *Biotechnol Bioeng* 2009; **103**(1): 187–98.
25. Viravaidya K, Sin A, Shuler ML. Development of a microscale cell culture analog to probe naphthalene toxicity. *Biotechnol Prog* 2004; **20**(1): 316–23.
26. Sin A, Chin KC, Jamil MF, Kostov Y, Rao G, Shuler ML. The design and fabrication of three-chamber microscale cell culture analog devices with integrated dissolved oxygen sensors. *Biotechnol Prog* 2004; **20**(1): 338–45.
27. Mahler GJ, Esch MB, Glahn RP, Shuler ML. Characterization of a gastrointestinal tract microscale cell culture analog used to predict drug toxicity. *Biotechnol Bioeng* 2009; **104**(1): 193–205.
28. Tran NH, Cavalcante LL, Lubner SJ, et al. Precision medicine in colorectal cancer: The molecular profile alters treatment strategies. *Ther Adv Med Oncol* 2015; **7**(5): 252–62.
29. Miles G, Rae J, Ramalingam SS, Pfeifer J. Genetic testing and tissue banking for personalized oncology: Analytical and institutional factors. *Sem Oncol* 2015; **42**(5): 713–23.

30. Bando H, Takebe N. Recent innovations in the USA National Cancer Institute-sponsored investigator initiated Phase I and II anticancer drug development. *Jpn J Clin Oncol* 2015; **45**(11): 1001–6.
31. Hayes DF, Schott AF. Personalized medicine: Genomics trials in oncology. *Trans Am Clin Climatol Assoc* 2015; **126**: 133–43.
32. Cantrell MA, Kuo CJ. Organoid modeling for cancer precision medicine. *Genome Med* 2015; **7**(1): 32.
33. Gao D, Vela I, Sboner A, et al. Organoid cultures derived from patients with advanced prostate cancer. *Cell* 2014; **159**(1): 176–87.
34. Bersini S, Jeon JS, Dubini G, et al. A microfluidic 3D in vitro model for specificity of breast cancer metastasis to bone. *Biomaterials* 2014; **35**(8): 2454–61.
35. Bersini S, Jeon JS, Moretti M, Kamm RD. In vitro models of the metastatic cascade: From local invasion to extravasation. *Drug Discov Today* 2014; **19**(6): 735–42.
36. Niu Y, Bai J, Kamm RD, Wang Y, Wang C. Validating antimetastatic effects of natural products in an engineered microfluidic platform mimicking tumor microenvironment. *Mol Pharm* 2014; **11**(7): 2022–9.
37. Polacheck WJ, German AE, Mammoto A, Ingber DE, Kamm RD. Mechanotransduction of fluid stresses governs 3D cell migration. *Proc Natl Acad Sci U S A* 2014; **111**(7): 2447–52.
38. Kim C, Kasuya J, Jeon J, Chung S, Kamm RD. A quantitative microfluidic angiogenesis screen for studying anti-angiogenic therapeutic drugs. *Lab Chip* 2015; **15**(1): 301–10.
39. Karakiulakis G, Papanikolaou C, Jankovic SM, et al. Increased type IV collagen–degrading activity in metastases originating from primary tumors of the human colon. *Invasion Metastasis* 1997; **17**(3): 158–68.
40. Franci C, Zhou J, Jiang Z, et al. Biomarkers of residual disease, disseminated tumor cells, and metastases in the MMTV-PyMT breast cancer model. *PLoS One* 2013; **8**(3): e58183.

14 Cancer Metastasis-on-a-Chip

Ran Li, Michelle B. Chen,* and Roger D. Kamm*

CONTENTS

14.1 CANCER METASTATIC CASCADE

Cancer metastasis, the spread of cancer cells from a primary tumor site to a secondary organ, accounts for 90% of deaths among cancer patients [1]. In order for cancer cells to transmigrate and colonize a distal organ, however, they must overcome several barriers in a series of steps termed the "metastatic cascade." In the first step, tumor cells of epithelial origin at the primary site undergo an epithelial-to-mesenchymal

* These authors contributed equally to this work.

transition [2], and acquire the capability to invade and migrate through the basement membrane as well as the stromal matrix surrounding the epithelial tissues. This enhanced migratory ability allows cancer cells to spread throughout the primary tumor site, and *cancer cell migration* can be further promoted by chemical (growth factors), cellular (tumor-associated stromal cells), and mechanical (interstitial flow) cues within the tumor microenvironment [3]. The growth of the tumor also initiates *tumor angiogenesis*, allowing blood vessels to extend into the tumor tissues [4]. Besides supplying cancer cells with nutrients, these tumor-associated blood vessels also provide cancer cells with a route of escape from the primary tumor. Once a migrating cancer cell comes into contact with a blood or lymphatic vessel, it can transmigrate through the endothelial wall and enter the circulation in a process called *intravasation*. In the blood stream, these circulating tumor cells can travel to the secondary site and exit the blood vessel by *extravasation*. These extravasated cancer cells can *recolonize* the secondary site [1]. Because of this, cancer metastasis is a complex process that involves a series of distinct steps (Figure 14.1) [94].

Recognizing the complexity of this process and the corresponding challenges associated with studying them *in vivo*, it is perhaps not surprising that a full understanding of the critical mechanistic details of metastatic disease is lacking. For similar reasons, there is a marked absence of drugs available to treat metastatic cancer by inhibiting one or more of the steps of the metastatic cascade. These key shortcomings have led in recent years to a rapid expansion of new *in vitro* technologies, especially in the field of microfluidics, to study metastasis and to create assays with the potential to screen for drugs with therapeutic value.

14.2 MICROFLUIDIC TECHNOLOGIES TO STUDY THE METASTATIC CASCADE

Although no *in vitro* model can fully replicate the complexity of the *in vivo* metastatic milieu, a considerable effort has been devoted to the development of more sophisticated *in vitro* technologies to facilitate our understanding of tumor cell invasion, migration, and interactions with the vasculature. Conventional cell culture assays including petri dishes and Transwell/Boyden chambers have proven to be pivotal in enhancing our understanding of cellular interactions in metastasis by recreating discrete events in a complex cascade and offering tight control of certain critical experimental parameters. However, there remain several needs that most traditional *in vitro* assays cannot yet satisfy. These include the recapitulation of spatial and temporal chemokine gradients, mechanical stresses, and relevant spatial organization of multiple cell types. Microfluidics has revolutionized the field of cell biology, allowing scientists to fabricate sophisticated 3D models in highly controlled microenvironments. Devices can easily accommodate different cell types which can be arranged in physiologically relevant configurations to mimic the *in vivo* tumor microenvironment [5]. Chemokine gradients and mechanical forces such as interstitial fluid pressure on tumor cells or shear stresses on endothelial cells can be readily applied, controlled, and quantified [6]. Cells can be embedded and cultured in 3D extracellular matrix (ECM) hydrogels that mimic the *in vivo* cell migration scenario, but can still be amenable to high-resolution imaging, as the small volume and/or thickness of cell

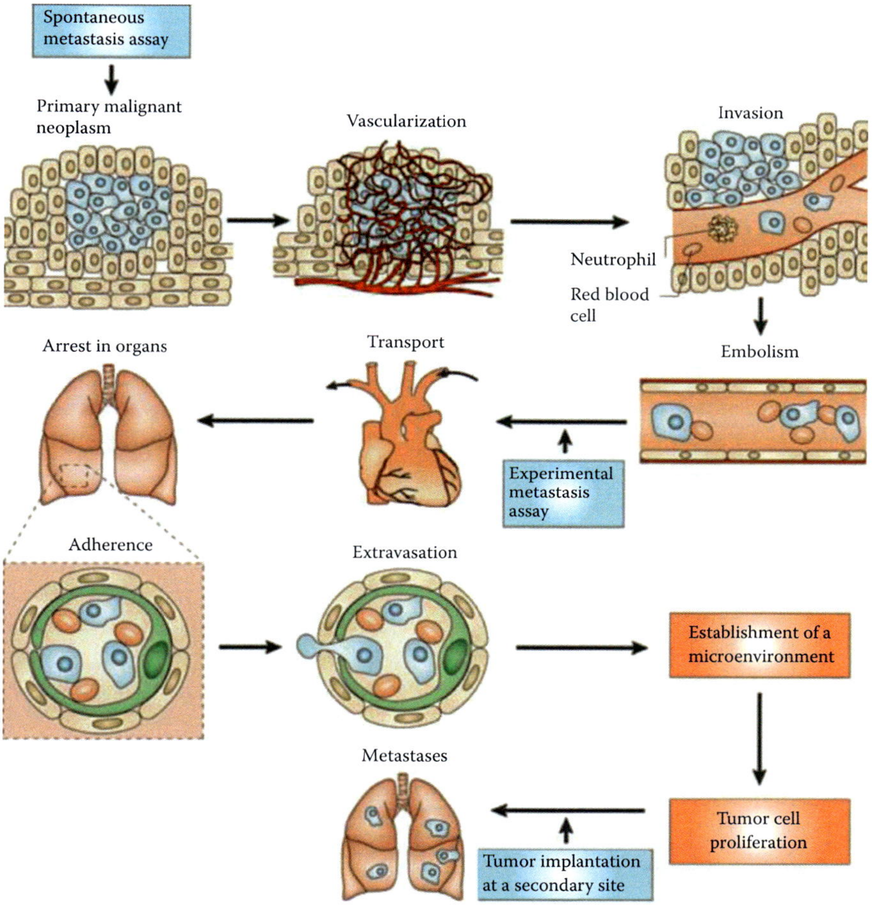

FIGURE 14.1 The metastatic cascade is a complex sequence of events beginning with the vascularization of the primary tumor mass, escape of tumor cells from the primary tumor, and migration toward the vasculature or lymphatic circulation. Cells interact with the endothelium and host stromal cells to enter the bloodstream and circulate throughout the body, a step termed *early metastatic seeding*. Circulating tumor cells that survive the transport process may arrest at a distant organ and extravasate. These cells have the potential to migrate into the surrounding stroma, proliferate, and recolonize, eventually forming a secondary metastasis. (Reproduced from Reymond, N., et al., *Nat. Rev. Cancer*, 13, 858–870, 2013. With permission.)

culture chambers generally provide excellent optical clarity [5]. High spatiotemporal resolution imaging in experimental metastasis assays is important as it enables characterization of the morphological details during tumor cell invasion and migration, and endothelial cells during transendothelial migration. As such, microfluidic technologies have emerged as a promising approach to develop intricate *in vitro* cancer models and recapitulate the different steps in the metastatic cascade.

14.3 MICROFLUIDIC TECHNOLOGIES TO STUDY CANCER CELL MIGRATION

A myriad of chemical, cellular, and biophysical cues in the tumor microenvironment can promote and direct cancer cell migration from the primary tumor tissue to the surrounding stroma [3]. Studying the effects of these microenvironmental cues on cancer cell migration using traditional cell-culture platforms is challenging, however. In the past few years, a variety of microfluidic devices have been reported to study the effects of these cues on cancer cell migration, providing important mechanistic insights into cancer cell metastasis.

14.3.1 Microfluidic Platforms to Apply Chemical Cues

Growth factors and cytokines within the tumor microenvironment form the chemical cues for cancer cell migration. These chemical cues are produced by cancer cells themselves as well as tumor-associated stromal cells within the tumor tissues. Diffusion of these soluble cues inside and around the tumor tissue creates concentration gradients, inducing cancer cells to migrate in the direction of increasing concentration, a process termed chemotaxis. These chemical stimuli attract cancer cells by promoting cellular polarization, protrusion formation, cellular contraction, and matrix metalloproteinase (MMP) production. Cancer cells use MMPs to break down ECM, which acts to impede cell migration. In addition to soluble chemical cues discussed earlier, oxygen is another important chemical stimulus for cancer cell migration. Within tumor tissues, an oxygen concentration gradient often forms due to the depletion of oxygen at the core of the tumor. This oxygen gradient could also guide the movement of cancer cells [7].

14.3.1.1 Growth Factors and Cytokines

Traditional chemotactic assays, such as the Boyden and Dunn chambers, are limited in their ability to generate steady, controlled, and quantifiable chemical gradients. Various microfluidic devices have recently been designed to address these shortcomings [8]. This new generation of chemotactic assays allows the establishment of user-defined, spatially controlled, complex gradients to study cancer cell chemotaxis in a quantitative and reproducible fashion [9]. To date, three classes of microfluidic chemotactic devices have been introduced: (1) laminar flow-based chemotactic devices, (2) diffusion-based 3D chemotactic devices with static reservoirs, and (3) diffusion-based 3D chemotactic devices with flow-based reservoirs [8,10].

The first class of microfluidic chemotactic devices is based on the absence of convective mixing in slow, laminar flow [11–14]. In this design, two or more streams of fluid having different concentrations of the same solute merge in a single, common channel. As they flow together, the solute diffuses between streams producing a continuous, relatively stable gradient in concentration across the channel. The laminar flow-based devices are usually composed of two regions. The first region is the gradient generating region where a network of microchannels serves to generate fluid streams of chemoattractants with different concentrations. The microchannels from the gradient generating region merge into a single large microchannel that serves as the gradient application region. Cancer cells can be seeded into the gradient application region, and their migration can be tracked using time-lapse microscopy (Figure 14.2a) [95]. User-defined gradient profiles, such as linear, parabolic, and polynomial profiles, can be established by changing the concentrations of the inlet stream to the gradient generating region. Using these versatile tools, Wang et al. showed that a nonlinear Epidermal growth factor (EGF) gradient was more likely to induce a chemotactic response from cancer cells than a linear EGF gradient [15]. Moreover, a detailed analysis showed that a steep EGF concentration profile induced faster migration, greater directional persistence, and a higher chemotactic index than a shallow one [15,16]. In a related study, Mosadegh et al. showed that a CXCL12 gradient alone could not induce chemotaxis of breast cancer cells. However, when a uniform concentration of EGF was superimposed on the gradient of CXCL12, a strong chemotactic response was observed [17]. The insights gained from these studies could not have been possible with traditional chemotactic assays,

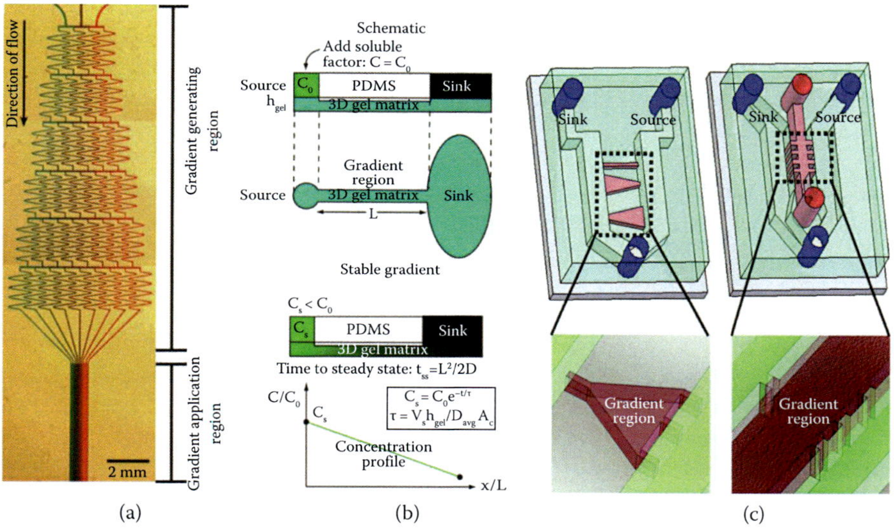

FIGURE 14.2 Microfluidic devices to study the effects of chemical cues on cancer cell migration. (a) Laminar flow-based chemotactic devices showing a gradient generating region and a gradient application region where the gradient of chemoattractants is applied to the cells migrating in 2D. (Reprinted with permission from Derringer, S.K.W., et al., *Anal. Chem.*, 73, 6, 1240–1246, 2001. Copyright [2001] American Chemical Society.) (b) Diffusion-based 3D chemotactic devices with static reservoirs. These devices consist of source and sink reservoirs connected by a 3D gel matrix (top). By supplying chemoattractant to the source reservoir, a quasi-steady linear concentration profile can be produced (bottom). (Abhyankar, V.V., et al., A platform for assessing chemotactic migration within a spatiotemporally defined 3D microenvironment, *Lab Chip*, 2008; 8: 1507–1515. Reproduced by permission of The Royal Society of Chemistry.) (c) Diffusion-based 3D chemotactic device with flow-based reservoirs. These devices consist of a 3D gel matrix (gradient region) flanked by source and sink channels. By providing continuous flow in the source and sink channels, constant boundary conditions can be established (right). Changing the shape of the gradient region can change the shape of the gradient profile established (left). (Reprinted with permission from Mosadegh, B., et al., *Langmuir*, 23, 10910–10912, 2007. Copyright [2007] American Chemical Society.)

which do not allow the generation of user-defined gradient profiles. The advantages of these laminar flow-based chemotactic devices over the traditional assays are clear: they allow for the generation of user-defined, long-term, stable gradients of complex shapes. However, these devices are not without their drawbacks. The cells in the devices are under the effects of continuous laminar flow, which could mechanically activate signaling pathways within the cells. In addition, most of these flow-based assays are 2D migration assays, which do not accurately mimic the *in vivo* condition [8].

Since tumor cells more often migrate in a 3D microenvironment, diffusion-based 3D migration devices have been developed and utilized for cancer research. These devices consist of sink and source reservoirs connected by a microfluidic gradient region containing a hydrogel (collagen I gel or Matrigel) (Figure 14.2b). Chemoattractants are introduced in the source reservoirs and allowed to passively diffuse through the hydrogel to establish a pseudo-steady state gradient [18–20]. Cells are seeded in the hydrogel, and the migration of these cells is tracked with time-lapse microscopy. Aside from providing a 3D environment, these devices have the advantage that the cells are not activated by flow. Abhyankar et al. used these diffusion-based devices to show that MTLn3 mouse breast carcinoma cells can perform chemotaxis when supplied with a gradient of EGF in a 3D collagen I gel [20]. Although these diffusion-based devices have been used for various cancer cell migration studies, they still have several key shortcomings. Since the source and sink reservoirs are not continuously replenished (static reservoirs), the boundary conditions for the diffusion are not kept constant with time. Hence, the gradients established in these devices are not stable with time. In addition, the lack of the control of the boundary conditions leads to inability to dynamically modulate the gradient [10].

To control the boundary conditions for diffusion, several novel microfluidic platforms have been designed to continuously replenish the source and the sink reservoirs. These devices consist of a hydrogel gradient region situated between two microchannels that act as source and sink reservoirs. To maintain a constant boundary condition for diffusion, the microchannels are replenished through the continuous flow of fluid. The microchannels are in some cases connected at the outlet to ensure that no pressure gradient is established across the hydrogel that could give rise to convective transport through the gel. This is to assure that mass transport through the gradient generating region is purely diffusion-based [21–23]. An added benefit of this arrangement is that the gradient can be dynamically modulated simply by changing the concentration of chemoattractant in one or both of the reservoirs, although one needs to allow for the time required to establish a new steady-state concentration gradient within the gel. Moreover, the shape of gradient profile generated using this method can be modified by changing the geometry of the gel region [21]. Therefore, this method of gradient generation combines many of the advantages of both flow-based gradient devices and diffusion-based gradient devices with static reservoirs (Figure 14.2c). It allows for the assessment of cancer cell migration in 3D ECM under the influence of dynamically controlled stable gradient of chemoattractants. Using this versatile tool, Kim et al. found that breast cancer cells migrate toward increasing concentration of CXCL12 in a 3D collagen I matrix [23]. However, this 3D chemotactic behavior was abrogated when the gradient of CXCL12 was superimposed with a uniform concentration of EGF. These results contradicted the results obtained from

2D flow-based assay [17], pointing to the possibility that there is a fundamental difference in how cancer cells sense and respond to gradients in 2D and in 3D.

14.3.1.2 Oxygen Tension

To date, no traditional chemotaxis or migration assay exists that is capable of investigating the effects of oxygen or hypoxic *gradient* on cancer cell movement. This is due to the fact that precise control over microenvironment cannot be achieved with these traditional assays. Utilizing the advantages of microfluidics, however, a variety of proof-of-concept microfluidic hypoxic gradient generators have recently been designed. These devices fall into two categories: (1) devices that use chemical reaction to generate oxygen gradient and (2) devices that use gas supply channels to generate oxygen gradient [24].

The first class of microfluidic oxygen gradient generators utilizes oxygen-generating chemicals and oxygen-scavenging chemicals to generate the oxygen gradient. These devices typically consist of a cell chamber flanked by two microchannels (Figure 14.3a). In one of the microchannels, oxygen generators, such as hydrogen peroxide and

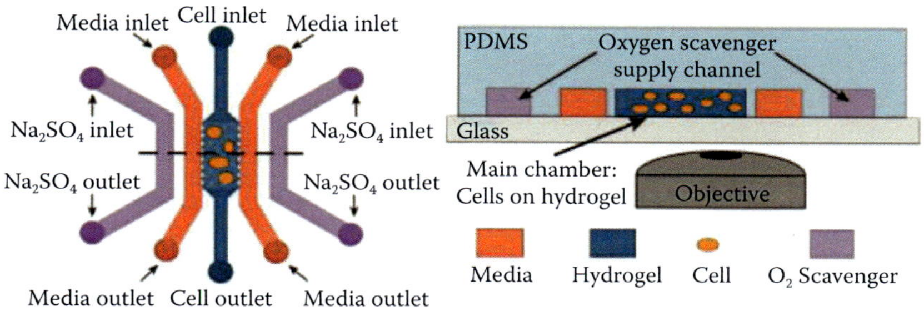

(a)

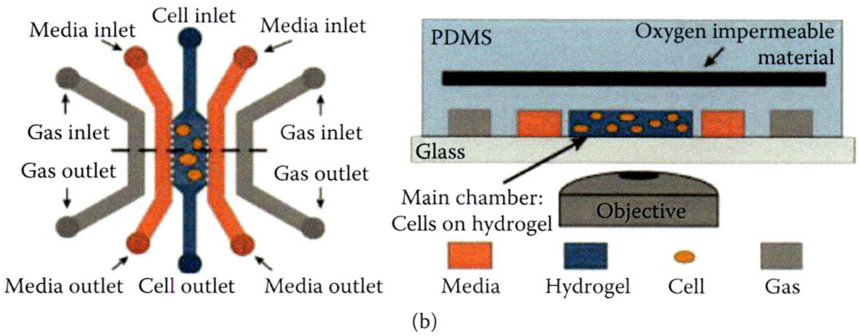

(b)

FIGURE 14.3 Microfluidic devices to generate hypoxia gradient, and microfluidic devices for co-culture studies. (a) Schematics of microfluidic platforms that utilize chemical reactions to generate an oxygen gradient. (b) Schematics of microfluidic platforms that utilize gas of defined oxygen composition to generate an oxygen gradient. (From Byrne, M.B., et al., *Trends Biotechnol.*, 32, 556–563, 2014. Reproduced by permission of the Cell Press.)

(Continued)

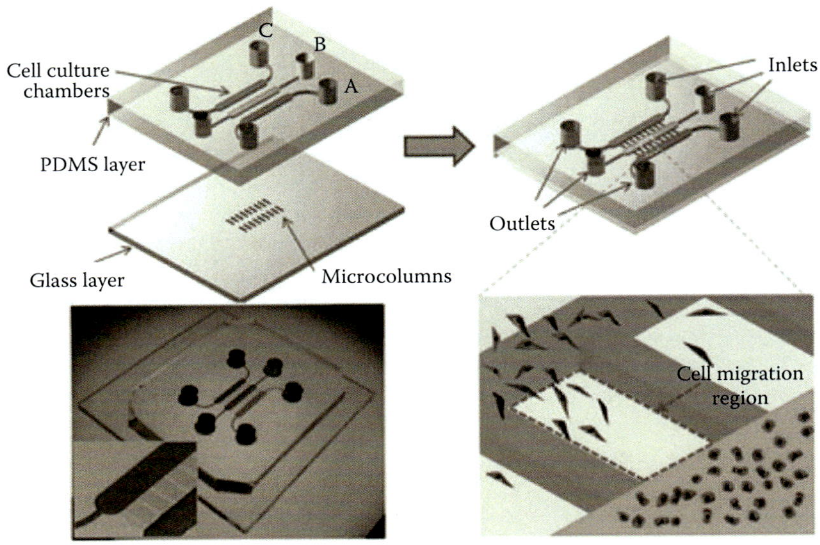

(c)

FIGURE 14.3 (Continued) Microfluidic devices to generate hypoxia gradient, and microfluidic devices for co-culture studies. (c) Schematics of representative microfluidic co-culture platforms showing cancer cells and fibroblasts cultured in interconnecting cell microchambers. (H. Ma, et al.: Characterization of the interaction between fibroblasts and tumor cells on a microfluidic co-culture device, *Electrophoresis*. 2010. 31. 1599–605. Copyright Wiley-VCH Verlag GmbH & Co. KGaA. Reproduced with permission.)

sodium hypochlorite, are introduced. Oxygen scavengers, such as pyrogallol, sodium hydroxide, and sodium sulfite, are flowed into the other microchannel. By controlling the flow rate of the oxygen generators and scavengers, user-defined boundary conditions for oxygen diffusion can be maintained, and a stable gradient of oxygen can be established across the cell chamber. These oxygen gradient generators have been used to study the efficacy of chemotherapeutic agents under hypoxic conditions [25,26].

The second class of microfluidic oxygen gradient generators utilizes gas with specified oxygen composition to generate oxygen gradients [27,28] (Figure 14.3b). Gas with high oxygen concentration flows through the source channel, while the gas with low oxygen concentration is introduced to the sink channel. By setting the boundary conditions of oxygen constant in the source and sink channels, an oxygen gradient can be maintained in the cell chamber. Gas-impermeable films, made of PC or PMMA, are patterned on top of the cell chamber to prevent the atmospheric oxygen from interfering with the gradient generation. Using these gradient generators, Funamoto et al. demonstrated that breast cancer cells migrate with greater persistence and speed when subjected to hypoxic conditions [29]. More recently, Mosadegh et al. introduced a micro-invasion assay that showed, for the first time, that cancer cells could undergo chemotaxis in an oxygen gradient [7].

Although a variety of these proof-of-concept microfluidic gradient generators have been introduced in the past few years, a detailed mechanistic study of cancer cell chemotaxis under a hypoxic gradient has yet to be achieved, highlighting the need to improve usability of these gradient generators.

14.3.2 Microfluidic Platforms to Apply Cellular Cues

Tumor-associated stromal cells can promote cancer cell migration and invasion through the stromal tissues, and various microfluidic co-culture platforms have been designed to investigate this process. Rather than assessing how a single chemical factor affects migration, as in the microfluidic chemotaxis assays, these co-culture devices investigate how interactions between cancer cells and tumor-associated stromal cells affect cancer cell migration. These devices could also help in the discovery of important signaling molecules that are involved in cancer cell-stromal cell interaction.

Most microfluidic co-culture platforms utilize a common design, usually consisting of two hydrogel regions arranged adjacent to each other. Cancer cells are seeded inside one of the hydrogel regions, and tumor-associated stroma cells are seeded in the other. By observing the 3D migration of cancer cells and stromal cells by phase and/or fluorescent microscopy, the interaction between these two cell types can be investigated (Figure 14.3c). Using one of these co-culture platforms, Ma et al. discovered that fibroblasts migrated toward cancer cells [30]. Moreover, Liu et al. discovered, using a similar platform, that cancer-associated fibroblasts (CAFs) could promote cancer cell migration, but normal fibroblasts lacked this ability [31]. This study further identified the role of MMPs in CAF-promoted cancer cell migration. Finally, Sung et al. developed a microfluidic co-culture assay to investigate the effects of fibroblasts on the progression of ductal carcinoma *in situ* to invasive ductal carcinoma [32]. This co-culture assay has recently been automated, highlighting a further advantage of microfluidics [33].

A different microfluidic co-culture system has been adopted by Hsu et al. to investigate the effects of fibroblasts and macrophages on cancer cell migration on a 2D substrate [34]. In this device, macrophages, fibroblasts, and cancer cells were each cultured in separate, interconnecting chambers. The mixing of the media in the chambers was controlled by pneumatic microvalves. By controlling the flow of media from fibroblast and/or macrophage chambers to the cancer cell chamber, the effects of stromal cell conditioned media on cancer cell migration were evaluated. Hsu et al. found that fibroblasts could enhance cancer cell migration speed. However, when these fibroblasts were pretreated with macrophage-conditioned media, the speed-enhancing effects of fibroblasts were diminished.

14.3.3 Microfluidic Platforms to Apply Mechanical Cues

The biophysical microenvironment of tumor tissues is fundamentally different from that of normal tissues. For example, cancer-associated fibroblasts can lay down collagen matrix inside the tumor, making it much denser and stiffer than normal tissues. In addition, fibroblasts can modify the ECM in the vicinity of the tumor by inducing bundling and alignment of collagen I fibers. Finally, the growth of the solid tumor often results in the buildup of interstitial fluid pressure inside the tumor, which subsequently drives an interstitial flow from the center of the tumor to the surrounding stroma. All of these mechanical cues (increased stiffness, altered ECM structure, and elevated interstitial flow) have recently been identified as promoting factors in tumor cell proliferation and migration [35]. To mimic these pathophysiological mechanical cues and study the effects of these cues on cell migration, researchers have designed a range of microfluidic systems to apply controlled mechanical stimuli to cells [10].

14.3.3.1 Stiffness

A variety of different methods have been utilized to generate a stiffness gradient in a microfluidic device. For instance, a microfluidic platform was developed to generate stiffness gradient along the surface of a 2D polymethylsiloxane (PDMS) membrane by integrating a layer of micropatterned PDMS micropillars underneath the membrane. By varying the pattern of the micropillars, the gradient in stiffness of the membrane could be modulated. Using this device, Palamà et al. discovered that stiffness gradients affect the elongation of HeLa cells, as well as the activation of focal adhesion kinases inside these cells [36].

Zaari et al. used an alternative strategy for generating stiffness gradients in their microfluidic device. This device uses a laminar flow-based chemical gradient generator described in the previous section to generate a gradient of crosslinker for polyacrylamide (PAAM) gel (Figure 14.4a). By photopolymerizing the PMMA gel in the presence of a gradient of crosslinkers, a gradient of stiffness can be produced [37]. In addition, a microfluidic device to apply both chemical gradients and stiffness gradients has recently been described [38].

14.3.3.2 Interstitial Flow

Elevated interstitial flow inside the tumor tissue is an indicator of poor prognosis. Through various studies performed with transwell assays, researchers discovered that elevated interstitial flow promotes cancer cell invasion via autologous chemotaxis or glycocalyx-mediated mechanotransduction [39–41, 6]. However, since the transwell assay is an end-point assay, it cannot be used to understand how interstitial flow affects the dynamics of cancer cell migration. To address this question, researchers have employed microfluidic flow assays [6]. These assays have a basic design of a hydrogel sandwiched between two microchannels. A pressure gradient is established across the hydrogel to impose an interstitial flow through the hydrogel seeded with cancer cells (Figure 14.4b). Using these devices, Polacheck et al. discovered a second mechanism that competes with CCR7-dependent autologous chemotaxis to influence cancer cell migration under interstitial flow. When cell-seeding density is low, cancer cells tend to migrate with the flow based on the CCR7-dependent mechanism. When cell-seeding density is high, cancer cells tend to migrate against the direction of the flow due to a CCR7-independent mechanotransduction mechanism [42]. Further studies from the same group showed that this mechanotransduction mechanism depended on flow-induced activation of β1 integrin on the upstream side of the cell. This polarized activation led to the accumulation of vinculin, focal adhesion kinase, and paxillin on the upstream side of the cells, cumulating in protrusion formation against the direction of the flow [43]. Haessler et al., using a similar

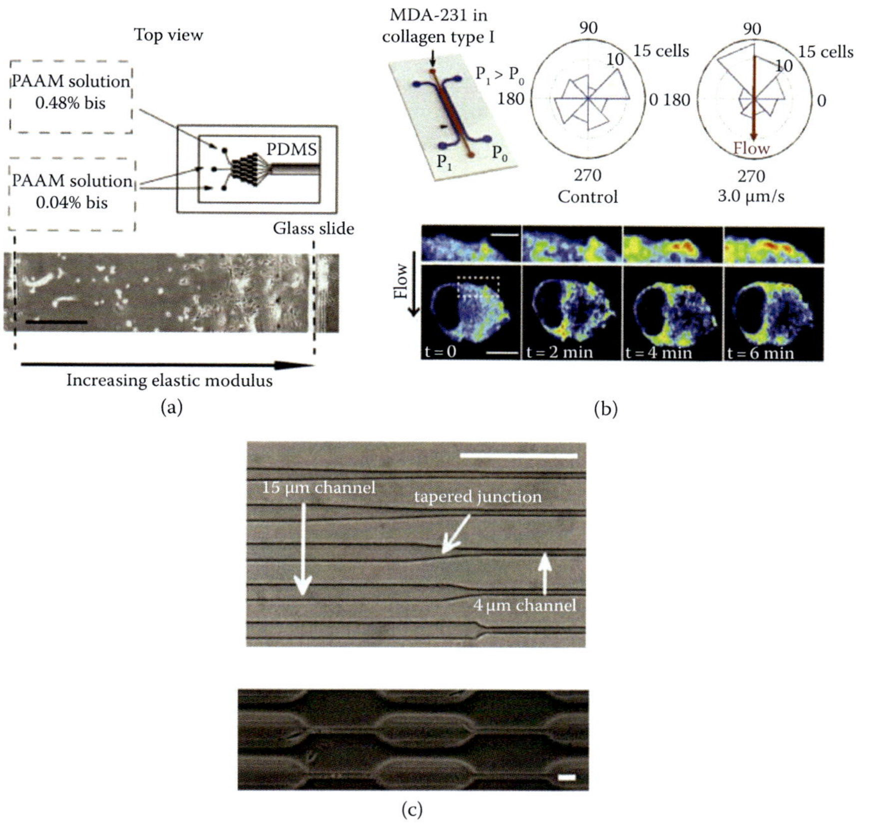

FIGURE 14.4 Microfluidic devices to study the effects of mechanical cues on cancer cell migration. (a) Schematics of microfluidic devices to generate stiffness gradients using the laminar flow-based chemical gradient generator. PAAM solutions with different concentration of cross-linkers were flowed into the gradient generator to create a gradient of cross-linkers, which translated into a gradient of stiffness. (From Zaari, N., et al.: *Adv. Mateorol.* 2133–2137, 2004. Copyright Wiley-VCH Verlag GmbH & Co. KGaA. Reproduced with permission.) (b) Microfluidic device to study the effects of interstitial flow on cancer cell migration. The device consists of a collagen gel sandwiched between to microchannels. A pressure gradient is established across the gel ($P_1 > P_0$) (top left). This device was used to demonstrate that cancer cells migrate against the direction of the flow (upstream) according to the rose plot (top right). This device also revealed that vinculin accumulated on the upstream side of the cells within 6 minutes after the initiation of the flow (bottom). (From Polacheck, W.J., et al., *Proc. Natl. Acad. Sci. U. S. A.*, 108, 11115–11120, 2011; and Polacheck, W. J., et al., *Proc. Natl. Acad. Sci. U. S. A.*, 111, 2447–52, 2014. With permission.) (c) Microfluidic devices to study the effects of physical confinement on cancer cell migration. These devices consist of channels with dimensions comparable to, or smaller than, the size of a cell (1 μm to 15 μm) to simulate the pore size of the ECM. (From Mak, M., et al., *PLoS One*, 6, e20825, 2011. Reproduced by permission of the Public Library of Science; and Pathak, A. and Kumar, S., *Proc. Natl. Acad. Sci. U. S. A.*, 109, 10334–10339, 2012. Reproduced by permission of the United States National Academy of Science.)

flow assay, demonstrated that interstitial flow affected different subpopulations of cancer cells differently. Specifically, cancer cells that migrated against the direction of the flow moved with greater speed but less persistence compared to the cells that migrated with the flow [44]. These results highlight the fact that the effects of interstitial flow on cancer cell migration is complex, and more studies are needed to elucidate the mechanotransduction mechanisms involved in this process.

14.3.3.3 Physical Confinement

Tumor tissue consists of a complex mix of cancer cells and other cell types, along with the ECM, which can present a steric hindrance to migration. To migrate through the dense ECM surrounding the tumor tissue, cancer cells can enzymatically degrade the ECM by secreting MMPs. Alternatively, cancer cells can also squeeze through the narrow pores of the ECM through an MMP-independent process [35]. This MMP-independent migration has recently received much attention due to the failure of anti-MMP therapy in treating metastasis. Microfluidic devices provide useful experimental platforms to mimic the migration of cancer cells through the narrow pores of the ECM. These devices incorporate microchannels with width on the order of 5–10 μm to mimic the pores of the ECM (Figure 14.4c) [97]. Cancer cells are seeded into these channels, and the migratory behaviors (speed, persistence, and morphology) of these cells are analyzed. Using these physical confinement devices, Irimia et al. found that physical confinement promoted fast and persistent migration of cancer cells that was independent of extracellular chemical cues [45]. Rolli et al. discovered that cancer cells that were transmigrating through narrow microchannels (7 μm wide) migrated with a smooth sliding motion, in stark contrast to the intermittent motion of cancer cells migrating in 2D [46]. More recently, devices with an array of physical confinement channels have also been developed to study how cancer cells deform and migrate through pores that are smaller than the nucleus [47–49]. In addition, the interplay between matrix stiffness and physical confinement has been investigated using microfluidic confinement devices with varying stiffness. Pathak et al. demonstrated that both physical confinement and stiffness could synergistically enhance cancer cell migration speed [50].

Microfluidic platforms have also been designed to investigate the migration of cancer cells on aligned and bundled collagen I fibers. Collagen fiber alignment has been found in solid tumor tissues, especially in breast tumor tissues. To study the effects of collagen fiber alignment on cancer cell migration, Riching et al. encapsulated a collagen hydrogel in a microchamber. By stretching the microchamber, the collagen fibers in the hydrogel were forced to align. Studies in this system demonstrated that collagen fiber alignment increased cancer cell migration persistence, but not speed [51].

14.4 MICROFLUIDIC TECHNOLOGIES TO STUDY TUMOR ANGIOGENESIS

Sprouting angiogenesis is the formation of microvasculature from pre-existing blood vessels. The growth of the tumor cells creates a hypoxic environment, inducing tumor cells to produce angiogenic growth factors that subsequently lead to vascularization of the tumor [4]. Indeed, angiogenesis is an integral part of metastatic cascade, since it provides tumor cells with not only nutrients to grow and migrate, but also with access to the circulation. A variety of microfluidic platforms have been devised to study the effects of chemical and mechanical cues on angiogenesis.

14.4.1 Chemical Cues

As mentioned earlier, the tumor microenvironment contains a wealth of growth factors (vascular endothelial cell growth factor [VEGF], fibroblast growth factor [FGF], and sphingosine-1-phosphate [S1P]) that promote angiogenesis [4]. These can be produced by tumor cells and tumor-associated stromal cells, and comprise the chemical cues for angiogenesis. Since the formation of nascent vessels requires the interaction of endothelial cells (ECs) with 3D ECM, microfluidic angiogenesis assays all follow a similar design consisting of a 3D collagen gel flanked by two microchannels. In one of the microchannels, the EC monolayers are formed on the sidewall of the collagen gel. Angiogenic factors (i.e., VEGF, FGF, and S1P) are introduced into either both microchannels or only the other microchannel to produce a gradient across the collagen gel, which induces a directed growth of the microvasculature into the collagen gel [52] (Figure 14.5a). In the past few years, numerous microfluidic devices have been designed to study the effects of growth factors on angiogenesis [8,10]. For example, Vickerman et al. designed and utilized such a device to study the effects of VEGF and S1P on endothelial cell sprouting angiogenesis. They found that VEGF and S1P could induce the formation of complex, 3D capillary networks with functional lumens [52]. Wood et al., using a similar device, found an inverse relationship between vessel elongation rate and vessel diameter during VEGF-induced sprout formation. The roles of MT1-MMP, MMP-1, MMP-2, and MMP-9 in the angiogenic process were systematically investigated using this microfluidic device [52]. Since microfluidic devices allow for high-resolution time-lapse tracking of sprout formation (Figure 14.5b), they also serve as a useful tool for the development of computational models of angiogenesis. Wood et al. and Farahat et al. used data generated from the microfluidic angiogenic assays to inform their computational models of angiogenesis [53,54]. More recently, a microfluidic angiogenesis platform that is comprised of channels micropatterned in a 3D collagen type I gel scaffold has been used to study the efficacy of angiogenic inhibitors. Endothelial cells are seeded in one of the channels to mimic the blood vessel. This platform has an advantage over the others since the blood vessels formed using this assay are fully encased within gel, mimicking the *in vivo* condition. Using this platform, Nguyen et al. discovered that endothelial cells could switch between VEGF-dependent and VEGF-independent angiogenesis [55].

Microfluidic devices have also been used to analyze the effects of cancer cells, epithelial cells, and stromal cells on sprouting angiogenesis. For instance, Chung et al. designed a microfluidic co-culture platform to assess the effects of cancer cells and smooth muscle cells on angiogenesis. They found that MTLn3 breast cancer cells induced sprouting angiogenesis of human microvascular endothelial cells (HMVECs), while U87MG glioblastoma cells had no observable effect on angiogenesis (Figure 14.5a). Interestingly, 10T½ smooth muscle cells seemed to repel the formation of angiogenic sprouts from the EC monolayer [56]. In addition, Sudo et al. used a similar microfluidic platform to investigate the interaction between hepatocytes and ECs [57]. Finally, Barkefors et al. used a microfluidic angiogenesis assay to study directional sprouting of endothelial-like cells from embryonic mouse kidneys under the effects of VEGF gradient [58].

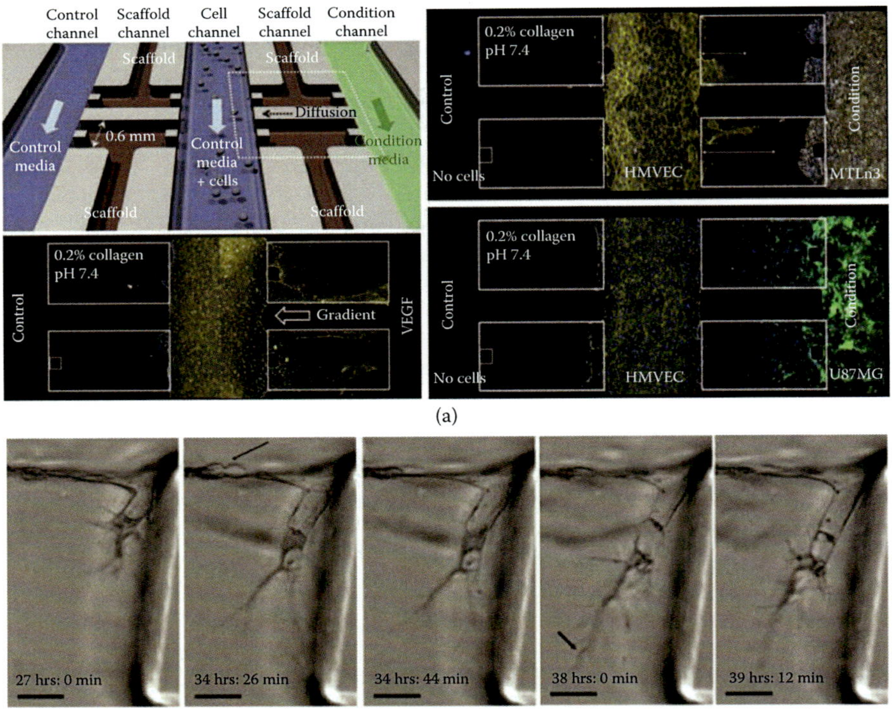

FIGURE 14.5 Microfluidic devices to study angiogenesis. (a) Schematics of microfluidic devices for angiogenesis. (From Chung, S., et al., *Lab Chip*, 9, 269–275, 2009. Reproduced by permission of The Royal Society of Chemistry.) Cells are seeded in the middle channel to form monolayers on the side of the gel scaffold. Cancer cells or angiogenic factors are introduced to the condition channel on the right. Control channel on the left contains only medium (top left). Endothelial cells underwent sprouting angiogenesis and invaded the collagen gel under the effects of a VEGF gradient (bottom left). Endothelial cells (HMVEC) sprout into the collagen gel under the effects of MTLn3 (top right); however, no sprouting was observed when endothelial cells were co-cultured with U87MG cancer cells (bottom right). (b) Time-lapse images of sprout formation in the microfluidic devices when endothelial cells were subjected to a gradient of VEGF. (From Vickerman, V., et al., *Lab Chip*, 8, 1468–1477, 2008. Reproduced by permission of The Royal Society of Chemistry.) *(Continued)*

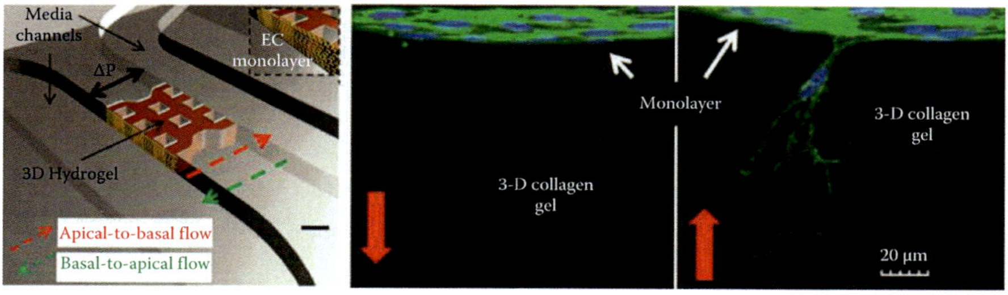

(c)

FIGURE 14.5 (Continued) Microfluidic devices to study angiogenesis. (c) Schematics of microfluidic devices used for studying the effects of transendothelial flow on angiogenesis. (From Vickerman, V. and Kamm, R.D., *Integr. Biol. (Camb)*, 48, 863–874, 2012. Reproduced by permission of The Royal Society of Chemistry.) Cells are seeded on the sidewall of the 3D hydrogel and a pressure gradient is established across the gel to induce apical-to-basal or basal-to-apical flow (left). Immunofluorescent images showing that apical-to-basal flow does not induce sprout formation (middle), while basal-to-apical does (right).

14.4.2 Mechanical Cues

Elevated fluid pressure in the vicinity of a growing tumor could also influence transmural flow in blood vessels near the tumor mass, which could induce angiogenesis through mechanotransduction pathways, and various microfluidic flow assays have been designed to investigate this hypothesis [6]. These assays are similar to those used to study the effects of chemical cues on angiogenesis (presented in previous section), but instead of a chemical gradient, a pressure gradient is applied across the collagen gel to create a flow across an EC monolayer seeded on the sidewall of the gel [59] (Figure 14.5c). Microfluidic technologies offer a convenient means to study the effects of transendothelial flow on angiogenesis, since they allow for the establishment of user-defined mechanical stimuli and real-time high-resolution imaging of EC sprouting, two of the advantages that traditional angiogenic assays lack. Using similar microfluidic platforms, Song et al. and Vickerman et al. reported that basal-to-apical transendothelial flow induced sprouting angiogenesis from the EC monolayer, while apical-to-basal flow had a much lesser effect on angiogenesis [59,60]. Vickerman et al. also proposed that tension imparted on the EC by basal-to-epical flow could activate integrin signaling, which could then lead to angiogenesis through the down-stream mechanotransduction events involving FAK, Src, ROCK, NO, and glycocalyx signaling [59]. Song et al. further reported that RhoA was also involved in this mechanically activated angiogenesis [61]. The observation that basal-to-apical transendothelial flow enhances sprouting angiogenesis is consistent with the *in vivo* observation that nascent sprouts often form from venules [6]. Recently, a shear stress threshold for activating sprouting angiogenesis has been identified using microfluidic devices of similar design [62].

14.5 MICROFLUIDIC TECHNOLOGIES TO STUDY TUMOR-ENDOTHELIAL INTERACTIONS: INTRAVASATION AND EXTRAVASATION

The first steps in the metastatic cascade consist of tumor cell detachment from the primary tumor and subsequent interactions with the extracellular matrix and stromal cells as they invade into the surrounding parenchyma. As the cells migrate they encounter nearby blood vessels or lymphatics, and some percentage of them begin interactions with the endothelium in order to enter the lymphatic or blood vasculature, or "intravasate." Upon successful entry, tumor cells circulate throughout the body, and of those that survive, some small fraction might arrest at a distant site, migrate out of the vessel (extravasate) and recolonize to form a secondary metastasis. In order to model these steps of the metastatic cascade, both tumor-ECM and tumor-endothelial interactions must be accurately recapitulated.

Intravasation is considered one of the early rate-limiting steps in metastasis. Increased tumor cell intravasation rates result in higher circulating tumor cell numbers and have been found to correlate with an increase in the incidence of secondary metastases [1]. It is known that entry of tumor cells can occur via both lymphatic and blood vessels; however, the mechanisms regulating these early steps are largely unknown. Questions arise regarding the mechanistic differences between blood and lymphatic intravasation, whether tumor cells migrate actively toward blood vessels via chemotactic responses or are shed passively into the circulation, and the role of immune cells in the intravasation microenvironment [63].

Extravasation involves a cascade of events consisting of (1) tumor cell arrest on the endothelium (2) tumor cell transendothelial migration (TEM), and (3) subsequent invasion. There have been two general hypotheses regarding the mode of arrest of circulating tumor cells (CTCs) in a blood vessel: mechanical trapping of cells and/or active preferential adhesion of the tumor cell onto the endothelium in a distant organ. James Ewing proposed that metastasis at distant organs is dictated by the anatomy of the blood and lymphatic vessels and blood circulatory paths between primary and secondary tumor sites [64]. CTCs, due to their relatively large size (~20 um diameter), may become physically trapped in the microcirculation, become activated, and eventually transmigrate [65]. However, Steven Paget's "seed and soil" hypothesis suggests that there exist interactions between different tumor cell types and specific organ microenvironments that guide their metastatic spread. The vasculature of such organs may be primed with surface receptors/molecules or secrete chemokines that cause specific tumor cell types to preferentially "home and seed" at that particular tissue environment. This is supported by the observation that breast cancer cells frequently metastasize to the bone [66]. Although two distinct phenomena have been described, is it clear now that the two theories may not be mutually exclusive; in fact, it has been suggested that tissue/organ tropism can be influenced by both physical trapping and interactions between "seed cells" and "receptive soils" [67].

14.5.1 *In Vivo* Models of Tumor-Endothelial Interactions

Conventional studies of intravasation and extravasation have been mostly limited to *in vivo* mouse models, either via live intravital microscopy or end-point assays with fixed tissues. These platforms have allowed the investigation of genes involved in the metastatic cascade and proteins that mediate cancer invasion [68]. Commonly used mouse models of metastasis include rat-tail vein, orthotopic, intracardial, and subcutaneous injections. For instance, tumor cells can be injected intravenously and intravital microscopy can be employed to visualize the circulation of tumor cells in the vasculature and interactions with the metastatic organ [69]. Although these models allow for a high degree of physiological relevance, there remain a number of challenges associated with studying the underlying mechanisms of tumor cell intravasation and extravasation. For example, probing the molecular mechanisms and cellular interactions is complicated due to the adaptive response of the tumor microenvironment, such as the recruitment of immune cells. Furthermore, real-time intravital imaging requires custom setups, which may perturb tumor pathophysiology [70,71]. Many *in vivo* assays are limited to end-point data collection and do allow for real-time high-resolution dynamic visualization of extravasation events. More recently, intravital microscopy performed on CAMs [72] and optically transparent transgenic zebra fish has enabled high spatial and temporal resolution imaging [73,74]; however, as with most *in vivo* platforms, the ability to perform parametric high-throughput studies in a tightly regulated microenvironment is restricted.

14.5.2 Conventional *In Vitro* Assays to Study Tumor-Endothelial Interactions

In vitro tools are invaluable for investigation of cellular interactions as they allow discretization of individual events in a complex cascade and offer tight control of experimental parameters. The most commonly used and accepted *in vitro* model to study tumor-endothelial interactions in the context of metastasis is the Boyden chamber/Transwell assay, which provides a relatively simple and high-throughput method for parametric cell migration studies, and overcomes some limitations of *in vivo* experiments by increasing throughput, ease of quantification of transmigration events, and the ability to employ human cell culture [75]. Here, an endothelial cell layer is grown to confluence on a porous membrane insert and tumor cells are seeded on top of the layer and allowed to transmigrate. The number of cells transmigrated can then be quantified by collecting the cells on the underside of the membrane. More recently, similar assays incorporating a 3D ECM matrix beneath the endothelial monolayer have increased the physiological relevance by recapitulating the subendothelial ECM into which tumor cells transmigrate [76]. A subluminal-to-luminal transendothelial migration assay has also been developed to model intravasation, which features a thin layer of ECM on the bottom of a Transwell filter with layer of endothelial cells seeded onto this basement membrane–like matrix [77]. However, "top-down" assays where the endothelium and matrix are layered on top of each other in the x-y plane usually do not allow for real-time and high-resolution monitoring of the entire processes of intravasation and extravasation.

14.5.3 Microfluidic Assays to Study Tumor Endothelial Interactions

Recently, microfluidic cell culture platforms have emerged to address the need for greater resolution, higher throughput, control of the cellular microenvironment and enhanced physiological relevance compared to traditional cell culture methods (Figure 14.6). The motivation for applying microfluidic technologies to study tumor-endothelial interactions include, but are not limited to:

- Isolation and discretization of specific steps in the intravasation and extravasation cascade (adhesion, transendothelial migration, basement membrane breaching) to facilitate understanding of the mechanism of action of different cellular proteins or drugs
- Ability to recapitulate physiologically relevant spatial patterning and organization of multiple cell types at the metastatic site
- Facilitation of 3D cell culture in ECM or ECM-like hydrogels
- High spatial resolution imaging of tumor-endothelial interactions during intravasation and extravasation via conventional microscopy techniques
- Facilitation of real-time visualization transmigration events, allowing close observation of the dynamics of the endothelium and tumor cells
- Ability for highly controlled and tunable microenvironments, such as precise application of relevant spatiotemporal chemical gradients, mechanical stresses, and complex interactions between different cell types [5,78].
- Relatively lower cost and higher throughput than conventional *in vivo* models
- Relatively lower reagent volume compared to conventional *in vitro* cell culture methods

14.5.3.1 Microfluidic Devices for Investigating Intravasation

One of the challenges in studying intravasation and extravasation is the complexity of the interface between the tumor microenvironment and vascular system. Recently, groups have developed *in vitro* intravasation platforms by positioning tumor cells next to engineered vascular beds or individual vascular tubes, which are formed either by natural angiogenic/vasculogenic-like processes or via monolayer formation on a hydrogel surface. For instance, Wong and Searson et al. formed cylindrical tubes inside a collagen type I matrix using Nitinol rods that are preseeded with tumor cells as a template. After solidification of the gel, the cylindrical tube is seeded with endothelial cells to form an intact lumen, and allowing close monitoring of tumor endothelial interactions during intravasation is possible via live-cell fluorescence imaging [79] (Figure 14.6a). Recently, other microfluidic devices have been developed featuring hydrogel regions flanked by media channels [5,80,81]. Here, a single lumen can be modeled by seeding endothelial cells into the flanking media channels, which forms an intact monolayer on the hydrogel and micropost surfaces. Tumor cells are seeded in the opposing media channel and transendothelial

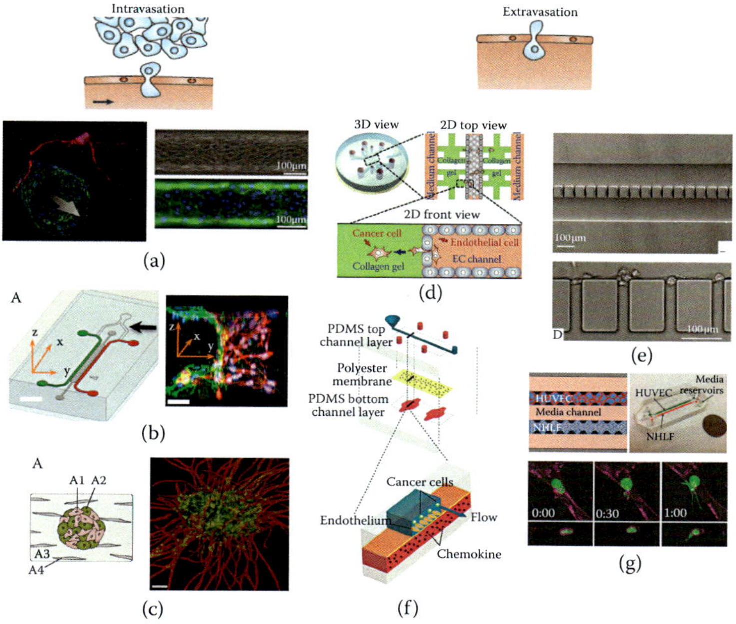

FIGURE 14.6 Microfluidic models of intravasation and extravasation. (a) A single endothelialized (HUVEC) tube is created inside an ECM hydrogel and tumor cells are injected into the surrounding gel to recapitulate intravasation events. (Reproduced from Wong, A.D. and Searson, P.C., *Cancer Res.*, 74, 4937–4945, 2014. With permission.) (b) An endothelial (HMVEC) layer is grown on a hydrogel-PDMS post surface to mimic the vascular barrier. Tumor cells are seeded in the opposing media channel and allowed to migrate across the collagen type 1 gel and past the endothelium. (Reproduced from Zervantonakis, I.K., et al., *Proc. Natl. Acad. Sci. U. S. A.*, 2012. With permission.) (c) A prevascularized tumor (PVT) model is created by co-culturing endothelial cells with tumor spheroids, resulting in an interconnected radial vascular network. Tumor cells can be observed to migrate intravascularly inside the lumens. (Reproduced from Ehsan, S.M., et al., *Integr. Biol.*, 6, 603–610, 2015. With permission.) (d) An endothelial monolayer is grown to confluence on the side of a gel-region to mimic the vascular barrier. Tumor cells are seeded in the "lumen" and extravasation events can be observed in the X-Y plane of view. (Reproduced from Jeon, J.S., et al., *PLoS One*, 8, e56910, 2013. With permission.) (e) A multi-step microfluidic device where micro-gaps between posts are coated with Matrigel and seeded with human microvascular endothelial cells to mimic the endothelium. Tumor cells can then be observed to transmigrate past the small EC-lined gaps (Reproduced from Chaw, K.C., et al., *Lab Chip*, 7, 1041–1047, 2007. With permission). (f) An extravasation model featuring an endothelial monolayer on top of a porous membrane sandwiched between two microchannels. The endothelium can be subject to physiologically relevant shear stresses while tumor cells are allowed to transmigrate from the top to the bottom microchannel. (Reproduced from Song, J.W., et al., *PLoS One*, 4, e5756, 2009. With permission.) (g) A perfusable interconnected microvascular network is formed by co-culturing HUVECs and fibroblasts into a fibrin gel matrix inside a microfluidic device. Tumor cells can be perfused into the networks via induction of a pressure drop and extravasation events can be observed in high resolution via confocal microscopy. (Reproduced from Chen, M.B., et al., *Integr. Biol.*, 5, 1262–1271, 2013. With permission.)

migration events can be observed in high resolution and real-time as cells migrate largely in the x-y plane of observation (Figure 14.6b). Using this model, it was found that endothelial signaling with macrophages in the intravasation microenvironment via secretion of tumor necrosis factor alpha results in increased endothelial barrier impairment and subsequent tumor cell transmigration [80]. Such planar single-lumen models greatly facilitate the imaging and quantification processes, as the geometry of the "vascular wall" is well defined and controlled.

Alternatively, the vasculature has also been modeled via spontaneous network formation by endothelial cells seeded in ECM hydrogels. Ehsan et al. developed an *in vitro* platform to recapitulate both the tumor neovascularization and intravasation stages of the metastatic cascade (Figure 14.6c). Termed the prevascularized tumor (PVT) model, the platform features PVT spheroids formed through the direct co-culture of primary human endothelial cells and human tumor cells, and surrounding fibroblasts. After 7 days of culture, endothelial cells are seen to sprout radially from and into the tumor spheroid, forming an intricate vascular bed. Via fluorescence microscopy, the percentage of tumor cells intravasated from the spheroid into the surrounding vasculature can be analyzed. It was further observed that decreased oxygen tension increased the intravasation efficiency of SW620 cells [82]. Similarly, Khuon et al. fabricated a vascular network in 3D collagen gels via a vasculogenic-like process followed by the introduction of MDA-MB-231 tumor cells into the surrounding gel either by co-assembly or multispot injection. Tumor cells could then be observed invading into the vascular network, and confocal and electron microscopy revealed that these cells are capable of transmigrating via the transcellular route [83].

14.5.3.2 Microfluidic Devices for Investigating Extravasation

Much effort has also been devoted to the design of microfluidic assays for observing the early stage of metastatic seeding. Extravasation is similar to intravasation as it also involves complex tumor-endothelial interactions. However, the extravasation microenvironment also differs greatly in terms of the types of chemical cues, auxiliary cell types, and mechanical forces present. Similar methods of engineering vasculature can be applied in this scenario, including self-organized formation of microvascular networks and endothelial monolayer formation on hydrogel surfaces or porous membranes. Recently, the Takayama group described a microfluidic chip consisting of two chambers separated by a thin membrane onto which an endothelial monolayer is cultured [84] (Figure 14.6f). The design resembles a traditional Transwell assay but with the added advantages of microfluidic systems including high resolution imaging and application of relevant fluid shear stresses. Tumor cell adhesion and arrest on the endothelium can be observed dynamically and in high resolution, and it was demonstrated that breast cancer cell receptors CXCR4 and CXCR7 are involved in the adhesion of tumor cells to the endothelium. Similarly, Shin et al. developed a microfluidic platform to recapitulate the metastatic cascade on a single chip, beginning with intravasation followed by arrest and adhesion on an endothelial monolayer. It was demonstrated that E-selectin expression and shear stress values dictated the adhesion of colon cancer cells onto the endothelium [85]. These types of "top-down" devices (where the endothelium lies in the x-y plane and tumor cell extravasation occurs in the z-plane) are particularly amenable to visualizing the morphological dynamics of the endothelium (e.g., endothelial junction behaviors).

In a different approach, microfluidic devices consisting of microchannels connected by 3D ECM hydrogels, tumor cells are seeded in one channel, arrest onto and extravasate across an EC monolayer oriented perpendicular to the image plane and migrate into a collagen gel (Figure 14.6d). Using a system of this design, tumor cell transmigration events were shown to occur in the first 24 hours and were accompanied with an increase in endothelial monolayer permeability [81]. This type of assay is particularly amenable to high resolution imaging of tumor cell morphological dynamics during extravasation as migration largely occurs in the x-y plane. However, a potential challenge is that the use of PDMS posts to contain the gel sometimes prevents the formation of a continuous, low-permeable EC monolayer. It is also difficult to predict how changes in stiffness between the PDMS and gel affect extravasation mechanisms. A similar type of microfluidic device was designed by Zhang et al. where one main endothelialized channel is connected vertically with five separated matrix-filled channels. Human Umbilical Vein Endothelial Cells (HUVECs) are grown to confluence on the hydrogels at the ends of the matrix channels, and tumor aggregates are seeded into the endothelial channel where extravasation events can be observed [86] (Figure 14.6e). Another microfluidic device recapitulating the extravasation microenvironment was devised by Chaw et al. [87]. In this multistep device, tumor cells must first deform past 10 μm gaps between PDMS posts, which mimic the narrow capillaries through which tumor cells must traverse in the circulation. A portion of these deformed cells were then collected and transferred to a "transmigration chamber" which consists of an endothelialized channel connected to Matrigel-filled channels, similar to that described in Zhang et al. [86].

It is important to note that most studies of transendothelial migration across planar EC monolayers are limited in their physiological relevance. For instance, the barrier permeability values achieved are usually 1 to 2 orders of magnitude higher than those found *in vivo* [80,81,88] and the planar geometry does not allow an accurate reproduction of physiological phenomena such as tumor cell trapping in the circulation. Recently, the ability to generate self-organized microvascular networks via suspension of endothelial cells inside collagen type 1 or fibrin hydrogels has enabled high-resolution visualization of tumor endothelial interactions while increasing physiological relevance compared to flat monolayers [89–91]. Chen et al. [92] applied this methodology to the study of tumor cell extravasation from within microvascular networks inside microfluidic devices. HUVECs and normal human lung fibroblasts are seeded in separate hydrogel regions held in place by microposts and flanked by media channels (Figure 14.6g). Within 4 days HUVECs form intricate interconnected networks that are perfusable via application of a pressure drop across the endothelial gel region. It was found via high-resolution confocal microscopy that the transmigrating tumor cell undergoes large cytoplasmic and nuclear deformations (subnuclear in dimension) when breaching the endothelium. This platform is particularly amenable to high-resolution imaging, as most extravasation events occur in a thin z-plane since the vascular network lies mainly in a pseudo-2D plane. Furthermore, microvessels have dimensions on the order of *in vivo* capillaries to arterioles (~5 to 80 μm), which can recapitulate relevant phenomena such as tumor cell trapping and arrest. This is not possible in flat monolayer assays and it is also difficult to achieve such small lumen diameters (~15 μm) via needle-templating. Using the same concept, Jeon and Bersini [93] created a bone extravasation microenvironment in an effort to recapitulate the complex chemical cues and signaling present when circulating tumor cells arrest and form secondary metastases in bone, a common occurrence during breast and prostate cancer progression [66].

14.6 CONCLUSION AND FUTURE CHALLENGES

It is essential to acquire a deeper understanding of the cellular and molecular mechanisms behind the metastatic cascade in the quest for new cancer therapeutics. Traditional *in vitro* assays including 2D petri dish and Transwell chamber cultures have played a central role in dissecting the extrinsic signals and cell-autonomous programs involved in metastasis, especially when used in conjunction with *in vivo* experiments. In recent years, microfluidics has emerged as a new approach to *in vitro* modeling that aims to address some of the key limitations in conventional platforms, enabling multiparametric, high-resolution studies on cell-cell and cell-matrix interactions in a precisely controlled yet more complex and physiologically relevant microenvironment.

While much progress has been made, these "metastasis-on-a-chip" platforms still have the potential to be more useful by overcoming remaining challenges. First, many devices are still complex to fabricate, limiting their accessibility to the wider community, as well as the throughput of their production and scalability. The ability to perform quantification and analysis of RNA, DNA, and protein levels in the cells cultured in microfluidic chambers is also crucial in understanding cellular mechanisms (e.g., protein expression, transcriptional level changes, etc.), beyond the information provided by microscopy and immunohistochemistry. Currently, the small number of cells in microdevices limits the amount of protein than can be collected and analyzed, and removing cells from enclosed microfluidic channels can be challenging. Finally, a key feature of metastasis is the tendency for specific tumor cells to home to distinct secondary organs. For instance, it has been observed that breast and prostate cancers frequently metastasize to bone [66]; however, the underlying mechanisms behind this "homing" behavior are largely unknown. As such, there is great potential in the development of "organ-specific" metastasis models, with the aim to understand the complex signaling that occurs between host cells and tumor cells in the extravasation and recolonization microenvironment. This information may pave the way to identifying new therapeutic targets to prevent secondary metastatis for certain tumor types. In summary, the development of microfluidic technologies has paved the way for a new generation of powerful *in vitro* experimental metastasis assays, which, combined with *in vivo* validation, can deepen our fundamental understanding of cancer biology and enable the discovery of new drug targets to combat metastatic progression.

REFERENCES

1. D. X. Nguyen, P. D. Bos, and J. Massagué, Metastasis: From dissemination to organ-specific colonization, *Nat. Rev. Cancer*, vol. 9, no. 4, pp. 274–84, 2009.

2. D. Hanahan and R. A. Weinberg, Hallmarks of cancer: The next generation, *Cell*, vol. 144, no. 5, pp. 646–74, 2011.
3. J. A. Joyce and J. W. Pollard, Microenvironmental regulation of metastasis, *Nat. Rev. Cancer*, vol. 9, no. 4, pp. 239–52, 2009.
4. P. Carmeliet and R. K. Jain, Molecular mechanisms and clinical applications of angiogenesis, *Nature*, vol. 473, no. 7347, pp. 298–307, 2011.
5. Y. Shin, S. Han, J. S. Jeon, K. Yamamoto, I. K. Zervantonakis, R. Sudo, R. D. Kamm, and S. Chung, Microfluidic assay for simultaneous culture of multiple cell types on surfaces or within hydrogels, *Nat. Protoc.*, vol. 7, no. 7, pp. 1247–59, 2012.
6. W. J. Polacheck, R. Li, S. G. M. Uzel, and R. D. Kamm, Microfluidic platforms for mechanobiology, *Lab Chip*, vol. 13, no. 12, pp. 2252–67, 2013.
7. B. Mosadegh, M. R. Lockett, K. T. Minn, K. A. Simon, K. Gilbert, S. Hillier, D. Newsome, et al., A paper-based invasion assay: Assessing chemotaxis of cancer cells in gradients of oxygen, *Biomaterials*, vol. 52, pp. 262–71, 2015.
8. S. Kim, H. J. Kim, and N. L. Jeon, Biological applications of microfluidic gradient devices, *Integr. Biol. (Camb.)*, vol. 2, no. 11–12, pp. 584–603, 2010.
9. J. Li and F. Lin, Microfluidic devices for studying chemotaxis and electrotaxis, *Trends Cell Biol.*, vol. 21, no. 8, pp. 489–97, 2011.
10. H. Ma, H. Xu, and J. Qin, Biomimetic tumor microenvironment on a microfluidic platform, *Biomicrofluidics*, vol. 7, no. 1, p. 11501, 2013.
11. S. Takayama, E. Ostuni, P. LeDuc, K. Naruse, D. E. Ingber, and G. M. Whitesides, Selective chemical treatment of cellular microdomains using multiple laminar streams, *Chem. Biol.*, vol. 10, no. 2, pp. 123–130, 2003.
12. C. W. Frevert, G. Boggy, T. M. Keenan, and A. Folch, Measurement of cell migration in response to an evolving radial chemokine gradient triggered by a microvalve, *Lab Chip*, vol. 6, no. 7, pp. 849–56, 2006.
13. D. Irimia, S.-Y. Liu, W. G. Tharp, A. Samadani, M. Toner, and M. C. Poznansky, Microfluidic system for measuring neutrophil migratory responses to fast switches of chemical gradients, *Lab Chip*, vol. 6, no. 2, pp. 191–8, 2006.
14. N. Li Jeon, H. Baskaran, S. K. W. Dertinger, G. M. Whitesides, L. Van de Water, and M. Toner, Neutrophil chemotaxis in linear and complex gradients of interleukin-8 formed in a microfabricated device, *Nat. Biotechnol.*, vol. 20, no. 8, pp. 826–30, 2002.
15. S.-J. Wang, W. Saadi, F. Lin, C. Minh-Canh Nguyen, and N. Li Jeon, Differential effects of EGF gradient profiles on MDA-MB-231 breast cancer cell chemotaxis, *Exp. Cell Res.*, vol. 300, no. 1, pp. 180–9, 2004.
16. W. Saadi, S.-J. Wang, F. Lin, and N. L. Jeon, A parallel-gradient microfluidic chamber for quantitative analysis of breast cancer cell chemotaxis, *Biomed. Microdevices*, vol. 8, no. 2, pp. 109–18, 2006.
17. B. Mosadegh, W. Saadi, S.-J. Wang, and N. L. Jeon, Epidermal growth factor promotes breast cancer cell chemotaxis in CXCL12 gradients, *Biotechnol. Bioeng.*, vol. 100, no. 6, pp. 1205–13, 2008.
18. I. Zervantonakis, S. Chung, R. Sudo, M. Zhang, J. Charest, and R. Kamm, Concentration gradients in microfluidic 3D matrix cell culture systems, *Int. J. Micro-Nano Scale Transp.*, vol. 1, pp. 27–36, 2010.
19. S. Chung, R. Sudo, V. Vickerman, I. K. Zervantonakis, and R. D. Kamm, Microfluidic platforms for studies of angiogenesis, cell migration, and cell-cell interactions. Sixth International Bio-Fluid Mechanics Symposium and Workshop March 28–30, 2008 Pasadena, California, *Ann. Biomed. Eng.*, vol. 38, no. 3, pp. 1164–77, 2010.
20. V. V Abhyankar, M. W. Toepke, C. L. Cortesio, M. A. Lokuta, A. Huttenlocher, and D. J. Beebe, A platform for assessing chemotactic migration within a spatiotemporally defined 3D microenvironment, *Lab Chip*, vol. 8, no. 9, pp. 1507–15, 2008.

21. B. Mosadegh, C. Huang, J. W. Park, H. S. Shin, B. G. Chung, S.-K. Hwang, K.-H. Lee, H. J. Kim, J. Brody, and N. L. Jeon, Generation of stable complex gradients across two-dimensional surfaces and three-dimensional gels, *Langmuir*, vol. 23, no. 22, pp. 10910–12, 2007.
22. W. Saadi, S. W. Rhee, F. Lin, B. Vahidi, B. G. Chung, and N. L. Jeon, Generation of stable concentration gradients in 2D and 3D environments using a microfluidic ladder chamber, *Biomed. Microdevices*, vol. 9, no. 5, pp. 627–35, 2007.
23. B. J. Kim, P. Hannanta-anan, M. Chau, Y. S. Kim, M. A. Swartz, and M. Wu, Cooperative roles of SDF-1α and EGF gradients on tumor cell migration revealed by a robust 3D microfluidic model, *PLoS One*, vol. 8, no. 7, p. e68422, 2013.
24. M. B. Byrne, M. T. Leslie, H. R. Gaskins, and P. J. A. Kenis, Methods to study the tumor microenvironment under controlled oxygen conditions, *Trends Biotechnol*, vol. 32, no. 11, pp. 556–63, 2014.
25. Y.-A. Chen, A. D. King, H.-C. Shih, C.-C. Peng, C.-Y. Wu, W.-H. Liao, and Y.-C. Tung, Generation of oxygen gradients in microfluidic devices for cell culture using spatially confined chemical reactions, *Lab Chip*, vol. 11, no. 21, pp. 3626–33, 2011.
26. L. Wang, W. Liu, Y. Wang, J. Wang, Q. Tu, R. Liu, and J. Wang, Construction of oxygen and chemical concentration gradients in a single microfluidic device for studying tumor cell-drug interactions in a dynamic hypoxia microenvironment, *Lab Chip*, vol. 13, no. 4, pp. 695–705, 2013.
27. C.-C. Peng, W.-H. Liao, Y.-H. Chen, C.-Y. Wu, and Y.-C. Tung, A microfluidic cell culture array with various oxygen tensions, *Lab Chip*, vol. 13, no. 16, pp. 3239–45, 2013.
28. H. E. Abaci, R. Devendra, Q. Smith, S. Gerecht, and G. Drazer, Design and development of microbioreactors for long-term cell culture in controlled oxygen microenvironments, *Biomed. Microdevices*, vol. 14, no. 1, pp. 145–52, 2012.
29. K. Funamoto, I. K. Zervantonakis, Y. Liu, C. J. Ochs, C. Kim, and R. D. Kamm, A novel microfluidic platform for high-resolution imaging of a three-dimensional cell culture under a controlled hypoxic environment, *Lab Chip*, vol. 12, no. 22, pp. 4855–63, 2012.
30. H. Ma, T. Liu, J. Qin, and B. Lin, Characterization of the interaction between fibroblasts and tumor cells on a microfluidic co-culture device, *Electrophoresis*, vol. 31, no. 10, pp. 1599–605, 2010.
31. T. Liu, B. Lin, and J. Qin, Carcinoma-associated fibroblasts promoted tumor spheroid invasion on a microfluidic 3D co-culture device, *Lab Chip*, vol. 10, no. 13, pp. 1671–7, 2010.
32. K. E. Sung, N. Yang, C. Pehlke, P. J. Keely, K. W. Eliceiri, A. Friedl, and D. J. Beebe, Transition to invasion in breast cancer: A microfluidic in vitro model enables examination of spatial and temporal effects, *Integr. Biol. (Camb.)*, vol. 3, no. 4, pp. 439–50, 2011.
33. S. I. Montanez-Sauri, K. E. Sung, J. P. Puccinelli, C. Pehlke, and D. J. Beebe, Automation of three-dimensional cell culture in arrayed microfluidic devices, *J. Lab. Autom.*, vol. 16, no. 3, pp. 171–85, 2011.
34. T.-H. Hsu, Y.-L. Kao, W.-L. Lin, J.-L. Xiao, P.-L. Kuo, C.-W. Wu, W.-Y. Liao, and C.-H. Lee, The migration speed of cancer cells influenced by macrophages and myofibroblasts co-cultured in a microfluidic chip, *Integr. Biol. (Camb.)*, vol. 4, no. 2, pp. 177–82, 2012.
35. D. Wirtz, K. Konstantopoulos, and P. C. Searson, The physics of cancer: The role of physical interactions and mechanical forces in metastasis, *Nat. Rev. Cancer*, vol. 11, no. 7, pp. 512–22, 2011.
36. I. E. Palamà, S. D'Amone, A. M. L. Coluccia, M. Biasiucci, and G. Gigli, Cell self-patterning on uniform PDMS-surfaces with controlled mechanical cues, *Integr. Biol. (Camb.)*, vol. 4, no. 2, pp. 228–36, 2012.
37. N. Zaari, P. Rajagopalan, S. K. Kim, A. J. Engler, and J. Y. Wong, Photopolymerization in microfluidic gradient generators: Microscale control of substrate compliance to manipulate cell response, *Adv. Mater.*, vol. 16, no. 23–24, pp. 2133–7, 2004.

38. S. García, R. Sunyer, A. Olivares, J. Noailly, J. Atencia, and X. Trepat, Generation of stable orthogonal gradients of chemical concentration and substrate stiffness in a microfluidic device, *Lab Chip,* vol. 15, no. 12, pp. 2606–14, 2015.
39. J. D. Shields, M. E. Fleury, C. Yong, A. A. Tomei, G. J. Randolph, and M. A. Swartz, Autologous chemotaxis as a mechanism of tumor cell homing to lymphatics via interstitial flow and autocrine CCR7 signaling, *Cancer Cell*, vol. 11, no. 6, pp. 526–38, 2007.
40. Z.-D. Shi, H. Wang, and J. M. Tarbell, Heparan sulfate proteoglycans mediate interstitial flow mechanotransduction regulating MMP-13 expression and cell motility via FAK-ERK in 3D collagen, *PLoS One*, vol. 6, no. 1, p. e15956, 2011.
41. A. C. Shieh and M. A. Swartz, Regulation of tumor invasion by interstitial fluid flow, *Phys. Biol.*, vol. 8, no. 1, p. 015012, 2011.
42. W. J. Polacheck, J. L. Charest, and R. D. Kamm, Interstitial flow influences direction of tumor cell migration through competing mechanisms, *Proc. Natl. Acad. Sci. U. S. A.*, vol. 108, no. 27, pp. 11115–20, 2011.
43. W. J. Polacheck, A. E. German, A. Mammoto, D. E. Ingber, and R. D. Kamm, Mechanotransduction of fluid stresses governs 3D cell migration, *Proc. Natl. Acad. Sci. U. S. A.*, vol. 111, no. 7, pp. 2447–52, 2014.
44. U. Haessler, J. C. M. Teo, D. Foretay, P. Renaud, and M. A. Swartz, Migration dynamics of breast cancer cells in a tunable 3D interstitial flow chamber, *Integr. Biol. (Camb.)*, vol. 4, no. 4, pp. 401–9, 2012.
45. D. Irimia and M. Toner, Spontaneous migration of cancer cells under conditions of mechanical confinement, *Integr. Biol. (Camb.)*, vol. 1, no. 8–9, pp. 506–12, 2009.
46. C. G. Rolli, T. Seufferlein, R. Kemkemer, and J. P. Spatz, Impact of tumor cell cytoskeleton organization on invasiveness and migration: A microchannel-based approach, *PLoS One*, vol. 5, no. 1, p. e8726, 2010.
47. M. Mak, C. A. Reinhart-King, and D. Erickson, Elucidating mechanical transition effects of invading cancer cells with a subnucleus-scaled microfluidic serial dimensional modulation device, *Lab Chip*, vol. 13, no. 3, pp. 340–8, 2013.
48. M. Mak and D. Erickson, Mechanical decision trees for investigating and modulating single-cell cancer invasion dynamics, *Lab Chip*, vol. 14, no. 5, pp. 964–71, 2014.
49. M. Mak, C. A. Reinhart-King, and D. Erickson, Microfabricated physical spatial gradients for investigating cell migration and invasion dynamics, *PLoS One*, vol. 6, no. 6, p. e20825, 2011.
50. A. Pathak and S. Kumar, Independent regulation of tumor cell migration by matrix stiffness and confinement, *Proc. Natl. Acad. Sci. U. S. A.*, vol. 109, no. 26, pp. 10334–9, 2012.
51. K. M. Riching, B. L. Cox, M. R. Salick, C. Pehlke, A. S. Riching, S. M. Ponik, B. R. Bass, et al., 3D collagen alignment limits protrusions to enhance breast cancer cell persistence, *Biophys. J.*, vol. 107, no. 11, pp. 2546–58, 2014.
52. V. Vickerman, J. Blundo, S. Chung, and R. Kamm, Design, fabrication and implementation of a novel multi-parameter control microfluidic platform for three-dimensional cell culture and real-time imaging, *Lab Chip*, vol. 8, no. 9, pp. 1468–77, 2008.
53. W. A. Farahat, L. B. Wood, I. K. Zervantonakis, A. Schor, S. Ong, D. Neal, R. D. Kamm, and H. H. Asada, Ensemble analysis of angiogenic growth in three-dimensional microfluidic cell cultures, *PLoS One*, vol. 7, no. 5, p. e37333, 2012.
54. L. B. Wood, A. Das, R. D. Kamm, and H. H. Asada, A stochastic broadcast feedback approach to regulating cell population morphology for microfluidic angiogenesis platforms, *IEEE Trans. Biomed. Eng.*, vol. 56, no. 9, pp. 2299–303, 2009.
55. D.-H. T. Nguyen, S. C. Stapleton, M. T. Yang, S. S. Cha, C. K. Choi, P. A. Galie, and C. S. Chen, Biomimetic model to reconstitute angiogenic sprouting morphogenesis in vitro, *Proc. Natl. Acad. Sci. U. S. A.*, vol. 110, no. 17, pp. 6712–17, 2013.

56. S. Chung, R. Sudo, P. J. Mack, C.-R. Wan, V. Vickerman, and R. D. Kamm, Cell migration into scaffolds under co-culture conditions in a microfluidic platform, *Lab Chip*, vol. 9, no. 2, pp. 269–75, 2009.
57. R. Sudo, S. Chung, I. K. Zervantonakis, V. Vickerman, Y. Toshimitsu, L. G. Griffith, and R. D. Kamm, Transport-mediated angiogenesis in 3D epithelial coculture, *FASEB J.*, vol. 23, no. 7, pp. 2155–64, 2009.
58. I. Barkefors, S. Thorslund, F. Nikolajeff, and J. Kreuger, A fluidic device to study directional angiogenesis in complex tissue and organ culture models, *Lab Chip*, vol. 9, no. 4, pp. 529–35, 2009.
59. V. Vickerman and R. D. Kamm, Mechanism of a flow-gated angiogenesis switch: Early signaling events at cell-matrix and cell-cell junctions, *Integr. Biol. (Camb.)*, vol. 4, no. 8, pp. 863–74, 2012.
60. J. W. Song and L. L. Munn, Fluid forces control endothelial sprouting, *Proc. Natl. Acad. Sci. U. S. A.*, vol. 108, no. 37, pp. 15342–7, 2011.
61. J. W. Song, J. Daubriac, J. M. Tse, D. Bazou, and L. L. Munn, RhoA mediates flow-induced endothelial sprouting in a 3-D tissue analogue of angiogenesis, *Lab Chip*, vol. 12, no. 23, pp. 5000–6, 2012.
62. P. A. Galie, D.-H. T. Nguyen, C. K. Choi, D. M. Cohen, P. A. Janmey, and C. S. Chen, Fluid shear stress threshold regulates angiogenic sprouting, *Proc. Natl. Acad. Sci. U. S. A.*, vol. 111, no. 22, pp. 7968–73, 2014.
63. M. Bockhorn, R. K. Jain, and L. L. Munn, Active versus passive mechanisms in metastasis: Do cancer cells crawl into vessels, or are they pushed?, *Lancet. Oncol.*, vol. 8, no. 5, pp. 444–8, 2007.
64. B. Psaila and D. Lyden, The metastatic niche: Adapting the foreign soil, *Nat. Rev. Cancer*, vol. 9, pp. 285–93, 2009.
65. J. D. Crissman, J. S. Hatfield, D. G. Menter, B. Sloane, and K. Y. Honn, Morphological study of the interaction of intra vascular tumor cells with endothelial cells and subendothelial matrix1, *Cancer Res.*, vol. 48, no. 14, pp. 4065–72, 1988.
66. K. M. Bussard, C. V Gay, and A. M. Mastro, The bone microenvironment in metastasis; what is special about bone?, *Cancer Metastasis Rev.*, vol. 27, no. 1, pp. 41–55, 2008.
67. I. J. Fidler, The role of the organ microenvironment in brain metastasis, *Semin. Cancer Biol.*, vol. 21, no. 2, pp. 107–12, 2011.
68. G. Francia, W. Cruz-munoz, S. Man, P. Xu, and R. S. Kerbel, Mouse models of advanced spontaneous metastasis for experimental therapeutics, *Nat. Rev. Cancer*, vol. 11, no. 2, pp. 135–41, 2011.
69. S. I. J. Ellenbroek and J. van Rheenen, Imaging hallmarks of cancer in living mice, *Nat. Rev. Cancer*, vol. 14, no. 6, pp. 406–18, 2014.
70. D. Kedrin, B. Gligorijevic, J. Wyckoff, V. V. Verkhusha, J. Condeelis, J. E. Segall, and J. van Rheenen, Intravital imaging of metastatic behavior through a mammary imaging window, *Nat. Methods*, vol. 5, no. 12, pp. 1019–21, 2008.
71. J. Condeelis and R. Weissleder, In vivo imaging in cancer, *Cold Spring Harb. Perspect. Biol.*, vol. 2, no. 12, p. a003848, 2010.
72. H. S. Leong, A. E. Robertson, K. Stoletov, S. J. Leith, C. A. Chin, A. E. Chien, M. N. Hague, et al., Invadopodia are required for cancer cell extravasation and are a therapeutic target for metastasis, *Cell Rep.*, vol. 8, no. 5, pp. 1558–70, 2014.
73. K. Stoletov, H. Kato, E. Zardouzian, J. Kelber, J. Yang, S. Shattil, and R. Klemke, Visualizing extravasation dynamics of metastatic tumor cells, *J. Cell Sci.*, vol. 123, no. Pt 13, pp. 2332–41, 2010.
74. K. Stoletov, V. Montel, R. D. Lester, S. L. Gonias, and R. Klemke, High-resolution imaging of the dynamic tumor cell vascular interface in transparent zebrafish, *Proc. Natl. Acad. Sci. U. S. A.*, vol. 104, no. 44, pp. 17406–11, 2007.

75. E. T. Roussos, J. S. Condeelis, and A. Patsialou, Chemotaxis in cancer, *Nat. Rev. Cancer*, vol. 11, no. 8, pp. 573–87, 2011.
76. C. T. Mierke, Cancer cells regulate biomechanical properties of human microvascular endothelial cells, *J. Biol. Chem.*, vol. 286, no. 46, pp. 40025–37, 2011.
77. E. T. Roussos, M. Balsamo, S. K. Alford, J. B. Wyckoff, B. Gligorijevic, Y. Wang, M. Pozzuto, et al., Mena invasive (MenaINV) promotes multicellular streaming motility and transendothelial migration in a mouse model of breast cancer, *J. Cell Sci.*, vol. 124, no. Pt 13, pp. 2120–31, 2011.
78. J. Lii, W.-J. Hsu, H. Parsa, A. Das, R. Rouse, and S. K. Sia, Real-time microfluidic system for studying mammalian cells in 3D microenvironments, *Anal. Chem.*, vol. 80, no. 10, pp. 3640–7, 2008.
79. A. D. Wong and P. C. Searson, Live-cell imaging of invasion and intravasation in an artificial microvessel platform, *Cancer Res.*, vol. 74, no. 17, pp. 4937–45, 2014.
80. I. K. Zervantonakis, S. K. Hughes-Alford, J. L. Charest, J. S. Condeelis, F. B. Gertler, and R. D. Kamm, Three-dimensional microfluidic model for tumor cell intravasation and endothelial barrier function, *Proc. Natl. Acad. Sci. U. S. A.*, vol. 109, no. 34, pp. 13515–20, 2012.
81. J. S. Jeon, I. K. Zervantonakis, S. Chung, R. D. Kamm, and J. L. Charest, In vitro model of tumor cell extravasation, *PLoS One*, vol. 8, no. 2, p. e56910, 2013.
82. S. M. Ehsan, K. M. Welch-reardon, M. L. Waterman, C. W. Christopher, S. C. George, and M. Genetics, A three-dimensional in vitro model of tumor cell intravasation, *Integr. Biol.*, vol. 6, no. 6, pp. 603–10, 2015.
83. S. Khuon, L. Liang, R. W. Dettman, P. H. S. Sporn, R. B. Wysolmerski, and T.-L. Chew, Myosin light chain kinase mediates transcellular intravasation of breast cancer cells through the underlying endothelial cells: A three-dimensional FRET study, *J. Cell Sci.*, vol. 123, no. Pt 3, pp. 431–40, 2010.
84. J. W. Song, S. P. Cavnar, A. C. Walker, K. E. Luker, M. Gupta, Y.-C. Tung, G. D. Luker, and S. Takayama, Microfluidic endothelium for studying the intravascular adhesion of metastatic breast cancer cells, *PLoS One*, vol. 4, no. 6, p. e5756, 2009.
85. M. K. Shin, S. K. Kim, and H. Jung, Integration of intra- and extravasation in one cell-based microfluidic chip for the study of cancer metastasis, *Lab Chip*, vol. 11, no. 22, pp. 3880–7, 2011.
86. Q. Zhang, T. Liu, and J. Qin, A microfluidic-based device for study of transendothelial invasion of tumor aggregates in realtime, *Lab Chip*, vol. 12, no. 16, pp. 2837–42, 2012.
87. K. C. Chaw, M. Manimaran, E. H. Tay, and S. Swaminathan, Multi-step microfluidic device for studying cancer metastasis, *Lab Chip*, vol. 7, no. 8, pp. 1041–7, 2007.
88. K. M. Chrobak, D. R. Potter, and J. Tien, Formation of perfused, functional microvascular tubes in vitro, *Microvasc. Res.*, vol. 71, no. 3, pp. 185–96, 2006.
89. S. Kim, H. Lee, M. Chung, and N. L. Jeon, Engineering of functional, perfusable 3D microvascular networks on a chip, *Lab Chip*, vol. 13, no. 8, pp. 1489–500, 2013.
90. J. A. Whisler, M. B. Chen, and R. D. Kamm, Control of perfusable microvascular network morphology using a multiculture microfluidic system, *Tissue Eng. Part C. Methods*, vol. 20, no. 7, pp. 543–52, 2014.
91. Y.-H. Hsu, M. L. Moya, P. Abiri, C. C. W. Hughes, S. C. George, and A. P. Lee, Full range physiological mass transport control in 3D tissue cultures, *Lab Chip*, vol. 13, no. 1, pp. 81–9, 2013.
92. M. B. Chen, J. A. Whisler, J. S. Jeon, and R. D. Kamm, Mechanisms of tumor cell extravasation in an in vitro microvascular network platform, *Integr. Biol.*, vol. 5, no. 10, pp. 1262–71, 2013.
93. J. S. Jeon, S. Bersini, M. Gilardi, G. Dubini, J. L. Charest, M. Moretti, and R. D. Kamm, Correction for Jeon et al., Human 3D vascularized organotypic microfluidic assays to study breast cancer cell extravasation, *Proc. Natl. Acad. Sci. U. S. A.*, vol. 112, no. 7, p. E818, 2015.

94. N. Reymond, B. B. D'Água, and A. J. Ridley, Crossing the endothelial barrier during metastasis, *Nat. Rev. Cancer*, vol. 13, no. 12, pp. 858–70, 2013.
95. S. K. W. Dertinger, D. T. Chiu, N. L. Jeon, and G. M. Whitesides, Generation of gradients having complex shapes using microfluidic networks, *Anal. Chem.*, vol. 73, no. 6, pp. 1240–6, 2001.
96. W. J. Polacheck, A. E. German, A. Mammoto, D. E. Ingber, and R. D. Kamm, Mechanotransduction of fluid stresses governs 3D cell migration, *Proc. Natl. Acad. Sci. U. S. A.*, vol. 111, no. 7, pp. 2447–52, 2014.
97. M. Mak, C. A. Reinhart-King, and D. Erickson, Microfabricated physical spatial gradients for investigating cell migration and invasion dynamics, *PLoS One*, vol. 6, no. 6, p. e20825, 2011.

15 Breast Cancer-on-a-Chip

Pierre-Alexandre Vidi and Sophie A. Lelièvre

CONTENTS

15.1 INTRODUCTION

"Organs-on-chips" represent the most advanced implementation of three-dimensional (3D) cell culture for research and development purposes. Three decades of fruitful work by biologists to establish the extracellular conditions mimicking physiologically relevant tissue phenotypes are merging with state-of-the-art technologies that provide scaffolding and other microenvironmental cues essential for the cells to optimally express their identity. This marriage between biology and engineering should last; indeed, there are many aspects of organs and their diseases to investigate and potentially treat, as is the case in the breast. On-a-chip culture permits the mimicry of functional and structural portions of organs or systems on devices amenable for investigation in the laboratory. For the breast it means reproducing portions of the ductal tree in the normal situation as well as environments in which diseases, notably cancers, develop and progress (Vidi, Leary, and Lelièvre 2013).

Undoubtedly, normal and cancerous phenotypes of the mammary gland have been among the precursor models to establish 3D cell culture methods. This is likely due to the enormous impact of breast cancer in the world as well as a strong will to decrease the burden that this disease has bestowed on many populations. In the breast cancer field it has been recognized early on that cancer cannot be understood unless we also decipher the homeostasis of the normal tissue (Bissell 1981). Moreover, to this date preventing tumors from developing is one of the highest challenges in cancer research; as such, "primary prevention" is the only way to reduce a globally rising incidence of breast cancer (Lelièvre and Weaver 2013). Therefore, recapitulating normal phenotypes and notably, the polarity axis, a fundamental feature of epithelia, was emphasized early on (Barcellos-Hoff et al. 1989; Plachot and Lelièvre 2004). The mammary gland is made of a monolayer of luminal epithelial cells delineated by a layer of myoepithelial cells lined against a specialized type of extracellular matrix (ECM), the basement membrane. This epithelium is arranged into ducts of decreasing diameter as they branch out to ultimately lead to terminal ductal glandular units containing multiple acini where the milk is secreted (Figure 15.1a). Milk flows through the ductal tree to reach the opening to the external environment at the nipple. The most frequent breast disease is cancer with tumors that usually grow within the lumen of the duct before cutting through the basement membrane and expanding inside the stroma as cells become invasive (Hodges et al. 2014). Lymph vessels and blood vessels are plentiful within the breast parenchyma and provide two modes of dissemination of breast cancer cells to colonize other tissues and organs.

While branching of the ductal system could be reproduced in the murine model upon treatment with epimorphine (Hirai et al. 1998), the classic models for the human mammary gland mimic the terminal, spheroidal structures of the breast glandular system referred to as acini, with one layer of polarized luminal epithelial cells surrounding a lumen (Petersen et al. 1992; Plachot et al. 2009). In more complete epithelial models, myoepithelial cells lining externally the luminal epithelium are included (Gudjonsson et al. 2002).

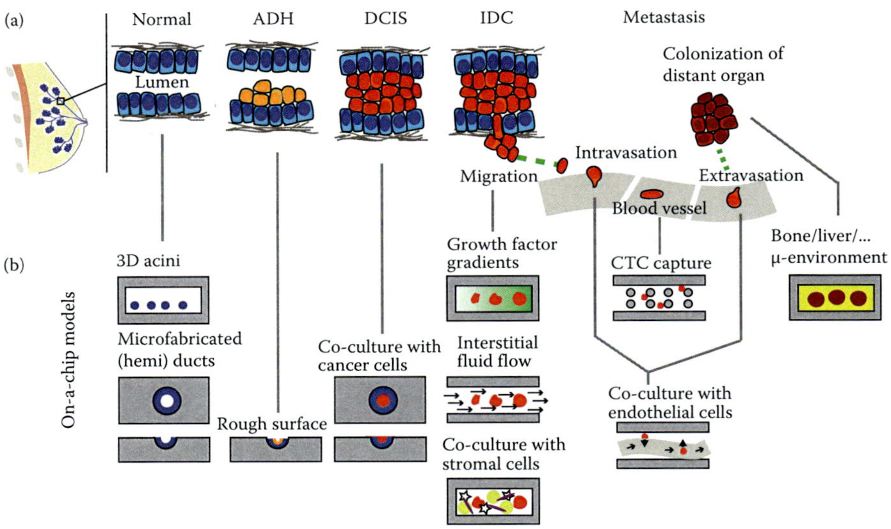

FIGURE 15.1 The different stages of breast cancer progression (a) and existing breast-cancer-on-a-chip models, described in the main text (b). Normal epithelial cells are shown in blue, cancer cells from the primary tumor in red, and metastasized cells in brown. ADH, atypical ductal hyperplasia; CTC, circulating tumor cells; DCIS, ductal carcinoma *in situ*; IDC, invasive ductal carcinoma.

The different steps in the progression of breast cancer have been fairly well described; it is important to reproduce these steps in order to study and prevent advancement of the disease. Although there is an established progression from hyperplasia to carcinoma *in situ* and invasive carcinoma, these steps are not obligatory passages; indeed, the current concept in breast oncology is that some of these steps might be bypassed, and that the multistep progression pathway depends on the molecular subtype of cancer (Lopez-Garcia et al. 2010; Bombonati and Sgroi 2011). The noncancerous stage of atypical hyperplasia has been associated with an increased risk for breast cancer development (Hartmann et al. 2014; Buckley et al. 2015); it is characterized by a strong heterogeneity in the appearance of nuclei and often associated with cell multilayering; the loss of a monolayered luminal epithelium is linked with an alteration of the polarity axis (Bradbury, Arno, and Edwards 1993; Lesko et al. 2015). The next stage is the cancerous preinvasive phenotype of carcinoma *in situ* in which cells proliferate within the ductal lumen while maintaining a certain degree of basal polarity, with an almost-continuous basement membrane surrounding the tumor. The final stage of local breast tumors is invasive carcinoma in which aggressive behavior is characterized by an extensive degradation of the basement membrane associated with loss of basal polarity and the formation of multicellular expansions that invade the environment surrounding the tumor (Bombonati and Sgroi 2011).

Mimicking the different stages of cancer progression with standard 3D culture has been successful to a certain extent. A progression model from partially differentiated MCF10A non-neoplastic cells to ductal carcinoma *in situ* (DCIS) was first reported with cells cultured in the presence of ECM enriched in basement membrane components from Engelbreth-Holm-Swarm (EHS) murine sarcoma tumors in a drip 3D system. Progression resulted from the induction of ErbB2 receptor dimerization in MCF10A cells, hence triggering the sustained epidermal growth factor receptor (EGFR) signaling observed in certain types of breast cancers (Muthuswamy et al. 2001). Another progression model with MCF10A cells reproduces very early stages of tumor formation, like hyperplasia (with MCF10A-NeoT cells) and atypical hyperplasia (with MCF10A-AT1 cells) in drip 3D culture; in this case stable NeoT cell lines were generated via transfection with *ras* oncogene derived from the human T24 bladder carcinoma cell line (Li et al. 2008). Progression from DCIS to invasive ductal carcinoma (IDC) is recapitulated by the neoplastic cell lines S2 and T4-2 that were derived from HMT-3522 non-neoplastic S1 cells by adapting them to a defined medium lacking epidermal growth factor (EGF) (Briand and Lykkesfeldt 2001; Briand, Petersen, and Van Deurs 1987). When placed in EHS-based 3D culture, the S2 cell line reproduces DCIS-like tumor nodules of different sizes. Upon analysis it was concluded that small, intermediate size, and large nodules represent different degrees of aggressiveness, with the largest nodules being possibly precursors to IDC mimicked by the T4-2 cells (Rizki et al. 2008). Hence the HMT-3522 S2 cell line is an interesting model that recapitulates some of the different tumor grades identified by pathologists for DCIS of the breast. The invasive stage of local breast tumors has been reproduced in 3D culture with different cell lines often with the use of collagen I as the matrix for embedding. An important characteristic of these 3D cell cultures is their potential to reproduce the level of aggressiveness of the cells, measured notably via the proliferation index and invasive protrusions (Koch et al. 2012). These relatively simple models illustrate that 3D cultures performed with a single cell type can approximate the normal phenotype of the breast epithelium and the different neoplastic stages, depending on the selected cell lines and an appropriate environment. However, there are specific cell behaviors or specific research questions that require the combination of different cell types in 3D culture.

Co-culture in standard 3D cell culture has been explored, notably for advanced stages of tumors where interactions between cancer and stromal cells influence aggressiveness and possibly metastasis (Cheung and Ewald 2014). Co-culture methods are being investigated in particular, to reproduce the metastatic stage of the disease with models of invasion within lymph vessels (Pisano et al. 2015) and blood vessels (Nangia-Makker et al. 2010). Co-culture models are also required to understand homing mechanisms of cancer cells, as shown for bone metastasis (Subia et al. 2015), and metastatic seeding behavior, as illustrated with liver and brain metastasis (Yates et al. 2007; Choi et al. 2014).

The 3D cell culture models have been instrumental for the identification of pathways driving breast cancer development (Vidi, Bissell, and Lelièvre 2013), such as the coupling between the EGFR and the β1-integrin pathways (Weaver et al. 1997) and the role of E-cadherin in epithelial cell migration (Shamir and Ewald 2015). They have increasingly convinced the scientific community that anticancer drug assessment in classical 2D culture is obsolete (Vidi et al. 2012; Howes et al. 2014; Bray et al. 2015; Imamura et al. 2015; Lovitt, Shelper, and Avery 2015). However, standard 3D culture is not sufficient to readily serve research on breast glandular development and disease; this is mainly due to the difficulty of mimicking the complex level of multitissue interactions and attaining sufficient reproducibility of the conditions sought. On-a-chip models are emerging as obvious alternatives to standard 3D culture. Notably, on-a-chip designs have been used to improve the control of the cellular environment, not only for assays requiring homogeneity of multicellular structures (Dolega et al. 2015), but also to more closely reproduce the anatomy of the organ in a controlled manner. Indeed, the geometry of the support on which tissues are grown appears to influence cell behavior and the 3D organization of multicellular structures, which is of high importance for organs containing curved structures, like channels or ducts, as their main feature (Vidi et al. 2014; Hribar et al. 2015).

The discovery of molecular subtypes of breast cancer that respond differently to treatments and correspond to distinct prognoses emphasizes the importance of using cell lines related to specific categories, in addition to specific stages, of breast cancer. These cell lines have been categorized into known molecular subtypes of breast cancer, including luminal A, luminal B, HER2, triple negative, or basal-like (Neve et al. 2006). Thus, models used in standard and organ-on-a-chip 3D cultures can be validated based on how faithfully they recapitulate the characteristics of each subtype of breast cancer. Models of triple-negative breast cancers are among the most studied due to the aggressiveness of the disease. They are characterized by the absence of expression of estrogen receptor, progesterone receptor, and HER2/NEU receptor. Furthermore, to become a model for basal-like cancer, triple negative tumor nodules should also be positive for EGFR and CK5/6 (Hodges et al. 2014). Studies with 3D cultures of triple negative breast cancer cells demonstrated the potentials of JAK/STAT pathways and inhibitors of IκB kinase (IKK)-related kinase IKBKE expression for the treatment of these cancers (Barbie et al. 2014). Cell lines corresponding to another aggressive form of neoplasia—the inflammatory breast cancer—are also available and have been used for instance to study the formation of cancer cell emboli (Lehman et al. 2013) that permit rapid metastasis progression via lymph vessels.

The identification of heterogeneity within individual tumors has shed light on a major reason for resistance to treatment. In particular, intratumor heterogeneity might explain why cancers respond differently to given therapeutic regimens from one individual to another, although they were given for the same pathological classification. Such heterogeneity occurs at the individual cell level due to a plethora of genetic and epigenetic features that determine cell behaviors as we have detailed elsewhere (Lelièvre 2014). Recapitulating heterogeneity at the cellular level is essential for research on the mechanisms of cancer onset and progression. For instance, the current concept is that the difference in proportion of stem cells between DCIS and IDC of triple negative nature participates in the progression to invasiveness (Shah et al. 2013; Uchoa Dde et al. 2014). The presence of cancer cells in quiescence as well as stem cells within tumors dramatically affects sensitivity to drugs (Oliveras-Ferraros et al. 2012). Moreover, the extracellular milieu modulates cell behavior, and thus the level of aggressiveness and the sensitivity to drugs. Indeed, basal polarity remnant in tumor nodules controls the sensitivity of cancer cells to chemotherapy and radiotherapy (Weaver et al. 2002) and modulates DNA repair efficacy (Vidi et al. 2012). The evolving levels of hypoxia in tumors, controlled notably by the stroma, also influence the response to treatment (Schmaltz et al. 1998; Strese et al. 2013).

Overall, unraveling the complexity of the organization of breast tumors and identifying the anatomical regions where cancers initiate and progress have made the scientific community aware that sophisticated 3D culture models making use of engineered scaffolds and fluidics are warranted. These models are particularly useful in research aiming at identifying molecular pathways that control specific cell phenotypes and behaviors, and at developing and testing compounds with therapeutic potential. Yet specific criteria need to be taken into account in order to use organ-on-a-chip systems rather than standard 3D culture.

In the first part of this chapter we present on-a-chip models that recapitulate normal and neoplastic breast tissues along with the justification for favoring such models compared to standard 3D culture. The second part of this chapter explores a central application of on-a-chip models for drug screening in which we discuss the advantages of these models compared to other 3D culture systems as well as their limits. Finally, we assess the possibility of breast-on-a-chip systems to serve the purpose of precision medicine.

15.2 ON-A-CHIP MODELS OF THE NORMAL BREAST AND OF MAMMARY TUMORS

The mammary gland is an organ that alternates phases of proliferation, differentiation, and involution during reproductive and menstrual cycles (Ferguson et al. 1992). Therefore, it is an ideal model to study basic cellular mechanisms of tissue differentiation and homeostasis. Understanding these mechanisms is key to the development of novel cancer prevention approaches. The growing emphasis on breast cancer prevention indeed stimulates the development of models of the normal breast epithelium to study risk factors and their mitigation (Vidi, Leary, and Lelièvre 2013). In addition, physiologically relevant models of the diseased breast epithelium are needed to identify and understand drivers of cancer initiation and progression, to develop therapeutic approaches, and to test new drugs or new drug combinations. On-a-chip models have been engineered that reproduce all major stages of breast cancer initiation and progression (Figure 15.1b). We expect that this emerging technology will drive important advances in basic and translational breast cancer research.

15.2.1 The Normal Breast Epithelium

One approach to reproduce the normal breast epithelium is to refine established 3D cell culture of mammary epithelial cells based on hydrogels by integrating these models in microfluidic devices. Perfusion systems in microchambers enable long-term cultures and time-lapse imaging under tightly controlled levels of nutrients, gas, and temperature (Meyvantsson et al. 2008; Chen, Hung, and Lee 2011). In these systems, perfusion can be driven by differential surface tension between growth medium inlets and outlets, which circumvents the need for pumps and complex tubing equipment. These microchamber devices have proven adequate for acinar differentiation, as shown by MCF10A and HMT-3522 S1 acini formation (Meyvantsson et al. 2008), and hold promise for future developments, notably for high-throughput screening of chemopreventive agents.

Microfabricated devices mimicking mammary ducts represent a complementary approach to perfusion systems containing 3D hydrogel cultures. These devices shift the classic 3D culture paradigm in that epithelial cells expand on engineered scaffolds, which creates closed (Bischel et al. 2013; Bischel, Beebe, and Sung 2015) or open (Grafton et al. 2011; Vidi et al. 2014) ductal microenvironments. Closed ductal systems are achieved using a viscous finger patterning method (Bischel, Lee, and Beebe 2012), whereas hemichannels can be produced in polydimethylsiloxane [PDMS] using soft lithography (Grafton et al. 2011) or on acrylic substrates with laser micromachining (Vidi et al. 2014). In both closed and open configurations, coating the plastic substrates with basement membrane proteins stimulates epithelial differentiation as evidenced by cell polarization. While optimizing hemichannel systems, we noticed that the shape and roughness of the substrate strongly influenced cell behavior: non-neoplastic mammary epithelial cells formed multilayers in the angles of squared-section channels and at the bottom of v-shaped channels. Similarly, these cells piled up if the acrylic surfaces generated by laser micromachining were rough (Vidi et al. 2014). Although multiple layers of non-neoplastic cells may prove useful to model hyperplasia, most applications for on-a-chip models of mammary ducts require a single layer of epithelial cells. Smoothing the growth substrate by spin-coating polymethyl methacrylate [PMMA] solves the issue of cell multilayering in hemichannels (Vidi et al. 2014). In the closed-channel conformation, the smooth, spherical cross-section of ECM hydrogel is achieved by the flow of culture medium prior to polymerization of the ECM hydrogel (viscous finger patterning; Bischel, Lee, and Beebe 2012).

15.2.2 Ductal Carcinoma *In Situ*

A key element of the ductal models described earlier is that they recapitulate morphogenesis of a lumen or a lumen microenvironment in which cancer cells can be seeded to mimic *in situ* breast cancer. Bischel and colleagues (Bischel, Beebe, and Sung 2015) used a cell line model of DCIS (MCF10A-DCIS.com; Miller et al. 2000) to produce tumors lined with non-neoplastic (MCF10A) cells in the closed channel system. Their study showed that human mammary fibroblasts co-cultured in parallel channels promote invasiveness, consistent with their previous work with compartmentalized co-cultures (Sung et al. 2011). This on-a-chip model of DCIS is appropriate to study the transition from *in situ* to invasive carcinoma because, in contrast to the classic Boyden chambers or EHS-based invasion assays, it recapitulates physiological cell-cell interactions between DCIS cells and the epithelial layer.

Our group developed a model with an open "hemichannel" conformation to study the influence of the mammary duct geometry on tumor cell morphology and drug responses. The results revealed more circular nuclei in tumor nodules developing in the ducts compared to their counterparts on flat surfaces (Vidi et al. 2014). Nuclear shape is an important component of nuclear grading used by pathologists to estimate tumor aggressiveness and classify cancers; the fact that the geometry of the tumor microenvironment produced on-a-chip influences this key morphologic parameter underscores the importance of the design of the chip.

15.2.3 Invasive Ductal Carcinoma

The concept of surface tension-driven perfusion of microchannels developed by the Beebe laboratory was used to study the transition from DCIS to IDC, specifically the effect of stromal cells on this transition (Sung et al. 2011). The critical role played by tumor-associated fibroblasts in cancer progression is well recognized, thanks notably to 3D co-culture systems (Cirri and Chiarugi 2011; Paulsson and Micke 2014). Sung and colleagues used a compartmentalized microchannel design to co-culture side-by-side MCF10A-DCIS.com cells and fibroblasts in 3D matrix, hence enabling the study of distance-dependent interactions between stromal and neoplastic epithelial cells (Sung et al. 2011). The results indicate that, in addition to a paracrine effect, either direct contact between epithelial cells and fibroblasts or localized matrix remodeling by the fibroblasts is a necessary step for transition to an invasive phenotype.

A key cellular feature enabling the progression from DCIS to invasive carcinoma is cancer cell migration; microfluidic systems have been developed to study this phenomenon in physiologically relevant microenvironments (Polacheck, Zervantonakis, and Kamm 2013). A multitude of intrinsic and extrinsic factors including epithelial-to-mesenchymal transition signaling programs, autocrine and paracrine signals, and physical properties of ECM fibers interdependently influence migration. Animal models are often too complex to tease apart the individual contributions of these factors, whereas classic assays such as wound healing and Boyden chambers are too reductionist. On-a-chip systems are now starting to fill this complexity gap. Also, cell migration is a highly stochastic phenomenon and lab-on-a-chip systems enabling time-lapse imaging of single live cells can capture speed, directionality, and persistence of cell movements. The Kamm and Chung laboratories have developed highly versatile on-a-chip platforms to study, among other things, cell migration under tightly controlled conditions (Chung et al. 2009; Shin et al. 2012). The microfluidic designs consist of a central cell culture channel connected to flanking ECM-filled channels via micropoles. Peripheral channels, also in direct contact with the ECM, can be used to introduce well-defined stimuli—notably gradients of growth factors. Peripheral channels may also be used for co-cultures with endothelial or stromal cells. Using a similar microfluidic device, Liu and colleagues showed matrix metalloproteinase (MMP)-dependent migration of MCF-7 cells toward a gradient of EGF (Liu et al. 2009). Migration of invasive mammary adenocarcinoma (MTLn3) cells in collagen gels toward an EGF gradient was also measured using a microfluidic device with source and sink reservoirs (Abhyankar et al. 2008). In the complex tumor microenvironment, cancer cells integrate multiple stimuli that cooperatively modulate cell behaviors. Hence, it is important to develop *in vitro* systems capable of generating complex gradients with multiple chemokines. Microfluidic models have already been engineered that combine one chemokine gradient with different basal levels of a second chemokine (e.g., EGF and the stromal-derived growth factor 1 [SDF-1/CXCL12] also implicated in cancer cell invasion and metastasis (Guo et al. 2016; Mosadegh et al. 2008; Kim et al. 2013). Both studies reported cooperativity between the two chemokines, but with apparently conflicting results: Mosadegh and colleagues showed that EGF promotes a directional response of MDA-MB-231 cells to SDF-1 gradients (Mosadegh et al. 2008), whereas in the second study, EGF prevented SDF-1 chemoattraction (Kim et al. 2013). This discrepancy highlights that differences in the amplitude and shape of the chemokine gradient and, more importantly, in the cell culture conditions (2D vs. 3D for the two studies discussed earlier) strongly affect the cellular integration of external stimuli. At the moment these types of on-a-chip system are limited by the lack of biological knowledge; indeed, determining physiologically relevant chemokine gradients is a challenge given the highly heterogeneous microenvironment of tumors and the technological gap to measure with precision local chemokine levels *in vivo*.

It is well established that chemical stimuli crosstalk with mechanical cues from the microenvironment (Bissell et al. 2002; Pickup, Mouw, and Weaver 2014). Matrices of defined stiffness spanning a range corresponding to soft breast tissues to stiff breast tumors can be engineered by crosslinking collagen fibers or by using acrylamide scaffolding (Wang and Pelham 1998; Levental et al. 2009). We foresee that incorporating stiffness gradients to existing and to new 3D culture microfluidic devices will facilitate studies aimed at understanding the mechanisms by which increased matrix stiffness leads to higher breast cancer risk and stimulates cancer progression (Butcher, Alliston, and Weaver 2009). In addition to mechanical properties dictated by ECM scaffolding molecules, interstitial fluid flow greatly influences the pressure and shear forces exerted on normal and malignant tissues. Interstitial flow establishes chemokine gradients promoting cancer cell migration toward lymph vessels, a common path for metastasis from breast cancer (Shields et al. 2007). Microfluidic systems are particularly well suited to expose cancer cells to well-defined flow rates, and studies using these systems are providing insight on how interstitial flow influences cancer cell migration. As intuitively expected, interstitial flow increases overall migration speed and directionality: a subpopulation of MDA-MB-231 cells migrated with high directionality toward the flow; this process was dependent on chemotaxis via the beta chemokine receptor CCR7 (Polacheck, Charest, and Kamm 2011; Haessler et al. 2012), confirming previous studies with 3D cell culture (Shields et al. 2007). Interestingly, a distinct population of cells migrated upstream (i.e., against the flow) with less directionality but higher velocity (Polacheck, Charest, and Kamm 2011; Haessler et al. 2012). This effect was mediated by the integration of the tension generated by the flow via β1-integrin/vinculin/focal adhesion kinase (FAK) signaling (Polacheck et al. 2014). These findings suggest that a subtle balance between rheo- and chemotaxis may define the direction of cell migration. In addition to the relatively well-characterized effects of interstitial flow on tumor cells, the biomechanical stimuli produced by lymph flow across and within lymph capillaries may also influence cancer cell migration. Experiments using a flow chamber system integrating tunable transmural flow (medium flow through the ECM matrix, perpendicular and across lymphatic endothelial cells), as well as tunable luminal flow (flow of medium along a lymphatic endothelial cell monolayer) show that both mechanical stimuli increase MDA-MB-231 cell migration across lymphatic endothelial cells. This model system appears to reproduce lymph vessel intravasation occurring *in vivo*. Similarly to interstitial flow, the effect of transmural and luminal flow rates on cancer cell migration is also in part mediated by CCR7 and its ligand CCL21 (Pisano et al. 2015).

15.2.4 Breast Cancer Metastasis

Breast cancer metastasizes predominantly to the bones, brain, lungs, and liver. Rather than the primary tumors, it is the irreversible damage to these essential organs induced by metastatic lesions that often leads to high morbidity of the disease. Whereas the journey for metastatic cells usually starts in the lymphatic system, long-distance dissemination occurs via the bloodstream. A higher frequency of circulating tumor cells (CTC) has been reported for advanced (stage III) cancers and negatively correlates with survival (Cristofanilli et al. 2004). Hence, detecting and isolating CTC is clinically relevant but challenging due to the scarcity of these cells. Microfluidic devices have been developed for the affinity capture of tumor cells from blood samples (Esmaeilsabzali et al. 2013; Hyun and Jung 2013). One important limitation of these approaches is the achievable flow rate; progress in the design of the chips is resolving this bottleneck. For example, the OncoBean chip (Murlidhar et al. 2014) can operate at a flow rate of 10 ml/h (about 10 times higher than predecessor systems without blood preprocessing). It relies on a radial flow design across microspots coated with antibodies against the epithelial cell adhesion molecule (EpCAM) for affinity capture. An alternative approach for the detection and isolation of CTC has been implemented in which CTC are immobilized on a chip with microperforations using magnetic beads functionalized with antibodies against CTC antigens (Chang et al. 2015).

Understanding the mechanisms by which cancer cells penetrate blood vessels (intravasation) and exit the bloodsteam (extravasation) may lead to new approaches to prevent metastasis. There is a strong momentum and rapid progress in the field of regenerative medicine to recapitulate blood vessels *in vitro*, and metastasis research will likely benefit greatly from it. A microfluidic system has been developed by Zervantonakis and colleagues that recreates the interface between invasive tumor cells and the vasculature (Zervantonakis et al. 2012). Experiments with this system showed that the presence of macrophages leads to decreased endothelial barrier function, as measured using fluorescent dextran permeability, and increased cancer cell intravasation. These data indicate that intravasation predominantly occurs at cell-cell junctions rather than transcellularly (i.e., across endothelial cell bodies). Blocking tumor necrosis factor alpha (TNFα) partially restored endothelial barrier function and reduced intravasation, highlighting the prominent role of inflammatory cytokines from the tumor microenvironment in cancer cell dissemination. Consistent with these observations, TNFα disrupted junctional cell-cell complexes and increased cancer cell intravasation in a microfluidic system containing self-assembled microvessels in a microchannel (Lee et al. 2014).

Microfluidic systems of the vasculature are also important to study breast cancer cell adhesion to endothelial walls and extravasation. Such a model with tunable flow rates was used to mimic circulating cancer cells and to follow their behavior in the presence or absence of defined levels of a chemokine, CXCL12, situated basally from the endothelium (Song et al. 2009). Metastatic breast cancer cells (MDA-MB-231) were found to adhere preferentially to regions of the endothelium stimulated by SDF-1/CXCL12, which matches the preferential homing *in vivo* to organs with high levels of this chemokine. The results also suggest that expression of the SDF-1 receptor on the endothelium plays an important role in cancer cell adhesion. Bone-specific extravasation of circulating breast cancer cells was further studied using perfusable microfluidic models of the vasculature incorporating osteoblasts (Bersini et al. 2014; Jeon et al. 2015). Increased extravasation of MDA-MB-231 cells from endothelia surrounded by a bone-like microenvironment was dependent on the cytokine CXCL5 and its receptor CXCR2 expressed in breast cancer cells (Bersini et al. 2014).

The next generation of microfluidic devices developed by the Kamm laboratory integrates endothelial cells within different microenvironments, including a bone-mimicking microenvironment recapitulated by a meshwork of microvasculature surrounded by differentiated osteoblasts and mesenchymal stem cells (Jeon et al. 2015). In this model, extravasation of the metastatic breast cancer cells was more pronounced in the bone-mimicking microenvironment compared to acellular matrices or to matrices containing myoblasts mimicking the muscle microenvironment; these differences did not correlate with the variation in endothelial cell permeability across the different matrices. Rather, expression of adenosine by myoblasts and adenosine signaling via the adenosine A_3A receptor in cancer cells reduced extravasation in the muscle-like microenvironment—an effect reproduced by treating the bone-like microenvironment with adenosine. Importantly, conditioning the microvasculature with shear force generated by medium flow increased endothelial permeability and accordingly decreased cancer cell extravasation, highlighting the influential contribution of mechanical stimuli to endothelial and cancer cell physiology.

In addition to the bones, the liver and brain are important metastatic sites for breast cancer cells. Efforts in regenerative medicine have yielded precise organoid models of the liver (Peloso et al. 2015; Clark et al. 2014), and 3D hepatic cell culture models have been combined with metastatic breast cancer cells, revealing phenotypic changes of these cells toward more epithelial characteristics and increased chemoresistance (Clark et al. 2014). On-a-chip neurovascular platforms developed for the study of neurodegenerative disorders (e.g., Achyuta et al. 2013) may also be adapted as models of brain metastasis.

In conclusion, breast- and breast cancer-on-a-chip models recapitulating all steps of cancer initiation and progression—from increased risk to cancer cell homing in new microenvironments—advance breast cancer research. These devices offer numerous advantages, including precise control and rapid equilibration of growth factor concentration and gradients, mechanical constraints of the matrices, flow of fluids, and even gas concentrations. Among the multitude of future developments for applications in basic research, integrative on-a-chip models recapitulating oxygen gradients will be particularly relevant to study how hypoxia induced by the tumor microenvironment promotes genomic instability and changes leading to metastatic progression (Gilkes, Semenza, and Wirtz 2014). Systems enabling spatiotemporal control of oxygen tension are being developed (Funamoto et al. 2012; Acosta et al. 2014). We also envisage future microfluidic models integrating tunable pH gradients to study the impact of acidosis on tumor progression, as well as new organ-specific models to study homing in the brain, liver, and lungs, and finally diversification of breast cancer cell lines used in these systems to derive knowledge for specific breast cancer subtypes. The compatibility of most on-a-chip designs with high-resolution, time-lapse imaging is extremely advantageous as it enables experiments with high temporal resolution, and hence, to tease apart stochastic from directed cell behaviors.

15.3 IMPROVED MODELS FOR DRUG DISCOVERY AND SCREENING

The development of small molecules that can exquisitely target specific regulatory pathways in cells requires *in vitro* models that faithfully reproduce the different subtypes of breast cancers. Indeed, the classification of breast cancers corresponds to distinct molecular pathways that provide tumor strength for progression and the potential to resist standard chemotherapy targeting general cell proliferation and survival mechanisms. Models based on traditional 3D cell culture are being investigated to recapitulate the characteristics of specific cancer subtypes. Some of these models include interactions with other cell types that might modulate drug sensitivity. For instance, a model of ER+ breast cancer cells (MCF7) and mammary fibroblasts (Hs 578Bst) has been designed to study the impact of drugs that modulate steroidogenesis in the context of paracrine interactions among these cell types (Wang et al. 2015). A combination of endothelial cells and triple negative breast cancer cells was used to explore therapies that inhibit angiogenesis and the cytokine-driven growth of immune-activated subsets of triple negative breast cancers (Barbie et al. 2014). However, the identification of new drug candidates requires screening of libraries with thousands of chemical and natural compounds, which also necessitates the production of tumor models that are high-throughput, with the possibility to harness the needed phenotype in many replicates and with as much automation as possible. The difficult task is to find a compromise between sophisticated on-a-chip systems necessary to recapitulate tumor organization and highly reproducible culture models that are easy to handle.

15.3.1 Models Used for Drug Screening in Breast Cancer

For high-throughput screening of anticancer drugs, 3D cell culture models that permit fast formation of tumor nodules are in vogue. Most of the models developed involve aggregation of cells either in a hanging drop (Kelm et al. 2003), a nonadhesive hydrogel (Chen, Gupta, and Cheung 2010), or a hydrogel millibeads (Pradhan, Chaudhury, and Lipke 2014) system. Yet inclusion of these systems into platforms such as 96- or 384-well plates provides high-throughput assays that may not necessitate using on-a-chip characteristics (Charoen et al. 2014). In nonadherent culture systems, cells are usually seeded at very high concentrations in order to rapidly produce spheroids of several hundred microns in diameter. Here, there is no developmental process in which cells progressively organize their tumor environment (e.g., particular interactions among cells, deposition of specific ECM molecules, transient areas of hypoxia). It has been claimed that tumors formed in this manner within two to three days display the characteristic organization of a tumor nodule *in vivo*, with metabolically active and proliferating cells mostly at the periphery of the spheroid and necrosis at the center (Pradhan, Chaudhury, and Lipke 2014; Charoen et al. 2014). More recent methods suggest embedding spheroids, first formed without exogenous ECM contact, in collagen I for instance, in order to mimic the mechanical properties of the tumor microenvironment (Charoen et al. 2014). Nevertheless, this onion-peel view of tumor organization is somewhat reductionist as tumors have a much more complex intranodular structure than a central hypoxic and necrotic core surrounded by dividing cells. It is understandable that cells forced to form large nodules via aggregation will indeed have increased tendency to divide at the periphery where nutrients are easily accessed, and will die by necrosis in the center where nutrients and oxygen cannot accumulate in high enough concentration. Moreover, the fact that a tumor looks like a spheroid under different culture conditions does not mean that the spheroids produced under these different conditions function similarly. An elegant study in which MCF7 tumor spheroids formed using hanging drops or in PDMS microwells, two methods that prevent cellular adhesion to a substratum to favor the aggregation of cells, revealed significant differences in respiratory activity, rate of senescence, and glucose metabolism (Zhou et al. 2013). All of these parameters influence the response to drugs. Notably, rates of proliferation and metabolism were higher in spheroids formed in PDMS structures compared to the hanging drop method. Gene expression analysis identified markers specific for each population (hanging drop–formed or PDMS-formed) of spheroids. Yet, results suggested that a small proportion of the spheroids formed with the hanging drop method had a specific set of markers similar to those corresponding to spheroids formed via the PDMS method, which Zhou and colleagues interpreted as a possible sign of instability of the spheroids produced by the hanging drop method.

Another system for fast formation of tumors is that of magnetic levitation (Jaganathan et al. 2014). This system permits the production of heterogeneous tumors in which interaction with the stromal environment, such as fibroblasts, can be reproduced, in particular to study the impact of the microenvironment on drug efficacy. This system relies on the rapid self-assembly of cells and allows the production of tumors in the millimeter range and the possibility to control the composition and, to a certain extent, the organization of tumors. The mimicry of the tumor microenvironment encompasses, depending on the type and ratio of fibroblasts and cancer cells, the formation of a fibrotic capsule and entrapment of fibroblasts within the tumor as well as deposition of ECM molecules such as fibronectin, collagen, and laminin. Findings revealed that the mixture of fibroblasts and cancer cells affected the response to doxorubicin compared to cancer cells alone, involving an impact on both tumor area and cell density as seen *in vivo*. Although appealing, drug testing performed on models based on the formation of cell aggregates described in the precedent paragraphs is likely to be affected by the lack of control over the metabolism and over the origin of heterogeneity within tumor nodules. For hydrogel-based and PDMS-based systems, drug testing might be influenced by the lack of control of drug access to the cells as well (indeed drug molecules might be trapped in these substrata) (Chen, Gupta, and Cheung 2010).

On-a-chip systems are required to produce multicellular interactions in a controlled manner by influencing locally the growing tumors, especially via the design of the culture device that influences the spatial organization of cells and the delivery of environmental factors. Interactions among stem and cancer cells or fibroblasts and cancer cells exist in different proportions depending on the subtype of breast cancer; this situation can be easily reproduced by mixing cells or culturing cells in separate compartments allowing paracrine interaction in 3D culture. However, the interaction involving contact of tumors with other tissue structures such as vessels made of endothelial cells, lymph endothelial cells, or non-neoplastic epithelial cells organized into ducts for instance, requires patterning. A star-shaped polyethylene glycol (starPEG)-heparin based hydrogel was used to reproduce breast tumor vascularization (Bray et al. 2015). In this matrix RGD motifs provide cell attachment, and the sensitivity to MMPs permits the modification of mechanical properties within a culture encompassing breast cancer cells, endothelial cells, and mesenchymal stromal cells. Specific molecules such as angiogenic factors can also be added to the gel. In the study by Bray and colleagues, the use of drugs that target cancer cells did not reveal a significant difference in sensitivity depending on the presence or absence of vessel mimicry. Instead, based on the preliminary data, this model is likely to bring information regarding the use of inhibitors of angiogenesis. Co-culture of breast cancer cells and primary adipose-derived stromal cells has been used in a microfluidics model developed to assess the efficacy of photodynamic therapy (PDT) with therapeutic agents such as photosensitizers in an approach based on gold nanoparticles (Yang et al. 2015). The system enabled modulation of the microenvironment, with appropriate depth of tissue for PDT assessment, and creation of conditions with relevant circulation flow, like interstitial diffusion for mass transport, drug delivery in a perfusion-like manner, and nutrient supply and removal of metabolic waste. The model allowed time for the tissue environment to form (up to two weeks). Once completed, the chip included a succession of culture chambers that could be irradiated by light simultaneously with live imaging. In this particular study, the model was useful to show that gold nanoparticles improved the use of the photosensitizer studied by allowing PDT to induce cell death throughout the depth of the tumor.

The concept of 3D cell culture integrating breast tumors within the non-neoplastic epithelium of hemichannels that we presented briefly in Section 15.2.2 is based on the assumption that the curved geometry of the breast ducts in which tumors originally grow might have an impact on cell behavior. Indeed, the idea of continuous structure from the ECM to the cell nucleoskeleton (Lelièvre, Weaver, and Bissell 1996; Lelièvre et al. 1998) underlies the possibility for extracellular constraints to influence gene expression, and thus potentially the response to drugs. We have compared triple negative tumor nodules formed by proliferation of breast cancer cells over five days, either on a flat support or within a curved geometry, for their sensitivity to doxorubicin and bleomycin. In both cases tumor cells showed a different response depending on the geometrical context in which they thrived (Vidi et al. 2014). Thus, in addition to the influence of non-neoplastic cells on cancer cells' drug sensitivity already documented, a simple flat versus curved geometry for the growth of tumors also matters. Whether differences in drug sensitivity are linked to measured alterations in the circularity and area of the cell nucleus depending on the tissue geometry remains to be investigated (Vidi et al. 2014; and unpublished data).

Another aspect of on-a-chip systems is the possibility of studying drug sensitivity within specific organ or tissue niches, even those located far from the initial tumor site. A bone-breast cancer cell system has been developed using silk fibroin as a scaffold for the distribution model (Subia et al. 2015). This system permits the investigation of the impact of drugs on normal functions and features of the host tissue. In this study there was no recorded impact on bone mineralization, but osteogenic markers like osteocalcin and alkaline phosphatase were decreased. It is proposed that this type of model might ease the development of targeted drugs by taking into account the environment of the tumor cells. Obviously, there is more to producing effective drug screening systems than simple high-throughput tumor production.

The LiverChip model has been used to mimic micrometastases and test their sensitivity to drugs (Clark et al. 2014). In this system, breast cancer cell lines have been tested for their integration into the hepatic niche. Importantly, the MDA-MB-231 aggressive cells have shown a reversion to a more epithelioid phenotype associated with resistance to chemotherapy as observed in patients. Resistance was found to be conferred by the re-expression of E-cadherin. In contrast to the fast-assembling tumors described earlier, this promising system uses time as the essence for best mimicry of a physiologically relevant situation. Specifically, it encompasses continuous live assessment of liver function and the potential for studying drug efficacy based on circadian rhythm and the burst of mild inflammatory states.

15.3.2 Integration of Biosensors within on-a-Chip Models

Drug screening requires measuring either cell death (for cytotoxic drugs) or cell proliferation (for cytostatic drugs), or even sometimes the expression level of specific targets and response pathways (e.g., DNA repair activity, etc.). Although these measurements can be easily performed by immunostaining and observation under fluorescence microscopy, this method is time consuming and often relies on subjective assessment. Instead, automated imaging and the integration of biosensors for the analysis of specific readouts will facilitate high-throughput and continuous screening. Ideally, biosensors should work for live cell analysis and might encompass probes to measure extracellular events such as oxygen, ions, metabolic factors, and cytokines as well as intracellular events including specific signaling pathways, death, metabolic activity, stress, etc. An example of imaging-based analysis is the combination of nanoculture plates (without addition of ECM or gel) for the formation of tumor nodules and the imaging device BioStation CT that has been employed to assess drug sensitivity of breast cancer cells representing luminal A and luminal B subtypes in 96-well plates. A correlation was made between proliferation, measured via ATP levels, and spheroid velocity and fusion of spheroids both in cell lines and primary breast cancer cells when comparing control and anticancer drug-treated samples (Sakamoto et al. 2015). Whether nodule migration as measured here represents a physiologically relevant phenomenon remains to be confirmed.

A biosensor relying on fluorescence energy-transfer (FRET) has been designed to detect apoptosis in breast tumor nodules in culture with the capability of measurements in real time and over long culture periods (Anand et al. 2015). The GFP-caspase sensors are first transfected in the cell lines to be used for drug screening. Stable transfection enables the constant availability of sensors and permits the assessment of drug sensitivity in specific areas of the tumors (e.g., at the periphery vs. in the center of the nodule). These capabilities developed with standard 3D cultures are relatively easy to use with on-a-chip systems since sensors are part of the cell lines.

A relatively underexplored marker for biosensing is nuclear architecture, with features as easy to investigate as circularity, a shape descriptor. Using a simple DAPI staining, as mentioned earlier, we identified a correlation between nuclear circularity (i.e., how round the cell nucleus is) and drug sensitivity (Vidi et al. 2014). Previous work had revealed a link between p53 expression level and nuclear circularity (Mijovic et al. 2013), illustrating the importance of taking nuclear shape into account in drug sensitivity assays, and suggesting a mechanism by which nuclear shape might influence drug sensitivity. This type of marker is particularly meaningful in cancer since tumors are characterized by different degrees of heterogeneity in nuclear shape, and pathological assessment based on morphometric analyses of nuclei seems to improve the evaluation of tumor aggressiveness (Hannen et al. 1998; Okudela et al. 2010; Mijovic et al. 2013). We showed in the disease-on-a-chip system that there was less heterogeneity when tumors were grown in the physiologically relevant curved geometry of the duct compared to tumors grown on a flat support (Vidi et al. 2014). *In vivo*, each subtype of breast cancer appears to have its own average nuclear circularity (Lelièvre laboratory, unpublished results), which supports further efforts to investigate the potential for nuclear circularity to be used as a prognostic marker for drug sensitivity. Notably, "sensing" nuclear circularity would be particularly important with certain groups of drugs, the target of which would be influenced, like p53, by nuclear circularity.

Other biosensors relate to the cells' environment, as is the case for hypoxia, which influences cell behavior and potentially drug sensitivity (Vaupel 2008). Thus, it is necessary to monitor and possibly modify oxygenated regions in tumors, as hypoxia can be transient (Matsumoto et al. 2010), in order to improve the outcomes of drug screening.

It has been argued that microfluidics permits spatial and temporal control necessary to modulate the oxygenation within tumors (Grist et al. 2015). Notably, the short diffusion distance between channels in a microfluidic device leads to quick equilibrium and thus, permits the completion of several cycles of oxygenation for a particular tumor region per hour. In the three-layer microfluidic platform designed by Grist and colleagues, gas control channels are on three of the sides of the cell culture channel; there are also hydration channels perfused with liquid between the gas and cell culture channels that reduce evaporation in the cell culture channels. Ratiometric optical oxygen sensors, instead of intensity sensors, are integrated into the microfluidics device to provide a map of oxygen levels.

Other sensors used so far to measure behavioral features of cells might also be useful for drug screening within an on-a-chip platform. Electrical cell-substrate impedance sensing (ECIS) has been integrated in a microfluidics platform to measure, with high sensitivity, the migration of single breast cancer cells in 3D matrices (Nguyen et al. 2013). The device detects rapid changes of impedance as invasive MDA-MB-231 cells leave the electrode and migrate in the hydrogel toward chemoattractants. This approach might provide readouts based on phenotypical responses to measure the impact of antimetastatic drugs.

15.3.3 High-Throughput Screening with Organs-on-Chips

High-throughput drug screening encompasses several aspects. It requires multiplicity, with screening of many drugs and/or cell types at once. It also requires rapid and reliable assessment of results. Thus, the biosensors used for such assessment should be easily readable, with highly reproducible results, and placed within structures that allow testing of possibly hundreds of cell cultures on-a-chip. The cancer-on-a-chip might not reproduce every single aspect of the real situation, but rather the essential aspects to mimic and measure a specific tumor phenotype.

If targeting invasive capabilities is not the purpose for drug screening, a model such as the disease-on-a-chip described earlier, with tumors growing within the confines of a duct lined with non-neoplastic cells, might be sufficient (Vidi et al. 2014). This model can be produced quickly, adapted for multiwell plates, and used with automated cell culture and biological assessment. If invasiveness is a characteristic of interest in the drug screening process, a more sophisticated model would be necessary, such as that designed based on finger patterning of channels through ECM hydrogels, allowing breast ductal carcinoma cells to display invasion from inside the channel throughout the gel toward fibroblasts located in adjacent compartments (Bischel, Beebe, and Sung 2015). The high-throughput nature of this system is based on the fact that the channel patterning is created using a micropipette; the authors indicate that manufacturing could be done with automated pipetting and liquid handling systems. Another system that recapitulates invasive potential has been developed in a titer-plate format with patterning of adjacent lanes of gels and liquid without the need for physical separation, thanks to the use of phase guides (i.e., geometric features acting as pressure barriers). Thus, it can reproduce heterogeneous tissues. Here high-throughput capabilities include compatibility with automation and high-content screening equipment (Trietsch et al. 2013). A clever system transforms 3D

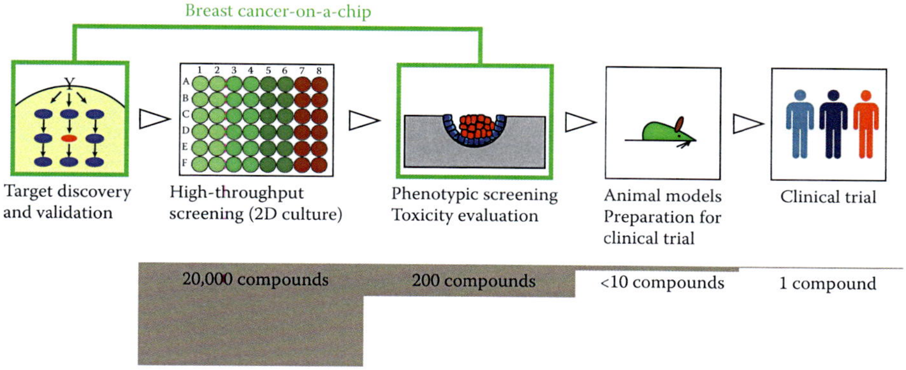

FIGURE 15.2 Integration of lab-on-a-chip technologies into the drug discovery and screening pipeline. First, breast-cancer-on-a-chip systems can be used to identify new targets with tumor models specific of cancer subtypes (target discovery). Second, following bioassays to screen a vast array of compounds (high-throughput screening), breast-cancer-on-a-chip can be used as a filter to identify drugs with real phenotypic impact (phenotypic screening) (e.g., reduction of tumor size) from those selected via the high-throughput screening, hence permitting the selection of a reduced number of drugs with the best therapeutic potential for animal testing before clinical trials.

cultures into 200 μm-thick sections for easy imaging by peeling layers of stacks of 200 μm-thick papers containing 96 "spots" with cell-including Matrigel™ integrated in the culture vessel. Each tissue is separated from tissues produced in neighboring wells, enabling testing of different treatments, although it does not prevent lateral diffusion of molecules through the hydrophobic wax. High-throughput is possible for production and handling via robots (Deiss et al. 2013). Ultimately, it is the combination of high-throughput reading of the sensors or parameters and high-throughput layout of on-a-chip models that will provide optimal conditions for fast and reliable drug screening systems. Once these models have been tested for reproducibility of the results, we envision a bright future enabling screening to filter out drugs that would not work well once cells are in their 3D context, consequently reducing animal use and costs for drug testing (Figure 15.2).

15.4 PERSPECTIVES FOR PRECISION MEDICINE

Breast cancer is a heterogeneous disease, or perhaps more accurately stated, it is a collection of diseases with different histopathological characteristics, genomic instability levels, prognoses, and relative occurrences in different ethnic groups. The identification of breast cancer subtypes described in the introduction, with distinct patterns of expression for hormonal and epidermal growth factor receptors, has led to the development of targeted therapies, such as tamoxifen and aromatase inhibitors for ER+ tumors, and Herceptin and Lapatinib for HER2-overexpressing cancers. These targeted therapies have opened the path toward precision breast oncology, i.e., breast cancer treatments tailored to individual patients. Subsequently, molecular breast cancer subtypes have been refined based on global gene expression profiles, leading to the discovery of new subgroups with unique prognoses (Perou et al. 2000; Curtis et al. 2012). Every cancer develops a unique set of genetic and epigenetic anomalies, some of which are (or will be) useful for therapeutic decisions. Precisely targeting these specific anomalies is expected to guarantee the best possible outcomes. This assumption is being tested for metastatic breast cancer in a large multinational prospective study (Zardavas et al. 2014). Thanks to the rapid progress in DNA sequencing techniques, clinical precision medicine programs now offer the possibility to sequence entire tumor genomes or, more frequently, panels of genes that are unambiguously linked to cancer progression (Stover and Wagle 2015). For example, the Foundation One panel (Frampton et al. 2013) that consists of 315 cancer-related genes complemented with 28 genomic regions frequently rearranged in cancer enables the detection of substitutions, deletions, rearrangements, and copy number alterations in small subsets of cells within solid tumors. Knowledge from such analyses is helping clinicians and patients select targeted therapies. Indeed, current testing identifies treatable genomic defects in about 75% of solid tumors tested (Frampton et al. 2013). Genomic profiling can also reveal potential new treatment approaches, as recently shown for metaplastic breast carcinoma (Ross et al. 2015). Indeed, in addition to frequently altered genes (e.g., TP53 and PIK3CA, mutated in more than 30% of breast cancers), large-scale breast cancer sequencing projects have detected many mutations occurring at lower frequency that may also contribute to cancer progression, notably by defining the pharmacogenetic landscape of individual tumors. These rare yet specific alterations hold promise for precision medicine (Curtis 2015; Stover and Wagle 2015). One major limitation to precision medicine is that many gene-based alterations linked to cancer progression and detectable in patients cannot (yet) be targeted with drugs. However, current research focused on cancer-related molecular pathways is broadening the palette of drugs to eventually match the range of profiles of tumor anomalies; the increase in clinical trials with specific genomic alterations as a recruitment criterion (Roper et al. 2015) attests to the potential of targeted therapies.

An application for microfluidic devices will be to facilitate tumor DNA collection from blood samples ("liquid biopsies") for genomic analyses, with the goal to employ less invasive procedures while increasing sampling efficacy and "time resolution" compared to currently used biopsies (Kidess and Jeffrey 2013). The microfluidics developments described in Section 15.2 for the isolation of circulating cancer cells constitute one approach. The capture of circulating cell-free tumor DNA (ctDNA) from the plasma represents another approach. Indeed, studies have revealed that ctDNA is a sensitive biomarker to monitor breast cancer progression and to detect resistant subclones (reviewed in Curtis 2015).

Whereas genomic data enable cancer classification, prognosis prediction, and increasingly effective guidance of therapeutic choices, phenotypical outcomes in tumors such as drug responses cannot be fully predicted from a list of genomic anomalies. Epigenetic modulation of gene expression in cancer cells, mechanical properties of the tissues, cancer-stroma and cancer-epithelium interactions, as well as the level of tumor heterogeneity, are only a few factors critically influencing tumor phenotypes. A promising prospect for precision medicine is the use of on-a-chip technologies to directly assess tumor responses to different drug regimens in order to predict effectiveness and resistance. More distantly, we may envisage this approach to screen libraries for compounds specifically tailored to a particular cancer. Such devices should (1) enable a rapid turnover (up to a few weeks, as currently achieved for genomic testing); (2) place tumor cells in a physiologically relevant microenvironment, similar to their site of origin; (3) require minimal amounts of tumor samples (typically less than 1 cm^3 of resected tissue is available after breast cancer surgeries), and (4) faithfully sample and maintain intratumor heterogeneity. Heterogeneity within breast tumors is largely fueled by genomic instability characteristic of tumor cells, in particular in basal-like tumors. Stochastic interconversion between non-stem and stem stages as well as the nonhomogeneous tumor microenvironment (e.g., hypoxia gradients) may also contribute to heterogeneity among cells of a given cancer (Marusyk and Polyak 2010; Lelièvre, Hodges, and Vidi 2014; Martelotto et al. 2014; Skibinski and Kuperwasser 2015). Taking into account intratumor heterogeneity is important because this cancer characteristic constitutes a challenge for both classic cytotoxic drugs and targeted therapies. Indeed, tumors with diverse cancer cells have higher chances to develop treatment resistances and cancer progression (Almendro, Cheng, et al. 2014; Almendro, Kim, et al. 2014). Tumor cells once placed in culture tend to loose heterogeneity. A likely explanation for this phenomenon is the abrupt change from complex (and heterogeneous) microenvironments *in vivo* to monotonous conditions *in vitro*, which at best approximate the chemical and physical cues against which the tumor has evolved. Only the cancer cells able to adapt and to proliferate rapidly in the new environment can be tested, leading to biased pharmacological readouts.

Patient-derived mouse xenograft (PDX) models, also called tumorografts, are currently used to maintain and amplify human tumors while preventing phenotypical drift (DeRose et al. 2011; Zhang et al. 2013; Zhang et al. 2014). By placing cancer cells in a physiological microenvironment and maintaining the tumor architecture, PDX models constitute a useful resource for preclinical drug assessment. However, these models take time to establish and may not be appropriate for large-scale use as clinical predictors for the treatment of cancer patients. In addition, the physiology of the mice may introduce a bias.

Short-term cultures in 3D matrices from primary solid breast tumors or from pleural effusions have been used to validate drug efficacy predicted based on gene expression signatures (Cohen et al. 2011). More recently, the SpheroNEO study (Halfter et al. 2015) used spheroid assays to predict responses to neoadjuvant therapy. Spheroids produced from dissociated breast tumors were treated with the same drugs used for the neoadjuvant cycles. The study revealed that cell survival *in vitro* is a good predictor for the pathological complete response (pCR) in the surgical excision specimens, defined in this study by the absence of "vital tumor." We envision the generalization of similar short-term approaches with miniaturized microfabricated devices that will maintain tumor behaviors and heterogeneity and use minimal amounts of samples for testing drug efficacy. Microfluidic systems are already being tested for drug screening with patient samples. For example, chemotherapeutic dosage and combinations have been assessed on lung tumor–derived cancer cells in 3D culture using a microfluidic platform (Xu et al. 2013). Large discrepancies in drug sensitivity between 3D cultures from primary tumor cells and immortalized cell lines were observed, as expected. It will be essential to systematically assess the value of this and other microfluidic platforms for predicting treatment outcomes in clinical studies.

15.5 CONCLUDING REMARKS

Microfluidic technology is lifting 3D cell culture to the next level of sophistication. The increasing use of organ-on-a-chip models is very encouraging given the paramount influence of the microenvironmental contexts that require engineering for the making of features such as culture surfaces with a given geometry, extracellular matrices with specific mechanical properties, and sensors adapted to the minute pieces of tissues created.

Reproducing environments with malleable physical properties is the next frontier for the breast-on-a-chip. Indeed the breast is one of the original organs for which mechanical properties, such as stiffness of the stroma, have been shown to be modified with cancer development (Paszek et al. 2005; Acerbi et al. 2015). It is also known that breast density, in part linked to the increased presence of collagen I in the stroma, is associated with heightened breast cancer risk (Seewaldt 2012). Hence, being able to modulate stiffness via engineered matrices such as those developed based on modulating Type I collagen oligomers and collagen precursors (Bailey et al. 2011; Whittington et al. 2013) is tremendously appealing. These matrices have been primarily used in standard 3D culture and for tissue engineering, yet they should be usable regardless of the culture vessel, and thus for organs-on-chips as well.

Another exciting challenge with breast-on-a-chip models is to reproduce tissue complexity, not so much in terms of multiple cell types and matrices, as these elements are amenable for bioprinting if needed, but in terms of fluctuating heterogeneity in disease. The advent of organs-on-chips was mainly initially meant for reproducing the normal situation. With the first few disease-on-chips, notably pioneered by the breast cancer model (Vidi et al. 2014), the question of reproducing the fast-changing conditions within tumors has emerged. For instance, hypoxia of breast tumors influenced by the stroma can vary greatly over a 24-hour period (Acosta et al. 2014), which can make *in vitro* drug testing inaccurate unless such heterogeneity associated with monitoring via sensors is achieved. Another example of heterogeneity is that of epigenetic profiles (Lelièvre, Hodges, and Vidi 2014; Brooks, Burness, and Wicha 2015). Here, controlling the microenvironment on-a-chip seems an obvious first approach in order to induce a mixed population of cancer cells to recreate physiologically relevant situations in which the extent/size of groups of cells with specific epigenetic traits changes with the breast cancer subtype, as well as with the malignant progression of a particular cancer subtype. An example of this situation is that of high-grade breast cancers, like triple negative breast cancers, that include a high proportion of $CD44^{+}$/$CD24^{-/low}$ cancer stem cells possibly generated via epigenetic regulation (Kagara et al. 2012; Lim et al. 2013). Yet, once the tumor becomes invasive, the percentage of cancer stem cells greatly diminishes; this is likely due to the acquisition of a certain differentiation level of these cells. In addition to controlling triggers of phenotypic alterations, such as oxidative stress and inducers of inflammatory responses, being able to read epigenetic marks at the single-cell level on-a-chip is paramount for this type of models.

There is a long road ahead to create the many cell culture models necessary for the next level of biological and biomedical physiologically relevant discoveries. Collaboration is a very attractive way to work with organs-on-chips (Benam et al. 2015) as team science projects regrouping engineering, basic science, and clinical perspectives are most likely to significantly contribute to our understanding and reproduction of the extreme complexity of breast cancer onset and progression in the laboratory. Then, integration of models to mimic processes linked with the absorption of oral medicine and food, and cancer models on-a-chip (Imura, Sato, and Yoshimura 2010) will lead to exquisite pipelines for assessing therapies within defined environmental contexts. We consider nanosensing within breast-on-a-chip platforms as the next frontier where parameters such as hypoxia, stiffness, pH, and cellular and nuclear morphology will have to be managed on a per-cell basis, within an undisturbed multicellular organization. Mechanical engineers and electrical engineers have managed to design state-of-the-art nanosensors, many of which are awaiting applications within cells or within complex groups of cells. We are hopeful that biologists are ready to take on the collaborative challenge.

ACKNOWLEDGMENTS

We acknowledge colleagues whose work could not be cited in this chapter due to space limitation and the presentation of work that we restricted to studies focusing on breast cancer progression. Support was provided by the National Health Institutes (CA163957 to PAV and CA171704 to SAL) and the Trask Innovation Fund (to SAL). SAL is a member of International Breast Cancer and Nutrition (IBCN).

REFERENCES

Abhyankar, V. V., M. W. Toepke, C. L. Cortesio, M. A. Lokuta, A. Huttenlocher, and D. J. Beebe. 2008. A platform for assessing chemotactic migration within a spatiotemporally defined 3D microenvironment. *Lab Chip* 8 (9): 1507–15. doi: 10.1039/b803533d.

Acerbi, I., L. Cassereau, I. Dean, Q. Shi, A. Au, C. Park, Y. Y. Chen, J. Liphardt, E. S. Hwang, and V. M. Weaver. 2015. Human breast cancer invasion and aggression correlates with ECM stiffening and immune cell infiltration. *Integr Biol (Camb)* 7 (10): 1120–34. doi: 10.1039/c5ib00040h.

Achyuta, A. K., A. J. Conway, R. B. Crouse, E. C. Bannister, R. N. Lee, C. P. Katnik, A. A. Behensky, J. Cuevas, and S. S. Sundaram. 2013. A modular approach to create a neurovascular unit-on-a-chip. *Lab Chip* 13 (4): 542–53. doi: 10.1039/c2lc41033h.

Acosta, M. A., X. Jiang, P. K. Huang, K. B. Cutler, C. S. Grant, G. M. Walker, and M. P. Gamcsik. 2014. A microfluidic device to study cancer metastasis under chronic and intermittent hypoxia. *Biomicrofluidics* 8 (5): 054117. doi: 10.1063/1.4898788.

Almendro, V., Y. K. Cheng, A. Randles, S. Itzkovitz, A. Marusyk, E. Ametller, X. Gonzalez-Farre, et al. 2014. Inference of tumor evolution during chemotherapy by computational modeling and in situ analysis of genetic and phenotypic cellular diversity. *Cell Rep* 6 (3): 514–27. doi: 10.1016/j.celrep.2013.12.041.

Almendro, V., H. J. Kim, Y. K. Cheng, M. Gonen, S. Itzkovitz, P. Argani, A. van Oudenaarden, S. Sukumar, F. Michor, and K. Polyak. 2014. Genetic and phenotypic diversity in breast tumor metastases. *Cancer Res* 74 (5): 1338–48. doi: 10.1158/0008-5472.CAN-13-2357-T.

Anand, P., A. Fu, S. H. Teoh, and K. Q. Luo. 2015. Application of a fluorescence resonance energy transfer (FRET)-based biosensor for detection of drug-induced apoptosis in a 3D breast tumor model. *Biotechnol Bioeng* 112 (8): 1673–82. doi: 10.1002/bit.25572.

Bailey, J. L., P. J. Critser, C. Whittington, J. L. Kuske, M. C. Yoder, and S. L. Voytik-Harbin. 2011. Collagen oligomers modulate physical and biological properties of three-dimensional self-assembled matrices. *Biopolymers* 95 (2): 77–93. doi: 10.1002/bip.21537.

Barbie, T. U., G. Alexe, A. R. Aref, S. Li, Z. Zhu, X. Zhang, Y. Imamura, et al. 2014. Targeting an IKBKE cytokine network impairs triple-negative breast cancer growth. *J Clin Invest* 124 (12): 5411–23. doi: 10.1172/JCI75661.

Barcellos-Hoff, M. H., J. Aggeler, T. G. Ram, and M. J. Bissell. 1989. Functional differentiation and alveolar morphogenesis of primary mammary cultures on reconstituted basement membrane. *Development* 105 (2): 223–35.

Benam, K. H., S. Dauth, B. Hassell, A. Herland, A. Jain, K. J. Jang, K. Karalis, et al. 2015. Engineered in vitro disease models. *Annu Rev Pathol* 10: 195–262. doi: 10.1146/annurev-pathol-012414-040418.

Bersini, S., J. S. Jeon, G. Dubini, C. Arrigoni, S. Chung, J. L. Charest, M. Moretti, and R. D. Kamm. 2014. A microfluidic 3D in vitro model for specificity of breast cancer metastasis to bone. *Biomaterials* 35 (8): 2454–61. doi: 10.1016/j.biomaterials.2013.11.050.

Bischel, L. L., D. J. Beebe, and K. E. Sung. 2015. Microfluidic model of ductal carcinoma in situ with 3D, organotypic structure. *BMC Cancer* 15: 12. doi: 10.1186/s12885-015-1007-5.

Bischel, L. L., S. H. Lee, and D. J. Beebe. 2012. A practical method for patterning lumens through ECM hydrogels via viscous finger patterning. *J Lab Autom* 17 (2): 96–103. doi: 10.1177/2211068211426694.

Bischel, L. L., E. W. Young, B. R. Mader, and D. J. Beebe. 2013. Tubeless microfluidic angiogenesis assay with three-dimensional endothelial-lined microvessels. *Biomaterials* 34 (5): 1471–7. doi: 10.1016/j.biomaterials.2012.11.005.

Bissell, M. J. 1981. The differentiated state of normal and malignant cells or how to define a "normal" cell in culture. *Int Rev Cytol* 70: 27–100.

Bissell, M. J., D. C. Radisky, A. Rizki, V. M. Weaver, and O. W. Petersen. 2002. The organizing principle: Microenvironmental influences in the normal and malignant breast. *Differentiation* 70 (9–10): 537–46.

Bombonati, A., and D. C. Sgroi. 2011. The molecular pathology of breast cancer progression. *J Pathol* 223 (2): 307–17. doi: 10.1002/path.2808.

Bradbury, J. M., J. Arno, and P. A. Edwards. 1993. Induction of epithelial abnormalities that resemble human breast lesions by the expression of the neu/erbB-2 oncogene in reconstituted mouse mammary gland. *Oncogene* 8 (6): 1551–8.

Bray, L. J., M. Binner, A. Holzheu, J. Friedrichs, U. Freudenberg, D. W. Hutmacher, and C. Werner. 2015. Multi-parametric hydrogels support 3D in vitro bioengineered microenvironment models of tumour angiogenesis. *Biomaterials* 53: 609–20. doi: 10.1016/j.biomaterials.2015.02.124.

Briand, P., and A. E. Lykkesfeldt. 2001. An in vitro model of human breast carcinogenesis: Epigenetic aspects. *Breast Cancer Res Treat* 65 (2): 179–87.

Briand, P., O. W. Petersen, and B. Van Deurs. 1987. A new diploid nontumorigenic human breast epithelial cell line isolated and propagated in chemically defined medium. *In Vitro Cell Dev Biol* 23 (3): 181–8.

Brooks, M. D., M. L. Burness, and M. S. Wicha. 2015. Therapeutic implications of cellular heterogeneity and plasticity in breast cancer. *Cell Stem Cell* 17 (3): 260–71. doi: 10.1016/j.stem.2015.08.014.

Buckley, E., T. Sullivan, G. Farshid, J. Hiller, and D. Roder. 2015. Risk profile of breast cancer following atypical hyperplasia detected through organized screening. *Breast* 24 (3): 208–12. doi: 10.1016/j.breast.2015.01.006.

Butcher, D. T., T. Alliston, and V. M. Weaver. 2009. A tense situation: Forcing tumour progression. *Nat Rev Cancer* 9 (2): 108–22. doi: 10.1038/nrc2544.

Chang, C. L., W. Huang, S. I. Jalal, B. D. Chan, A. Mahmood, S. Shahda, B. H. O'Neil, D. E. Matei, and C. A. Savran. 2015. Circulating tumor cell detection using a parallel flow micro-aperture chip system. *Lab Chip* 15 (7): 1677–88. doi: 10.1039/c5lc00100e.

Charoen, K. M., B. Fallica, Y. L. Colson, M. H. Zaman, and M. W. Grinstaff. 2014. Embedded multicellular spheroids as a biomimetic 3D cancer model for evaluating drug and drug-device combinations. *Biomaterials* 35 (7): 2264–71. doi: 10.1016/j.biomaterials.2013.11.038.

Chen, M. C., M. Gupta, and K. C. Cheung. 2010. Alginate-based microfluidic system for tumor spheroid formation and anticancer agent screening. *Biomed Microdevices* 12 (4): 647–54. doi: 10.1007/s10544-010-9417-2.

Chen, S. Y., P. J. Hung, and P. J. Lee. 2011. Microfluidic array for three-dimensional perfusion culture of human mammary epithelial cells. *Biomed Microdevices* 13 (4): 753–8. doi: 10.1007/s10544-011-9545-3.

Cheung, K. J., and A. J. Ewald. 2014. Illuminating breast cancer invasion: Diverse roles for cell-cell interactions. *Curr Opin Cell Biol* 30: 99–111. doi: 10.1016/j.ceb.2014.07.003.

Choi, Y. P., J. H. Lee, M. Q. Gao, B. G. Kim, S. Kang, S. H. Kim, and N. H. Cho. 2014. Cancer-associated fibroblast promote transmigration through endothelial brain cells in three-dimensional in vitro models. *Int J Cancer* 135 (9): 2024–33. doi: 10.1002/ijc.28848.

Chung, S., R. Sudo, P. J. Mack, C. R. Wan, V. Vickerman, and R. D. Kamm. 2009. Cell migration into scaffolds under co-culture conditions in a microfluidic platform. *Lab Chip* 9 (2): 269–75. doi: 10.1039/b807585a.

Cirri, P., and P. Chiarugi. 2011. Cancer associated fibroblasts: The dark side of the coin. *Am J Cancer Res* 1 (4): 482–97.

Clark, A. M., S. E. Wheeler, D. P. Taylor, V. C. Pillai, C. L. Young, R. Prantil-Baun, T. Nguyen, et al. 2014. A microphysiological system model of therapy for liver micrometastases. *Exp Biol Med (Maywood)* 239 (9): 1170–9. doi: 10.1177/1535370214532596.

Cohen, A. L., R. Soldi, H. Zhang, A. M. Gustafson, R. Wilcox, B. E. Welm, J. T. Chang, et al., 2011. A pharmacogenomic method for individualized prediction of drug sensitivity. *Mol Syst Biol* 7: 513. doi: 10.1038/msb.2011.47.

Cristofanilli, M., G. T. Budd, M. J. Ellis, A. Stopeck, J. Matera, M. C. Miller, J. M. Reuben, et al. 2004. Circulating tumor cells, disease progression, and survival in metastatic breast cancer. *N Engl J Med* 351 (8): 781–91. doi: 10.1056/NEJMoa040766.

Curtis, C. 2015. Genomic profiling of breast cancers. *Curr Opin Obstet Gynecol* 27 (1): 34–9. doi: 10.1097/GCO.0000000000000145.

Curtis, C., S. P. Shah, S. F. Chin, G. Turashvili, O. M. Rueda, M. J. Dunning, D. Speed, et al. 2012. The genomic and transcriptomic architecture of 2,000 breast tumours reveals novel subgroups. *Nature* 486 (7403): 346–52. doi: 10.1038/nature10983.

Deiss, F., A. Mazzeo, E. Hong, D. E. Ingber, R. Derda, and G. M. Whitesides. 2013. Platform for high-throughput testing of the effect of soluble compounds on 3D cell cultures. *Anal Chem* 85 (17): 8085–94. doi: 10.1021/ac400161j.

DeRose, Y. S., G. Wang, Y. C. Lin, P. S. Bernard, S. S. Buys, M. T. Ebbert, R. Factor, et al. 2011. Tumor grafts derived from women with breast cancer authentically reflect tumor pathology, growth, metastasis and disease outcomes. *Nat Med* 17 (11): 1514–20. doi: 10.1038/nm.2454.

Dolega, M. E., F. Abeille, N. Picollet-D'hahan, and X. Gidrol. 2015. Controlled 3D culture in Matrigel microbeads to analyze clonal acinar development. *Biomaterials* 52: 347–57. doi: 10.1016/j.biomaterials.2015.02.042.

Esmaeilsabzali, H., T. V. Beischlag, M. E. Cox, A. M. Parameswaran, and E. J. Park. 2013. Detection and isolation of circulating tumor cells: Principles and methods. *Biotechnol Adv* 31 (7): 1063–84. doi: 10.1016/j.biotechadv.2013.08.016.

Ferguson, J. E., A. M. Schor, A. Howell, and M. W. Ferguson. 1992. Changes in the extracellular matrix of the normal human breast during the menstrual cycle. *Cell Tissue Res* 268 (1): 167–77.

Frampton, G. M., A. Fichtenholtz, G. A. Otto, K. Wang, S. R. Downing, J. He, M. Schnall-Levin, et al. 2013. Development and validation of a clinical cancer genomic profiling test based on massively parallel DNA sequencing. *Nat Biotechnol* 31 (11): 1023–31. doi: 10.1038/nbt.2696.

Funamoto, K., I. K. Zervantonakis, Y. Liu, C. J. Ochs, C. Kim, and R. D. Kamm. 2012. A novel microfluidic platform for high-resolution imaging of a three-dimensional cell culture under a controlled hypoxic environment. *Lab Chip* 12 (22): 4855–63. doi: 10.1039/c2lc40306d.

Gilkes, D. M., G. L. Semenza, and D. Wirtz. 2014. Hypoxia and the extracellular matrix: Drivers of tumour metastasis. *Nat Rev Cancer* 14 (6): 430–9. doi: 10.1038/nrc3726.

Grafton, M. M., L. Wang, P. A. Vidi, J. Leary, and S. A. Lelievre. 2011. Breast on-a-chip: Mimicry of the channeling system of the breast for development of theranostics. *Integr Biol (Camb)* 3(4): 451–9. doi: 10.1039/c0ib00132e.

Grist, S. M., J. C. Schmok, M. C. Liu, L. Chrostowski, and K. C. Cheung. 2015. Designing a microfluidic device with integrated ratiometric oxygen sensors for the long-term control and monitoring of chronic and cyclic hypoxia. *Sensors (Basel)* 15(8): 20030–52. doi: 10.3390/s150820030.

Gudjonsson, T., L. Ronnov-Jessen, R. Villadsen, F. Rank, M. J. Bissell, and O. W. Petersen. 2002. Normal and tumor-derived myoepithelial cells differ in their ability to interact with luminal breast epithelial cells for polarity and basement membrane deposition. *J Cell Sci* 115 (Pt 1): 39–50.

Guo, F., Y. Wang, J. Liu, S. C. Mok, F. Xue, and W. Zhang. 2016. CXCL12/CXCR4: A symbiotic bridge linking cancer cells and their stromal neighbors in oncogenic communication networks. *Oncogene* 35 (7): 816–26. doi: 10.1038/onc.2015.139.

Haessler, U., J. C. Teo, D. Foretay, P. Renaud, and M. A. Swartz. 2012. Migration dynamics of breast cancer cells in a tunable 3D interstitial flow chamber. *Integr Biol (Camb)* 4 (4): 401–9. doi: 10.1039/c1ib00128k.

Halfter, K., N. Ditsch, H. C. Kolberg, H. Fischer, T. Hauzenberger, F. E. von Koch, I. Bauerfeind, et al. 2015. Prospective cohort study using the breast cancer spheroid model as a predictor for response to neoadjuvant therapy—The SpheroNEO study. *BMC Cancer* 15: 519. doi: 10.1186/s12885-015-1491-7.

Hannen, E. J., J. A. van der Laak, J. J. Manni, M. M. Pahlplatz, H. P. Freihofer, P. J. Slootweg, R. Koole, and P. C. de Wilde. 1998. An image analysis study on nuclear morphology in metastasized and non-metastasized squamous cell carcinomas of the tongue. *J Pathol* 185 (2): 175–83. doi: 10.1002/(SICI)1096-9896(199806)185:2<175::AID-PATH69>3.0.CO;2-U.

Hartmann, L. C., D. C. Radisky, M. H. Frost, R. J. Santen, R. A. Vierkant, L. L. Benetti, Y. Tarabishy, K. Ghosh, D. W. Visscher, and A. C. Degnim. 2014. Understanding the premalignant potential of atypical hyperplasia through its natural history: A longitudinal cohort study. *Cancer Prev Res (Phila)* 7 (2): 211–17. doi: 10.1158/1940-6207.CAPR-13-0222.

Hirai, Y., A. Lochter, S. Galosy, S. Koshida, S. Niwa, and M. J. Bissell. 1998. Epimorphin functions as a key morphoregulator for mammary epithelial cells. *J Cell Biol* 140 (1): 159–69.

Hodges, K. B., D. Bazzoun, K. McDole, R. Talhouk, and S. A. Lelievre. 2014. The role of epigenetics in mammary gland development and breast cancer risk, in mammary glands: anatomy, development and diseases. In *Veterinary Sciences and Medicine*, edited by E. Rucker, 63–112. Nova Science, New York.

Howes, A. L., R. D. Richardson, D. Finlay, and K. Vuori. 2014. 3-Dimensional culture systems for anti-cancer compound profiling and high-throughput screening reveal increases in EGFR inhibitor-mediated cytotoxicity compared to monolayer culture systems. *PLoS One* 9 (9): e108283. doi: 10.1371/journal.pone.0108283.

Hribar, K. C., D. Finlay, X. Ma, X. Qu, M. G. Ondeck, P. H. Chung, F. Zanella, et al. 2015. Nonlinear 3D projection printing of concave hydrogel microstructures for long-term multicellular spheroid and embryoid body culture. *Lab Chip* 15 (11): 2412–18. doi: 10.1039/c5lc00159e.

Hyun, K. A., and H. I. Jung. 2013. Microfluidic devices for the isolation of circulating rare cells: A focus on affinity-based, dielectrophoresis, and hydrophoresis. *Electrophoresis* 34 (7): 1028–41. doi: 10.1002/elps.201200417.

Imamura, Y., T. Mukohara, Y. Shimono, Y. Funakoshi, N. Chayahara, M. Toyoda, N. Kiyota, et al. 2015. Comparison of 2D- and 3D-culture models as drug-testing platforms in breast cancer. *Oncol Rep* 33 (4): 1837–43. doi: 10.3892/or.2015.3767.

Imura, Y., K. Sato, and E. Yoshimura. 2010. Micro total bioassay system for ingested substances: Assessment of intestinal absorption, hepatic metabolism, and bioactivity. *Anal Chem* 82 (24): 9983–8. doi: 10.1021/ac100806x.

Jaganathan, H., J. Gage, F. Leonard, S. Srinivasan, G. R. Souza, B. Dave, and B. Godin. 2014. Three-dimensional in vitro co-culture model of breast tumor using magnetic levitation. *Sci Rep* 4: 6468. doi: 10.1038/srep06468.

Jeon, J. S., S. Bersini, M. Gilardi, G. Dubini, J. L. Charest, M. Moretti, and R. D. Kamm. 2015. Human 3D vascularized organotypic microfluidic assays to study breast cancer cell extravasation. *Proc Natl Acad Sci U S A* 112 (1): 214–19. doi: 10.1073/pnas.1417115112.

Kagara, N., K. T. Huynh, C. Kuo, H. Okano, M. S. Sim, D. Elashoff, K. Chong, A. E. Giuliano, and D. S. Hoon. 2012. Epigenetic regulation of cancer stem cell genes in triple-negative breast cancer. *Am J Pathol* 181 (1): 257–67. doi: 10.1016/j.ajpath.2012.03.019.

Kelm, J. M., N. E. Timmins, C. J. Brown, M. Fussenegger, and L. K. Nielsen. 2003. Method for generation of homogeneous multicellular tumor spheroids applicable to a wide variety of cell types. *Biotechnol Bioeng* 83 (2): 173–80. doi: 10.1002/bit.10655.

Kidess, E., and S. S. Jeffrey. 2013. Circulating tumor cells versus tumor-derived cell-free DNA: Rivals or partners in cancer care in the era of single-cell analysis? *Genome Med* 5 (8): 70. doi: 10.1186/gm474.

Kim, B. J., P. Hannanta-anan, M. Chau, Y. S. Kim, M. A. Swartz, and M. Wu. 2013. Cooperative roles of SDF-1alpha and EGF gradients on tumor cell migration revealed by a robust 3D microfluidic model. *PLoS One* 8 (7): e68422. doi: 10.1371/journal.pone.0068422.

Koch, T. M., S. Munster, N. Bonakdar, J. P. Butler, and B. Fabry. 2012. 3D Traction forces in cancer cell invasion. *PLoS One* 7 (3): e33476. doi: 10.1371/journal.pone.0033476.

Lee, H., W. Park, H. Ryu, and N. L. Jeon. 2014. A microfluidic platform for quantitative analysis of cancer angiogenesis and intravasation. *Biomicrofluidics* 8 (5): 054102. doi: 10.1063/1.4894595.

Lehman, H. L., E. J. Dashner, M. Lucey, P. Vermeulen, L. Dirix, S. Van Laere, and K. L. van Golen. 2013. Modeling and characterization of inflammatory breast cancer emboli grown in vitro. *Int J Cancer* 132 (10): 2283–94. doi: 10.1002/ijc.27928.

Lelievre, S. A. 2014. Taking a chance on epigenetics. *Front Genet* 5: 205. doi: 10.3389/fgene.2014.00205.

Lelievre, S. A., and C. M. Weaver. 2013. Global nutrition research: Nutrition and breast cancer prevention as a model. *Nutr Rev* 71 (11): 742–52. doi: 10.1111/nure.12075.

Lelievre, S. A., V. M. Weaver, J. A. Nickerson, C. A. Larabell, A. Bhaumik, O. W. Petersen, and M. J. Bissell. 1998. Tissue phenotype depends on reciprocal interactions between the extracellular matrix and the structural organization of the nucleus. *Proc Natl Acad Sci U S A* 95 (25): 14711–16.

Lelievre, S., K. B. Hodges, and P. A. Vidi. 2014. Application of theranostics to measure and treat cell heterogeneity in cancer. In *Cancer Theranostics*, edited by X. Chen and S. Wong, 493–516. Academic Press, Elsevier.

Lelievre, S., V. M. Weaver, and M. J. Bissell. 1996. Extracellular matrix signaling from the cellular membrane skeleton to the nuclear skeleton: A model of gene regulation. *Recent Prog Horm Res* 51: 417–32.

Lesko, A. C., K. H. Goss, F. F. Yang, A. Schwertner, I. Hulur, K. Onel, and J. R. Prosperi. 2015. The APC tumor suppressor is required for epithelial cell polarization and three-dimensional morphogenesis. *Biochim Biophys Acta* 1854 (3): 711–23. doi: 10.1016/j.bbamcr.2014.12.036.

Levental, K. R., H. Yu, L. Kass, J. N. Lakins, M. Egeblad, J. T. Erler, S. F. Fong, et al. 2009. Matrix crosslinking forces tumor progression by enhancing integrin signaling. *Cell* 139 (5): 891–906. doi: S0092-8674(09)01353-1 [pii] 10.1016/j.cell.2009.10.027.

Li, Q., S. R. Mullins, B. F. Sloane, and R. R. Mattingly. 2008. p21-Activated kinase 1 coordinates aberrant cell survival and pericellular proteolysis in a three-dimensional culture model for premalignant progression of human breast cancer. *Neoplasia* 10 (4): 314–29.

Lim, Y. Y., J. A. Wright, J. L. Attema, P. A. Gregory, A. G. Bert, E. Smith, D. Thomas, et al. 2013. Epigenetic modulation of the miR-200 family is associated with transition to a breast cancer stem-cell-like state. *J Cell Sci* 126 (Pt 10): 2256–66. doi: 10.1242/jcs.122275.

Liu, T., C. Li, H. Li, S. Zeng, J. Qin, and B. Lin. 2009. A microfluidic deviee for characterizing the invasion of cancer cells in 3-D matrix. *Electrophoresis* 30 (24): 4285–91. doi: 10.1002/elps.200900289.

Lopez-Garcia, M. A., F. C. Geyer, M. Lacroix-Triki, C. Marchio, and J. S. Reis-Filho. 2010. Breast cancer precursors revisited: Molecular features and progression pathways. *Histopathology* 57 (2): 171–92. doi: 10.1111/j.1365-2559.2010.03568.x.

Lovitt, C. J., T. B. Shelper, and V. M. Avery. 2015. Evaluation of chemotherapeutics in a three-dimensional breast cancer model. *J Cancer Res Clin Oncol* 141 (5): 951–9. doi: 10.1007/s00432-015-1950-1.

Martelotto, L. G., C. K. Ng, S. Piscuoglio, B. Weigelt, and J. S. Reis-Filho. 2014. Breast cancer intra-tumor heterogeneity. *Breast Cancer Res* 16 (3): 210. doi: 10.1186/bcr3658.

Marusyk, A., and K. Polyak. 2010. Tumor heterogeneity: causes and consequences. *Biochim Biophys Acta* 1805 (1): 105–17. doi: 10.1016/j.bbcan.2009.11.002.

Matsumoto, S., H. Yasui, J. B. Mitchell, and M. C. Krishna. 2010. Imaging cycling tumor hypoxia. *Cancer Res* 70 (24): 10019–23. doi: 10.1158/0008-5472.CAN-10-2821.

Meyvantsson, I., J. W. Warrick, S. Hayes, A. Skoien, and D. J. Beebe. 2008. Automated cell culture in high density tubeless microfluidic device arrays. *Lab Chip* 8 (5): 717–24. doi: 10.1039/b715375a.

Mijovic, Z., M. Kostov, D. Mihailovic, N. Zivkovic, M. Stojanovic, and M. Zdravkovic. 2013. Correlation of nuclear morphometry of primary melanoma of the skin with clinicopathological parameters and expression of tumor suppressor proteins (p53 and p16(INK4a)) and bcl-2 oncoprotein. *J BUON* 18 (2): 471–6.

Miller, F. R., S. J. Santner, L. Tait, and P. J. Dawson. 2000. MCF10DCIS.com xenograft model of human comedo ductal carcinoma in situ. *J Natl Cancer Inst* 92 (14):1185–6.

Mosadegh, B., W. Saadi, S. J. Wang, and N. L. Jeon. 2008. Epidermal growth factor promotes breast cancer cell chemotaxis in CXCL12 gradients. *Biotechnol Bioeng* 100 (6): 1205–13. doi: 10.1002/bit.21851.

Murlidhar, V., M. Zeinali, S. Grabauskiene, M. Ghannad-Rezaie, M. S. Wicha, D. M. Simeone, N. Ramnath, R. M. Reddy, and S. Nagrath. 2014. A radial flow microfluidic device for ultra-high-throughput affinity-based isolation of circulating tumor cells. *Small* 10 (23): 4895–904. doi: 10.1002/smll.201400719.

Muthuswamy, S. K., D. Li, S. A. Lelièvre, M. J. Bissell, and J. S. Brugge. 2001. ErbB2, but not ErbB1, reinitiates proliferation and induces luminal repopulation in epithelial acini. *Nat Cell Biol* 3 (9): 785–92. doi: 10.1038/ncb0901-785 ncb0901-785 [pii].

Nangia-Makker, P., Y. Wang, T. Raz, L. Tait, V. Balan, V. Hogan, and A. Raz. 2010. Cleavage of galectin-3 by matrix metalloproteases induces angiogenesis in breast cancer. *Int J Cancer* 127 (11): 2530–41. doi: 10.1002/ijc.25254.

Neve, R. M., K. Chin, J. Fridlyand, J. Yeh, F. L. Baehner, T. Fevr, L. Clark, et al. 2006. A collection of breast cancer cell lines for the study of functionally distinct cancer subtypes. *Cancer Cell* 10 (6): 515–27. doi: 10.1016/j.ccr.2006.10.008.

Nguyen, T. A., T. I. Yin, D. Reyes, and G. A. Urban. 2013. Microfluidic chip with integrated electrical cell-impedance sensing for monitoring single cancer cell migration in three-dimensional matrixes. *Anal Chem* 85 (22): 11068–76. doi: 10.1021/ac402761s.

Okudela, K., T. Woo, H. Mitsui, T. Yazawa, H. Shimoyamada, M. Tajiri, N. Ogawa, M. Masuda, and H. Kitamura. 2010. Morphometric profiling of lung cancers-its association with clinicopathologic, biologic, and molecular genetic features. *Am J Surg Pathol* 34 (2): 243–55. doi: 10.1097/PAS.0b013e3181c79a6f.

Oliveras-Ferraros, C., B. Corominas-Faja, S. Cufi, A. Vazquez-Martin, B. Martin-Castillo, J. M. Iglesias, E. Lopez-Bonet, A. G. Martin, and J. A. Menendez. 2012. Epithelial-to-mesenchymal transition (EMT) confers primary resistance to trastuzumab (Herceptin). *Cell Cycle* 11 (21): 4020–32. doi: 10.4161/cc.22225.

Paszek, M. J., N. Zahir, K. R. Johnson, J. N. Lakins, G. I. Rozenberg, A. Gefen, C. A. Reinhart-King, et al. 2005. Tensional homeostasis and the malignant phenotype. *Cancer Cell* 8 (3): 241–54. doi: 10.1016/j.ccr.2005.08.010.

Paulsson, J., and P. Micke. 2014. Prognostic relevance of cancer-associated fibroblasts in human cancer. *Semin Cancer Biol* 25: 61–8. doi: 10.1016/j.semcancer.2014.02.006.

Peloso, A., A. Dhal, J. P. Zambon, P. Li, G. Orlando, A. Atala, and S. Soker. 2015. Current achievements and future perspectives in whole-organ bioengineering. *Stem Cell Res Ther* 6: 107. doi: 10.1186/s13287-015-0089-y.

Perou, C. M., T. Sorlie, M. B. Eisen, M. van de Rijn, S. S. Jeffrey, C. A. Rees, J. R. Pollack, et al. 2000. Molecular portraits of human breast tumours. *Nature* 406 (6797): 747–52. doi: 10.1038/35021093.

Petersen, O. W., L. Ronnov-Jessen, A. R. Howlett, and M. J. Bissell. 1992. Interaction with basement membrane serves to rapidly distinguish growth and differentiation pattern of normal and malignant human breast epithelial cells. *Proc Natl Acad Sci U S A* 89 (19): 9064–8.

Pickup, M. W., J. K. Mouw, and V. M. Weaver. 2014. The extracellular matrix modulates the hallmarks of cancer. *EMBO Rep* 15 (12): 1243–53. doi: 10.15252/embr.201439246.

Pisano, M., V. Triacca, K. A. Barbee, and M. A. Swartz. 2015. An in vitro model of the tumor-lymphatic microenvironment with simultaneous transendothelial and luminal flows reveals mechanisms of flow enhanced invasion. *Integr Biol (Camb)* 7 (5): 525–33. doi: 10.1039/c5ib00085h.

Plachot, C., L. S. Chaboub, H. A. Adissu, L. Wang, A. Urazaev, J. Sturgis, E. K. Asem, and S. A. Lelièvre. 2009. Factors necessary to produce basoapical polarity in human glandular epithelium formed in conventional and high-throughput three-dimensional culture: Example of the breast epithelium. *BMC Biol* 7 (1): 77. doi: 1741-7007-7-77 [pii] 10.1186/1741-7007-7-77.

Plachot, C., and S. A. Lelièvre. 2004. DNA methylation control of tissue polarity and cellular differentiation in the mammary epithelium. *Exp Cell Res* 298 (1): 122–32.

Polacheck, W. J., J. L. Charest, and R. D. Kamm. 2011. Interstitial flow influences direction of tumor cell migration through competing mechanisms. *Proc Natl Acad Sci U S A* 108 (27): 11115–20. doi: 10.1073/pnas.1103581108.

Polacheck, W. J., A. E. German, A. Mammoto, D. E. Ingber, and R. D. Kamm. 2014. Mechanotransduction of fluid stresses governs 3D cell migration. *Proc Natl Acad Sci U S A* 111 (7): 2447–52. doi: 10.1073/pnas.1316848111.

Polacheck, W. J., I. K. Zervantonakis, and R. D. Kamm. 2013. Tumor cell migration in complex microenvironments. *Cell Mol Life Sci* 70 (8): 1335–56. doi: 10.1007/s00018-012-1115-1.

Pradhan, S., C. S. Chaudhury, and E. A. Lipke. 2014. Dual-phase, surface tension-based fabrication method for generation of tumor millibeads. *Langmuir* 30 (13): 3817–25. doi: 10.1021/la500402m.

Rizki, A., V. M. Weaver, S. Y. Lee, G. I. Rozenberg, K. Chin, C. A. Myers, J. L. Bascom, et al. 2008. A human breast cell model of preinvasive to invasive transition. *Cancer Res* 68 (5): 1378–87. doi: 68/5/1378 [pii] 10.1158/0008-5472.CAN-07-2225.

Roper, N., K. D. Stensland, R. Hendricks, and M. D. Galsky. 2015. The landscape of precision cancer medicine clinical trials in the United States. *Cancer Treat Rev* 41 (5): 385–90. doi: 10.1016/j.ctrv.2015.02.009.

Ross, J. S., S. Badve, K. Wang, C. E. Sheehan, A. B. Boguniewicz, G. A. Otto, R. Yelensky, et al. 2015. Genomic profiling of advanced-stage, metaplastic breast carcinoma by next-generation sequencing reveals frequent, targetable genomic abnormalities and potential new treatment options. *Arch Pathol Lab Med* 139 (5): 642–9. doi: 10.5858/arpa.2014-0200-OA.

Sakamoto, R., M. M. Rahman, M. Shimomura, M. Itoh, and T. Nakatsura. 2015. Time-lapse imaging assay using the BioStation CT: A sensitive drug-screening method for three-dimensional cell culture. *Cancer Sci* 106 (6): 757–65. doi: 10.1111/cas.12667.

Schmaltz, C., P. H. Hardenbergh, A. Wells, and D. E. Fisher. 1998. Regulation of proliferation-survival decisions during tumor cell hypoxia. *Mol Cell Biol* 18 (5): 2845–54.

Seewaldt, V. L. 2012. Cancer: Destiny from density. *Nature* 490 (7421): 490–1. doi: 10.1038/490490a.

Shah, S. N., L. Cope, W. Poh, A. Belton, S. Roy, C. C. Talbot, Jr., S. Sukumar, D. L. Huso, and L. M. Resar. 2013. HMGA1: A master regulator of tumor progression in triple-negative breast cancer cells. *PLoS One* 8 (5): e63419. doi: 10.1371/journal.pone.0063419.

Shamir, E. R., and A. J. Ewald. 2015. Adhesion in mammary development: Novel roles for E-cadherin in individual and collective cell migration. *Curr Top Dev Biol* 112: 353–82. doi: 10.1016/bs.ctdb.2014.12.001.

Shields, J. D., M. E. Fleury, C. Yong, A. A. Tomei, G. J. Randolph, and M. A. Swartz. 2007. Autologous chemotaxis as a mechanism of tumor cell homing to lymphatics via interstitial flow and autocrine CCR7 signaling. *Cancer Cell* 11 (6): 526–38. doi: 10.1016/j.ccr.2007.04.020.

Shin, Y., S. Han, J. S. Jeon, K. Yamamoto, I. K. Zervantonakis, R. Sudo, R. D. Kamm, and S. Chung. 2012. Microfluidic assay for simultaneous culture of multiple cell types on surfaces or within hydrogels. *Nat Protoc* 7 (7): 1247–59. doi: 10.1038/nprot.2012.051.

Skibinski, A., and C. Kuperwasser. 2015. The origin of breast tumor heterogeneity. *Oncogene.* 34 (42): 5309–16. doi: 10.1038/onc.2014.475.

Song, J. W., S. P. Cavnar, A. C. Walker, K. E. Luker, M. Gupta, Y. C. Tung, G. D. Luker, and S. Takayama. 2009. Microfluidic endothelium for studying the intravascular adhesion of metastatic breast cancer cells. *PLoS One* 4 (6): e5756. doi: 10.1371/journal.pone.0005756.

Stover, D. G., and N. Wagle. 2015. Precision medicine in breast cancer: Genes, genomes, and the future of genomically driven treatments. *Curr Oncol Rep* 17 (4): 15. doi: 10.1007/s11912-015-0438-0.

Strese, S., M. Fryknas, R. Larsson, and J. Gullbo. 2013. Effects of hypoxia on human cancer cell line chemosensitivity. *BMC Cancer* 13: 331. doi: 10.1186/1471-2407-13-331.

Subia, B., T. Dey, S. Sharma, and S. C. Kundu. 2015. Target specific delivery of anticancer drug in silk fibroin based 3D distribution model of bone-breast cancer cells. *ACS Appl Mater Interfaces* 7 (4): 2269–79. doi: 10.1021/am506094c.

Sung, K. E., N. Yang, C. Pehlke, P. J. Keely, K. W. Eliceiri, A. Friedl, and D. J. Beebe. 2011. Transition to invasion in breast cancer: A microfluidic in vitro model enables examination of spatial and temporal effects. *Integr Biol (Camb)* 3 (4): 439–50. doi: 10.1039/c0ib00063a.

Trietsch, S. J., G. D. Israels, J. Joore, T. Hankemeier, and P. Vulto. 2013. Microfluidic titer plate for stratified 3D cell culture. *Lab Chip* 13 (18): 3548–54. doi: 10.1039/c3lc50210d.

Uchoa Dde, M., M. S. Graudenz, S. M. Callegari-Jacques, C. R. Hartmann, B. P. Ferreira, M. Fitarelli-Kiehl, and M. I. Edelweiss. 2014. Expression of cancer stem cell markers in basal and penta-negative breast carcinomas—A study of a series of triple-negative tumors. *Pathol Res Pract* 210 (7): 432–9. doi: 10.1016/j.prp.2014.03.005.

Vaupel, P. 2008. Hypoxia and aggressive tumor phenotype: Implications for therapy and prognosis. *Oncologist* 13 (Suppl 3): 21–6. doi: 10.1634/theoncologist.13-S3-21.

Vidi, P. A., M. J. Bissell, and S. Lelievre. 2013. Three dimensional culture of human breast epithelial cells: The how and the why. *Methods in Molecular Biology*, vol. 945, pp. 193–219.

Vidi, P. A., G. Chandramouly, M. Gray, L. Wang, E. Liu, J. J. Kim, V. Roukos, M. J. Bissell, P. V. Moghe, and S. A. Lelievre. 2012. Interconnected contribution of tissue morphogenesis and the nuclear protein NuMA to the DNA damage response. *J Cell Sci* 125 (Pt 2): 350–61. doi: jcs.089177 [pii] 10.1242/jcs.089177.

Vidi, P. A., J. F. Leary, and S. A. Lelievre. 2013. Building risk-on-a-chip models to improve breast cancer risk assessment and prevention. *Integr Biol (Camb)* 5 (9): 1110–18. doi: 10.1039/c3ib40053k.

Vidi, P. A., T. Maleki, M. Ochoa, L. Wang, S. M. Clark, J. F. Leary, and S. A. Lelievre. 2014. Disease-on-a-chip: Mimicry of tumor growth in mammary ducts. *Lab Chip* 14 (1): 172–7. doi: 10.1039/c3lc50819f.

Wang, X., X. Sang, C. Diorio, S. X. Lin, and C. J. Doillon. 2015. In vitro interactions between mammary fibroblasts (Hs 578Bst) and cancer epithelial cells (MCF-7) modulate aromatase, steroid sulfatase and 17beta-hydroxysteroid dehydrogenases. *Mol Cell Endocrinol* 412: 339–48. doi: 10.1016/j.mce.2015.05.032.

Wang, Y. L., and R. J. Pelham, Jr. 1998. Preparation of a flexible, porous polyacrylamide substrate for mechanical studies of cultured cells. *Methods Enzymol* 298: 489–96.

Weaver, V. M., S. A. Lelièvre, J. N. Lakins, M. A. Chrenek, J. C. Jones, F. Giancotti, Z. Werb, and M. J. Bissell. 2002. beta4 integrin-dependent formation of polarized three-dimensional architecture confers resistance to apoptosis in normal and malignant mammary epithelium. *Cancer Cell* 2 (3): 205–16. doi: S1535610802001253 [pii].

Weaver, V. M., O. W. Petersen, F. Wang, C. A. Larabell, P. Briand, C. Damsky, and M. J. Bissell. 1997. Reversion of the malignant phenotype of human breast cells in three-dimensional culture and in vivo by integrin blocking antibodies. *J Cell Biol* 137 (1): 231–45.

Whittington, C. F., E. Brandner, K. Y. Teo, B. Han, E. Nauman, and S. L. Voytik-Harbin. 2013. Oligomers modulate interfibril branching and mass transport properties of collagen matrices. *Microsc Microanal* 19 (5): 1323–33. doi: 10.1017/S1431927613001931.

Xu, Z., Y. Gao, Y. Hao, E. Li, Y. Wang, J. Zhang, W. Wang, Z. Gao, and Q. Wang. 2013. Application of a microfluidic chip-based 3D co-culture to test drug sensitivity for individualized treatment of lung cancer. *Biomaterials* 34 (16):4109–17. doi: 10.1016/j.biomaterials.2013.02.045.

Yang, Y., X. Yang, J. Zou, C. Jia, Y. Hu, H. Du, and H. Wang. 2015. Evaluation of photodynamic therapy efficiency using an in vitro three-dimensional microfluidic breast cancer tissue model. *Lab Chip* 15 (3): 735–44. doi: 10.1039/c4lc01065e.

Yates, C., C. R. Shepard, G. Papworth, A. Dash, D. Beer Stolz, S. Tannenbaum, L. Griffith, and A. Wells. 2007. Novel three-dimensional organotypic liver bioreactor to directly visualize early events in metastatic progression. *Adv Cancer Res* 97: 225–46. doi: 10.1016/S0065-230X(06)97010-9.

Zardavas, D., M. Maetens, A. Irrthum, T. Goulioti, K. Engelen, D. Fumagalli, R. Salgado, et al. 2014. The AURORA initiative for metastatic breast cancer. *Br J Cancer* 111 (10): 1881–7. doi: 10.1038/bjc.2014.341.

Zervantonakis, I. K., S. K. Hughes-Alford, J. L. Charest, J. S. Condeelis, F. B. Gertler, and R. D. Kamm. 2012. Three-dimensional microfluidic model for tumor cell intravasation and endothelial barrier function. *Proc Natl Acad Sci U S A* 109 (34): 13515–20. doi: 10.1073/pnas.1210182109.

Zhang, H., A. L. Cohen, S. Krishnakumar, I. L. Wapnir, S. Veeriah, G. Deng, M. A. Coram, et al. 2014. Patient-derived xenografts of triple-negative breast cancer reproduce molecular features of patient tumors and respond to mTOR inhibition. *Breast Cancer Res* 16 (2): R36. doi: 10.1186/bcr3640.

Zhang, X., S. Claerhout, A. Prat, L. E. Dobrolecki, I. Petrovic, Q. Lai, M. D. Landis, et al. 2013. A renewable tissue resource of phenotypically stable, biologically and ethnically diverse, patient-derived human breast cancer xenograft models. *Cancer Res* 73 (15): 4885–97. doi: 10.1158/0008-5472.CAN-12-4081.

Zhou, Y., T. Arai, Y. Horiguchi, K. Ino, T. Matsue, and H. Shiku. 2013. Multiparameter analyses of three-dimensionally cultured tumor spheroids based on respiratory activity and comprehensive gene expression profiles. *Anal Biochem* 439 (2): 187–93. doi: 10.1016/j.ab.2013.04.020.

16 Disease Modeling

Uta Grieshammer and Kelly A. Shepard

CONTENTS

16.1 INTRODUCTION

Modeling of disease phenotypes in the laboratory has long been recognized as both a means to better understand a pathological process and a tool for discovering potential treatments. The analysis of animal models exhibiting induced or naturally occurring features similar to human disease is an important preclinical activity, and the ability to genetically engineer animals, especially mice, to harbor mutations or to simulate cellular defects known to drive human disease has been a powerful tool for investigating disease mechanisms and test novel drug candidates. However, animal models, while reflective of whole-body physiology, do not truly recapitulate the human condition and often point to treatments that prove ineffective when translated to human patients. In addition, they tend to be expensive, time and labor intensive, and may pose ethical challenges. The ability to procure and culture both animal and human cells, including the use of immortalized cell lines, has been fundamental for our understanding of cell biology and has led to the development of *in vitro* disease models that have fueled decades of academic research and have enabled the pharmaceutical industry to discover drugs based on molecular targets or other disease-relevant defects observable in cellular models. While these approaches continue to have use, they suffer from limitations. Human cell models overcome species-related issues inherent in animal models but are also limited in their ability to accurately reflect a relevant phenotype. Models based on primary cells directly isolated from humans, even those that can be expanded and maintained for a period in culture, suffer from sample variability and limitations in supply, while immortalized or transformed cell lines have unlimited expansion potential but no longer represent a cell state found *in vivo*. Another significant hurdle for *in vitro* disease modeling is the complexity of human disease that depends on interactions between multiple cell types within an organ or is manifested in a systemic manner. While this latter hurdle is being addressed by the development of organ- or body-on-a-chip approaches, human pluripotent stem cell (hPSC) technology makes available unlimited quantities of disease-relevant human cell types. In this chapter, we discuss the advances that have been made in hPSC-based disease modeling and how combining these two technologies, hPSCs and organ-on-a-chip, have the potential to significantly accelerate progress in understanding and discovering treatments for human disease.

16.2 CURRENT USES OF hPSC DISEASE MODELS

The *in vitro* derivation of embryonic stem cells (ESCs) from blastocysts, first in mice[1,2] and then in humans,[3] unlocked an exciting array of opportunities for disease modeling based on two key properties of these cells: their ability to self renew, which enables ESC cultures to be expanded and maintained in their early embryonic state indefinitely, and their pluripotency, or ability to be converted into any cell type of the body via appropriate biological cues, thus offering renewable sources of human cells and tissues that are normally difficult to obtain, such as those of the heart or brain (Figure 16.1). A second major development was the generation of induced pluripotent stem cells (iPSCs), which resemble ESCs but can be derived by introducing certain exogenous factors, e.g., OCT4, SOX2, KLF4, and c-MYC, into easily accessible somatic cells, such as those from blood or skin, from individual donors[4–6] (Figure 16.1). Such "reprogrammed" cells harbor the donor's unique genetic signature, offering the potential to model phenotypes based on authentic human genomic backgrounds from diseased individuals and to analyze the genetic contribution to those disease phenotypes. In the decade since their discovery, human (h)iPSCs, along with human (h)ESCs have shed new light on a variety of disease mechanisms and are increasingly used for identifying drug candidates and assessing drug toxicities in both academia and industry.[7]

16.2.1 Exploring Mechanisms of Pathogenesis

16.2.1.1 Monogenic Diseases

Amongst the first hPSC-based approaches for elucidating disease mechanisms were those explored in the context of monogenic disorders associated with highly penetrant phenotypes. There are several ways to establish such models: (1) hESC can be derived from blastocysts that carry disease-causing mutations, e.g., those

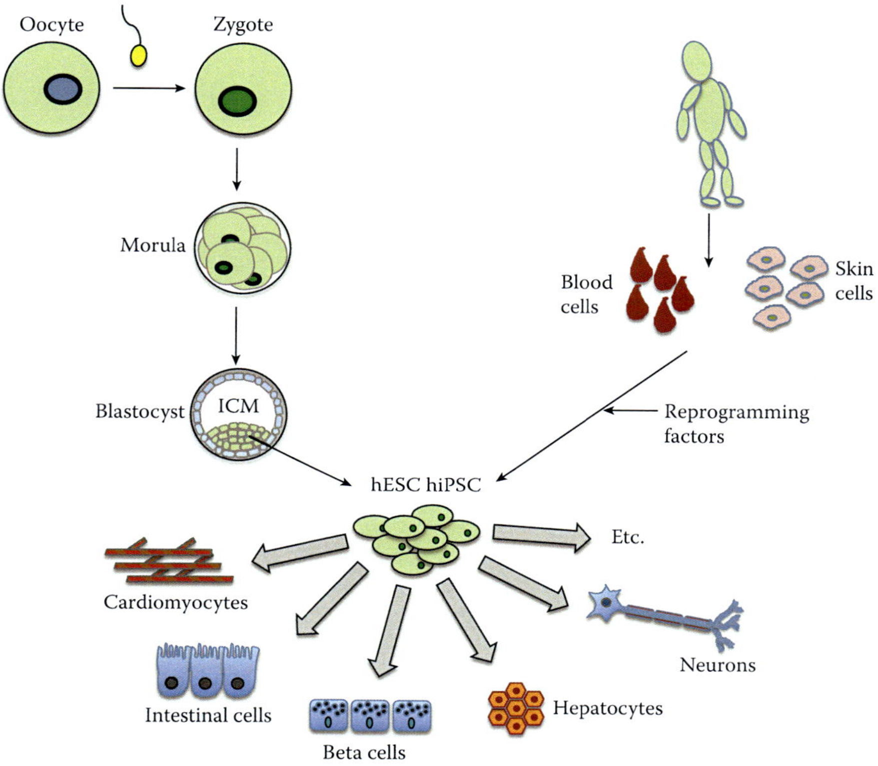

FIGURE 16.1 Generation and utility of human pluripotent stem cells. *Human embryonic stem cells (hESC)*—During *in vitro* fertilization, human sperm are used to fertilize human eggs (oocytes), resulting in the formation of a zygote that continues to develop *in vitro* through the 8–16 cell morula stage to the blastocyst stage, when the pluripotent inner cell mass (ICM) has formed. Blastocysts not needed for reproductive purposes can be donated to research for the derivation of hESC: when ICM cells are cultured under appropriate conditions, they become self-renewing, i.e., have unlimited ability to expand, and remain pluripotent, illustrated here by 5 of many possible differentiation outcomes. *Human induced pluripotent stem cells (hiPSC)*: Somatic cells from tissue donors, including easily obtained blood or skin cells, can be converted to cells resembling hESC by forced expression of a few genes encoding reprogramming factors. Like hESC, hiPSC are self-renewing and pluripotent.

that were donated to research after preimplantation diagnostics revealed the presence of such a mutation; (2) hiPSC can be derived from individuals with genetic diseases, whether the disease-causing mutation is known or not; or (3) relevant mutations can be introduced into normal hPSC through gene editing (Figure 16.2). There have now been many reports of cell-type specific phenotypes observed in differentiated derivatives of such hESCs or hiPSCs, for conditions ranging from neurodevelopmental and neurodegenerative disorders such as spinal muscular atrophy and Huntington's Disease, to syndromes associated with

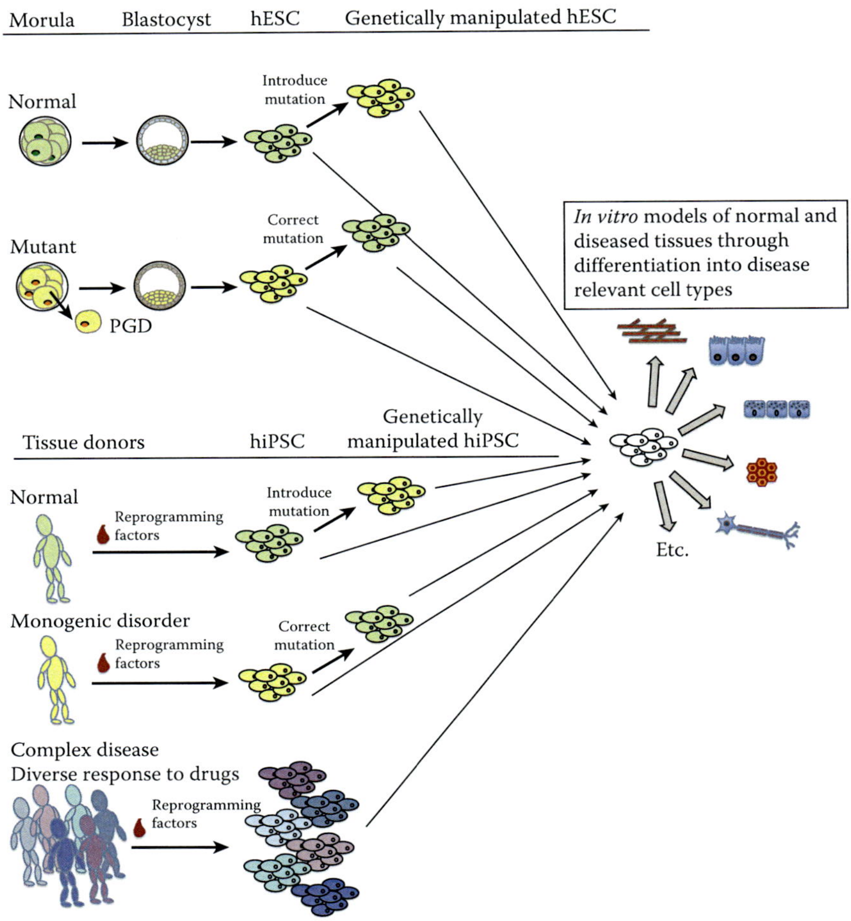

FIGURE 16.2 Genetic background of human pluripotent stem cells. *Human embryonic stem cells (hESC)*—hESC can be generated from apparently normal (indicated by the color green) blastocysts, or from blastocysts that carry mutations known to cause monogenic disorders (yellow). Such embryos are identified through preimplantation genetic diagnosis (PGD). hESC from normal and mutant embryos can be used to compare phenotypes in *in vitro* models of disease-relevant cell types. Furthermore, gene editing technology can be used to either introduce mutations known to cause monogenic disease into normal hESC or to correct the disease-causing mutation in mutant hESC, thereby creating isogenic hPSC lines that differ only at the manipulated locus. *Human induced pluripotent stem cells (hiPSC)*—hiPSC can be derived from normal subjects and those suffering from monogenic disorders, and they can be genetically manipulated to introduce or correct the relevant mutations, respectively, to generate *in vitro* models of normal and diseased tissues. Since hiPSC can be generated from any individual who manifests with a disease, they can also be derived from populations of individuals with complex disease for which the genetic underpinnings may be manifold (different shades of blue), and most often not fully understood.

cardiac arrhythmia, defects in metabolism, and blood disease.[7] Disease-specific phenotypes encompass a diverse spectrum of cell behaviors such as reduced survival or degeneration of specific neurons in culture, arrhythmic beating of cardiomyocytes, aggregation of misfolded proteins, and abnormalities in cell signaling, to name only a few.[7]

While most disease modeling applications of hPSC to date have focused on a single disease-relevant cell type, more complex cellular constructs, such as self-assembling organoids with organ-like tissue architecture (see Section 16.3.2), are beginning to be employed for this purpose. For instance, hiPSC derived from a patient with microcephaly and directed to form cerebral organoids displayed decreased numbers of neural progenitors and increased numbers of neurons consistent with the hypothesis that small brain size is due to a failure of progenitor expansion and to premature neuronal differentiation. Interestingly, spindle orientations during neural progenitor division were often found to be oblique and vertical in microcephaly organoids, while they were exclusively horizontal in control organoids. This provides a possible explanation for the reduction in neural progenitor numbers in the diseased tissues, since precise horizontal spindle orientation is necessary for early symmetric expansion of neural stem cells.[8]

In addition to a rapidly growing body of literature on the use of hPSC to identify and study disease phenotypes in a dish, there is also increasing evidence that hPSC-based phenotypes, representing a wide range of disorders, can be reversed or ameliorated by correcting the genetic defect or by treating cells with drugs known to affect disease outcome, thereby validating the *in vitro* models and fueling hope they can be further developed for identifying novel treatments or diagnostics.[7,9]

16.2.1.2 Complex Diseases

While rare, monogenic disorders might provide the most straightforward path for linking cellular and molecular pathways to a given disease, stem cell–based models are also useful for studying much more common complex disorders where the underlying genetic basis is not or only partially known, and the combined effects of different genes, environmental exposure, and/or lifestyle factors leads to a pathological outcome.[10] Furthermore, the extent and composition of genetic contributions to complex disease may differ amongst diseased individuals, and hiPSC offer the opportunity to study cellular phenotypes in the presence of different genomes. The usefulness of hiPSC models is illustrated by several recent investigations of neurodegenerative disease. While specific alleles at a few genetic loci have been linked to rare familial forms of Parkinson's and Alzheimer's diseases, the majority of cases are sporadic, with variable disease onset and progression due to unknown factors, only some of which are heritable. It has been possible to identify similar defects in hiPSC-derived dopaminergic neurons from patients with either sporadic or genetic forms of Parkinson's disease,[11] and in hiPSC-derived neurons of patients with one of the several genetic or sporadic forms of Alzheimer's disease.[12,13] These studies illustrate that hiPSCs enable disease modeling when the underlying genetic and other causes remain unknown (Figure 16.2), and representative cellular or animal models can thus not be engineered. Similarly, in diseases where some of the genetic risk factors have been identified through genome-wide association studies (GWASs), engineering of appropriate cellular or animal models to investigate their functional significance is hampered by the fact that most of these genetic variants lie in non-coding regions and they themselves may not be causal variants. hiPSC-based disease models offer the opportunity to probe the effects and functional significance of such risk alleles in the context of patient genomes and cell type-specific phenotypes. As a case in point, a recent hiPSC-based study investigated the molecular mechanism by which a genomic region, marked by several single nucleotide polymorphisms (SNPs), increases the risk for sporadic Alzheimer's disease. The SNPs are located in the 5′ non-coding region of the SORL1 gene, whose expression has previously been shown to have a protective effect against Alzheimer's disease. Consistent with an elevated risk for the disease, the presence of the risk variants in patient hiPSC-derived neurons blunted the induction of SORL1 gene expression in response to brain-derived neurotrophic factor (BDNF), providing evidence that disease risk is linked to altered regulation of gene expression.[14] Tremendous effort has gone into identifying genetic risk variants for common diseases through GWASs, but it remains a challenge to define the variants' roles in disease manifestation. hiPSC-based models provide a tool to probe the mechanism by which they alter disease risk, especially in conjunction with recently developed facile approaches for genome editing that allow examination of the functional significance of individual SNPs.[10,15]

16.2.1.3 Infectious Diseases

Disease course and susceptibility to disease by infectious agents is another area of research in which hiPSC modeling may break ground. Several groups have successfully modeled productive infection of hPSC-derived hepatocytes by the hepatitis C virus,[16–19] and by mosquito-borne malaria pathogens such as *Plasmodium vivax*.[20] hPSC models have also been useful for studying infection of hESC-derived neurons by Varicella zoster,[21] and exploring mechanisms of infection and latency of the human cytomegalovirus in hPSC-derived hematopoietic and neural cells, providing novel insights as to how this virus may lead to congenital defects.[22–24] hiPSC-derived cardiomyocytes were infected with coxsackievirus B3 strain, and this model was able to mimic known effects of antiviral compounds.[25] In an example of organoid-based disease modeling (see Section 16.3.2), *Heliobacter pylori* bacteria, which cause ulcers, were injected into the lumen of hPSC-derived gastric organoids and were shown to infect the epithelium and trigger known molecular responses, thus establishing a model for *H. pylori*–mediated gastric disease.[26]

16.2.2 hPSC Models in Drug Discovery and Development

Just as hPSC models represent tools for elucidating mechanisms of disease, they also provide a novel means to screen for drugs that alleviate or cure those diseases. Many drug candidates that show promise in preclinical studies ultimately prove ineffective in costly late-stage clinical trials, often due to the fact that they were discovered and tested in a model system that did not accurately reflect relevant disease features in humans. This phenomenon, together with toxicities first observed when tested in humans, contributes to significant attrition of drug candidates and is a major driver of the high cost of drug development.

In addition to poor correspondence between preclinical models and human subjects, some blame high drug attrition rates, at least in part, on limitations of target-based drug discovery, the main approach used by the pharmaceutical industry in recent decades to identify new disease-modifying compounds.[27,28] The notion is that poor choice of molecular targets contributes to the lack of drug efficacy often encountered in late stage phase II and III clinical trials, while well-designed phenotypic screens that aim to correct an abnormal cellular process rather than a molecular target's activity may deliver better results. As with target-based approaches, the appropriate design of phenotypic screens still depends on understanding the disease biology sufficiently to identify *in vitro* phenotypes that, if reversed, would change disease outcomes. If hiPSCs deliver on the promise of providing access to more authentic disease-relevant, patient-derived cellular models than previously possible, phenotypic screening, or target-based screening in disease-relevant cell types, may indeed revitalize the drug discovery process.

In an early proof-of-principle experiment, hiPSCs prepared from patients with a rare monogenic disorder, the peripheral neuropathy familial dysautonomia (FD), were differentiated into neural crest progenitors, the cells that give rise to sensory and autonomic neurons affected in that disease, and found to display tissue-specific phenotypes. Those phenotypes included decreased expression level of the normal form of the IKBKAP gene, which is due to a splicing error, the molecular basis of the disease, and two cellular phenotypes, namely defects in autonomic neuron differentiation and cell migration.[29] Kinetin, which had been identified in a screen of 1040 compounds as a molecule that can increase levels of normal IKBKAP in an FD lymphoblast cell line,[30] had a similar effect in FD hiPSC-derived neural crest progenitors. A comparable target-based screen of 6912 compounds was performed in these disease-relevant cells to identify additional molecules. Interestingly, two of the eight hits were able to reverse the differentiation defect in FD hiPSC-derived neural crest progenitors while kinetin did not,[31] illustrating the utility of investigating cellular phenotypes in disease-relevant cell types. However, none of the eight hits, nor kinetin, reversed the migration defect,[31] an observation that merits further investigation.

In subsequent years, others have described library screens in which potentially active drugs were identified that reverse phenotypes in disease-relevant cell types, mostly neurons, derived from either normal hPSCs or from patient-specific hiPSCs (reviewed in[9,32]). For instance, screens of hundreds to thousands of compounds were performed in normal hPSC-derived neurons to identify molecules that prevent β-amyloid 1–42–induced cell death,[33] that modulate the Wnt/beta-catenin signaling pathway,[34] or that regulate a transcriptional repressor[35] in search of potential therapeutic compounds for Alzheimer's disease, neuropsychiatric disorders, or Huntington's disease, respectively. Similarly, hiPSC-derived neurons from patients with amyotrophic lateral sclerosis (ALS) were screened for molecules that reverse TDP-43 aggregation,[36] while hiPSC-derived hepatocyte-like cells from patients with the liver disorder alpha-1 antitrypsin (AAT) deficiency were used to identify compounds that reduce AAT protein accumulation.[37] While there are not yet any clinical trials based on drugs discovered through such a platform, there is general optimism that with further advancements, this approach will prove fruitful in the near future.

In addition to the challenge of identifying appropriate molecular targets or cellular phenotypes for drug screening campaigns, another critical hurdle encountered in drug development is that some drugs are only effective in specific subpopulations of patients. Since in many cases the basis for these differences is unknown, appropriate patient cohorts likely to respond to the drug candidate cannot be identified prior to starting clinical trials. Therefore, drug candidates may fail to show efficacy even if some patients respond well to the drug, because others don't. A promising opportunity lies in the creation of panels of hiPSC lines that reflect the genetic diversity of the human population, which may prove useful for stratifying patient populations into likely responders and nonresponders prior to initiating clinical trials (Figure 16.2). In an even more advanced scenario, hiPSCs from clinical trial participants could be used in assays that mirror the human trial and serve to interrogate mechanisms of differential efficacy observed during the trial. Such findings could be used to identify predictive biomarkers that can be validated in samples from clinical trial participants, and inform future clinical trial design to e.g., enable effective subgroup analyses.[38,39]

Investigators are optimistic that with further advances, it may even be possible to expand use of hPSC-based models to other areas of drug development that have traditionally relied on the use of animal models for predicting human physiological responses, such as the pharmacokinetic/pharmacodynamic characterization of novel drugs. In general, the entry of hPSC-based models into the drug development process will require validation of their predictive capability across broad drug classes and clinical outcomes, and much progress has already been made in this regard for the use of hPSC-derived cardiomyocytes in predicting drug toxicities. In addition, though, they will likely contribute important new tools for other steps in the drug development pipeline, lowering drug attrition rates during late-stage clinical trials by improving the accuracy of preclinical predictions.

16.3 CURRENT LIMITATIONS AND CHALLENGES IN hPSC-BASED DISEASE MODELING

16.3.1 Complexity of Human Biology

Although hPSCs overcome certain limitations of previous cellular and animal models by enabling, at least in principle, the production of large quantities of human cells with genomes and cellular identities relevant to disease, a number of significant challenges remain before this technology can prove broadly transformative. As with any model, the design of hPSC-based disease models relies on incomplete understanding as to which particular cell type or cell types are involved in and affected by disease, and whether a phenotype observed *in vitro* accurately represents a relevant disease phenotype at the *in vivo* level. This latter issue is exacerbated by the fact that cells grown on a flat, hard surface in a cell culture dish experience a very different environment than cells that reside in a living organ, and although some disease states may be truly cell autonomous, many disease phenotypes involve interactions between multiple cell types, and/or reflect a complex interplay of cellular, extracellular, systemic, environmental, and other factors. Consistent with this notion, in some hPSC-based disease studies, certain phenotypes were revealed or exacerbated under conditions of cellular stress or experimentally induced aging. For instance, hiPSC-derived dopaminergic neurons from Parkinson's disease patients were more prone to cell death in response to oxidative stress than those from normal individuals, while genetically-induced aging was necessary to reveal dendrite degeneration and an inability to rescue dopaminergic neuron loss in a mouse model of the disease.[7,40] Inducing aging is of general interest for late-onset diseases where genotype and age conspire to trigger disease, especially since there is evidence that hiPSC generation rejuvenates various characteristics of aged cells.[40–42] In another example, a mechanical cue was needed to reveal a difference between normal and diseased hiPSCs, as exposure to shear stress to simulate hemodynamic forces, which are known to protect heart valves from calcification, triggered expression of anti-osteogenic gene networks, as expected, in wild type hiPSC-derived endothelial cells but not in those with NOTCH1 haploinsufficiency, a condition that causes aortic valve calcification.[43]

To model the role of cell-cell interactions in disease course, investigators co-culture two cell types typically found juxtaposed in an organ. For instance, in addition to defects intrinsic to motor neurons, their degeneration in ALS has been shown to involve effects exerted by neighboring astrocytes, at least in the case of the rare familial form of ALS caused by mutations in the gene that encodes superoxide dismutase (SOD). When normal hPSC-derived motor neurons are cultured in the presence of astrocytes from SOD mutant mice they die, whereas astrocytes from wild type mice do not have this toxic effect (reviewed in[44]). Interestingly, in a study of familial ALS caused by a mutation in TDP-43, hiPSC-derived astrocytes displayed cell-autonomous phenotypes including reduced survival, but did not cause death of normal hiPSC-derived motor neurons.[45] Clearly, much remains to be learned about the pathogenic mechanisms of familial and sporadic ALS, and co-culture of hiPSC-derived motor neurons and astrocytes will continue to serve as an effective tool.

In order to provide human cellular models with a more authentic physiological environment than a culture dish, researchers have long transplanted them into immune-deficient mice. Prominent examples include tumor xenografts and the reconstitution of the hematopoietic system of irradiated mice with human hematopoietic stem cells (HSCs). HSCs can be obtained from cord blood or through mobilization into the periphery in adults. Other human cell types are procured from fetal or cadaveric tissues to replace, e.g., mouse liver or thymus to create humanized mice, thereby generating valuable models for the study of human cells in the context of whole animal physiology. Such *in vivo* platforms can now take advantage of hiPSC-derived cells that harbor genetic predispositions to disease—for instance mature human intestinal epithelium developed from hPSC-derived intestinal cells[46] following transplantation under the kidney capsule of immune-deficient mice—setting the stage for the analysis of intestinal disease in future studies utilizing patient hiPSCs. Similar approaches can be envisioned for diseases that are known to involve systemic interactions, such as the autoimmune disease type 1 diabetes, where mice harboring hiPSC-derived pancreatic beta cells, the cells destroyed in disease, and also hiPSC-derived thymus and HSCs to model the defective immune system, could recapitulate the disease etiology, using patient cells.[47]

16.3.2 Functionality and Diversity of *In Vitro* hPSC-Derived Cell Types

Another major hurdle toward creating faithful disease models in a dish is the current inability to produce functional adult cell types with authentic, mature phenotypes from hPSCs *in vitro*. Molecular and functional analyses of *in vitro*–differentiated hPSC products show that the cells typically acquire a fetal rather than adult state. The reasons for this inability to reach maturity remain largely unknown, although evidence exists that PSCs have the capacity to mature into functional adult cell types. In mice, this is best illustrated by the ability of ESCs and iPSCs to generate entire fertile animals through tetraploid complementation, a technique in which PSCs are injected into a blastocyst that was manipulated such that host cells are excluded from contributing to the embryo proper, and thus the resulting mice are derived entirely from injected PSCs.[48–51] Similar experiments cannot be ethically performed using human PSCs, but when injected into adult immune-deficient mice, hPSCs form teratomas, a form of tumor that contains many different tissue types. Interestingly, it has now been demonstrated that teratomas derived from hiPSCs that were co-injected with hematopoietic support cells, with or without hematopoietic cytokines, give rise to functional repopulating human HSCs.[52,53] This finding is evidence that hPSCs can acquire the HSC state, while attempts at directed differentiation toward this, and so many other adult functional cell states *in vitro*, remain unsuccessful. Furthermore, there are examples of hPSC-derived immature cell populations that, once transplanted into animal models, mature and become functional, such as the maturation of beta cell progenitors,[54] hepatocyte progenitors,[55] or oligodendrocyte progenitors[56] *in vivo*.

Attempts to develop protocols for *in vitro* directed differentiation often have the goal of creating a pure population of a particular cell type, and are typically guided by what is known about the cues cells respond to during embryonic development. This knowledge is likely incomplete, especially as it pertains to later stages of tissue maturation. Not knowing all the cues that cells need to acquire a desired cell state, and possibly not even knowing the identity of all cell types involved in disease progression, makes the *in vitro* generation of disease-relevant cellular models from hPSCs challenging.

One way to mimic the *in vivo* environment of developing cells, rather than attempting to systematically provide all relevant cellular, extracellular, and other cues, is to allow cells to re-create cellular complexity normally present *in vivo* by promoting organoid development. Organoids are derived from hPSCs or adult stem cells placed in a three-dimensional (3D) environment that enables recapitulation of endogenous developmental, homeostatic, or regenerative processes, such that they self-organize into 3D structures containing multiple cell types that are arranged in local patterns resembling those found in the corresponding organs.[57,58] For instance, during the development of intestinal organoids from hPSCs, epithelial–mesenchymal boundaries formed, suggesting that crosstalk between these cell layers, which is known to be important in normal development, may have contributed to the successful generation of various intestinal cell types in the organoids.[59] Similarly, cerebral and retinal organoids undergo developmental processes that include formation of juxtaposed regions that resemble signaling boundaries known to participate in local cellular interactions in developing embryos.[8,60,61]

Although organoids appear to facilitate the development of various cell types that are otherwise difficult to obtain in isolation, they generally remain, similar to their two-dimensional (2D) counterparts, functionally immature.[57] For instance, gastric and lung organoids generate multiple cell types typical of the respective organ, but remain in an immature state.[26,62] Similarly, hPSC-derived retinal organoids formed a multilayered neural retina with multiple fetal-like cell types, and photoreceptors did not become light sensitive.[60] Interestingly, comparing retinal organoids developing from mouse and human PSC showed species-specific timing differences that could be altered by molecular manipulation,[60] suggesting that long culture times may be needed for some human cell types to reach maturity based on intrinsic developmental timing, or specific manipulations may be needed to accelerate it.

Not all organ-resident cell types form in organoids, partly because they normally arrive in developing organs through cell migration, rather than being induced by local cell-cell interactions. Their absence may affect the developmental progression of local cell types. For instance, innervation or signals released from immune and other blood cells can contribute to tissue maturation during development. Similarly, lack of vasculature, and associated lack of oxygen and nutrient supply, has been recognized as a limitation of organoid development, especially as it pertains to overall growth potential, but it may also have effects on cell diversity and maturation within a given organ, as the endothelium is known to contribute important signaling functions during development. As a case in point, Takebe et al.[55] combined hiPSC-derived hepatic endoderm cells with human endothelial cells and human mesenchymal stem cells, which together self-organized into liver buds that were still immature but exhibited higher level liver functions than those generated without the added cells.

Although it remains unknown which exact maturation status is needed for any given hPSC-based model to best represent a disease phenotype, and interesting deficiencies consistent with known disease manifestations are observed using current differentiation protocols, it is generally recognized that for hPSCs to reach their full potential as predictive tools, researchers need to devise differentiation protocols that result in authentic cell types. Systematic efforts, employing comprehensive transcriptomic, epigenomic, and other omics analyses to compare *in vitro*–derived cell states to those found in developing human tissues, are under way to help devise protocols that guide cells to acquire desired cell fates. Investigators continue to attempt to mimic normal developmental signaling cues, while some screen small molecule libraries to identify compounds that direct differentiation steps. Since there is ample evidence that hPSCs can reach mature states when placed *in vivo* and that they can assume diverse cell fates when provided with a 3D environment *in vitro*, this goal appears to be reachable.

16.4 ORGAN-ON-A-CHIP TECHNOLOGY IN DISEASE MODELING

As described in detail in previous chapters, organ-on-a-chip technology holds tremendous promise for improving the authenticity of *in vitro*–generated cell-based models through generation of biomimetic microsystems containing multiple organ-relevant cell types that interact with each other through regulated fluid flow. Unlike 2D cultures or 3D organoids, organs-on-chips can be designed to enable interaction of tissue resident cells with circulating immune and other blood cells, and engineered tissue-tissue interfaces can mimic tissue barrier functions such as those between endothelium and e.g., alveoli (alveolar-capillary interface of the lung) or astrocytes (blood-brain barrier) to analyze transcellular transport, and absorption and secretion of endogenously produced or administered molecules. Precise control of cell culture parameters, such as fluid flow, coupled with microsensing capabilities enables real-time control and analysis of drug responses, tissue barrier integrity, cell migration, fluid pressure, and mechanical activity of contractile tissues to provide a comprehensive view of biological processes.[63]

16.4.1 Improving Differentiation

The validity of organ-on-a-chip models depends on the availability of authentic cell types that represent those of the organ in question, and current studies mainly make use of cell lines and cells isolated from primary tissues. As with 2D or organoid-based modeling of human biology, including disease biology, organ-on-a-chip models would greatly benefit from the use of hPSCs because they provide easy access to a variety of human cell types with genomic backgrounds representative of apparently healthy individuals and those predisposed or destined to develop disease. While the functionality and maturity of hPSC-derived cell types represents a hurdle, there is evidence that organ-on-a-chip technology may alleviate some of the differentiation barriers.

As described earlier, there is growing evidence that hPSCs progressing through developmental stages within organoids create cellular and extracellular environments that guide cells to adopt different cell fates normally found in an organ. However, there are additional factors, such as mechanical forces, that are known to affect cell fate decisions. Such forces do not occur in developing organoids but can be simulated in organs-on-chips. For instance, the various cell types that line kidney tubules are normally exposed to fluid sheer stress, and organ-on-a-chip studies, using primary rat or human kidney cells, have shown that application of physiologically relevant fluid sheer stress improved renal cell functionality, such as epithelial cell polarization, primary cilia formation, and molecular transport activity.[64,65] Similarly, dynamic hydraulic compression improved osteogenic differentiation of human mesenchymal stem cells[66] and trickling flow and cyclic mechanical distortions that mimic peristaltic motions triggered a human intestinal epithelial cell line to form a polarized columnar epithelium with multiple intestinal cell types that grows into folds, recapitulating the structure of intestinal villi and producing a better mimic of whole intestine than achieved with conventional culture.[67,68] Similar to this intestine-on-a-chip, functionally complex intestinal models have also been obtained from hPSCs in organoid culture.[59] Aside from comparing these two different models it would be interesting to determine whether application of trickling flow and cyclic mechanical distortions further improves maturation of cell types in the hPSC-derived organoid system. While such organoids would have to be miniaturized for incorporation into chips, a similar approach has already been implemented for liver microtissues that were 3D-engineered from hepatocytes and fibroblasts and introduced into chips to study drug responses under flow.[69] This approach took advantage of the greater complexity of a 3D microenvironment, which served to stabilize liver-specific function over several weeks.

Which exact approach is chosen to optimize hPSC-containing organs-on-chips will depend on the demands of the model in terms of functionality of cell types and intended readouts as well as the need for scale-up and automation. Parallel advances in other technologies, such as 3D printing and biomaterial design, may further increase the complexity of tissues that could be incorporated within chips and the throughput at which they could be produced and analyzed. When designing organ-on-a-chip experiments, it is important, however, to keep in mind that the choice of approach may impact the differentiation state of hPSC-derived cell types.

16.4.2 Organ-on-a-Chip Disease Models

Several studies that describe the use of biomimetic microdevices for the generation of disease models have now been published. A great example that illustrates the transformative potential of this approach is the simulation of drug-induced pulmonary edema in a lung-on-a-chip.[70] This model consists of alveolar and vascular channels, seeded with a human cell line and primary cells, respectively, that are apposed, to create tissue-tissue interfaces. Air and fluid flow as well as cyclic mechanical strain, mimicking breathing motions, are applied to recreate lung biology. Perfusion of the vascular channel with IL-2, which is administered systemically in patients to treat certain cancers, elicited vascular leakage and thus fluid accumulation, or edema, in the alveolar air space, a known side effect of IL-2. While mechanical stretch alone did not compromise barrier integrity in this model, the response to IL-2 was much more pronounced in the presence of cyclic mechanical strain as compared to IL-2 treatment alone,[70] showcasing the power of the microdevice approach in mimicking organ biology.

In an example of hPSC-based disease modeling, drug responsiveness in a myocardium-on-a-chip system showed better concordance with clinical observations and large-scale animal models than that observed in 2D hiPSC-based models.[71] Although 2D cultures of hPSC-derived cardiomyocytes are routinely used to query cardiac conditions, this microdevice system better mimicked myocardial organization through creation of aligned tissue structure during differentiation and it also simulated microcirculation, allowing for continuous exchange of nutrients, metabolites, and drugs, as occurs *in vivo*. In a different study, a monogenic mitochondrial myopathy called Barth syndrome was modeled using a heart-on-a-chip approach to investigate abnormalities in sarcomere arrangement and contractility in patient-specific hiPSC-derived cardiomyocytes and to assess the effectiveness of potential drug interventions.[72] While conventional 2D cultures use measurements of contractile function in single cardiomyocytes, the heart-on-a-chip model enabled analysis of impaired contractile stress generation in Barth syndrome myocardial tissue constructs.

As hPSC differentiation protocols improve over the next few years, possibly aided by mechanical and other characteristics that can be mimicked in microdevices, the incorporation of hiPSC-derived cell types will empower organs-on-chips to model disease in the context of human disease genetics.

16.4.3 Organ-on-a-Chip Models for Drug Discovery

The role that organ-on-a-chip models will play in drug discovery will partly depend on the extent to which their production can be scaled up and automated. While this technology may not support high throughput screening campaigns, it will likely provide more authentic models of human organ biology for validation and optimization of hits from primary screens than offered by current cell-based assays and animal models. To date, organ-on-a-chip models have illustrated a remarkable ability to report on many aspects of human organ biology, presenting effective new tools for de-risking the drug development process. Additionally, they may offer other efficiencies such as reduced supply costs due to their micro scale and reduced animal use.

As with any model, the predictive ability of organs-on-chips needs to be validated, and readouts must represent the clinical endpoints they are intended to model.[38] Much work is needed to reach that point but as an example, the lung-on-a-chip study of pulmonary edema described earlier tested two molecular interventions predicted to alleviate the IL-2–induced vascular leakage. The investigators showed that a molecule known to stabilize endothelial intercellular junctions, Ang-1, and a compound that blocks TRPV4, an ion channel known to cause increased permeability when stimulated by mechanical strain, reverted the effects of IL-2. Importantly, the investigators were able to replicate the ameliorating effects of these two molecules in murine whole lungs that suffered from increased barrier permeability due to IL-2 exposure.[70]

The impact of biomimetic microdevices on the drug development process will be greatest in cases where no good cellular or animal models currently exist. In addition though, the organ-level functionality displayed by organ-on-a-chip models will likely prove superior to many current cell-based assays, as long as they are amenable to reproducible production and also to adaptation to automatic robotic platforms and to scale up as needed for the stage of drug development they are intended to serve. Their impact will be further augmented when combined with hiPSC-derived cell types to access diverse human genomic backgrounds, especially as they represent the genetic basis of complex disease traits or the diversity in drug response and metabolism among future patients.

REFERENCES

1. Evans, M.J. & Kaufman, M.H. Establishment in culture of pluripotential cells from mouse embryos. *Nature* 292 no. 5819 (1981): 154–156.
2. Martin, G.R. Isolation of a pluripotent cell line from early mouse embryos cultured in medium conditioned by teratocarcinoma stem cells. *Proc Natl Acad Sci U S A* 78 no. 12 (1981): 7634–7638.

3. Thomson, J.A., Itskovitz-Eldor, J., Shapiro, S.S. et al. Embryonic stem cell lines derived from human blastocysts. *Science* 282 no. 5391 (1998): 1145–1147.
4. Takahashi, K. & Yamanaka, S. Induction of pluripotent stem cells from mouse embryonic and adult fibroblast cultures by defined factors. *Cell* 126 no. 4 (2006): 663–676.
5. Takahashi, K., Tanabe, K., Ohnuki, M. et al. Induction of pluripotent stem cells from adult human fibroblasts by defined factors. *Cell* 131 no. 5 (2007): 861–872.
6. Yu, J., Vodyanik, M.A., Smuga-Otto, K. et al. Induced pluripotent stem cell lines derived from human somatic cells. *Science* 318 no. 5858 (2007): 1917–1920.
7. Sterneckert, J.L., Reinhardt, P. & Scholer, H.R. Investigating human disease using stem cell models. *Nat Rev Genet* 15 no. 9 (2014): 625–639.
8. Lancaster, M.A., Renner, M., Martin, C.A. et al. Cerebral organoids model human brain development and microcephaly. *Nature* 501 no. 7467 (2013): 373–379.
9. Ko, H.C. & Gelb, B.D. Concise review: Drug discovery in the age of the induced pluripotent stem cell. *Stem Cells Transl Med* 3 no. 4 (2014): 500–509.
10. Grieshammer, U. & Shepard, K.A. Proceedings: Consideration of genetics in the design of induced pluripotent stem cell-based models of complex disease. *Stem Cells Transl Med* 3 no. 11 (2014): 1253–1258.
11. Sanchez-Danes, A., Richaud-Patin, Y., Carballo-Carbajal, I. et al. Disease-specific phenotypes in dopamine neurons from human iPS-based models of genetic and sporadic Parkinson's disease. *EMBO Mol Med* 4 no. 5 (2012): 380–395.
12. Israel, M.A., Yuan, S.H., Bardy, C. et al. Probing sporadic and familial Alzheimer's disease using induced pluripotent stem cells. *Nature* 482 no. 7384 (2012): 216–220.
13. Kondo, T., Asai, M., Tsukita, K. et al. Modeling Alzheimer's disease with iPSCs reveals stress phenotypes associated with intracellular Abeta and differential drug responsiveness. *Cell Stem Cell* 12 no. 4 (2013): 487–496.
14. Young, J.E., Boulanger-Weill, J., Williams, D.A. et al. Elucidating molecular phenotypes caused by the SORL1 Alzheimer's disease genetic risk factor using human induced pluripotent stem cells. *Cell Stem Cell* 16 no. 4 (2015): 373–385.
15. Merkle, F.T. & Eggan, K. Modeling human disease with pluripotent stem cells: From genome association to function. *Cell Stem Cell* 12 no. 6 (2013): 656–668.
16. Wu, X., Robotham, J.M., Lee, E. et al. Productive hepatitis C virus infection of stem cell-derived hepatocytes reveals a critical transition to viral permissiveness during differentiation. *PLoS Pathog* 8 no. 4 (2012): e1002617.
17. Yoshida, T., Takayama, K., Kondoh, M. et al. Use of human hepatocyte–like cells derived from induced pluripotent stem cells as a model for hepatocytes in hepatitis C virus infection. *Biochem Biophys Res Commun* 416 no. 1–2 (2011 late): 119–124.
18. Schwartz, R.E., Trehan, K., Andrus, L. et al. Modeling hepatitis C virus infection using human induced pluripotent stem cells. *Proc Natl Acad Sci U S A* 109 no. 7 (2012): 2544–2548.
19. Roelandt, P., Obeid, S., Paeshuyse, J. et al. Human pluripotent stem cell-derived hepatocytes support complete replication of hepatitis C virus. *J Hepatol* 57 no. 2 (2012): 246–251.
20. Ng, S., Schwartz, R.E., March, S. et al. Human iPSC-derived hepatocyte-like cells support plasmodium liver-stage infection in vitro. *Stem Cell Reports* 4 no. 3 (2015): 348–359.
21. Markus, A., Grigoryan, S., Sloutskin, A. et al. Varicella-zoster virus (VZV) infection of neurons derived from human embryonic stem cells: Direct demonstration of axonal infection, transport of VZV, and productive neuronal infection. *J Virol* 85 no. 13 (2011 missed): 6220–6233.
22. Belzile, J.P., Stark, T.J., Yeo, G.W. & Spector, D.H. Human cytomegalovirus infection of human embryonic stem cell-derived primitive neural stem cells is restricted at several steps but leads to the persistence of viral DNA. *J Virol* 88 no. 8 (2014): 4021–4039.

23. Penkert, R.R. & Kalejta, R.F. Human embryonic stem cell lines model experimental human cytomegalovirus latency. *MBio* 4 no. 3 (2013): e00298–00213.
24. D'Aiuto, L., Di Maio, R., Heath, B. et al. Human induced pluripotent stem cell-derived models to investigate human cytomegalovirus infection in neural cells. *PLoS One* 7 no. 11 (2012): e49700.
25. Sharma, A., Marceau, C., Hamaguchi, R. et al. Human induced pluripotent stem cell-derived cardiomyocytes as an in vitro model for coxsackievirus B3-induced myocarditis and antiviral drug screening platform. *Circ Res* 115 no. 6 (2014): 556–566.
26. McCracken, K.W., Cata, E.M., Crawford, C.M. et al. Modelling human development and disease in pluripotent stem-cell-derived gastric organoids. *Nature* 516 no. 7531 (2014): 400–404.
27. Zheng, W., Thorne, N. & McKew, J.C. Phenotypic screens as a renewed approach for drug discovery. *Drug Discov Today* 18 no. 21–22 (2013): 1067–1073.
28. Swinney, D.C. & Anthony, J. How were new medicines discovered? *Nat Rev Drug Discov* 10 no. 7 (2011): 507–519.
29. Lee, G., Papapetrou, E.P., Kim, H. et al. Modelling pathogenesis and treatment of familial dysautonomia using patient-specific iPSCs. *Nature* 461 no. 7262 (2009): 402–406.
30. Slaugenhaupt, S.A., Mull, J., Leyne, M. et al. Rescue of a human mRNA splicing defect by the plant cytokinin kinetin. *Hum Mol Genet* 13 no. 4 (2004): 429–436.
31. Lee, G., Ramirez, C.N., Kim, H. et al. Large-scale screening using familial dysautonomia induced pluripotent stem cells identifies compounds that rescue IKBKAP expression. *Nat Biotechnol* 30 no. 12 (2012): 1244–1248.
32. Engle, S.J. & Vincent, F. Small molecule screening in human induced pluripotent stem cell-derived terminal cell types. *J Biol Chem* 289 no. 8 (2014): 4562–4570.
33. Xu, X., Lei, Y., Luo, J. et al. Prevention of beta-amyloid induced toxicity in human iPS cell-derived neurons by inhibition of Cyclin-dependent kinases and associated cell cycle events. *Stem Cell Res* 10 no. 2 (2013): 213–227.
34. Zhao, W.N., Cheng, C., Theriault, K.M. et al. A high-throughput screen for Wnt/beta-catenin signaling pathway modulators in human iPSC-derived neural progenitors. *J Biomol Screen* 17 no. 9 (2012): 1252–1263.
35. Charbord, J., Poydenot, P., Bonnefond, C. et al. High throughput screening for inhibitors of REST in neural derivatives of human embryonic stem cells reveals a chemical compound that promotes expression of neuronal genes. *Stem Cells* 31 no. 9 (2013): 1816–1828.
36. Burkhardt, M.F., Martinez, F.J., Wright, S. et al. A cellular model for sporadic ALS using patient-derived induced pluripotent stem cells. *Mol Cell Neurosci* 56 no. (2013): 355–364.
37. Choi, S.M., Kim, Y., Shim, J.S. et al. Efficient drug screening and gene correction for treating liver disease using patient-specific stem cells. *Hepatology* 57 no. 6 (2013): 2458–2468.
38. Esch, E.W., Bahinski, A. & Huh, D. Organs-on-chips at the frontiers of drug discovery. *Nat Rev Drug Discov* 14 no. 4 (2015): 248–260.
39. Chen, Z., Cheng, K., Walton, Z. et al. A murine lung cancer co-clinical trial identifies genetic modifiers of therapeutic response. *Nature* 483 no. 7391 (2012): 613–617.
40. Miller, J.D., Ganat, Y.M., Kishinevsky, S. et al. Human iPSC-based modeling of late-onset disease via progerin-induced aging. *Cell Stem Cell* 13 no. 6 (2013): 691–705.
41. Mahmoudi, S. & Brunet, A. Aging and reprogramming: A two-way street. *Curr Opin Cell Biol* 24 no. 6 (2012): 744–756.
42. Freije, J.M. & Lopez-Otin, C. Reprogramming aging and progeria. *Curr Opin Cell Biol* 24 no. 6 (2012): 757–764.

43. Theodoris, C.V., Li, M., White, M.P. et al. Human disease modeling reveals integrated transcriptional and epigenetic mechanisms of NOTCH1 haploinsufficiency. *Cell* 160 no. 6 (2015): 1072–1086.
44. Veyrat-Durebex, C., Corcia, P., Dangoumau, A. et al. Advances in cellular models to explore the pathophysiology of amyotrophic lateral sclerosis. *Mol Neurobiol* 49 no. 2 (2014): 966–983.
45. Serio, A., Bilican, B., Barmada, S.J. et al. Astrocyte pathology and the absence of non-cell autonomy in an induced pluripotent stem cell model of TDP-43 proteinopathy. *Proc Natl Acad Sci U S A* 110 no. 12 (2013): 4697–4702.
46. Watson, C.L., Mahe, M.M., Munera, J. et al. An in vivo model of human small intestine using pluripotent stem cells. *Nat Med* 20 no. 11 (2014): 1310–1314.
47. Melton, D.A. Using stem cells to study and possibly treat type 1 diabetes. *Philos Trans R Soc Lond B Biol Sci* 366 no. 1575 (2011): 2307–2311.
48. Zhao, X.Y., Li, W., Lv, Z. et al. iPS cells produce viable mice through tetraploid complementation. *Nature* 461 no. 7260 (2009): 86–90.
49. Nagy, A., Rossant, J., Nagy, R., Abramow-Newerly, W. & Roder, J.C. Derivation of completely cell culture-derived mice from early-passage embryonic stem cells. *Proc Natl Acad Sci U S A* 90 no. 18 (1993): 8424–8428.
50. Kang, L., Wang, J., Zhang, Y., Kou, Z. & Gao, S. iPS cells can support full-term development of tetraploid blastocyst-complemented embryos. *Cell Stem Cell* 5 no. 2 (2009): 135–138.
51. Boland, M.J., Hazen, J.L., Nazor, K.L. et al. Adult mice generated from induced pluripotent stem cells. *Nature* 461 no. 7260 (2009): 91–94.
52. Amabile, G., Welner, R.S., Nombela-Arrieta, C. et al. In vivo generation of transplantable human hematopoietic cells from induced pluripotent stem cells. *Blood* 121 no. 8 (2013): 1255–1264.
53. Suzuki, N., Yamazaki, S., Yamaguchi, T. et al. Generation of engraftable hematopoietic stem cells from induced pluripotent stem cells by way of teratoma formation. *Mol Ther* 21 no. 7 (2013): 1424–1431.
54. Kroon, E., Martinson, L.A., Kadoya, K. et al. Pancreatic endoderm derived from human embryonic stem cells generates glucose-responsive insulin-secreting cells in vivo. *Nat Biotechnol* 26 no. 4 (2008): 443–452.
55. Takebe, T., Sekine, K., Enomura, M. et al. Vascularized and functional human liver from an iPSC-derived organ bud transplant. *Nature* 499 no. 7459 (2013): 481–484.
56. Keirstead, H.S., Nistor, G., Bernal, G. et al. Human embryonic stem cell-derived oligodendrocyte progenitor cell transplants remyelinate and restore locomotion after spinal cord injury. *J Neurosci* 25 no. 19 (2005): 4694–4705.
57. Lancaster, M.A. & Knoblich, J.A. Organogenesis in a dish: Modeling development and disease using organoid technologies. *Science* 345 no. 6194 (2014): 1247125.
58. Sato, T. & Clevers, H. Growing self-organizing mini-guts from a single intestinal stem cell: Mechanism and applications. *Science* 340 no. 6137 (2013): 1190–1194.
59. Spence, J.R., Mayhew, C.N., Rankin, S.A. et al. Directed differentiation of human pluripotent stem cells into intestinal tissue in vitro. *Nature* 470 no. 7332 (2011): 105–109.
60. Nakano, T., Ando, S., Takata, N. et al. Self-formation of optic cups and storable stratified neural retina from human ESCs. *Cell Stem Cell* 10 no. 6 (2012): 771–785.
61. Kadoshima, T., Sakaguchi, H., Nakano, T. et al. Self-organization of axial polarity, inside-out layer pattern, and species-specific progenitor dynamics in human ES cell-derived neocortex. *Proc Natl Acad Sci U S A* 110 no. 50 (2013): 20284–20289.
62. Dye, B.R., Hill, D.R., Ferguson, M.A. et al. In vitro generation of human pluripotent stem cell derived lung organoids. *Elife* 4 no. (2015): e05098.
63. Bhatia, S.N. & Ingber, D.E. Microfluidic organs-on-chips. *Nat Biotechnol* 32 no. 8 (2014): 760–772.

64. Jang, K.J., Mehr, A.P., Hamilton, G.A. et al. Human kidney proximal tubule-on-a-chip for drug transport and nephrotoxicity assessment. *Integr Biol (Camb)* 5 no. 9 (2013): 1119–1129.
65. Jang, K.J. & Suh, K.Y. A multi-layer microfluidic device for efficient culture and analysis of renal tubular cells. *Lab Chip* 10 no. 1 (2010): 36–42.
66. Park, S.H., Sim, W.Y., Min, B.H. et al. Chip-based comparison of the osteogenesis of human bone marrow- and adipose tissue-derived mesenchymal stem cells under mechanical stimulation. *PLoS One* 7 no. 9 (2012): e46689.
67. Kim, H.J., Huh, D., Hamilton, G. & Ingber, D.E. Human gut-on-a-chip inhabited by microbial flora that experiences intestinal peristalsis-like motions and flow. *Lab Chip* 12 no. 12 (2012): 2165–2174.
68. Kim, H.J. & Ingber, D.E. Gut-on-a-chip microenvironment induces human intestinal cells to undergo villus differentiation. *Integr Biol (Camb)* 5 no. 9 (2013): 1130–1140.
69. Li, C.Y., Stevens, K.R., Schwartz, R.E. et al. Micropatterned cell-cell interactions enable functional encapsulation of primary hepatocytes in hydrogel microtissues. *Tissue Eng Part A* 20 no. 15–16 (2014): 2200–2212.
70. Huh, D., Leslie, D.C., Matthews, B.D. et al. A human disease model of drug toxicity-induced pulmonary edema in a lung-on-a-chip microdevice. *Sci Transl Med* 4 no. 159 (2012): 159ra147.
71. Mathur, A., Loskill, P., Shao, K. et al. Human iPSC-based cardiac microphysiological system for drug screening applications. *Sci Rep* 5 no. (2015): 8883.
72. Wang, G., McCain, M.L., Yang, L. et al. Modeling the mitochondrial cardiomyopathy of Barth syndrome with induced pluripotent stem cell and heart-on-chip technologies. *Nat Med* 20 no. 6 (2014): 616–623.

17 *In Vivo*, *In Vitro*, and Stem Cell Technologies to Predict Human Pharmacology and Toxicology*

Harry Salem, Russell Dorsey, Daniel Carmany and Thomas Hartung

CONTENTS

17.1 PHARMACOLOGY AND TOXICOLOGY

"Pharmacology and toxicology" describes the beneficial and adverse effects, respectively, of chemicals on living tissue and organisms. It also includes mechanisms of action, such as the manner in which the drug or chemical acts such as blocking of receptors, enzymes, or stimulating hormone production, etc., or the effects of the drug on the physiologic or biochemical processes within the body to produce a given effect. *Dorland's Medical Dictionary* defines pharmacology as the science that deals with the origin, nature, chemistry, effects, and uses of drugs; it includes pharmacognosy (natural drugs), pharmacokinetics (adsorption, distribution, metabolism, excretion; the body's effects on the chemical), pharmacodynamics (the chemical's effects on the body), pharmacotherapeutics (therapeutic uses and effects of drugs), and toxicology (adverse effects of drugs, poisons, chemicals, detection, antidotes).[1] Drug toxicology has often been considered as exaggerated ("excess") pharmacology. Other terms that should be defined[2] include toxidrome, which is a constellation of clinical

* Disclaimer: The views and opinions expressed in this chapter are those of the authors and should not be construed as an official U.S. Department of the Army position, policy, or decision, unless so designated by other official documentation. Citation of trade names in this chapter does not constitute an official U.S. Department of the Army endorsement or approval of the use of such commercial items.

effects (signs and symptoms) characteristic of a given type of chemical. Toxicity is the intrinsic capacity of a substance to produce injury, while hazard is the capacity of a substance to produce injury under conditions of use, and risk is the probability of a substance to produce injury or harm under specified conditions of use or exposure. Safety, on the other hand, is the probability that injury or harm will not occur under specified conditions of use or exposure.

17.1.1 *In Vivo*

Safety has been paramount ever since this country was formed. Since the founding of our nation and the first half of the 1800s, drugs were not regulated by the federal government. Impure or bogus drugs were usually contained within a state or region. The manufacture and trade of drugs were governed by individual states. Following the Mexican-American War (1846–1848), although only 1773 Americans were killed in action, over 13,000 were killed by collateral causes such as food, poor living conditions, and infections. Public outrage focused on the poor medical care given to the soldiers, and the public concluded that the weak and adulterated drugs supplied to the soldiers caused the large number of deaths in the army. Lewis Caleb Beck's book, *Adulteration of Various Substances Used in Medicine and the Arts*, published in 1846, enraged the public, and the outcry forced Congress to pass the Drug Importation Act of 1848.[3] This was the first federal drug law signed into law by President James K. Polk. It was very limited in scope, and addressed only the purity of drugs. Customs was charged with enforcing the law.[4] This Act required that imported drugs meet the standards for strength and purity established in the United States Pharmacopeia (USP), which was prepared by committees of physicians and pharmacists. The Act was ineffective and short-lived, but it did help to solidify the USP's status as a national compendium, and set a precedent for future federal drug laws.[5]

Upton Sinclair's book *The Jungle* was the impetus that stimulated the passage of the Pure Food and Drug Act of 1906.[6] The book stressed the important issue of the unsanitary meat processing plants. This Act was the forerunner of the Food and Drug Administration (FDA). The main purpose of the Act was to ban foreign and interstate traffic in adulterated or mislabeled food and drug products, and it directed the U.S. Bureau of Chemistry in the Department of Agriculture to inspect products and refer offenders to prosecutors. It required that active ingredients be placed on the label of the drug's packaging and that drugs could not be below purity levels established by the United States Pharmacopeia or the National Formulary. The Pure Food and Drug Act of 1906 was signed by President Theodore Roosevelt on the same day that he signed the Federal Meat Inspection Act. The Bureau of Chemistry was renamed the U.S. Food and Drug Administration (FDA) in 1930.

The U.S. Food, Drug, and Cosmetic Act (FFDCA, FDCA, or FD&C) passed in 1938 gave the authority to the FDA to oversee the safety of food, drugs, and cosmetics. It also subjected new drugs to pre-market safety evaluation for the first time. This Act was influenced by the 100 deaths that occurred in 1937 due to the preparation Elixir Sulfanilamide (S.E. Massengill Co.) Although the sulfanilamide was safe, the diethylene glycol used to dissolve it was not. Thus the final drug preparation was now part of the evaluation along with both preclinical and clinical test results for new drugs.

The 1962 amendment is the Kefauver Harris Amendment, or the Drug Efficacy Amendment to the Federal Food, Drug, and Cosmetic Act. It followed the thalidomide events and required that the clinical tests demonstrated efficacy and safety prior to going to market. The thalidomide tragedy was that thousands of children were born with birth defects from mothers who had taken thalidomide for morning sickness during pregnancy. Thalidomide had not been approved for use in the United States, but was available in other countries, where the birth defects occurred. Francis Oldham Kelsey was the FDA reviewer who refused to approve thalidomide for use in the United States. This Act first introduced "proof of efficacy" that was not required previously. It also required that the drug advertising disclose accurate information on side effects and efficacy. Cheap drugs could not be marketed as expensive drugs under a new trade name as a new breakthrough medication as they were previously. President John F. Kennedy signed this law October 10, 1962.

Today, regulatory requirements for preclinical studies include pharmacology and toxicology studies in animals. These *in vivo* studies are conducted in live animals and consist of acute, subacute, subchronic, and chronic studies, as well as specialized tests such as carcinogenic, irritation, birth defects, or any other adverse effect. In rodents, acute tests are usually dosed once, or the dose is given in divided doses over a 24-hour period. The animals are observed for up to 14 days for overt signs and mortality. Food intake, hematology, and body weights are also observed. The dead animals are necropsied, pathology observed, and an LD_{50}, i.e., the lethal dose where 50% of animals (typically rats) die, is calculated. A vehicle control group should also be used. These studies can be performed in mice, rats, rabbits, and guinea pigs.

Subacute studies can be conducted where the animals are dosed for up to 30 days. These studies can be conducted in mice, rats, rabbits, guinea pigs, and dogs. Typically, groups of five animals of both sexes are dosed with increasing log doses. The animals are dosed daily and observed daily for overt signs, body weight, food consumption, clinical pathology, pathology and histopathology, as well as organ weights. A vehicle control group should also be used.

Subchronic studies can be conducted where the animals are dosed for up to 90 days. The rest of the study should proceed as in the other studies. In the chronic study the animals should be dosed for up to 2 years, and in the carcinogenic studies the animals are dosed for their lifespan.

In all of the aforementioned animal studies, the animals should be dosed by the same route as the humans who will be taking the drug. That is, orally, dermally, inhalation, parenteral, or as eyedrops.

From these data, the doses will be extrapolated to humans. Extrapolations can be made on body weight, body surface area, metabolism, or by other mathematical manipulations. Guidelines are available for any of these type of studies. One of the earliest documents published by the FDA that discussed nonclinical toxicity testing of drugs was "Appraisal of the Safety of Chemicals in Foods, Drugs, and Cosmetics."[7] This document provided guidance and protocols for various toxicity tests.

In May 1968, Goldenthal's article in the FDA Papers Current Views on Safety Evaluation of Drugs provided general guidelines for drug testing.[8] Concurrently, an FDA document titled Guidelines for Reproduction Studies for Safety Evaluation of

Drugs for Human Use (D'Aguanno) was published in which more specific information regarding teratological testing was provided.[9]

A group of pharmacologists and toxicologists from the Environmental Protection Agency (EPA), Consumer Product Safety Commission (CPSC), National Institute of Occupational Safety and Health (NIOSH), and the FDA formed a subcommittee on toxicology within the Interagency Regulatory Liaison Group (IRLG) and began writing acute and chronic toxicity testing guidelines. These guidelines were eventually discussed at an open meeting with representatives of other agencies, academia, and food, chemical, and pharmaceutical companies. The guidelines were subsequently incorporated into several drafts of FDA's book of toxicology guidelines, *Toxicological Principles for the Safety Assessment of Direct Food Additives and Color Additives used in Food*, also known as the *Redbook*.[10]

Since then, FDA pharmacologists and toxicologists have participated in international activities such as the Organization for Economic cooperation and Development (OECD) guideline development, Interagency Coordinating Committee on the Validation of Alternative Methods (ICCVAM) or European Center for the Validation of Alternative Methods (ECVAM) evaluations, and International Conference on Harmonization of Technical Requirements for Registration of Pharmaceuticals for Human Use (ICH) guidance development in efforts to modify existing toxicology protocols or to evaluate *in vitro* replacement methods.

Such methods can be used within the drug development area to generate acceptable nonclinical data to support human drug trials and accurate labeling statements. Because these are guidelines and not regulations, they can be modified as needed to reflect advances in the state of the art of toxicology.[11]

Toxicology studies can be conducted not only to determine safety or adverse effects but also for detection, protection, and decontamination, as well as for countermeasure development.

In the United States, most of the governmental agencies including FDA, EPA, CPSC, Department of Agriculture, Department of Transportation, and the National Toxicology Program require testing or conduct toxicity testing. Ninety six percent of the animals used in toxicology testing include mice, rats, birds, and cold-blooded animals, which are not covered by the Animal Welfare Act.

As of today, there are 97 million organic and inorganic substances in the Chemical Abstract Services Registry, the majority of which have not been evaluated for toxicity. Approximately 15,000 new substances are added each day. Current toxicity testing protocols rely primarily on animal models, which are very expensive and time consuming.

Human toxicity estimates are developed from animal toxicity studies. These are low-throughput, time consuming, and costly.[12] They are not scalable or automatable. To develop human estimates uncertainty factors are utilized in the analysis and extrapolation. They usually result in overestimation of hazard.[13] Olsen et al. (2000) estimated that rodents predict only about 43% of human toxicity, non-rodents predict around 63%, and non-rodents and rodents combined predict about 71%.[14] Andrew von Eschenbach, acting FDA Commissioner (2006–2009), stated, "consider just one stark statistic: today, nine out of ten compounds developed in the lab fail in human studies. They fail, in large part because they behave differently in people than they do in animal or laboratory tests."[15] Discordance among the species used is substantial; reliable extrapolation from animal data to humans is impossible,[16] and virtually all known human teratogens have so far been identified in spite of, rather than because of, animal-based methods. Despite the fact that animal-based teratology studies would fail strict validation criteria, three *in vitro* methods passed them.[17] Despite painstaking attempts to standardize them, animal protocols across laboratories are unable to provide consistent and reproducible results. Human estimate studies are based on animal models that may not accurately reflect the human response due to species-to-species differences in metabolism. These metabolic differences are related to genetics such as the P450 isoenzymes, and not to size. Table 17.1 shows species differences in P450 isoenzymes in the livers of different species. This table demonstrates that the human and rat only share three of the same P450 isoenzymes.[18]

Physiological and pharmacological reactions from drugs and chemicals vary enormously from species to species. For example, penicillin kills guinea pigs but is inactive in rabbits; aspirin kills cats and causes birth defects in rats, mice, guinea pigs, dogs, and monkeys, and causes bleeding and gastrointestinal irritation in humans; morphine is a depressant in humans, but stimulates goats, cats, and horses; dioxin demonstrates variability in oral toxicity in different species, and DDT shows variability due to age in the rat. The species differences in sensitivity to the dioxin TCDD is reported by Birnbaum,[19] and the oral LD_{50} are seen in Table 17.2. In the case of DDT, the oral LD_{50} for the newborn rat is greater than 4000 mg/Kg, which falls to 730 mg/Kg at 10 days, and then to 190 mg/Kg at 4 months, and is 220 mg/Kg in the adult.[19,20] Thus it appears that the neonates are not always more sensitive than the adult to chemical exposures.[20]

TABLE 17.1
Species-Specific Differences in the Most Abundant CYP Isozymes Expressed in the Liver

CYP	Human	Monkey	Rabbit	Rat	Mouse
1A	***1A1, 1A2***	***1A1, 1A2***	***1A1, 1A2***	***1A1, 1A2***	***1A1, 1A2***
2A	2A6, 2A7	–	2A10, 2A11	2A1, 2A2, 2A3	2A4, 2A5
2B	2B6	2B17	2B4, 2B5	2B1, 2B2, 2B3	2B9, 2B10, 2B13
2C	2C8, 2C9, 2C18, 2C19	2C20, 2C37	2C1, 2C2, 2C3, 2C4, 2C5, 2C14, 2C15, 2C16	2C6, 2C7, 2C11, 2C12, 2C13, 2C22, 2C23, 2C24	2C29
2D	2D6	2D17	–	2D1, 2D2, 2D3, 2D4, 2D5	2D9, 2D10, 2D11, 2D12, 2D13
2E	***2E1***	***2E1***	***2E1***, 2E2	***2E1***	***2E1***
3A	3A4, 3A5	3A8	3A6	3A1, 3A2, 3A9	3A11, 3A13, 3A16
4A	4A9, 4A11	–	4A4, 4A5, 4A6, 4A7	4A1, 4A2, 4A3, 4A8	4A10, 4A12, 4A14

Note: Bold, italicized isozymes are those that are shared between animals and humans.

TABLE 17.2
Acute Toxicity of Dioxin (TCDD)

Species	LD_{50} μg/kg po
Guinea pig	0.6–2.5
Mink	4
Rat	22–320
Monkey	<70
Rabbit	115–275
Mouse	114–280
Dog	>100–<3000
Hamster	1150–5000

Transgenic animal models, such as the humanized mouse models, have been in development for years, and appear to be a powerful tool for providing a more detailed understanding on the role of specific genes in biological pathways and systems. These models have been used in the field of toxicology, especially for the screening of mutagenic and carcinogenic potential, and for the characterization of toxic mechanisms of action. The future outlook for these models is in toxicology and risk assessment, and how transgenic technologies will likely be an integral tool for toxicity testing in the twenty-first century.[21]

17.1.2 *In Vitro*

In vitro toxicology is not a new field. Our interest in "the three Rs" goes back to its origins with William Russell and Rex Burch, who published *Principles of Humane Experimental Techniques* in 1959, which coined this term.[22] The three Rs are Refinement, Reduction, and Replacement of animal testing. It is impossible to summarize the plethora of current *in vitro* approaches in pharmacology and toxicology and therefore we use examples from our own laboratories to illustrate opportunities. Our interest at the U.S. Army Edgewood Chemical Biological Center (ECBC) goes back to the early 1980s, when we advocated a fourth R, Responsibility. At that time, in collaboration with other interested scientific organizations, we sponsored biennial DoD Alternative Symposia until the year 2000, and published four books on alternative methods.[23–26] In parallel, initiated originally by the cosmetics industry, the Center for Alternatives to Animal Testing (CAAT) was established in 1981 at the Johns Hopkins School of Public Health.[27] Over time, work has expanded across all industrial sectors and internationally, and since 2010 there is also a European branch. It remains after more than 35 years of existence the premier information hub for the three Rs in the United States and beyond, with its information portal AltWeb (http://altweb.jhsph.edu) and a number of programs promoting the three Rs. (http://caat.jhsph.edu). The leading journal ALTEX, official journal of CAAT, is similarly published for more than 30 years. (http://www.altex.ch/).

The European animal welfare legislation from 1986, Directive 86/609/EEC on the protection of animals used for scientific purposes, committed the European

Commission and the member states to promote alternatives to animal testing and use them whenever they are reasonably available. This led, among others, to the establishment of the European Centre for the Validation of Alternative Methods, which one of us (T.H.) led from 2002–2008. This first establishment of a validation body was followed by many countries (see below) and the international collaboration of these centers more than 50 alternative methods have been formally validated. Of note, European legislation has been considerably reinforced in 2010 with respect to the promotion of alternative approaches,[28] in line with the separate EU legislations for cosmetics and chemicals some years earlier.[29]

In 1993, in the United States, the National Defense Authorization Act (H.R. 10 2-346; 102d Congress, second session) directed the Secretary of Defense to establish aggressive and targeted programs to refine, reduce, and replace the current use of animals in the Department of Defense (DoD). In these symposia and books, we brought together the recent and relevant contributions of scientists from government, industry, and academia in North America, the United Kingdom, The Netherlands, Switzerland, France, Republic of China, Sweden, and Germany, to meet the needs for developing and validating alternatives such as refinement, reduction, and replacement of animal testing. The internationally recognized scientists presented what has been accomplished thus far in developing acceptable alternatives to traditional animal toxicological assessment and provided potentially new initiatives. This included validation and regulatory acceptance of alternatives in the United States, Canada, and the European Union. In addition, the history of humane treatment of animals was addressed, as were humane endpoints and pain and distress management in animal research and testing. Development of predictive methods based on the mechanisms of eye irritation at the ocular surface, dermal toxicity testing and molecular biomarkers, and transgenics for assessing neurotoxicity were also addressed. Other discussions included case studies in the use of alternatives to study the mechanisms of sulfur mustard action and the role of transgenics and toxicogenomics in the development of alternative toxicity tests. The most recent innovations in alternatives including archival data, and *in silico* techniques and the next generation of biochips were also addressed.

Our interest in alternatives was recognized by the invitation received to participate in the Interagency Regulatory Alternatives Group (IRAG) even though we were not a governmental regulatory agency. This ad hoc group was formed by Sidney Green, who at that time was with the FDA, and also included representatives from the EPA and the Consumer Product Safety Commission. This group ultimately evolved into the ad hoc Interagency Coordinating Committee for the Validation of Alternative Methods (ICCVAM) established by Bill Stokes of the National Institute of Environmental Health Sciences (NIEHS) and Dick Hill of the EPA in 1993. This committee resulted from a mandate in the National Institutes of Health (NIH) Revitalization Act of 1993 (Public Law 103-43 Section 1301), in which congress instructed the NIH to research replacement, reduction, and refinement alternatives and to establish the criteria for validation of regulatory acceptance, as well as to recommend a process for scientifically validated alternatives to be accepted for regulatory use. ICCVAM became established as a committee in 1997 and was made official in the year 2000. One of us (H.S.) has served as the official DoD representative on this committee since 1993.

Animal research has contributed significantly to many medical breakthroughs as well as to the database of biological knowledge. Since 1901, over 50 Nobel prizes in physiology and medicine have been awarded for discoveries based on the use of animals. Joseph Murray, the winner of the 1990 Nobel Prize for medicine, has stated, "there would not be a single person alive today as a result of an organ transplant without animal experimentation." There is much evidence that remarkable improvement of health and well-being, including quality of life, has been dependent on animal research. Diabetes, once a killing disease, took its toll with tragic rapidity until the summer of 1921, when two young Canadian scientists, Frederick Banting and Charles Best, succeeded in isolating insulin, which now enables millions of diabetics to lead normal lives. It was in the laboratories at the University of Toronto that Banting and Best kept the depancreatized dog Marjorie alive with injections of pancreatic extracts that contained what we now know as insulin. The development and safety testing of vaccines, including the vaccine that led to the eradication of polio, were also dependent on animal experimentation and safety testing. Animal research has resulted in many medical advances, from drug development for the treatment of life-threatening allergies, psychiatric disorders, heart disease, and infections to transplantation and transplant therapy, as well as surgery, including open-heart and coronary bypass procedures. When Steven Rosenberg of the NIH received the Research and Development Achievement Award for the development of a most promising breakthrough in the treatment of cancer using immunotherapy, he stated that none of this would have been possible without the use of animals. It is evident that animal experimentation has played a very important role in drug development and improving human health and well-being. It is hoped that by the use of *in vitro* systems, such as cell lines, isolated cells, micro-organisms, tissue cultures, invertebrates, quantitative structure-activity relationships, and physiologically based pharmacokinetic modeling, it may be possible to predict toxicity or a least gain insight about the possible target organs and mechanisms of action. These techniques should result in a reduction in the number of animals used in experimentation for safety and drug development. In fact, some validated alternative methods have shown to outperform the respective animal experiment, as exemplified by the whole blood pyrogen test for bacterial contaminations of drugs.[30]

Among the many *in vitro* techniques employed in the past two decades, a novel procedure which we have used very effectively is the one described by Albert Li et al.[31] Since the *in vitro* experiments attempt to simulate what occurs within the human body, it is necessary to assess the potential for hepatotoxicity, or the effects of metabolism on the toxicity of the parent compound being tested. The metabolism in the liver can activate or deactivate the parent compound, thus influencing the pharmacokinetics and pharmacodynamics of the material. Thus, *in vitro* systems that include a metabolic component would be more biologically relevant than plate-based experiments using single cell types.[32] Several similar systems have been described.[33] Although the Integrated Discrete Multiple Organ Co-Culture (IdMOC) system has never been used in combination with a high-content analysis instrument, we have combined them to provide the simple tool that allows for the incorporation of the metabolic component for *in vitro* screening and the high-content analysis, a quantitative, multiparametric tool that is powerful for the evaluation of toxicity *in vitro* and has been used to predict chemical-induced liver injury in both primary human hepatocyte and HepG2 models.[34,35] In our 2014 paper we reported on the results of the first use of high-content analysis on the IdMOC system.[32] This high-throughput method revealed four toxic endpoints that were consistent with the known detoxification of 4-aminophenol by hepatocytes, and the cytotoxicity of cyclophosphamide enhanced by co-culturing with hepatocytes as expected with its known metabolic activation. This first demonstration of the use of the multiparametric high-content analysis in combination with the IdMOC cell culture system provides the capability to quickly analyze the metabolic and toxicological effects of a large number of compounds in a simple co-culture system with multiple descriptive endpoints appropriate for the effects under consideration. We hope to adapt this system of using the idMOC combination of stem cell–derived organoids as a test bed for pharmacological and safety studies.

The field of *in vitro* pharmacology and toxicology is moving from the early use of mainly cancer cell lines and animal primary cells to human primary cells, co-cultures of different cells, three-dimensional (3D)[36,37] as well as organotypic cultures, and are increasingly combined with high-throughput (automation) and high-content (image analysis[38] and omics) approaches. This allows to base testing on understanding of mechanism of action and thereby make the models more human-relevant.

17.2 STEM CELLS

Although the use of stem cells to predict human pharmacology and toxicology is an *in vitro* technology, considering its importance to the future of this field we have given it specific attention in this chapter. The National Research Council in its "Toxicity Testing in the 21st Century: A Vision and a Strategy (Tox-21c)" (Figure 17.1) recommended the use of *in vitro* systems and pathways for detecting toxicants (NRC, 2007).[39] Since then it has been suggested that human stem cells be used, though the first major implementation activities of Tox-21c were focused more on high-throughput and high-content approaches.[40]

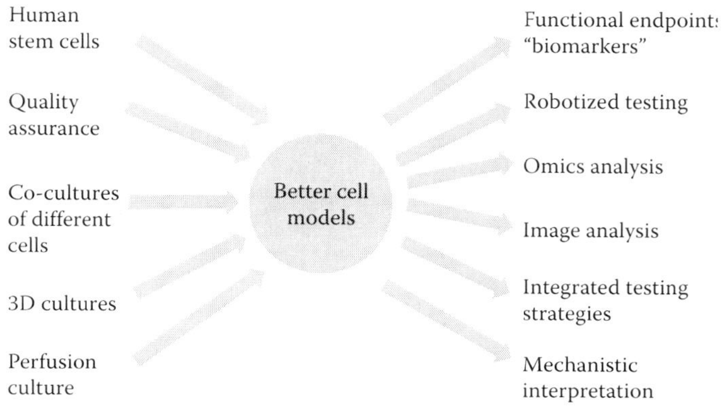

FIGURE 17.1 Twenty-first century cell culture for 21st century toxicology.

A parallel movement aims to develop organs on chip, i.e., human microphysiological systems, or their combinations in a single setup as "human-on-chip," predominantly based on stem-cell–derived models. This work was prompted by the U.S. DoD desire to develop medical countermeasures for bioterrorism and chemical warfare.[41] This could eliminate or reduce extrapolation errors from non-human to human. In addition it could reduce the time for testing as well as the cost, and most important, it could reduce the number of animals required for testing. Stem cells can divide and differentiate into almost any specific cell, have become in a short time a very important research tool in biology and medicine, and stem cell research has a definite place in pharmacology and medicine. 3D stem cell–derived organoids developed from humans onto scaffolds are currently being used.[42,43] Our own work at CAAT, for example, aims to develop a human brain microphysiological system from induced pluripotent stem cells. [44,45] Using cells from either healthy or diseased donors, we can test gene environment interactions. Using gyratory shaking and a regiment of growth factors, 300–400 μm diameter brain spheroids are formed over 10 weeks, which contain all neuronal cell types, astrocytes, and oligodendrocytes. The construct is spontaneously electrophysiologically active as shown in collaboration between our groups. The models allow for studies of neurodevelopmental and neurodegenerative diseases as well as neurotoxicity. For example, chemicals known to induce symptoms of Parkinson's disease in humans and animals selectively damaged the dopaminergic neurons in this model. Future work on infection and trauma can easily be envisioned.

Since 1917, the ECBC has been the nation's principal research and development resource for chemical and biological (CB) defense with capabilities in chemical warfare agent research, engineering, and testing, as well as full lifecycle test and evaluation of CB operational technologies. Thus ECBC is uniquely positioned to perform the task of CB evaluation challenge and evaluation of the body-on-a-chip platform under the X vivo Capability for Evaluation and Licensure (X.C.E.L.) program sponsored by the Defence Threat Reduction Agency (DTRA) through Wake Forest University under the Principal Investigator (PI) Anthony Atala. Wake Forest Institute for Regenerative Medicine (WFIRM), a leader in the translational medicine of implanting laboratory grown stem cell–derived organs into humans, will orchestrate the development of INGOTS (INtegrated Organoid Testing System) through collaboration with the Khademhosseini Lab at Harvard University, with strong experience in developing micro- and nanoscale bioengineering devices for controlling cellular behavior, with Shuichi Takayama, from the University of Michigan, who has experience in constructing microscale models of the body, expertise in biomicroelectromechanical and biomolecular devices, and has also developed technologies for high-throughput drug testing and multiplexed disease biomarker assays. Current safety and efficacy testing methods for the evaluation of medical products are not conducive to a flexible, rapid-response capability. The use of a 3D printed organoid system composed of human primary and/or induced pluripotent stem cells will provide a fast, flexible, multifunctional assay configurable for the elucidation of predictive metabolic profiles and presymptomatic biomarkers of exposure to both chemical and biological insults, as well as the effects of countermeasures used to treat them. While elements of this program have not yet been realized, it is advancing toward the combination of several organs by microfluidics combined with assessment of biomarkers of pharmacological and toxicological activity. This proposal includes several firsts: (1) bioprinted organoids into micro-reactors for a networked system; (2) using a universal blood surrogate for different organoids; and (3) testing organoids with surety agents for metabolic responses to mimic human/animal data. This system when developed will establish a comprehensive INGOTS-based plan for chemical/biological/toxicological agents and validate the system for proof-of-concept and predictive value for experimental and FDA-approved medical countermeasures. In the interim we are also working on human induced Pluripotent Stem Cells (iPSC)-derived organoids including the brain, heart, and lungs singly and with the liver.

In our *in vitro* toxicology program we are currently studying the cardio-specific effects of chemical warfare agents (CWAs) on stem cell–derived cardiomyocytes (SCDC). Using the ACEA Biosciences xCELLigence Real-time Analyzer Cardio System with noninvasive impedance readout for real-time monitoring of cell viability and functional beating activity of cardiomyocytes (see Figures 17.2 and 17.3).

The advantages of this system are that is fast, safe, inexpensive, and it is considered more predictive than other *in vitro* tests as well as being able to study direct effects that may be overlooked. This model is being studied to create a non-animal model that can be utilized to identify biomarkers for diagnostics and pathway identification leading to enhanced medical countermeasures. In these studies we have

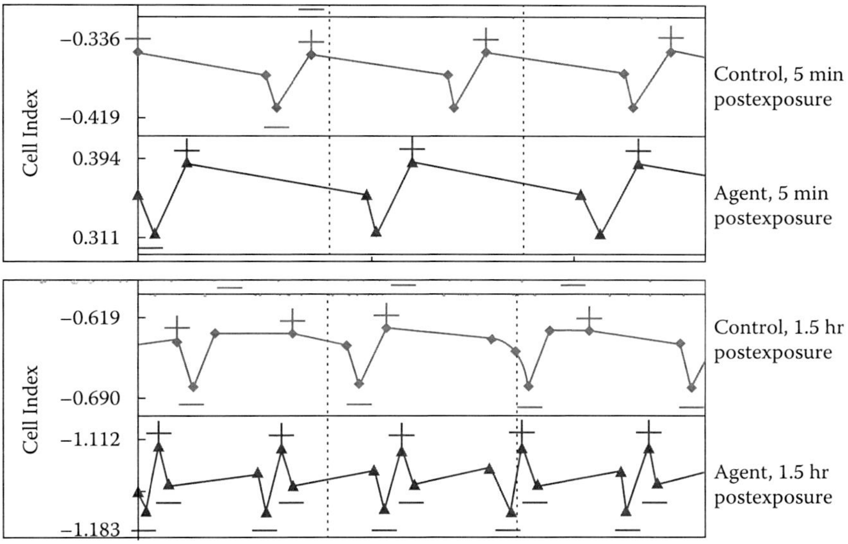

FIGURE 17.2 Effects of Agent on Cardiomyocyte Beating (Cell Index). The system uses 96-well plates with each well containing a number of electrodes to measure current, and the impedance (cell index) of the cell monolayer which is directly affected by the concentration and morphology of the cells. The impedance of each well is measured and recorded at multiple time points. A carbon dioxide incubator has been modified for the testing of CWAs within engineering controls.

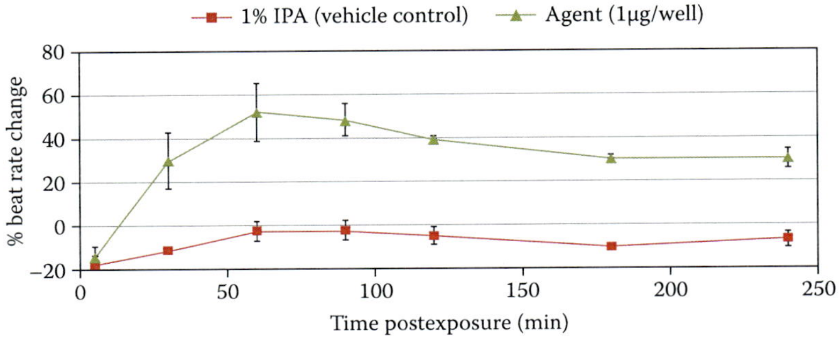

FIGURE 17.3 Stem cell-derived cardiomyocytes are exposed to a nerve agent, and beat rate was monitored. The beat rate of the exposed cell increased to a high of ~150% of the pre-exposed beat rate.

shown the direct effects of some CWAs. The elevation of the cardiomyocyte beat rate and the induced variation on the corresponding beat patterns establish that CWAs can have a direct effect on cardiac function. The mechanisms of action are being characterized using high-content analysis (HCA) to identify cellular changes during exposure and activity-based protein profiling (ABPP) to identify protein targets.

Stem cell technologies have brought about an avenue toward human cell models, which were before limited by the difficult access to human tissue and the shortcomings of cancer cell lines. Combined with advanced tissue culture technologies, they promise to make organotypic cultures possible that will reflect human (patho) physiology more closely.

COMPETING INTEREST STATEMENT

Thomas Hartung holds patents as inventor of the whole blood pyrogen test and the use of cryopreserved blood, which are licensed to Merck-Millipore; he receives royalties from Merck-Millipore from sales of the kit version.

A provisional patent has been filed by Johns Hopkins University for some of the mini-brain technologies described in this paper (inventors David Pamies, Helena Hogberg and Thomas Hartung), which was licensed to Organome, LLC. Thomas Hartung is cofounder of Organome LLC.

REFERENCES

1. Newman Dorland, WA. *Dorland's Illustrated Medical Dictionary*, Philadelphia, PA: Saunders, 2007.
2. Ferrario D, Hartung T. Glossary of reference terms for alternative test methods and their validation. *ALTEX* 2014;31:319–335.
3. Beck, LC., *Adulteration of Various Substances Used in Medicine and the Arts,* New York, S.S. and W. Wood, 1846.
4. McKinney DD. The Mexican-American War brings regulation on drug importation. *Frontline* 2002; Summer: 50–51.

5. 1848 Drug Importation Act, accessed September 19, 2016. http://www.usp.org/sites/default/files/fda-exhibit/legislation/1848.html
6. Sinclair, U., The Jungle. New York: Grosset and Dunlap, 1906.
7. The Editorial Committee. *Appraisal of the Safety of Chemicals in Foods, Drugs, and Cosmetics*. The Association of Food and Drug Officials of the United States; 1959.
8. Goldenthal, EL. Current views on safety evaluation of drugs. *FDA Papers* 1968:2–8.
9. D'Aguanno, W. *Guidelines for Reproduction Studies for Safety Evaluation of Drugs for Human Use*. U.S. FDA. Washington, DC: Bureau of Medicine; 1973, pp. 41–43.
10. U.S. FDA, Bureau of Foods. *Toxicological Principles for the Safety Assessment of Direct Food Additives and Color Additives Used in Food*. 1982 (*Redbook 1*). FDA, Washington, DC.
11. Hanig, JP, Osterberg RE. Drug safety toxicology. In: *Biological Concepts and Techniques in Toxicology: An Integrated Approach*. Reviere JE (Ed.), Boca Raton, FL: CRC Press; 2006. pp. 249–271.
12. Hartung T. Food for thought … on animal tests. *ALTEX* 2008;25:3–9.
13. Hartung T. Toxicology for the twenty-first century. *Nature* 2009;460:208–212.
14. Olsen H, et al. Concordance of the toxicity of pharmaceuticals in humans and in animals. *Regul Toxicol Pharmacol* 2000;32(1):56–67.
15. Prepared statement for FDA Teleconference: Steps to Advance the Earliest Phases of Clinical Research in the Development of Innovative Medical Treatments, January 12, 2006.
16. Hartung T. Look back in anger—What clinical studies tell us about preclinical work. *ALTEX* 2013;30:275–291.
17. Bailey J, Knight A, Balcombe J. The future of teratology research is in vitro. *Biogenic Amines* 2005;19(2):97–145.
18. Lewis DFV, Ioannides C, Parke DV. Cytochromes P450 and species differences in xenobiotic metabolism and activation of carcinogen. *Environ Health Perspect* 1998; 106(10):633–641.
19. Birnbaum LS. The mechanism of dioxin toxicity: Relationship of risk assessment. *Environ Health Perspect* 1944;102(Suppl 9):157–167.
20. Kacew S (Ed.). *Lu's Basic Toxicology: Fundamentals, Target Organs, and Risk Assessment*, 6th ed. CRC Press, Taylor & Francis Group, Boca Raton, FL. 2012.
21. Boverhof DR, et al. Transgenic animal models in toxicology: Historical perspectives and future outlook. *Toxicol Sci* 2011;121(2):207–233.
22. Russell, W. M. S. and Burch, R. L. (1959). *The Principles of Humane Experimental Technique*. London, UK: Methuen. 238pp.
23. Salem, H. *Animal Test Alternatives*. 1995. Marcel Dekker, New York.
24. Salem, H, Katz SA. *Advances in Animal Alternatives for Safety and Efficacy Testing*. 1998. Taylor and Francis, Philadelphia, PA.
25. Salem H, Katz SA. *Toxicity Assessment Alternatives; Methods, Issues, Opportunities*. 1999. Humana Press, Totowa, New Jersey.
26. Salem H, Katz SA. *Alternative Toxicological Methods*. 2003. CRC Press, Boca Raton, FL.
27. Goldberg AM. A history of the Johns Hopkins Center for Alternatives to Animal Testing (CAAT): The First 28 Years (1981–2009). *Applied in Vitro Toxicology* 2015;1(2):99–108. http://doi.org/10.1089/aivt.2015.0015
28. Hartung T. Comparative analysis of the revised Directive 2010/63/EU for the protection of laboratory animals with its predecessor 86/609/EEC—A t4 report. *ALTEX* 2010;27:285–303.
29. Hartung T. From alternative methods to a new toxicology. *Eur J Pharmaceut Biopharmaceut* 2011;77:338–349.
30. Hartung T. The human whole blood pyrogen test—Lessons learned in twenty years. *ALTEX* 2015;32:79–100.

31. Li AP, Bode C, Sakai YA. A novel in vitro system, the integrated discrete multiple organ cell culture (IDMOC) system, for the evaluation of human drug toxicity. *Chemico-Biol Inter* 2004;150(1):129–136
32. Cole SD, Madren-Whalley JS, Li AP, Dorsey R, and Salem H. High content analysis of an in vitro model for metabolic toxicity: results with the model toxicants 4-aminophenol and cyclophosphamide. *Journal Biomol Screening* 2014;19(10):1402–1408.
33. LeCluyse EL, Witek RP, Andersen ME, Powers MJ. Organotypic liver culture models: Meeting current challenges in toxicity testing. *Criti Rev Toxicol* 2012;42(6):501–548.
34. Xu JJ, et al. cellular imaging predictions of clinical drug-induced liver injury. *Toxicol Sci* 2008;105(1):97–105.
35. O'Brian PJ, et al. High concordance of drug induced human hepatotoxicity with in vitro cytotoxicity measured in a novel cell-based model using high content screening. *Arch Toxicol* 2006;80(9):580–604.
36. Hartung T. 3D—A new dimension of in vitro research. *Advanced Drug Delivery Reviews*, Preface Special Issue "Innovative tissue models for in vitro drug development" 2014;69:vi.
37. Alépée N, et al. State-of-the-art of 3D cultures (organs-on-a-chip) in safety testing and pathophysiology—A t4 report. *ALTEX* 2014;31:441–477.
38. van Vliet E, et al. Current approaches and future role of high content imaging in safety sciences and drug discovery. *ALTEX* 2014;31:479–493.
39. NRC (2007). Toxicity Testing in the 21st Century: A Vision and a Strategy. Washington, DC, The National Academies Press.
40. Bouhifd M, et al. The Human Toxome project. *ALTEX* 2015;32:112–124.
41. Hartung T, Zurlo J. Alternative approaches for medical countermeasures to biological and chemical terrorism and warfare. *ALTEX* 2012;29:251–260.
42. Mansfielf E, Oreskovic TL, Rentz, NS, Jeerage KM. Three-dimensional hydrogel constructs for exposing cells to nanoparticles. *Nanotoxicology* 2014;8(4):394–403. doi: 10.3109/17435390.2013.790998.
43. Movia D, Prina-Mello A, Bazou D, Volkov Y, Giordani S. Screening the cytotoxicity of single-walled carbon nanotubes using novel 3D tissue-mimetic models. *ACS Nano* 2011;5(11):9278–9290.
44. Hogberg HT, et al. Toward a 3D model of human brain development for studying gene/environment interactions. *Stem Cell Res Ther* 2013;4(Suppl 1):1–7.
45. Pamies D, Hartung T, Hogberg HT. Biological and medical applications of a brain-on-a-chip. *Exp Biol Med* 2014;239(9):1096–10107.

18 Personalized Medicine

Elisa Cimetta, Michael Lamprecht, Sarindr Bhumiratana, Nafissa Yakubova and Nina Tandon

CONTENTS

18.1 INTRODUCTION: WHY HAVEN'T WE "PERSONALIZED" MEDICINE BEFORE?

Personalized medicine has received much attention recently as an emerging approach for disease treatment and prevention that takes into account individual variability in genes, environment, and lifestyle. Yet although the notion of human individuality has strong roots in many human cultures, the central assumption—that pathology is related to human individuality (whether rooted in genetic or environmental causes)—has eluded applicability in medical science. Limitations in our ability to measure, modulate, and compute the relevant factors are largely to blame for that lag.

Not until very recently, for example, have we been able to directly measure genetic expression, let alone correlate it with clinical outcomes beyond the population level. (We've instead used proxies of ethnic groups or families with hereditary conditions.) Moreover, until the advent of combinatorial chemistry in the 1990s, the number of drug-like molecules that could be synthesized limited the size of available chemical libraries. Furthermore, despite the timelessness of the nature/nurture debate, our knowledge of the relevancy of various environmental factors has often come about anecdotally (e.g., cigarette smoking, radioactivity, pollution). Biomarkers also have often proved difficult and/or impossible to measure noninvasively and longitudinally (e.g., EKG, blood pressure, gene expression). Finally, even as more data become available, the multivariant datasets involved in correlating relevant longitudinal biomarkers with environmental and interventional regimes present non-trivial challenges to data scientists and data visualization experts.

Without the ability to systematically measure, modulate, or correlate relevant genetic and environmental data with individual medical outcomes, medical science has historically relied heavily on cell and animal testing methods prior to exposing human test subjects to experimental therapies. Then, upon commencement of human clinical trials, it has taken one of the two extreme approaches to developing treatment regimes:

- **Randomized trial:** in which two groups are assigned to different conditions and the overall group outcomes are studied. The first clinical trial was recorded in the Bible around 600 B.C. (see [1] for a great history of clinical trials).
- **N-of-one ad-hoc approach (patient treatment):** e.g., a doctor may prescribe one drug for hypertension and monitor its effect on a person's blood pressure before trying a different one.

Recent clinical trial innovations all aim to add refinement to the "gold standard" of the randomized clinical trial [2]. For instance, basket trials categorize patients into cohorts based on patient-specific molecular markers, umbrella trials test multiple drugs on a single disease, and adaptive trials vary the interventions on an ongoing basis based on patient responses.

These classic methods have yielded many positive results, such as penicillin for bacterial infections, smallpox vaccination, and insulin for insulin-independent diabetes [3]. Yet even with their recent innovations, standard approaches have been yielding diminishing results. Consider that the top 10 highest-grossing drugs in the United States have helped only between 4% and 25% of the people who take them [4]. Furthermore, with iterative approaches, valuable time and resources are wasted, and failures are not benign. These issues are only exacerbated with cancer treatments.

Where we aim to go with personalized medicine is mass customization of healthcare—with medical decisions, practices, and products tailored to the individual patient. In other words, we aim for an N-of-one approach that's both scalable and systematic.

18.1.1 The Human Genome Project: A Rosetta Stone for Understanding Disease?

The concept of personalized medicine started forming in the 1990s concurrently with the advent of the Human Genome Project, which underscored that we are all uniquely characterized by our distinct genetic makeup. A natural consequence has been the understanding that the same diseases might differ at a genetic level between individuals.

Throughout the ensuing decades, we gained significant knowledge of the genetic diversity of disease thanks to dramatic advances in the fields of systems biology, pharmacogenomics, and genomics technologies in general [5–7]. For example, it has long been known that that susceptibility to certain infectious diseases, such as tuberculosis, malaria, and HIV/AIDS, varies markedly with human genetics [8]. Yet more recently, we've also witnessed that variability at a molecular level, proving that within some types of tumors, such as leukemia and breast cancer, clinically significant differences in the genetic profiles can be identified [9,10]. Similar observations have been made for cardiovascular diseases such as long QT syndrome and familial hypertrophic cardiomyopathy [11–13].

Hopes grew high. A 2001 study predicted that, over the course of just a few years, advances in personalized medicine would revolutionize the way both drug development and clinical trials are conducted [14]. New drugs would be developed using patient-specific data and connecting all different stages of clinical development and patient treatment through feedback loops giving specific molecular and pharmacological information about patient response. Toxicogenomic markers might predict adverse drug reactions (ADRs) before human clinical trials. Clinical trials themselves would be conducted following novel paradigms in patient selection (based on biological and genomic data), grouping, and therapy. Advances in information technology, too, spurred hope for coordinating relevant health data among researchers, clinicians, and patients.

However, real progress proved slower than our aspirations. Translating genomic and other discoveries into personalized therapeutics, we found, requires overcoming a number of significant scientific, technical, economic, and social challenges.

18.1.2 Eroom's Law → Moore's Law in Reverse?

Despite the advances in genomic sequencing, combinatorial chemistry, and computing, there's been a kind of collective consternation at the paradox of improving inputs and declining output for developing therapies. In fact, so frustrating is the inefficiency that some observers have said drug development adheres to "Eroom's law"—Moore's law spelled in reverse [15].

A major limitation for pharma R&D in general, and especially the translation of genetic knowledge into the development pipeline, is the lack of reliable, high-throughput platforms to model the complex physiological function of the complete human. The development of the "organ-on-a-chip" technologies outlined in this volume means to address this shortcoming by supplementing preclinical and clinical testing (see Figure 18.1). Underlying technical issues, outlined in previous chapters, still need to be resolved. These include difficulties with reproducibly miniaturizing the relevant human organ systems into organoids, our limited ability to directly modify genes in living cells, combining multiple organoid systems (much less microbiome or epigenetic cues in the *in vivo* environment) into "human-on-chip" platforms, the scarce agreement on the marker set used to screen for disease [16–18], as well as difficulties with incorporating appropriate sensing systems into lab-on-chip systems. More regulatory, business, and logistical considerations will emerge that will influence drug R&D as scientists solve the technical problems and translate these systems out of the lab and into the drug development pipeline and clinic.

We're not there yet. However, it is undeniable that personalized medicine holds great promise in reducing the costs of the typical drug discovery and development processes, reducing the risks of drug toxicity, increasing drug efficacy, establishing preventive therapies, and ultimately relieving the healthcare system as a whole. We therefore need to take a step back and revise our technical, policy, and business aims to reach the goal of creating a novel, patient-centered standard of care, based on tailored approaches dictated by each individual's unique footprints.

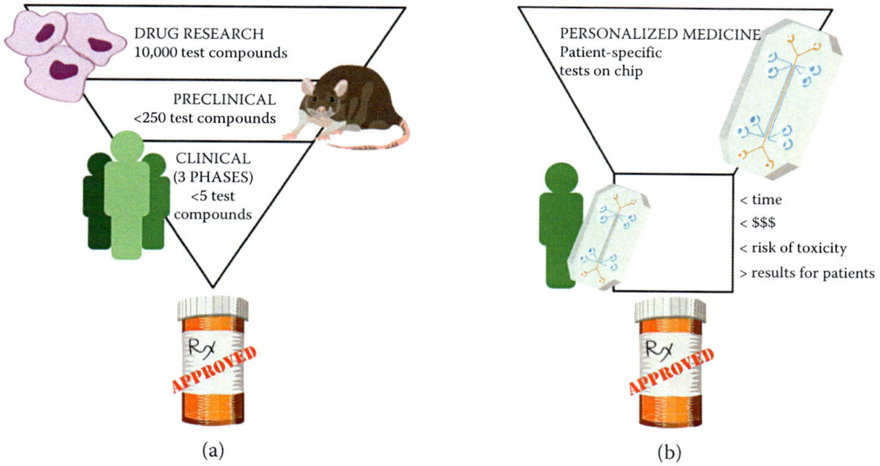

FIGURE 18.1 (a) The current drug development pipeline and (b) envisioned drug development pipeline incorporating "organ-on-a-chip" technologies which supplement preclinical and clinical testing. Patient-specific on-chip tests have the potential to reduce time, cost, and risks of toxicity, thereby improving patient outcomes.

18.1.3 Merging of *In Silico*, *In Vitro*, *In Vivo* → *In Proto*?

A critical tool for our continued progress is creating better systems to mimic human biology in the lab. This volume provides a comprehensive review of the technologies that have been applied to such "organ-on-a-chip" systems that simulate physiological function of human tissues and organs. We envision that systems like these (which we dare to call *in proto* systems) will reside somewhere on the spectrum of testing platforms in between *in silico* computer models, traditional "flat biology" *in vitro* systems, and animal and human *in vivo* models.

This approach has been made possible in recent years by the convergence of several technologies, including microfabrication, microelectronics, and microfluidics, which enable the precise fluid handling, sensing, and actuation required to measure and collect relevant data from modeled tissues. By combining these "lab-on-chip" systems with organoid cell culture, systems may be developed to mimic organ functionality *in vitro*, ranging from lung, liver, heart, and skin to multiple-tissue chips, as described in previous chapters. As *in proto* technologies get better at reproducibly recapitulating human *in vivo* biology, big data methods must then be applied to achieve the predictive analysis that we need to outdo traditional standard animal and cell culture testing, and perhaps even catch up to the utility of human clinical trials. In parallel, researchers can apply that same knowledge to the improvement of regenerative medicine technologies for use *in vivo*.

In the following sections we present a critical vision of several technological, regulatory, and business considerations, focusing on a series of case studies and giving a perspective view of where the regenerative medicine field will bring us in the next decades toward the goal of personalized medicine.

18.2 TECHNOLOGICAL CONSIDERATIONS

Today's aims of personalized medicine intersect with that of regenerative medicine in two main respects (see Figure 18.2): (1) "Organ-on-a-chip" systems that facilitate screening and optimizing traditional therapies (i.e., small molecule or biologic), as in the *in proto* systems that are the topic of this volume, and (2) development of individualized cell-based therapies. Technological efforts toward realizing these aims are parallel and interdependent:

- **Cell sources:** advanced cell sources, such as induced pluripotent stem cells, may serve as ingredients in cell therapies as well as *in proto* platforms for testing new molecular therapies
- ***In proto* niche:** advances in 3D cell culture, such as scaffolds and bioreactors, increase the fidelity of *in vitro* systems and help make engineered tissues more applicable in cell therapies.

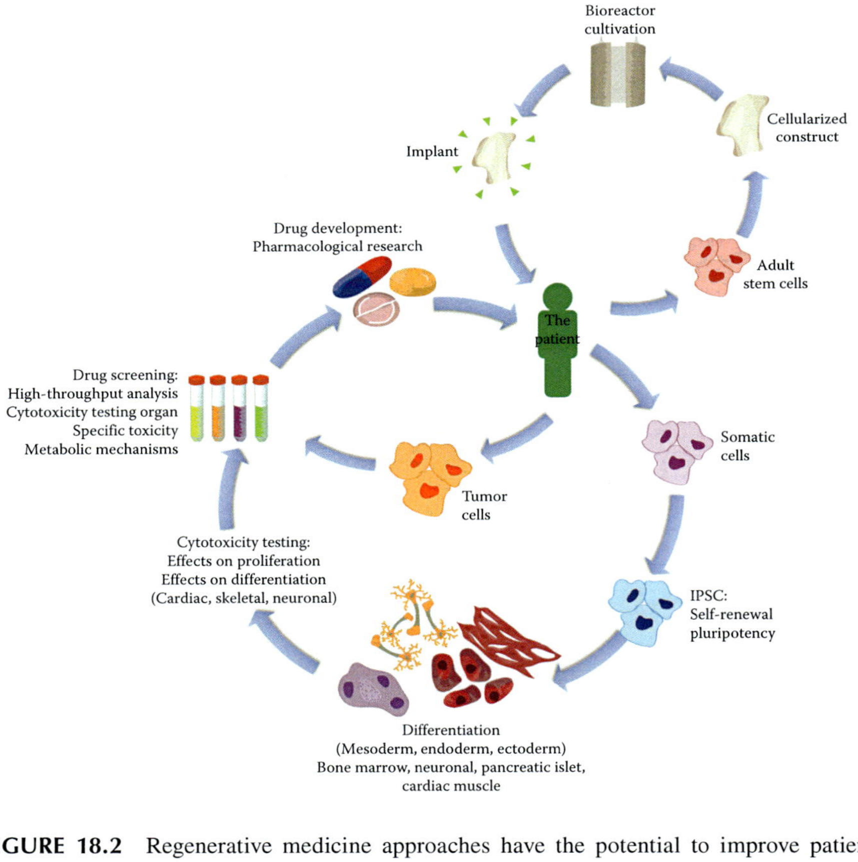

FIGURE 18.2 Regenerative medicine approaches have the potential to improve patient treatments in two main respects: left circle: "organ-on-a-chip" systems (i.e., *in proto* systems) derived from patient cells that facilitate screening and optimizing traditional therapies, and right circle: use of patient cells for deriving individualized cell-based therapies.

Supporting these efforts are next-generation genomics, which may aid in the direct correction of disease genes in addition to serving as a tool for researchers and clinicians to understand genetic diseases. New genome-editing technologies, such as CRISPR, have been applied to a range of organisms and applications including the correction of diseased genes in zygotes and human cell lines, and could prove critical to curing devastating diseases. Briefly, CRISPR stands for Clustered Regularly Interspaced Short Palindromic Repeats and defines sections of repeated base sequences in prokaryotic DNA. CRISPR and its associated proteins Cas, the CRISPR-Cas9 system, is an adaptive immune process used to fend off foreign DNA, and has now been harnessed to do the opposite—by delivering Cas9 with an appropriate guide RNAs, the cell's genome is specifically cut at a target location, allowing precise removal and/or insertion of genes.

The extent to which the aforementioned technologies succeed—allowing us to command the expansion, gene expression, and 3D architecture of cells and tissues—will help determine whether we find ourselves on the cusp of a truly new era in medicine as opposed to remaining in the trough of Eroom's law.

18.2.1 Cell Source

Pluripotent cells, such as adult and embryonic stem cells, are invaluable tools for research and can potentially serve as an ingredient for *in proto* systems as well as cell- and tissue-replacement therapies. Each cell source comes with its respective availability, differentiation potential, immunogenicity, oncogenecity, and so on—factors which influence its applicability for use in *in proto* or cell therapies, as outlined below.

18.2.1.1 Induced Pluripotent Stem Cells

Even with the recent advances in genomic technologies, genetic characterization of some diseases has been hindered by the limited availability of tissue samples carrying the disease, due to confined donor cell availability. The limited proliferation capacity and loss of functionality observed in *ex vivo* expanded cells compound the challenges. Yamanaka and colleagues were the first to report successful reprogramming of somatic cells into induced pluripotent stem cells (iPSCs) in 2006 [19,20]. Their technique can also be applied to reprogram various donor cell types, such as skin, neuronal, hematopoietic, and adipose stromal cells into iPSCs. Reprogramming of adult somatic cells into iPSCs is an approach that holds great promise to create patient-specific disease models for the high throughput screening necessary for development and testing of new therapeutic agents (see [21] for a review).

Banks and databases collecting disease- and patient-specific iPSCs are flourishing, constituting a great resource for studying individualized treatments and characterizing genetic traits of disease. However, when it comes to actually using those cells for *in vivo* implantation studies and/or cell therapies, some unresolved issues emerge. There is still an open debate on the quality of the obtained reprogrammed and redifferentiated cells, the resemblance to their natural counterparts in the human body, and the "somatic memory" that they potentially retain [22]. Terminally differentiated iPSCs typically reach a relatively low level of maturation, and thus can more faithfully be used as a model for the earlier stages of disease. Nonetheless, iPSCs are an invaluable means for obtaining relevant genetic and physiological information characterizing unique traits of disease, of its initiation and progression. They are also a powerful tool for performing toxicology screenings and drug- and therapy-development studies. To date, iPSCs have been successfully employed in studies of neurodegenerative diseases, hematopoietic disorders, metabolic conditions, and cardiovascular pathologies [23–30].

The use of iPSCs is currently mostly limited to *in vitro* studies, but recent publications have indicated that they tend to have a low immunogenicity and do not contain donor antigen–presenting cells [31]. This assumption does not hold for all cell types, however, and cardiomyocytes (CMs) represent one of the exceptions: iPSC-derived CMs show increased immunogenicity when implanted *in vivo*. However, for those iPSC-derived cells matching the low immunogenicity hypothesis, human leukocyte antigen (HLA) compatibility is sufficient to avoid acute rejections and has thus supported the creation of regional "haplobanks" [32].

Other than embryonic cells and iPSCs, adult stem cells also retain some tissue-regeneration potential. These populations have the capability to switch between homoeostasis or regeneration states [33] upon activation of defined signaling events. Among others, several studies proved how neural [34], cardiac [35,36], skeletal muscle [37], intestinal [38], and hepatic [39] cells could promote regeneration in adult injured tissues.

18.2.1.2 Autologous Adult Stem Cells

Multipotent adult stem cells, which are found throughout the body and do not involve embryos or raise as many safety concerns as pluripotent cells, also hold vast potential for regenerative medicine. The NIH's clinicaltrials.gov site lists some 4500 adult stem cell trials as compared with 27 for embryonic stem cells and 21 for iPSCs. It is therefore no surprise that adipose-derived stem cells (ASCs) have captured the attention of researchers, investors, physicians, patients, and—increasingly—regulators around the world.

Adult stem cells have been identified in many organs and tissues, including (but not limited to) brain, bone marrow, peripheral blood, blood vessels, skeletal muscle, skin, teeth, myocardium, gut, and liver [40–46]. These cells may serve as a source for creating living, functional tissues to repair or replace tissue or organ function lost due to age, disease, damage, or congenital defects. They may also serve as a cell source for high-throughput drug screening. Advantages of adult stem cells include the ease with which they are obtained, their autologous source, and their lack of political controversy. However, these cells are not without limitations. Most notably, their limited differentiation potential may pose limits for their use in certain applications (e.g., brain, heart), a constraint which may be exacerbated by certain disease states (e.g., cancer, diabetes). Nonetheless, adult stem cells hold great promise for shifting the paradigm away from a medical model that involves treatment of injured tissues and organs, and toward regeneration.

18.2.1.3 Autologous Cancer Cells

When available, autologous cells harvested from the patient's site of disease are still the source of choice for several applications, and remain the sole source in the case of cancer studies.

Cancer is characterized by a genetic diversity that has been known for decades, and patients would thus greatly benefit from a personalized approach in the choice of treatment. Only a few types of tumor have been linked to specific genetic traits that can be screened in healthy individuals in order to indicate appropriate preventive therapies. All other cancers are diagnosed after the onset of the disease, and therapy starts after diagnosis. Cells from tissue samples could be used to map the genetic identity of the tumor and thus determine the ideal therapeutic approach for each individual [47]. To reach the ultimate goal in personalizing the treatment of tumors, we would need to create large datasets containing tumor classifications based on their genetic traits, predicting clinical outcomes and suggesting the ideal personalized therapy. New data would be constantly added to these datasets, continually increasing the power of the included studies and contributing to greatly improved therapeutic outcomes. In addition, adding a supporting material (scaffold) to the cells in culture ensures an increased closeness to the *in vivo* tissue counterpart (see Section 18.2.2.1).

18.2.2 Gene Editing

Another constraint on the translation of genetic knowledge into therapy development has been our limited ability to directly modify genes in a living cell. With sickle-cell anemia or retinitis pigmentosa, for example, both diseases caused by a mutation in just one base pair, researchers have been helpless to correct them and halt their devastating effects. The CRISPR-Cas9 genome-editing system is a novel tool enabling precise and easy DNA modification of virtually any organism [48]. Several concerns and ethical debates still remain, but this technology may provide tools that researchers need to understand and cure some of our most deadly genetic and non-genetic diseases [49]. CRISPR-Cas9 technology can be applied to T-cell genome engineering, potentially greatly improving cell-based therapies for cancer, HIV, immune deficiencies, and autoimmune diseases [50–52].

Another tool is small interfering RNA (siRNA, known also as short interfering or silencing RNA). siRNA is used for short-term silencing of protein coding genes and has potential therapeutic use in selectively targeting and suppressing disease-causing genes [53] in cases ranging from viral diseases to cancer, hepatitis, and HIV. It also has the potential to play a role in target validation for animal disease models. siRNA is a synthetic RNA duplex that works by targeting specific Messenger RNA (mRNA) degradation. This tool is widely used to access individual gene contributions to different cellular phenotypes (i.e cell differentiation [54] and insulin signaling [55,56]), as well as in identifying novel disease pathways. However, delivery of siRNA *in vivo* presents challenges (poor stability, nontargeted biodistribution, adverse immune response), and the mechanism needs to be further studied before it can be used as a therapeutic tool [57].

18.2.2.1 *In Proto* Niche

A further bottleneck to the complete characterization of a disease *in vitro* is the lack of a proper design of the *in vivo* niche [58,59]—in other words, replication of the real-life environment within the body and relevant epigenetic events that are linked to the onset and progression of disease [60,61]. The re-creation of the *in vivo* niche has therefore been the focus of much research conducted over the last decade. These efforts are based on the understanding that traditional two-dimensional (2D) culture systems (i.e., Petri dishes and tissue flasks) fail to provide a number of real-life three-dimensional (3D) features: cell-cell and cell-matrix interactions, the spatial and temporal patterns of biochemical and physical signals, and crosstalk between different organ systems [62]. The latter is of particular importance in drug toxicity studies, given the incidence of ADRs in organs and tissues other than the target one [63,64].

There is thus a huge need for novel culture systems capable of capturing the whole organism's complexity and giving readouts more predictive of individual human physiology. Recent technological advances in the development of novel scaffolding and bioreactors, especially toward multitissue devices created from different patients' cells, could complete the missing step in the ladder from the dish to the patient and from the traditional "average" patient to real personalized medicine. We review key advances here.

18.2.2.1.1 Scaffolding Materials

Biomaterials have been used for biological applications for several decades, and the science behind their fabrication and optimization has advanced dramatically in recent years [65–67]. Scaffolding materials recapitulate several aspects of native tissues and give cells a 3D template guiding their attachment, growth, and fate specification. Scaffolds can be engineered to match defined sets of characteristics, from their composition (natural or synthetic, composite or mono-component, etc.) to their structure (shape and dimensions, porosity, pore size and shape, roughness, etc.) and mechanical properties (elastic modulus, surface stiffness) [68–72].

There are several approaches to creating a cell scaffold, ranging from decellularized tissues harvested from human or animal cadavers to synthetic 3D printed biomaterials [73–75]. Nature still provides the best scaffolds for cell growth; these are currently the gold standard for many applications. For instance, bone grafts derived from cow are commonly used in dental and spinal applications. Great care is taken to remove all of the cells and proteins from the bone source, leaving behind a scaffold with the correct mechanical strength, pore size, and microstructure for new bone to grow [76]. Using subtractive manufacturing, these scaffolds can also be shaped to an anatomically correct form. Such scaffolds are implanted directly to the site of injury, where they promote existing bone tissue to infiltrate and eventually replace the defect with the patient's own tissue (e.g., MedTronic's Infuse® Bone Graft). Decellularized tissues are effective but also carry the risk of introducing pathogens or initiating an immune response in the patient. To address that risk, tissue engineers have focused much attention on synthetic biomaterials.

One of the first biomaterials, alginate, is still used as a method to immobilize cells and is FDA-approved for use in cellular therapeutics [77]. However, synthetic biomaterials have improved markedly over the past 20 years. Determining the correct material is vital for a successful therapy. A wide range of biocompatible materials have been developed, including poly(lactic-co-glycolic acid), polycapralactone, and polyethylene glycol [78,79]. These materials are biologically inert and will degrade over time, an important characteristic for cell therapies if the goal is to completely replace an organ or tissue with the patient's own cells. Synthetic biomaterials are an attractive alternative due to low costs, the ability to change the materials' characteristics, and the ease of shaping the materials for an anatomical fit.

As 3D printers continue to evolve and become easily accessible to researchers, a new generation of 3D printed biomaterials is emerging [80,81]. There are many advantages to additive manufacturing. Some printers are capable of sizes less than 1 μm, allowing for intricate internal scaffold designs that can closely mimic the structures found in native tissues. To that end, software has continued to evolve to meet these needs. For instance, Within Labs has developed a program that can transfer the internal structure of a material as seen through a CT scan directly to a scaffold design [82]. Different regions can also contain different internal structures, making a true composite graft scaffold. However, this field is currently limited by the materials that are compatible with 3D printing. The development of new materials and perhaps combinations of materials will be necessary for true 3D printed scaffolds to become mainstream.

18.2.2.1.2 Bioreactors

Langer and Vacanti introduced the concept of the bioreactor in the 1990s [83], highlighting how culture devices designed to provide environmental control to the cells in culture—delivering nutrients and drugs and removing waste products—could bring to reality the idea of developing functional human tissue substitutes *in vitro*. Thus marked the starting point for the era of tissue engineering. Bioreactors typically have stirring and/or fluid perfusion elements, ensuring that mass transport by convection sustains provision of soluble factors to all cells in culture up to the centimeter scale [84–87]. Bioreactors allow us to surpass the limitation of standard culture techniques by breaking the barrier of the diffusion-limited development of 3D tissues. When cultured *in vitro* without the dense capillary network present *in vivo*, cells and tissues experience the fast formation of oxygen depletion zones (below threshold values for cell survival). Diffusion alone can only sustain the physiological supply of oxygen for depths up to 100–200 μm for dense tissues [88]. Thus without the contribution of convection, it is impossible to culture tissue constructs with relevant sizes and sound biological properties mimicking those of their *in vivo* counterpart. The establishment of a dynamic environment also more closely mimics the *in vivo* behavior of cells and tissues, which is characterized by multiple signals that are modulated in both space and time.

Overall, the key feature of bioreactors is that they allow tight control over operative parameters and culture conditions while reducing the variability that derives from the many operator-related interventions characterizing standard cultures. Bioreactors, if required by the tissue in culture, can also provide physiological stimulations such as shear stress–generated mechanical stimuli [85], compression and loading [89,90], and electrical stimuli [91,92]. These additional capabilities allow engineers to incorporate a tissue complexity that was lacking in constructs fabricated using standard *in vitro* culture techniques [93].

Bioreactors can be a fundamental tool in realizing the promise of personalized medicine [94]. Bioreactors are essential in ensuring tissue growth and maturation [95–100] and, when needed, in providing transportation capabilities for the cultured biological samples [101].

Miniaturization of the culture systems leads to changes in the fundamental forces determining the behavior of fluids and mass transport [102]. Flow is typically laminar (Reynolds numbers Re<100), with well-defined and predictable hydrodynamic profiles, and transport is regulated by molecular diffusion and finely tuned convective forces [103]. Overall, microscale platforms allow operating at time- and length-scales that are more representative of biological phenomena. Microfluidics and microscale technologies have broad applications, ranging from tissue engineering to scaffold fabrication to development of high-throughput assays [104–108]. In addition, they can play a major role in the development of personalized medicine approaches in both drug development and testing and disease modeling.

18.2.2.1.3 Scale

We introduced the concept of "scaling" in the previous paragraph and discuss it here in more depth.

When thinking about using tissues engineered *in vitro* for personalized medicine applications, there are several key considerations. First, it needs to be specified which level of similarity to the *in vivo* counterpart must be met. Human tissues and organs have a complexity that is almost impossible to fully re-create in *in vitro* models [109], and it is thus necessary to narrow the list of structural and biological properties that are necessary and sufficient to reproduce correct physiologic behaviors and responses. Clearly, to be efficacious, a tissue-mimic must develop measurable and repeatable responses when exposed to defined exogenous stimulations. The second fundamental consideration, consequently, is the identification of the so-called minimally functional unit of tissue [110] and its size. Tissue units will be radically different depending on the object under study, be it heart [111,112], liver [113,114], kidney [115], or vasculature [116]. Ultimately, since drugs always have downstream effects in organs other than the target one, the most efficient and predictive disease and drug testing platform will be patient-specific and composed of multiple types of interconnected tissues [117]. This is not an easy undertaking, and it thus also requires an interconnection of efforts between multiple laboratories sharing expertise and technologies to reach the ultimate goal of creating such a perfect platform. A consortium led by some of the strongest players in the field is currently carrying out such a project, and its aim is the completion of the so-called HeLiVa platform, an integrated system composed of heart, liver, and vascular microsystems [118] to be used for drug testing in human health and disease.

18.3 BUSINESS CONSIDERATIONS

The rise of personalized medicine is the result of many scientific advances. Yet as often happens with the emergence of any new paradigm, strong and powerful entrenched forces must be addressed to increase the likelihood that these advances

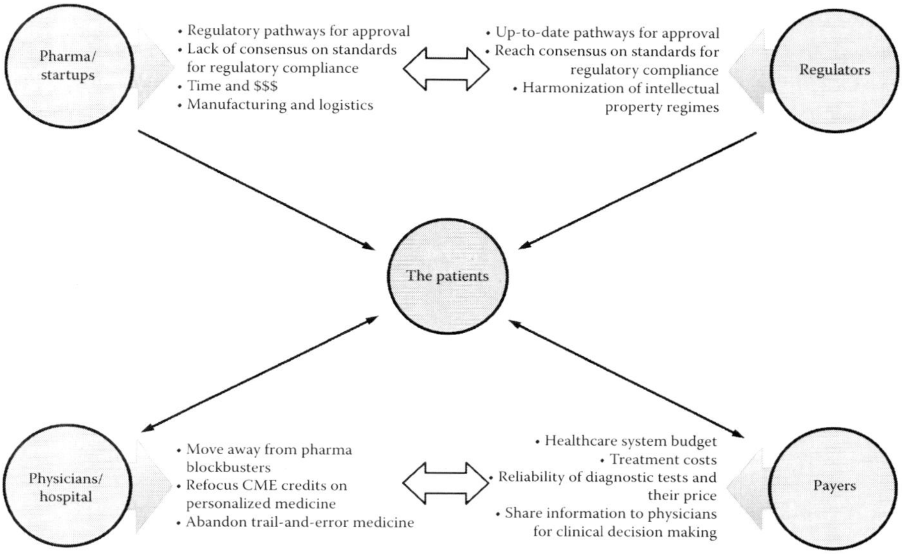

FIGURE 18.3 Schematic of stakeholders affecting the adoption of a patient-centered approach to personalized medicine into clinical practice.

are actually adopted into clinical practice. For regenerative medicine as it applies to personalized medicine, business interests from all major members of the healthcare ecosystem (i.e., regulators, industry, clinicians, payers), as well as environmental players (e.g., legal system and funding sources), are the forces to be contended with (see Figure 18.3).

18.3.1 Regulatory Considerations

One of the largest hurdles facing personalized medicine is evolving regulation. With the eventual end to "blockbuster" drugs and the growing number of cellular- and genetic-based therapeutics, quick, smart adaptation by regulatory agencies will become ever more urgent.

Personalized medicine generally involves the use of two types of medical products: (1) a diagnostic device, and (2) a therapeutic product. These two types of products map directly to the dual aims of *in proto* systems and regenerative medicine, as outlined in Section 18.2. Looking ahead, as each of these types of products is translated out of the laboratory toward use in the clinic, they will be subject to different regulatory pathways: that of molecular diagnostics, and those that apply to cell- and tissue-based therapies, respectively. Any effort to co-develop two or more medical products—such as an *in vitro* diagnostic and a drug—in tandem would face a number of regulatory, policy, and review management challenges, since such products are usually regulated by different regulatory centers and are often owned by separate companies.

Many experts believe that the global regulatory environment has not kept pace with the rapid advances in the field of personalized medicine. A major challenge has been that as the number of emerging biomarkers in the field proliferates, so too do the commensurate expenses for conducting trials. Until consensus emerges as to which biomarkers will constitute regulatory compliance, we run the risk that companies will be loathe to pursue clinical development.

18.3.1.1 Molecular Diagnostics

Personalized medicine relies on diagnostics to identify the best therapy for the patient at the right time: if the diagnostic test is inaccurate, then the treatment decision based on that test may not be optimal. Generally speaking, diagnostic techniques include both *in vitro* tests, such as assays used to identify genetic factors, and *in vivo* tests, such as electroencephalography (EEG), electrocardiography (EKG), or diagnostic imaging.

The *in proto* organ-on-a-chip systems described in this volume, if they are some day to be used clinically, are envisioned as next-generation *in vitro* tests. As such they will supplement the two technical categories of existing tests most relevant to personalized medicine: (1) pharmacogenomics tests, which study inter-individual variations in gene expression, and (2) pharmacogenetic tests, that study inter-individual variations in DNA sequences related to drug absorption and disposition [119].

In the United States, the *in vitro* tests that match a therapy to the patient fall into one of the two regulatory categories. *In vitro* diagnostic devices (IVDs) are diagnostic tests developed and manufactured by device manufacturers, regulated by the Food and Drug Administration (FDA). IVDs that direct treatment decisions are considered by the FDA to be high risk and therefore require pre-market approval. By contrast, diagnostic tests developed by a commercial laboratory, termed laboratory-developed tests (LDTs), are subject to oversight under the Clinical Laboratory Improvement Amendments of 1988 (CLIA) administered by the Centers for Medicare & Medicaid Services (CMS). The *in proto* organ-on-a-chip systems described in this volume, if used in such a setting, would fall under this regulatory jurisdiction.

For *in proto* systems, the expected level of regulatory oversight will no doubt affect the cost of providing laboratory services and, in turn, innovation, especially when the outcome of an LDT impacts pharmacological intervention. Yet the use of diagnostic testing to understand the molecular mechanisms of an individual patient's disease will be pivotal in the delivery of safe and effective therapy for many diseases in the future. It is therefore imperative that a risk-based regulatory approach sustains and promotes innovation while benefiting patient safety.

18.3.1.2 Cell-Based Therapeutics: A Short Primer on Regulatory Pathways

In all major global markets, different applications of a cell- or tissue-based therapy require different regulatory pathways for approval. In the United States, for cell-based therapeutics, these paths typically fall into one of the three categories: (1) devices, (2) biologics, or (3) combination products [120]. Devices and biologics have distinct regulatory hurdles for FDA approval, while a combination product involves first getting approval of both a device and a biologic individually before final approval of the combination product.

Between devices and biologics, devices have a shorter path to FDA approval. This is not surprising, as biologics present a broader set of challenges, which we discuss later. Devices are assigned as class I, II, or III, which determines the amount of research and clinical data necessary for approval [121]. Class I devices, for example, require submission of pre-market notification (PMN or 510[k]). These devices, due to their simplicity, are not required to gain FDA approval, but instead are "cleared" for use. After submission, a 510(k) is cleared or rejected within 90 days. A class II device requires pre-market notification and generally must show both safety and efficacy. The FDA determines requirements for approval on a case-by-case basis and may require human data. Finally, class III devices almost always require extensive testing including human clinical data for pre-market approval (PMA) and eventual device approval. A small number of devices that pose significant risk require extensive testing within a clinical trial. In order to use a device in a clinical trial, an investigational device exemption (IDE) must be submitted and approved. As with any clinical trial, it can take several years to collect the data necessary for submission. However, after the necessary data are collected and the PMA is submitted, the average approval time is approximately 1 year [122].

For biologics, clinical trials are always necessary. Before a clinical trial can start, an investigational new drug (IND) form must be submitted and approved [123]. To gain IND approval, it is typically necessary to show safety and efficacy in both a small and large animal study. This safety and efficacy study is first reviewed by the FDA in a pre-IND meeting for guidance on what is necessary for IND approval. When conducting the safety and efficacy study it is good practice to perform the study under good laboratory practices (GLPs)—the FDA guidelines for proper technique, data collection, analysis, and management of a research study [124]. Following IND approval, human clinical trials can begin. There are four phases: (I) safety and dosage, (II) efficacy and side effects, (III) a large-scale safety and efficacy study, and (IV) post-marketing surveillance [125]. If the biologic succeeds in the first three phases, developers submit a biologics license application (BLA) for approval, after which the therapy can be marketed and sold commercially [126]. The phase IV trial is conducted after the product is marketed and is aimed at studying the long-term effects of the drug. While the duration of clinical trials varies from case to case, the approval process can take 3–10 years to complete trials and come at substantial cost [127].

As personalized medicine typically requires genetic testing and/or manipulation of blood, cells, or tissue, these therapeutics are often placed in the PMA or BLA pathways. This distinction is important, as a PMA approval is typically shorter and requires less data than a BLA. Moreover, the distinction may be a subtle one, as the regulatory pathway is determined not just by the components of the product but also by the function of the human- or animal-derived cells within the product. If the cells serve as a homologous replacement to the application, the therapy is regulated under BLA. For example, Carticel® is a cellular-based therapy in which autologous chondrocytes are cultured and then used to repair a patient's cartilage defect. Because the cells within Carticel are used to replace missing cells in the patient, it was regulated under BLA [128]. If, instead, cells are used to produce a cell-derived product that results in the desired effect, it will be regulated under PMA. Apligraf®, engineered skin that is used for partial thickness ulcers, is an example of the latter. Instead, skin cells are cultured on a scaffold to produce a layer of collagen that enhances the patient's skin's natural ability to heal. Apligraf was regulated as a class III medical device that required PMA approval [129]

Interestingly, alternative use of the same product can result in a different FDA requirement. Apligraf (GINTUIT™), for example, when used for topical oral applications instead of ulcers of the skin, requires BLA approval. Because the mechanism of action is not well understood and it was unclear if allogenic cells remained at the site of injury, a PMA was not sufficient [130].

With the growing number of cellular- and genetic-based therapeutics, regulatory agencies are working hard to keep pace with these new technologies. However, these new therapeutics often raise a different set of problems during research and development than large-scale production of a chemical. For instance, certain concepts from drug development (e.g., pharmacokinetics, distribution, metabolism, and excretion) require adaptation to cell therapies. Studies that assess how the body responds to a therapy may need to explain engraftment, biodistribution, cell trafficking, and persistence of the cells *in vivo* [131].

Before a drug batch can be released, it must also pass expensive, time-consuming tests of sterility (often taking several weeks), including checks for mycoplasma, endotoxin, and viral infections. These types of tests may be inhibitive with cell-based therapies, which are inherently alive and have a short "shelf life" during which they may require nutrients, gas exchange, temperature control, or other care. Autologous cell therapies face further constraints, as each patient's product requires release testing. If these current standard practices remain rigid, with no adaptation to the nature of new therapies, they will significantly affect the cost to produce products based on autologous cells, inhibiting their widespread use and acceptance for reimbursement by insurance companies (payers).

It is easy to recognize the disparate needs of personalized medicine, but regulatory agencies have been slow to change. Significant advances in experimental techniques have reduced both the costs and time to test for sterility (e.g., the Fast MycoTOOL PCR-based test from Roche yields results within 6 hours as opposed to more traditional, 28-day cell-culture based tests). How the FDA treats such techniques will have an immediate impact on the personalized medicine market.

18.3.2 Dismantling Pharma's Blockbuster Model and Supporting a "Market of n = 1"

Regulatory regimes are not the only entrenched forces that have hindered the transition toward personalized medicine in the United States, and to varying degrees the rest of the world. Another strong force is the pharmaceutical industry's historically successful blockbuster model, which focuses on developing and marketing drugs for as broad a patient group as possible and discourages the development of therapies aimed at smaller subpopulations—and the diagnostic tests that can identify them. Furthermore, the new types of therapeutics coming online in the era of personalized medicine (such as cellular- and genetic-based therapies) require different manufacturing and logistics processes, which need to be disaggregated and thereby present their own issues. Again, our focus is the United States, but similar logic could also be applied in other countries.

18.3.2.1 What Personalized Medicine Can Learn from Orphan Diseases

Historically, pharma companies have looked to large disease populations as the biggest potential revenue streams. As personalized medicine gains increased attention, we may learn some economic lessons from the story of rare and orphan diseases.

The development of orphan drugs has been financially incentivized through U.S. law via the Orphan Drug Act (ODA) of 1983. Rising investment in treatments for rare diseases can be attributed in large part to the incentives provided by the ODA. Accordingly, development of therapies for rare diseases is on the rise (approximately one-third of the 39 drugs approved in the United States in 2012 carried orphan drug status) [132]. The success of the original Orphan Drug Act in the United States led to adoption of similar policies in other key markets, most notably in Japan in 1993, Australia in 1997, and the European Union in 2000 [204].

One of the ODA's key provisions is the Orphan Drug Tax Credit (ODTC), designed to promote research spending on orphan drug development. The ODTC allows orphan drug developers to receive a tax credit for 50% of qualified clinical trial costs for new orphan drugs. Without these tax credits in place, some estimates say, one-third fewer orphan drugs would likely have been developed over the past 30 years [133].

With President Obama's 2015 announcement of a $215 million Precision Medicine Initiative [134] meant to help speed the development of new individualized treatments and improve care, many of us are excited about the potential for personalized medicine to follow a similar path.

Since the ODA, other initiatives have promoted additional progress in the treatment of rare diseases. In 2002, legislation created what is today called the Office of Rare Disease Research within the National Institutes of Health (NIH) to coordinate research on treatments for rare diseases. As part of the FDA, the Office of Orphan Products Development (OOPD) oversees the provisions contained in the ODA and administers the Orphan Products Grant Program, which awards approximately $14 million in research grants each fiscal year. Today, many different networks and organizations, such as the National Organization for Rare Disorders (NORD), help facilitate research and patient support in the United States and other countries.

The Precision Medicine Initiative, too, seeks to address key issues in the broader incentives and hurdles influencing personalized therapies under development. It provides funding to the NIH, for example, not only for identifying genomic drivers to cancer but also to explore emerging issues in patient engagement and data sharing. It also allocates money for the FDA and the Office of the National Coordinator for Health Information Technology (ONC) to support the new databases and interoperability standards (along with their requisite privacy and security capabilities) that will be crucial for advancing and coordinating precision medicine and public health. With respect to the regenerative medicine technologies outlined in this volume, this measure will likely positively affect research and development in two ways: by directly increasing the demand for *in proto* cancer models, and more indirectly by highlighting the importance of other personalized therapy approaches, such as regenerative technologies (although not specifically mentioned in the initiative).

18.3.2.2 But if We Can Barely Afford to Pay for Orphan Diseases, How Can We Afford Personalized Medicine?

It is important to mention, however, that even the most comprehensive policies provide only a subset of the relevant incentives for development of personalized therapeutics. With orphan diseases, for example, rising investment may also be attributed to more standard economics: guaranteed market exclusivity and inelastic demand often mean that the fewer patients a drug helps, the more it costs. And so, as many manufacturers have discerned, a small patient group may yet prove extraordinarily lucrative at the right price.

The high cost of treatments for rare diseases is a topic of much controversy. Biotech companies justify their pricing to support the few patients who will ever need the therapies in question. Furthermore, although some targeted drugs, such as those for cancer, are expensive, their high price tag may also be justified because they only help a small number of patients for a few months.

However, it is also important to recognize that the costs of purchasing orphan drugs may mean that many patients risk being denied lifesaving treatments worldwide. An example of this can be found in a monoclonal antibody manufactured by Alexion and licensed in Canada, the United Kingdom, and the United States for the treatment of two rare blood diseases (paroxysmal nocturnal hemoglobinuria and atypical hemolytic uraemic syndrome). This drug, which is currently considered the highest priced drug in the world, is marketed under the name Soliris® and is the only treatment option available to give these patients the possibility of kidney transplantation and the chance of a normal life expectancy. However, it costs over $500,000 per patient per year, a figure that is not sustainable in public healthcare settings. When citing this example, critics have often noted that before testing Soliris for these diseases, Alexion also tested the drug for rheumatoid arthritis, which afflicts 1 million Americans. The trials failed, but if it had worked for arthritis, Alexion would likely have had to charge a much a lower price for this use, as it would have to compete against drugs that cost 10–20 times less.

Drug price controversies such as Soliris may be exacerbated when there is a lack of transparency on production costs. Despite the dual requirements of regulatory disclosure and (for the many publicly held pharmaceutical companies) financial transparency, little is often known about how much it costs to bring drugs to market and how manufacturers arrive at their selling prices (and even less so for privately held companies).

The same issue will likely apply to regenerative therapies, which generally have high R&D costs and high costs of goods sold. In the case of Alexion, the company refused to disclose production costs, preferring instead to state profit margins of 22%, to the disappointment of UK regulators [135].

At the time of this writing, cost transparency remains a quickly developing issue: in May 2015, New York joined five other states (California, Oregon, Massachusetts,

North Carolina, and Pennsylvania) in introducing legislation requiring pharmaceutical manufacturers to submit a report to the state outlining the total costs of the production of certain expensive drugs, with the information to be published on a public website [136].

The industry has rushed to develop orphan drugs in recent years because they cost their developers less to put through clinical trials and command higher prices when they do launch. Growth in the rare disease space is also partially due to the high number of biological orphan drugs, which are currently less susceptible to generic (biosimilar) competition as compared to small molecules. Their high economic value may thus continue beyond the end of patent expiration, a factor shared by regenerative therapies.

For this reason, there is often a combination of fear and promise that accompanies more personalized approaches to medicine beyond rare diseases. The increased understanding of gene mutations and disease-causing proteins, on the one hand, has resulted in the identification of subgroups within traditionally common illnesses. The drug ivacaftor (marketed as Kalydeco®), for example, treats just 4% of the 30,000 patients with cystic fibrosis in the United States and therefore meets the requirement for FDA orphan designation [137].

However, others worry about the potential for this "proliferation" of rare diseases to bankrupt healthcare systems, calculating that if an orphan drug were to be developed to treat each of the 6000 rare diseases in the United States at a cost of $300,000 per year, the cost of managing these conditions would be $1 trillion, the size of the global pharmaceutical market [138].

That calculation is an extreme example, but it cannot be denied that healthcare systems have finite budgets and so must at least to some degree weigh the costs and benefits of medicine. And so, thinking ahead toward regenerative medicine therapies in development, in order for therapies to move from the bench to the bedside we must acknowledge that beyond science, the economics, too, must work.

There is also a potential cost savings to consider with the *in proto* technologies described in this volume. To the extent that *in proto* systems enable us to better match patients with next-generation therapies with high effectiveness rates, or to customize dosage regimes of small-molecule therapies, developers of these systems will be able to make the case that they are helping to reduce the costs of "impersonalized medicine," especially those associated with administering ineffective drugs and treating side effects.

One illustrative example of this type of cost savings is warfarin, a widely prescribed anticoagulant. In the last 20 years, scientists determined that variations in two genes (CYP2C9 and VKORC1) could account for nearly half of the variability in patient response to warfarin, yet the $350 test had been administered to fewer than 5% of patients who started warfarin therapy. The FDA updated the label for warfarin in 2007 and 2010, informing doctors that patients with the variant genes might require non-standard warfarin doses when they start therapy [139]. However, genotyping strategies haven't caught on in clinical practice and guidelines still don't recommend genotyping patients ahead of prescribing them warfarin [140]. FDA members estimate that if diagnostic tests to detect certain gene variations were routinely administered to patients who need warfarin, the resulting reduction in serious bleeding events and strokes caused by under- and overdosing of the drug could save the U.S. health care system as much as $1.1 billion annually [141].

A lesson from this case for *in proto* systems is that, to the extent that a pharmaceutical company can demonstrate that its drug lowers the overall cost of treating a subpopulation with a disease, private and government insurers may influence willingness to pay for the relevant diagnostic test and to pay a higher price for the drug treatment. The same logic will also apply to regenerative therapies, as the benefits for patients (e.g., potentially fewer revision surgeries for living implants versus synthetic) may be realized over an individual's lifetime, while the reimbursement system controlled by these institutions pays for—and thus encourages—the performance of procedures rather than accurate diagnoses or prevention. In the United States, where the insurance markets are more fragmented than in other more centralized healthcare systems, these cases may be even more difficult to make, since the payer that paid for the original surgery might not the be entity that will realize the cost savings of fewer revision surgeries, etc.

18.3.2.3 An Assembly Line for the Pipeline? The Talent Shortage and Automated Cell Culture

Putting aside debates over which costs should be factored into drug development [133], one of the undeniable reasons for the high cost of goods sold in regenerative therapies is the high amount of expensive materials and operator handling required by modern biology techniques [205]. Furthermore, the costs of regulatory compliance involve additional expense in manufacturing, with its associated expensive GMP facilities, procedures, and materials.

Demand is rising for automated and quantitative cell culture technology, driven both by the intense activity in stem cell biology and by the emergence of systems biology. If you have to produce a large number of cells for high-throughput cell-based assays, automation can free up many hours of time compared to manual cell culture. The demands of scaling up into 3D tissues only make the need more apparent. Furthermore, automating cell culture could also make cell growth more consistent, because steps like mixing and pipetting are more tightly controlled. The cost and complexity of fully automated systems (many costing upward of $1 million) can be barriers for researchers used to manual cell culturing. The development of automated systems in the life-science laboratory has gone hand-in-hand with the information-management systems needed to handle the masses of data generated.

18.3.2.4 From Shelf Life toward Warm-Chain Distribution

An unbroken cold chain is an uninterrupted series of storage and distribution activities which maintain a given temperature range. Cold chains are common in the food and pharmaceutical industries and also in some chemical shipments, where vendors may provide temperature control and quality assurance for regulatory compliance. However, with regenerative medicine technologies certain samples may require chilling, while others may require room temperature or even incubation during transport. We may also anticipate that future shipping of live samples could require optimized sterile gas exchange, monitoring of key biomarkers (e.g., cellular viability, pH, glucose, CO_2, etc.), and high-precision tracking to ensure patient–product matching. Technologies to meet this challenge are being developed and range from precision temperature and ventilation control provided by vendors (e.g., Microq, http://www.microq.com/key-features.php) to proprietary biopreservation media (e.g., Biolifesolutions, http://www.biolifesolutions.com/biolife/markets/regenerative-medicine/), optical gas probes (e.g., PreSens, http://www.presens.de/) and clinical trial supply management software (e.g., Medidata Balance, http://www.mdsol.com/en/what-we-do/study-conduct/medidata-balance). Incorporation of these types of technologies (and the data they will produce) will be critical as cellular therapies continue to be developed for the clinic.

18.3.3 Legal Considerations

The $730.1 billion health and medical insurance industry represents, alongside construction and finance, one of the largest and most highly regulated sectors in the United States [142–143]. Moreover, on the therapeutic side, it is an industry that has historically been fueled by capitalization of intellectual property rights. For these reasons, the complex landscape of intellectual property and medical liability presents important legal considerations for future *in proto* and cell therapies.

18.3.3.1 Intellectual Property: Biosimilars, Gene Patenting, and Patent Trolls

Intellectual property rights refer to legal rights granted to inventors or creators to protect their inventions or creations for a certain period of time. This concept dates from the fourteenth century in Venice, Italy [144], and continues to play a vital role in the modern economy, enabling innovators to reap commercial benefits from their creative efforts.

The field of patent strategy is laced with metaphors that evoke medieval battles: in patent portfolios, patent "fences" may be built and decorated with "crown-jewel" patents. In patent "wars," patent "swords" may be employed offensively, while patent "shields" are used against others who file suit, and patent "trolls" abound as profiteers [145]. Yet intellectual property policy is also modern and dynamic and has been subject to significant changes in recent years. The 2013 America Invents Act, which contains the most significant changes to the U.S. patenting process in over 60 years, was intended to harmonize the American patent system with those in the rest of the world. This legislation provides for different standards of proof on intellectual property and different timing on protections, and it is having a profound influence on patent strategy in the life sciences [146].

Broadly speaking, patents are generally awarded for 20 years and are granted for products and processes that satisfy the criteria of novelty, non-obviousness, and utility [144]. The economic role that patents play in the medical field, however, ranges depending on the type of medical technology being developed. In the traditional model for small-molecule pharmaceutical development, for example, 10 to 12 of the patented years are typically spent developing the drug at a cost of about $1.5 billion to $2.5 billion, leaving only 8 to 10 years to make money before generic drug companies can sell bioequivalent formulations for a fraction of the price. At that point, sales for off-patent branded products quickly flatten by over 90% [147]. In contrast, although intellectual property is also often considered the core value generator for medical devices, this product class is not as much discovered as designed, and iteratively developed. The value of patents for marketed medical devices is therefore often outlived by the expiration date, as competitive forces tend to drive obsolescence before patent expiry [148].

The value of intellectual property in the class of biologic drugs, which increasingly encompasses regenerative therapies, bodes well for the retention of value even beyond the patent lifetime. Because biologic therapies (unlike small molecules, which are produced through relatively inexpensive biochemical processes) are made in living cells (and regenerative therapies are made of living cells), they require specialized and often proprietary production methods and incur high manufacturing costs [149]. More important, from an intellectual property perspective, they are impossible to copy with exactness. After biologic therapies lose patent, their generic equivalents are therefore not exact formulations but are instead dubbed "biosimilars" to denote their non-identical nature to their respective patented products. In order to demonstrate functional equivalency, these follow-on therapies are required to undergo more testing than ordinary generic drugs, all the while attempting to avoid infringing on the patents that litter the manufacturing process [150].

The combination of high manufacturing costs and high development costs for biologics has resulted in shifting for biosimilars the economics that have historically incentivized the production of generic drugs for small molecules. Generics typically cost less than their respective patented products because they are relatively inexpensive to produce, and generic manufacturers do not have to recoup the high drug development costs incurred by the molecules' originators. Biosimilars manufacturers, by contrast, must recoup not only high manufacturing costs but also the high costs of increasingly complex characterization and clinical evaluation. As a result, biosimilars' penetration rates have lagged those for small-molecule generics [151]. So far, the slow entry of biosimilars has been a boon to innovators, effectively extending the profitable lifetime of a biologic drug past the date of its patent expiration and incentivizing companies to prioritize biotech-based product development in coming years [152]. However, it is important to keep in mind that this is a developing issue, as the first biosimilar was only approved in Europe in 2006 [153] and in the United States in 2015 [150]. As of now, no biosimilars have to date been approved for regenerative therapies.

Meanwhile, innovation is progressing for *in proto* therapies and other molecular diagnostics, fueled by advances in genomics research. However, there has been some concern around the adequacy and appropriateness of current national patent regimes to address questions of DNA patenting and commercialization of intellectual property related to the human genome. In a landmark case, Myriad Genetics discovered that mutations in two human genes, BRCA1 and BRCA2, led to a significant increased risk of breast cancer: 55% to 65% of women who express these mutations have been estimated to develop breast cancer by age 70 [154]. The correlation is so dramatic that celebrity Angelina Jolie underwent a double mastectomy following a positive result by Myriad's diagnostic. By patenting these genes, Myriad guaranteed its position as the sole entity capable of testing for these mutations and priced its product accordingly.

Many customers sued Myriad Genetics over the high costs of the tests. The claim was that Myriad did not invent anything, despite having discovered a very important aspect of naturally occurring human DNA. Ultimately, the Supreme Court deliberated the case and found in favor of the plaintiffs [155]. With this ruling, a new era for genetic and personalized medicine was ushered in. No entity can patent naturally occurring genes, which means anyone can produce a test for the BRCA1 or BRCA2 mutations and sell it commercially.

However, the Supreme Court took a position that still incentivizes companies to discover gene-based therapies and diagnostics. While DNA cannot be directly patented, complementary DNA (cDNA) can. This nuance allows companies to produce a novel, patentable gene clone that can be used in healthcare for a variety of diagnostic or therapeutic applications. It is a small but far-reaching distinction that has shaped the current landscape for personalized medicine.

New science in the field of regenerative medicine holds enormous promise for improving lives. Yet its development will hinge on factors far beyond scientific discovery and technological innovation, including market needs and the costs of translating intellectual property into commercial ventures. Cost pressures from payers and longer development times are already squeezing medical innovators. Within that complex landscape, it's not yet clear how the recent dynamic changes in intellectual property will translate to *in proto* and regenerative cell therapies. Bitter fights are being waged as we write this manuscript over the CRISPR DNA editing technology, for one [156]. Neither is it entirely clear if most researchers (in either academia or industry) are violating patents as they develop and commercialize iPSC-based therapies [157]. And despite the lofty intentions of the Leahy-Smith America Invents Act (AIA), we are likely to continue to see the emergence of unintended consequences, such as the subversion of Patent Trial and Appeal Boards, which were meant to simplify appeals but after abuse from so-called "patent trolls" have been labeled "patent death squads" [144].

In sum, it remains to be seen which forces will prevail in the "patent wars," as "patent fences" rise and "patent swords" swing. But we hope and trust that, amid the field's many dynamic checks and balances, and given the high stakes, innovation will continue to survive.

18.3.3.2 Liability Concerns: New Knowledge Breeds New Duties (and Ideas of Negligence)

The era of personalized medicine is approaching—more slowly than many originally predicted, but nevertheless advancing in fits and starts. We have identified many specific challenges to the goal of integrating a personalized approach to clinical development and medical care, including technical hurdles, regulatory hurdles, and associated costs. One factor that has received relatively scant attention, yet has acted as a powerful driver for widespread behavioral change and the adoption of emerging technologies, is liability law.

Liability and healthcare go hand in hand. The concept behind medical liability dates from Roman times, the idea being that every person who enters into a learned profession undertakes to bring to the exercise a reasonable degree of care. And although we are no longer cutting off the hands of doctors who are held legally responsible for harm to patients (i.e., the Roman punishment; see [151] for an excellent review of malpractice law in the United States), punitive damages resulting from the U.S. medical malpractice system have been estimated at $55 billion annually [158].

The intent of the malpractice system rests on the assumption that the threat of a lawsuit will encourage providers to adhere to standards of care and manufacturers to manufacture safe and effective therapies, all of which should lead to better patient outcomes. Yet the malpractice system may also represent an obstacle to healthcare cost-effectiveness and quality, as provider fears over malpractice claims have been implicated in overuse of treatments and tests adding minimal value, as well as the avoidance of potentially risky technologies that may provide higher efficacies to patients [159].

Malpractice claims are costly and pervasive. An estimated one-quarter of American surgeons have been the subject of a malpractice suit over the preceding 2 years [158], and some estimate that by age 65, 99% of physicians in high-risk specialties will face a malpractice claim [160]. Such startling figures indicate that liability is likely to be a major driver of the future direction and implementation of personalized medicine, because as new tools such as genetic tests emerge, new commensurate responsibilities (and therefore concepts of negligence) arise.

The *in proto* systems presented in this book could help. The additional pharmacogenomic information that these systems provide has the potential to promote patient safety and efficacy while allowing patients to avoid unnecessary costs and side effects. However, discovering the legal balance between what is efficacious and necessary and what is superfluous will be difficult. We're already starting to see the effects: individuals injured by adverse drug effects are increasingly likely to bring lawsuits alleging that they have a polymorphism or biomarker conferring susceptibility to the drug that should have been identified and used to alter their drug treatment [161].

Likely targets of such lawsuits include drug manufacturers, third-party payers, physicians, and pharmacists. The handful of cases already decided by the courts involving clinical genetics—often prenatal testing—illustrate the particular liability threat that physicians face as compared to the other stakeholders in this realm. In addition to the traditional claims for negligence, genetic testing has given rise to new concepts in case law such as wrongful conception, wrongful birth, and wrongful life, while creating new applications for claims such as loss of chance and duty to third parties [162].

Looking ahead, these concepts are likely to apply to future *in proto* systems and imply that the false-positive and -negative rates of these tests (and the risks/efficacies of the therapeutic regimes for which they serve as diagnostics) will be of the utmost importance. Tricky questions may arise: if doctors fail to perform genetic tests that may have led to different drug regimes, for example, should they be held liable? If the treatment in question results in 100% success for the patient, it is easy to claim that the doctor should have performed the genetic test. However, as is the case with many cancer treatments, what if the difference in success is only 5%? Or if the treatment will have negative side effects? Or if the test results in false-positive and/or false-negative rates? Should insurance companies be held liable if they refuse to pay for such tests? Should manufacturers of future *in proto* systems be held liable for false positives and/or negatives? What if the test and corresponding drug regimen costs significantly more, leading to higher healthcare costs? The answers to these questions will likely have a direct impact on a patient's course of treatment, as continued cost increases in diagnostics, treatment, and healthcare insurance shape commercialization of personalized medicine.

In the realm of cell therapies, complex manufacturing and logistics procedures are likely to pose new product liability risks to manufacturers (e.g., those associated with cellular extractions from patients, transfer and laboratory manipulation of cellular material, and complex logistics for tracking personalized treatments), as well as physicians (who will need to learn new procedures and decide patient course of treatment).

The FDA plays an important role in defining liability. Product approvals and product labeling impact the division of responsibility between providers and product manufacturers. They also define the indications for which products may be used [163] and the distribution of liability between innovators and generic manufacturers [164].

Looking toward translation of future regenerative technologies, it will therefore be important to view clinical development and movement through the regulatory pathway through the lens of liability as it applies to stakeholders. The absence of many lawsuits today should not provide much comfort, given that the typical dynamics of litigation are that it starts slowly, then picks up momentum in a cascade that is very hard to stop once it starts [162]. Factors that may help reduce liability risks include physician education and a focus on proactively developing mechanisms commensurate with new technologies to ensure manufacturing quality.

18.3.4 Funding Sources: Fostering The "Entrepreneurial Ecosystem"

At present, we are by some measures witnessing a catastrophic decline in the productivity of the pharma industry. Fewer drugs than ever are being approved. Consider two numbers: 800,000 and 21 [165]. The first is the number of medical research papers published in 2008 [166]. The second is the number of new drugs [167] approved by the Food and Drug Administration the previous year.

Furthermore, it is unclear how to finance an increase in productivity: patent expirations are set to cost the pharmaceutical industry $100 billion from 2012–2016, while revenues expected from new drugs during the same time are estimated to be only one-third of that number [168]. So the pharmaceutical industry by itself is unlikely to provide the solution.

The United States government, meanwhile, has a double problem: despite the so-called pharma "patent cliff," healthcare spending is outpacing economic growth [169]. And since the movement to double the NIH budget ended in 2003, the number of funded grants has decreased by 50% even as grant applications explode, increasing by 20% in the same time frame [170]. Considering regenerative technologies, nascent in the lab and primed for development toward the clinic, the problem is so dire that some have even dubbed the gap between biomedical researchers and the patients who need their discoveries "the valley of death."

The reasons for this declining productivity are many and interdependent, including the explosion of molecular biology in the 1970s and the resultant emergence of biomedical research as a discipline in its own right, siloing clinical and basic research [171]. This same explosion has also likely contributed to development of more complex therapies, bottlenecking the drug development pipeline with complex clinical trials and the associated additional cost of drug development [166].

But a glimmer of hope may be hidden within another startling statistic. The federal government invests \$50 billion per year in academic research—\$23 billion in the life sciences alone—and yet the returns on investment, if measured by the translation of academic research toward the clinic, are uneven [168]. The San Francisco Bay Area and Baltimore, for example, each receive approximately \$2 billion annually in government R&D funding, while biotechnology startup productivity was 40-fold more in San Francisco [172]. It is therefore clear that getting a promising new treatment out of academia, which has become a main source of innovation in life sciences, and translated to market is reliant on many factors beyond R&D funding and innovation [168,171,176].

Many investors and policymakers believe that life science companies are most successful when located in the epicenter of what have been dubbed life science "clusters" or "hubs." Around the world, governments are making efforts to foster such hubs.

The qualities of a life-science hub include (please see [177] for an excellent review):

- Availability of affordable commercial lab space
- Availability of "bridging-the-gap funding" for proof-of-principle studies before the research would be of interest for licensing by established companies or serve as the foundation for a start-up
- High concentration of high-quality academic institutions, which represent a source of significant discovery and are the first adopters of new products and a source of talent for recruiting
- Location: proximity to three international airports, for easy participation in meetings and facilitated logistics—increasingly important in a global world
- Policies: incentives such as tax credits for R&D and/or employees, and safeguards such as regulatory and political transparency
- Access to financial markets: to translate science into successful businesses
- Pharma and startup companies: to provide a support network

Incubators, often located in life science hub cities, are key to fostering growth of technologies by allowing the sharing of real estate and facility expenses, which are among the industry's biggest costs. Furthermore, they may facilitate collaborations with other members of the "ecosystem" through proximity to local academic institutions and financial markets, and serve as a hub for cultivating talent through mentoring [172].

In recent years, life-science clusters have become more of a global phenomenon. Perhaps in part because of the focused attention and resources these clusters bring together, we are seeing an increasing share of non-U.S. patent applications and a rising number of doctorate and higher education degrees in the science and engineering fields [177]. More and more innovation is taking place in local pockets of small and medium sized organizations. This trend is also reflected in activity in the venture capital markets: initial rounds of funding (Series A) for novel drug R&D reached their highest levels in a decade in 2013, representing nearly 80% of venture capital for therapeutics [173,174], and new drugs approved in 2014 numbered 41 [175].

Because therapy development can take up to 20 years, the eventual impact of such efforts on the drug pipeline will only emerge with time. Success and advancements occur when the best minds are working in unison and, just like with bioreactors and cells, we cannot forget the importance of physical location and environmental cues for cultivating and maturing scientific breakthroughs.

18.3.5 New "Social Contract" toward Patients and Physicians as Partners

The word "patient" is derived from the Latin word *patientem*, meaning "bearing, supporting, suffering, enduring, permitting" [178]. The word conjures up a vision of quiet suffering, of someone lying patiently in a bed waiting for the doctor. It also conjures up a kind of unequal relationship in knowledge and power between the user of healthcare services and the provider (and, increasingly, the payer too).

We would also argue that the image of a passive patient is a barrier to realizing the true potential of personalized medicine, for several reasons. First, it's impractical. If, in the past, information traditionally flowed from doctor to patient, the Internet has now upended the relationship, and the change is here to stay. We seem to lack consensus about whether a more engaged patient population is always a good thing: on the one hand, using the web as a source of health information has been shown to be associated with anxiety reduction (and presumably therapeutic value) in patients [179]. On the other hand, critics cite the potential hassle and risks of low-quality information with so-called "armchair oncology" [180].

Nevertheless, Internet use has led to a host of increased expectations, and more and more, patients will continue to expect doctors to guide them through the frequently conflicting advice and opinions they read about online. As they learn more about potential treatments, patients also increasingly expect customized treatment plans [181]. The authors can personally attest to the engagement, curiosity, and enthusiasm of the patient community: as researchers in a small start-up company, with a product in preclinical development, we constantly receive emails from patients asking for treatments and wanting to volunteer for clinical trials. We hypothesize that if harnessed correctly, a patient-driven approach could be a powerful driver for the clinical adoption of personalized medicine.

Second, the passive patient poses serious risks to reaping the benefits of what may be the largest driver (and potential cost savings) of personalized health: understanding the effects of individual lifestyle choices. We can achieve this by performing screenings and empowering the patient and physician together to monitor, track, and maintain health (as opposed to treating an already manifested disease). Consider a future *in proto* screening test, which identifies hundreds or perhaps thousands of people with a particular problem—and treatment for the problem starts when symptoms have not yet surfaced. Assuming a molecular or genetic diagnostic test is reliable and relevant, the total cost of a test may only be understood in the context of the total cost of treatment. Without the coordinated engagement of patients and physicians we will not likely realize the benefits of such a test.

Test developers, vendors, and clinical labs must also be able to communicate the usefulness and reliability of their tests to regulators and payers, ensuring scientific and clinical acceptance of their tests as well as securing reimbursement. The first thing a payer wants to know is whether a particular molecular or genetic diagnostic test (and future *in proto* test) is reliable and therefore worth paying for as a covered health benefit. Does this test distinguish one treatment option from another? What are the test's false-positive and false-negative rates? Are the test results correlated with treatment options and health outcomes? Are there a sufficient number of qualified clinicians available to interpret the results? In essence, is this test ready for prime time? With regenerative therapies, similar questions abound, especially if the benefits of a particular treatment paradigm (e.g., a living implant versus a synthetic implant) might only be realized over a long period of time (i.e., reducing numbers of revision surgeries).

Third, the passive patient model poses risks to successful clinical development of personalized treatments. Blinding, for example, has been a mainstay of clinical trials since the mid-20th century, with research participants voluntarily forgoing knowledge about their treatment allocation in order to maintain scientific validity. However, several technologies are closing the gap between what patients and researchers may know. The growing set of wearable consumer technologies such as fitness trackers, smartphone apps, and sophisticated consumer medical devices have been shown to increase clinical trial compliance [182,183], and the proliferation of online communities (such as PatientsLikeMe) allows patients to collaborate online and share side effects [184]. These technologies are challenging the medical community to re-evaluate the power differential between patients and trialists to forge a "new social contract"—away from that of blinded test subjects and more toward true clinical partners in the development of new therapies [185]. Partnership with patients has many potential benefits, including facilitating patient recruitment and speeding up the process for gathering relevant data, as was done in a PatientsLikeMe analysis for amyotrophic lateral sclerosis (ALS) [186]. In the case of regenerative therapies, the patients themselves may even be the source of the key ingredients (i.e., cells) of the therapy itself.

Inclusion in the scientific process may also address another important, although less commonly acknowledged, "side effect" of illness, the more existential affliction that can plague those struggling with disease: that is, the underlying feeling of isolation and loneliness from the lack of a direct avenue to those in positions of power to make change. However, these possible benefits must be weighed against the potential risks of lack of standardization, to hone a model that maintains scientific rigor while respecting the patient autonomy required for personalized medicine.

The physician's role, which is still deeply rooted in trial-and-error medicine, will need to rapidly shift toward aiding patients in interpreting the many sources of health data that are becoming available. Clinicians must anticipate that patients will begin arriving at their office, for example, with the results of home genetic tests, or will expect to share the results of their consumer wearable devices, or will want their help to interpret the online medical information they have been searching. The word "doctor" is derived from the Latin word *docere*, "to show, teach, or cause to know" [187]. This is a role that, paradoxically, will likely fit clinicians more and more in a personalized medicine approach that will involve more challenges to the physician's expertise. Most doctors are already required to attend 12 to 50 hours of continuing medical education (CME) courses per year to maintain their licenses. To get physicians up to speed on personalized medicine, including developments in genomics, diagnostic testing, and targeted therapies, these courses will become increasingly important.

Interestingly, as scientific and precise as the process of personalized medicine can be, like any human process, it will take subjective and personal relationships to make it work.

18.4 CASE STUDIES

This section reviews personalized, cell-based products that are already in use, either in the United States or globally, as well as those undergoing experimentation and clinical trials, encompassing a range of complexity and human physiology (see Figure 18.4). We explore the nature of these treatments, their applications and challenges. We also review *in vitro* tissue regeneration (i.e., *in proto* systems) as personalized drug testing platforms.

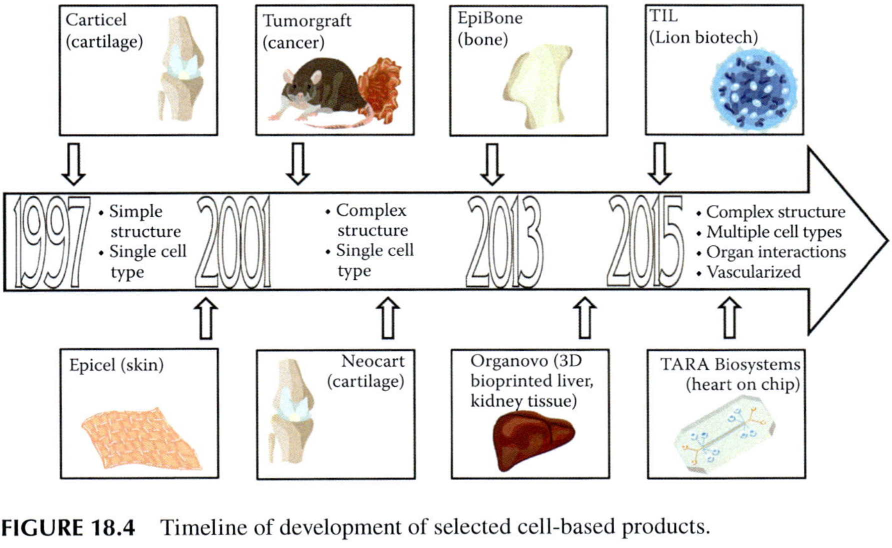

FIGURE 18.4 Timeline of development of selected cell-based products.

18.4.1 Personalized Cell Therapies and Products That are Already in Use

18.4.1.1 Cartilage: Carticel (Vericel)

Due to the avascular nature of cartilage tissue, it has very limited ability to regenerate. To overcome this hurdle, several companies have sought to grow cartilage tissue *in vitro* for subsequent *in vivo* implantation. Carticel offers an FDA-approved cell therapy for articular cartilage injuries of the knee in adults. A biopsy of healthy femoral cartilage provides autologous chondrocytes that are expanded *in vivo* [188]. When the cell numbers are sufficient, the cultured chondrocytes are then reintroduced to the damaged area. These newly implanted cells form hyaline-like cartilage that has similar properties to normal cartilage and helps improve the knee function and reduce pain. This individualized treatment aims to recreate and replace the damaged cartilage, offering long-lasting relief. While there are several companies focused on growing cartilage tissue *in vitro* (Histogenics, ProChon Biotech, CellGenix), Carticel is the only product currently FDA approved (1997) for use in the United States.

18.4.1.2 Skin: Epicel® (Vericel)

Epicel is an autologous dermal graft derived from a patient's own skin [189]. Starting with only a postage-stamp-size piece of skin, the skin cells are grown *in vitro* to large sheets approximately 2 to 8 cells thick. The cells are encouraged to grow by the use of mouse cells that secrete growth factors. Interestingly, even though Epicel is an autologous product derived from the patient's own cells, it is considered a xenograft, or a graft from one species to another, because it is grown with the help of mouse cells. These sheets are used to treat burn victims who have been burned over at least 30% of their body. While efficacy has still not been proven and the treatment is therefore not FDA approved in the United States, there is currently no alternative, so Epicel is sold as a humanitarian device only, restricting its commercial availability to extreme cases [190].

18.4.1.3 Cancer: TumorGraft® (Champion Oncology)

Dr. Manuel Hidalgo first described mouse avatars as an experimental method used to test individualized cancer therapies by implanting a patient's own tumor under the skin of immunodeficient mice [191]. Following growth of the tumor, different therapeutic approaches, such as chemotherapy, can be tested to determine which drugs are most effective. The same sort of method is now used in therapy with TumorGraft. Avatars are not without limitations. Human immune response to cancer plays a significant role in cancer development and treatment but is absent in mice. Also, because of the time necessary to complete an experiment, the patient may not be able to forego treatment or may succumb to the disease before results are delivered [192]. However, initial results of these tests look promising, and 87% of drugs identified as successful at reducing tumors in the mouse avatar model were also successful in the patient.

18.4.2 Personalized Cell Therapies and Products Already an Development

18.4.2.1 Bone: EpiBone

EpiBone uses customized bioreactor systems to grow anatomically precise, personalized living bone tissue in the lab [193]. From fat tissue, a patient's own mesenchymal stem cells are isolated and expanded *in vitro*. In parallel, a CT scan delivers the exact size and shape of the bone defect, and a personalized scaffold and bioreactor are prepared [96]. Once the scaffold is seeded and placed into the bioreactor, the patient's stem cells are coaxed into bone-generating cells. It takes approximately 3 weeks to produce a mature piece of bone. EpiBone is currently in the preclinical stage of development, testing its personalized bones in large-animal studies.

18.4.2.2 Cartilage: NeoCart® (Histogenics)

Similar to Carticel, NeoCart uses a biopsy of healthy cartilage from a patient and expands chondrocytes *in vitro*. However, NeoCart then seeds these chondrocytes onto a proprietary scaffold and cultures the graft under low oxygen, varying pressure and perfusion of nutrients. In a few weeks, the engineered tissue is sent to a physician, who implants the graft in the damaged area. NeoCart is currently in phase III clinical trials and has shown significant improvement in patients treated in phase II studies [194,195].

18.4.2.3 Cancer: Tumor-Infiltrating Lymphocytes (Lion Biotechnologies)

In early stage cancer, special immune cells known as tumor-infiltrating lymphocytes (TIL) migrate to the tumor site and attack it. But cancer cells adapt to such attacks and suppress immune response. Lion Biotechnologies is developing ready-to-infuse autologous T-cell therapy using TIL for the treatment of patients with Stage 4 metastatic melanoma [196]. The TIL treatment is currently in a phase II clinical trial.

18.4.3 Drug Testing Platforms

Besides growing living tissues for subsequent *in vivo* implantations, growing human tissues *in vitro* is also important for drug testing and research development. Animal models don't always work accurately for drug testing and evaluating human diseases [197]. Moreover, differences among individuals require personalized and customized drug testing platforms and disease models for more accurate results. While not personally "individualized" to each patient, these systems can represent subgroups of patients, such as those with certain diseases (e.g., diabetes) or a specific gene mutation. In this regard, they are a part of the personalized medicine revolution.

18.4.3.1 Liver and Kidney: Organovo

Organovo designs multicellular, dynamic, and functional human tissues using 3D bioprinting [198]. These tissues mimic the key aspects of the native tissues and could be used as drug testing platforms in medical research. For example, tissues inside the body contain multiple cell types and layers and have certain structure. Bioprinting enables the construction of these tissues layer by layer, including the cell types specific for each layer, resulting in living, 3D human tissue models. In addition to drug testing, diseases can be induced and studied in these microenvironments [199].

18.4.3.2 Cardiac: TARA Biosystems

TARA Biosystems develops an organ-on-a-chip platform to provide physiologically relevant heart-on-a-chip human tissue models for drug discovery and new therapy evaluations [200]. These micro-sized human heart tissues enable testing and offer more accurate readings on the safety and efficacy of the new therapies. They are developed *in vitro* using primary cardiac cells and are cultured under specific conditions that include perfusion and electrical current.

18.5 FUTURE DIRECTIONS: A COLLABORATIVE PATH

The coming era of personalized medicine is an exciting moment in our history, a kind of Gutenberg moment of the most personal kind. By helping us decipher and direct our own personal genetic programming (and, increasingly, that of the microorganisms and viruses that call our bodies home), it will help us live longer, healthy lives. The developments to date represent the cumulative efforts of countless scientists, engineers, clinicians, regulators, businesspeople, and generous patients, dedicated to learning about our collective selves and how to treat ourselves better when we're ill.

Yet as we've outlined in this chapter, the biggest hurdles for translating what we know about systems and molecular biology into personalized diagnostics and therapeutics include not only technical considerations and gaps in knowledge but also our human organizations. The explosion of molecular biology in the 1970s, culminating in the Human Genome Project, had the unanticipated side effects of siloing clinical and basic research and bottlenecking clinical development. New knowledge of polymorphisms and other biomarkers has led to breakthrough treatments, but also to frustrations from patent wars and physician paralysis in the face of potential liability to patients who are injured or who simply don't improve. We are, after all, human scientists, with human test subjects. As we look ahead toward the translation of *in proto* and *in vivo* regenerative technologies, we in the field of regenerative medicine will need to collaborate more and more, and with new communities.

A critical set of economic and scale-up challenges remain before cultured cell therapies can graduate from the laboratory to the production line. In order to become a widely accepted new standard of care, personalized medicine and point-of-care testing platforms must ameliorate existing techniques, offering improved sensitivity, time-to-results, ease of use, and cost-effectiveness. This will be difficult to accomplish when even now, obtaining precise longitudinal data for genetic expression over time remains difficult. In addition, there are likely to be issues around comparability for a range of relevant tests, including (but not limited to) release testing and potency assays. This is especially true for autologous therapies where physiochemical characterization alone is not adequate to demonstrate quality. Until these issues are addressed, biosimilars will likely not pose a financial threat in the way that generic drugs do. But in the meantime, quality and costs will nonetheless be a concern, especially when we consider rolling out technologies from the developed world to the developing world.

Unlocking these challenges successfully may require researchers to collaborate with partners we've never considered before—even non-scientists. Like crowd-funding and crowd-sourcing efforts in other fields, new avenues are opening up for citizen scientists to participate in cancer and genomics research (see Zooniverse and Foldit, and project funding through www.experiment.com, www.petridish.org, and others [201]). This type of collaboration could accomplish a dual purpose: amplifying the efforts of scientists in the lab and clinicians in clinical trials while also engendering public understanding and support for next-generation therapeutics outside the confines of government grants.

This technological dawn, like any other, is fraught with new possibilities as well as unanticipated consequences. We don't have to look far into the past for examples in which major disruptive technologies have been accompanied by serious ethical questions (e.g., organ donation, *in vitro* fertilization). If you talk to scientists in the field of regenerative medicine about culture and media, however, we will likely first think of DMEM and cytokines. But increasingly, the meanings of these words in the humanities will be ones that we cannot ignore, especially in the realm of bioethics [202]. Here again, we could seek unprecedented partners. The work of bioartists who incorporate biologically living matter within art installations and artifacts, such as those from SymbioticA in Australia [203], for example, provokes questions about scientific truths, what constitutes living, and the ethical and artistic implications of life manipulation. For those provocations alone, bioethics would benefit from opening to contributions from the arts. At EpiBone, in parallel with our scientific efforts, we have launched an artist residency program for the express purpose of reaching out to the bioart community early on.

New science is also emerging connecting our molecular selves to the quantum mechanics of the cosmos, helping to explain physiological processes as diverse as photosynthesis and cancer mutagenesis. Life arises from the complex interactions of organic molecules. As we reveal the molecular mechanisms of diseases through our current paradigms, we will reveal deeper mysteries and find ourselves collaborating even more widely with physicists and mathematicians. If the field of molecular biology made "coders of us all," perhaps tomorrow's discoveries will make us into physicists and mathematicians (and don't forget ethicists).

In Section 18.3.5, when we discussed the etymologies of the words patient (as sufferer) and doctor (as teacher), we left out that the word "physician" is derived from the Latin *physical*, or "things related to nature." Who we are, and what illnesses we suffer, depends not only on our genes but also on a complex intersection of environmental, genetic, social, and cultural factors. We have a great deal to learn about the biological, anatomical, and physiological mechanisms that underlie disease. Realizing a truly personalized approach to patient care will require fundamental advances in understanding each of these influences, as well as how they impact one another.

Erwin Schrödinger, the Austrian physicist and one of the founders of quantum mechanics in the 1920s, said: "The scientist only imposes two things, namely truth and sincerity, imposes them upon himself and upon other scientists."

As we look to a future in which we use science to lead the way toward better care of our fellow humans, may our most useful tool be our sincerest quest for truth. Or, in haiku form:

Sci-pothesizing
Probing, striving for proving
Rinsing; repeating

ACKNOWLEDGMENTS

The authors would like to thank the editors for their invitation to contribute to this volume; Gordana Vunjak-Novakovic for her mentorship; Grace Rubenstein for her copy polishing; Maia Yoshida for helping to produce portions of the illustrations; Nafissa Yakubova for her help with the manuscript; and Noah Keating for his multidisciplinary musings.

REFERENCES

1. A. Bhatt, Evolution of clinical research: A history before and beyond james lind, *Perspect Clin Res*, vol. 1, pp. 6–10, 2010.
2. N. Tandon, The new kind of clinical trial: It's personal, *Forbes*, pp. 1-6, 2015.
3. N. Black, Why we need observational studies to evaluate the effectiveness of health care, *BMJ*, vol. 312, pp. 1215–18, 1996.
4. N. J. Schork, Personalized medicine: Time for one-person trials, *Nature*, vol. 520, pp. 609–11, 2015.
5. A. D. Weston and L. Hood, Systems biology, proteomics, and the future of health care: Toward predictive, preventative, and personalized medicine, *J Proteome Res*, vol. 3, pp. 179–96, 2004.
6. M. J. Sorich and R. A. McKinnon, Personalized medicine: Potential, barriers and contemporary issues, *Curr Drug Metab*, vol. 13, pp. 1000–6, 2012.
7. R. Sabatier, A. Goncalves, and F. Bertucci, Personalized medicine: Present and future of breast cancer management, *Crit Rev Oncol Hematol*, vol. 91, pp. 223–33, 2014.
8. A. V. Hill, Genetics and genomics of infectious disease susceptibility, *Br Med Bull*, vol. 55, pp. 401–13, 1999.
9. T. R. Golub, D. K. Slonim, P. Tamayo, C. Huard, M. Gaasenbeek, J. P. Mesirov, et al., Molecular classification of cancer: Class discovery and class prediction by gene expression monitoring, *Science*, vol. 286, pp. 531–7, 1999.
10. I. Hedenfalk, D. Duggan, Y. Chen, M. Radmacher, M. Bittner, R. Simon, et al., Gene-expression profiles in hereditary breast cancer, *N Engl J Med*, vol. 344, pp. 539–48, 2001.
11. L. Fananapazir, Advances in molecular genetics and management of hypertrophic cardiomyopathy, *JAMA*, vol. 281, pp. 1746–52, 1999.
12. P. J. Schwartz, S. G. Priori, E. H. Locati, C. Napolitano, F. Cantu, J. A. Towbin, et al., Long QT syndrome patients with mutations of the SCN5A and HERG genes have differential responses to Na+ channel blockade and to increases in heart rate. Implications for gene-specific therapy, *Circulation*, vol. 92, pp. 3381–6, 1995.
13. W. Zareba, A. J. Moss, P. J. Schwartz, G. M. Vincent, J. L. Robinson, S. G. Priori, et al., Influence of genotype on the clinical course of the long-QT syndrome. International Long-QT Syndrome Registry Research Group, *N Engl J Med*, vol. 339, pp. 960–5, 1998.
14. G. S. Ginsburg and J. J. McCarthy, Personalized medicine: Revolutionizing drug discovery and patient care, *Trends Biotechnol*, vol. 19, pp. 491–6, 2001.

15. J. W. Scannell, A. Blanckley, H. Boldon, and B. Warrington, Diagnosing the decline in pharmaceutical R&D efficiency, *Nat Rev Drug Discov*, vol. 11, pp. 191–200, 2012.
16. J. M. Meyer and G. S. Ginsburg, The path to personalized medicine, *Curr Opin Chem Biol*, vol. 6, pp. 434–8, 2002.
17. P. C. Ng, S. S. Murray, S. Levy, and J. C. Venter, An agenda for personalized medicine, *Nature*, vol. 461, pp. 724–6, 2009.
18. M. U. Ahmed, I. Saaem, P. C. Wu, and A. S. Brown, Personalized diagnostics and biosensors: A review of the biology and technology needed for personalized medicine, *Crit Rev Biotechnol*, vol. 34, pp. 180–96, 2014.
19. K. Takahashi, K. Tanabe, M. Ohnuki, M. Narita, T. Ichisaka, K. Tomoda, et al., Induction of pluripotent stem cells from adult human fibroblasts by defined factors, *Cell*, vol. 131, pp. 861–72, 2007.
20. K. Takahashi and S. Yamanaka, Induction of pluripotent stem cells from mouse embryonic and adult fibroblast cultures by defined factors, *Cell*, vol. 126, pp. 663–76, 2006.
21. A. D. Ebert, P. Liang, and J. C. Wu, Induced pluripotent stem cells as a disease modeling and drug screening platform, *J Cardiovasc Pharmacol*, vol. 60, pp. 408–16, 2012.
22. D. A. Robinton and G. Q. Daley, The promise of induced pluripotent stem cells in research and therapy, *Nature*, vol. 481, pp. 295–305, 2012.
23. S. J. Chamberlain, X. J. Li, and M. Lalande, Induced pluripotent stem (iPS) cells as *in vitro* models of human neurogenetic disorders, *Neurogenetics*, vol. 9, pp. 227–35, 2008.
24. A. D. Ebert, J. Y. Yu, F. F. Rose, V. B. Mattis, C. L. Lorson, J. A. Thomson, et al., Induced pluripotent stem cells from a spinal muscular atrophy patient, *Nature*, vol. 457, pp. 277–U1, 2009.
25. S. Nishikawa, R. A. Goldstein, and C. R. Nierras, The promise of human induced pluripotent stem cells for research and therapy, *Nat Rev Molec Cell Biol*, vol. 9, pp. 725–9, 2008.
26. I. H. Park, N. Arora, H. Huo, N. Maherali, T. Ahfeldt, A. Shimamura, et al., Disease-specific induced pluripotent stem cells, *Cell*, vol. 134, pp. 877–86, 2008.
27. I. H. Park, R. Zhao, J. A. West, A. Yabuuchi, H. G. Huo, T. A. Ince, et al., Reprogramming of human somatic cells to pluripotency with defined factors, *Nature*, vol. 451, pp. 141–U1, 2008.
28. J. J. Unternaehrer and G. Q. Daley, Induced pluripotent stem cells for modelling human diseases, *Philos Trans Royal Soc B*, vol. 366, pp. 2274–85, 2011.
29. J. T. Dimos, K. T. Rodolfa, K. K. Niakan, L. M. Weisenthal, H. Mitsumoto, W. Chung, et al., Induced pluripotent stem cells generated from patients with ALS can be differentiated into motor neurons, *Science*, vol. 321, pp. 1218–21, 2008.
30. G. Lee and L. Studer, Modelling familial dysautonomia in human induced pluripotent stem cells, *Philos Trans R Soc Lond B Biol Sci*, vol. 366, pp. 2286–96, 2011.
31. D. Focosi, G. Amabile, A. Di Ruscio, P. Quaranta, D. G. Tenen, and M. Pistello, Induced pluripotent stem cells in hematology: Current and future applications, *Blood Cancer J*, vol. 4, p. e211, 2014.
32. A. Zimmermann, O. Preynat-Seauve, J. M. Tiercy, K. H. Krause, and J. Villard, Haplotype-based banking of human pluripotent stem cells for transplantation: Potential and limitations, *Stem Cells Dev*, vol. 21, pp. 2364–73, 2012.
33. A. Wabik and P. H. Jones, Switching roles: The functional plasticity of adult tissue stem cells, *EMBO J*, vol. 34, pp. 1164–79, 2015.
34. L. C. Fuentealba, S. B. Rompani, J. I. Parraguez, K. Obernier, R. Romero, C. L. Cepko, et al., Embryonic origin of postnatal neural stem cells, *Cell*, vol. 161, pp. 1644–55, 2015.
35. N. Smart, S. Bollini, K. N. Dube, J. M. Vieira, B. Zhou, S. Davidson, et al., De novo cardiomyocytes from within the activated adult heart after injury, *Nature*, vol. 474, pp. 640–4, 2011.

36. K. Turksen, Adult stem cells and cardiac regeneration, *Stem Cell Rev*, vol. 9, pp. 537–40, 2013.
37. D. Costamagna, E. Berardi, G. Ceccarelli, and M. Sampaolesi, Adult stem cells and skeletal muscle regeneration, *Curr Gene Ther*, vol. 15, pp. 348–63, 2015.
38. N. Barker, Adult intestinal stem cells: Critical drivers of epithelial homeostasis and regeneration, *Nat Rev Mol Cell Biol*, vol. 15, pp. 19–33, 2014.
39. W. Y. Lu, T. G. Bird, L. Boulter, A. Tsuchiya, A. M. Cole, T. Hay, et al., Hepatic progenitor cells of biliary origin with liver repopulation capacity, *Nat Cell Biol*, vol. 17, pp. 971–83, 2015.
40. N. Barker, J. H. van Es, J. Kuipers, P. Kujala, M. van den Born, M. Cozijnsen, et al., Identification of stem cells in small intestine and colon by marker gene Lgr5, *Nature*, vol. 449, pp. 1003–7, 2007.
41. Y. Dor, J. Brown, O. I. Martinez, and D. A. Melton, Adult pancreatic beta-cells are formed by self-duplication rather than stem-cell differentiation, *Nature*, vol. 429, pp. 41–6, 2004.
42. J. Gimble and F. Guilak, Adipose-derived adult stem cells: Isolation, characterization, and differentiation potential, *Cytotherapy*, vol. 5, pp. 362–9, 2003.
43. K. A. Jackson, S. M. Majka, H. Wang, J. Pocius, C. J. Hartley, M. W. Majesky, et al., Regeneration of ischemic cardiac muscle and vascular endothelium by adult stem cells, *J Clin Invest*, vol. 107, pp. 1395–402, 2001.
44. M. Korbling and Z. Estrov, Adult stem cells for tissue repair—A new therapeutic concept? *N Engl J Med*, vol. 349, pp. 570–82, 2003.
45. M. F. Pittenger, A. M. Mackay, S. C. Beck, R. K. Jaiswal, R. Douglas, J. D. Mosca, et al., Multilineage potential of adult human mesenchymal stem cells, *Science*, vol. 284, pp. 143–7, 1999.
46. J. G. Toma, M. Akhavan, K. J. Fernandes, F. Barnabe-Heider, A. Sadikot, D. R. Kaplan, et al., Isolation of multipotent adult stem cells from the dermis of mammalian skin, *Nat Cell Biol*, vol. 3, pp. 778–84, 2001.
47. L. J. van't Veer and R. Bernards, Enabling personalized cancer medicine through analysis of gene-expression patterns, *Nature*, vol. 452, pp. 564–70, 2008.
48. T. B. Meissner, P. K. Mandal, L. M. Ferreira, D. J. Rossi, and C. A. Cowan, Genome editing for human gene therapy, *Methods Enzymol*, vol. 546, pp. 273–95, 2014.
49. H. Ledford, CRISPR, the disruptor, *Nature*, vol. 522, pp. 20–4, 2015.
50. R. S. Leibman and J. L. Riley, Engineering t cells to functionally cure HIV-1 infection, *Mol Ther*, vol. 23, pp. 1149–59, 2015.
51. K. Schumann, S. Lin, E. Boyer, D. R. Simeonov, M. Subramaniam, R. E. Gate, et al., Generation of knock-in primary human T cells using Cas9 ribonucleoproteins, *Proc Natl Acad Sci U S A*, vol. 112, pp. 10437–42, 2015.
52. P. Tebas, D. Stein, W. W. Tang, I. Frank, S. Q. Wang, G. Lee, et al., Gene editing of CCR5 in autologous CD4 T cells of persons infected with HIV, *N Engl J Med*, vol. 370, pp. 901–10, 2014.
53. M. Dominska and D. M. Dykxhoorn, Breaking down the barriers: siRNA delivery and endosome escape, *J Cell Sci*, vol. 123, pp. 1183–9, 2010.
54. K. Kurisaki, A. Kurisaki, U. Valcourt, A. A. Terentiev, K. Pardali, P. Ten Dijke, et al., Nuclear factor YY1 inhibits transforming growth factor beta- and bone morphogenetic protein-induced cell differentiation, *Mol Cell Biol*, vol. 23, pp. 4494–510, 2003.
55. A. C. Hsieh, R. Bo, J. Manola, F. Vazquez, O. Bare, A. Khvorova, et al., A library of siRNA duplexes targeting the phosphoinositide 3-kinase pathway: Determinants of gene silencing for use in cell-based screens, *Nucleic Acids Res*, vol. 32, pp. 893–901, 2004.
56. Z. Y. Jiang, Q. L. Zhou, K. A. Coleman, M. Chouinard, Q. Boese, and M. P. Czech, Insulin signaling through Akt/protein kinase B analyzed by small interfering RNA-mediated gene silencing, *Proc Natl Acad Sci U S A*, vol. 100, pp. 7569–74, 2003.

57. K. Gavrilov and W. M. Saltzman, Therapeutic siRNA: Principles, challenges, and strategies, *Yale J Biol Med*, vol. 85, pp. 187–200, 2012.
58. D. Kaplan, R. T. Moon, and G. Vunjak-Novakovic, It takes a village to grow a tissue, *Nat Biotechnol*, vol. 23, pp. 1237–9, 2005.
59. G. Vunjak-Novakovic, L. Meinel, G. Altman, and D. Kaplan, Bioreactor cultivation of osteochondral grafts, *Orthodontics Craniofacial Res*, vol. 8, pp. 209–18, 2005.
60. C. A. Hamm and F. F. Costa, The impact of epigenomics on future drug design and new therapies, *Drug Discov Today*, vol. 16, pp. 626–35, 2011.
61. C. E. Romanoski, C. K. Glass, H. G. Stunnenberg, L. Wilson, and G. Almouzni, Epigenomics: Roadmap for regulation, *Nature*, vol. 518, pp. 314–6, 2015.
62. N. Tandon, D. Marolt, E. Cimetta, and G. Vunjak-Novakovic, Bioreactor engineering of stem cell environments, *Biotechnol Adv*, vol. 31, pp. 1020–31, 2013.
63. I. R. Edwards and J. K. Aronson, Adverse drug reactions: Definitions, diagnosis, and management, *Lancet*, vol. 356, pp. 1255–9, 2000.
64. U. A. Meyer, Pharmacogenetics and adverse drug reactions, *Lancet*, vol. 356, pp. 1667–71, 2000.
65. J. A. Burdick and G. Vunjak-Novakovic, Engineered microenvironments for controlled stem cell differentiation, *Tissue Eng Part A*, vol. 15, pp. 205–19, 2009.
66. M. P. Lutolf, P. M. Gilbert, and H. M. Blau, Designing materials to direct stem-cell fate, *Nature*, vol. 462, pp. 433–41, 2009.
67. M. P. Lutolf and J. A. Hubbell, Synthetic biomaterials as instructive extracellular microenvironments for morphogenesis in tissue engineering, *Nat Biotechnol*, vol. 23, pp. 47–55, 2005.
68. S. Gerecht, J. A. Burdick, L. S. Ferreira, S. A. Townsend, R. Langer, and G. Vunjak-Novakovic, Hyaluronic acid hydrogel for controlled self-renewal and differentiation of human embryonic stem cells, *PNAS*, vol. 104, pp. 11298–303, 2007.
69. S. Hofmann, H. Hagenmüller, A. M. Koch, R. Müller, G. Vunjak-Novakovic, D. L. Kaplan, et al., Control of *in vitro* tissue-engineered bone-like structures using human mesenchymal stem cells and porous silk scaffolds, *Biomaterials*, vol. 28, pp. 1152–62, 2007.
70. T. P. Kraehenbuehl, R. Langer, and L. S. Ferreira, Three-dimensional biomaterials for the study of human pluripotent stem cells, *Nat Methods*, vol. 8, pp. 731–6, 2011.
71. R. T. Tran, P. Thevenot, Y. Zhang, D. Gyawali, L. Tang, and J. Yang, Scaffold sheet design strategy for soft tissue engineering, *Nat Mater*, vol. 3, pp. 1375–89, 2010.
72. S. Bhumiratana, W. L. Grayson, A. Castaneda, D. N. Rockwood, E. S. Gil, D. L. Kaplan, et al., Nucleation and growth of mineralized bone matrix on silk-hydroxyapatite composite scaffolds, *Biomaterials*, vol. 32, pp. 2812–20, 2011.
73. I. Marcos-Campos, D. Marolt, P. Petridis, S. Bhumiratana, D. Schmidt, and G. Vunjak-Novakovic, Bone scaffold architecture modulates the development of mineralized bone matrix by human embryonic stem cells, *Biomaterials*, vol. 33, pp. 8329–42, 2012.
74. R. J. Miron, A. Sculean, Y. Shuang, D. D. Bosshardt, R. Gruber, D. Buser, et al., Osteoinductive potential of a novel biphasic calcium phosphate bone graft in comparison with autographs, xenografts, and DFDBA, *Clin Oral Implants Res*, vol. 26, pp. 668–75, 2016.
75. H. Seitz, W. Rieder, S. Irsen, B. Leukers, and C. Tille, Three-dimensional printing of porous ceramic scaffolds for bone tissue engineering, *J Biomed Mater Res B Appl Biomater*, vol. 74, pp. 782–8, 2005.
76. C. Gardin, S. Ricci, L. Ferroni, R. Guazzo, L. Sbricoli, G. De Benedictis, et al., Decellularization and delipidation protocols of bovine bone and pericardium for bone grafting and guided bone regeneration procedures, *PLoS One*, vol. 10, p. e0132344, 2015.
77. H. L. Kim, G. Y. Jung, J. H. Yoon, J. S. Han, Y. J. Park, D. G. Kim, et al., Preparation and characterization of nano-sized hydroxyapatite/alginate/chitosan composite scaffolds for bone tissue engineering, *Mater Sci Eng C Mater Biol Appl*, vol. 54, pp. 20–5, 2015.

78. D. W. Hutmacher, J. C. Goh, and S. H. Teoh, An introduction to biodegradable materials for tissue engineering applications, *Ann Acad Med Singapore*, vol. 30, pp. 183–91, 2001.
79. R. C. Thomson, M. J. Yaszemski, J. M. Powers, and A. G. Mikos, Fabrication of biodegradable polymer scaffolds to engineer trabecular bone, *J Biomater Sci Polym Ed*, vol. 7, pp. 23–38, 1995.
80. N. E. Fedorovich, J. R. De Wijn, A. J. Verbout, J. Alblas, and W. J. Dhert, Three-dimensional fiber deposition of cell-laden, viable, patterned constructs for bone tissue printing, *Tissue Eng* Part A, vol. 14, pp. 127–33, Jan 2008.
81. A. Khalyfa, S. Vogt, J. Weisser, G. Grimm, A. Rechtenbach, W. Meyer, et al., Development of a new calcium phosphate powder-binder system for the 3D printing of patient specific implants, *J Mater Sci Mater Med*, vol. 18, pp. 909–16, 2007.
82. Autodesk. *Autodesk Within*. 2015. Available from: http://www.withinlab.com/overview/new_index.php (Accessed on August 27, 2015).
83. R. Langer and J. P. Vacanti, Tissue engineering, *Science*, vol. 260, pp. 920–6, 1993.
84. E. Y. Fok and P. W. Zandstra, Shear-controlled single-step mouse embryonic stem cell expansion and embryoid body-based differentiation, *Stem Cells*, vol. 23, pp. 1333–42, 2005.
85. W. L. Grayson, D. Marolt, S. Bhumiratana, M. Fröhlich, X. E. Guo, and G. Vunjak-Novakovic, Optimizing the medium perfusion rate in bone tissue engineering bioreactors, *Biotechnol Bioeng*, vol. 108, pp. 1159–70, 2011.
86. D. Marolt, I. M. Campos, S. Bhumiratana, A. Koren, P. Petridis, G. Zhang, et al., Engineering bone tissue from human embryonic stem cells, *Proc Natl Acad Sci U S A*, vol. 109, pp. 8705–9, 2012.
87. Y. Martin and P. Vermette, Bioreactors for tissue mass culture: Design, characterization, and recent advances, *Biomaterials*, vol. 26, pp. 7481–503, 2005.
88. G. Vunjak-Novakovic and D. T. Scadden, Biomimetic platforms for human stem cell research, *Cell Stem Cell*, vol. 8, pp. 252–61, 2011.
89. B. M. Baker, R. P. Shah, A. H. Huang, and R. L. Mauck, Dynamic tensile loading improves the functional properties of mesenchymal stem cell-laden nanofiber-based fibrocartilage, *Tissue Eng* A, vol. 17, pp. 1445–55, 2011.
90. K. Bilodeau and D. Mantovani, Bioreactors for tissue engineering: Focus on mechanical constraints. A comparative review, *Tissue Eng*, vol. 12, pp. 2367–83, 2006.
91. N. Tandon, C. Cannizzaro, P. H. Chao, R. Maidhof, A. Marsano, H. T. Au, et al., Electrical stimulation systems for cardiac tissue engineering, *Nat Protoc*, vol. 4, pp. 155–73, 2009.
92. N. Tandon, A. Marsano, R. Maidhof, L. Wan, H. Park, and G. Vunjak-Novakovic, Optimization of electrical stimulation parameters for cardiac tissue engineering, *J Tissue Eng Regen Med*, vol. 5, pp. e115–25, 2011.
93. R. D. Abbott and D. L. Kaplan, Strategies for improving the physiological relevance of human engineered tissues, *Trends Biotechnol*, vol. 33, pp. 401–7, 2015.
94. L. G. Griffith and G. Naughton, Tissue engineering—Current challenges and expanding opportunities, *Science*, vol. 295, pp. 1009–14, 2002.
95. G. M. de Peppo, I. Marcos-Campos, D. J. Kahler, D. Alsalman, L. Shang, G. Vunjak-Novakovic, et al., Engineering bone tissue substitutes from human induced pluripotent stem cells, *Proc Natl Acad Sci U S A*, vol. 110, pp. 8680–5, 2013.
96. W. L. Grayson, M. Fröhlich, K. Yeager, S. Bhumiratana, M. E. Chan, C. Cannizzaro, et al., Engineering anatomically shaped human bone grafts, *Proc Nat Acad Sci U S A*, vol. 107, pp. 3299–304, 2010.
97. M. R. Ebrahimkhani, J. A. Neiman, M. S. Raredon, D. J. Hughes, and L. G. Griffith, Bioreactor technologies to support liver function *in vitro*, *Adv Drug Deliv Rev*, vol. 69–70, pp. 132–57, 2014.
98. P. Maghsoudlou, S. Eaton, and P. De Coppi, Tissue engineering of the esophagus, *Semin Pediatr Surg*, vol. 23, pp. 127–34, 2014.

99. S. Sundaram, J. One, J. Siewert, S. Teodosescu, L. Zhao, S. Dimitrievska, et al., Tissue-engineered vascular grafts created from human induced pluripotent stem cells, *Stem Cells Transl Med*, vol. 3, pp. 1535–43, 2014.
100. W. H. Zimmermann, I. Melnychenko, and T. Eschenhagen, Engineered heart tissue for regeneration of diseased hearts, *Biomaterials*, vol. 25, pp. 1639–47, 2004.
101. N. Tandon, A. Taubman, E. Cimetta, L. Saccenti, and G. Vunjak-Novakovic, Portable bioreactor for perfusion and electrical stimulation of engineered cardiac tissue, *Conf Proc IEEE Eng Med Biol Soc*, vol. 2013, pp. 6219–23, 2013.
102. E. Cimetta, E. Figallo, C. Cannizzaro, N. Elvassore, and G. Vunjak-Novakovic, Microbioreactor arrays for controlling cellular environments: Design principles for human embryonic stem cell applications, *Methods*, vol. 47, pp. 81–9, 2009.
103. T. M. Squires and S. R. Quake, Microfluidics: Fluid physics at the nanoliter scale, *Rev Mod Physics*, vol. 77, pp. 977–1016, 2005.
104. A. Khademhosseini, R. Langer, J. Borenstein, and J. P. Vacanti, Microscale technologies for tissue engineering and biology, *PNAS*, vol. 103, pp. 2480–7, 2006.
105. D. N. Breslauer, P. J. Lee, and L. P. Lee, Microfluidics-based systems biology, *Mol BioSyst*, vol. 2, pp. 97–112, 2006.
106. S. Haeberle and R. Zengerle, Microfluidic platforms for lab-on-a-chip applications, *Lab Chip*, vol. 7, pp. 1094–110, 2007.
107. S. K. Sia and G. M. Whitesides, Microfluidic devices fabricated in poly(dimethylsiloxane) for biological studies, *Electrophoresis*, vol. 24, pp. 3563–76, 2003.
108. M. Toner and D. Irimia, Blood-on-a-Chip, *Ann Rev Biomed Eng*, vol. 7, pp. 77–103, 2005.
109. P. M. Gilbert and H. M. Blau, Engineering a stem cell house into a home, *Stem Cell Res Ther*, vol. 2, p. 3, 2011.
110. E. Cimetta, A. Godier-Furnemont, and G. Vunjak-Novakovic, Bioengineering heart tissue for *in vitro* testing, *Curr Opin Biotechnol*, vol. 24, pp. 926–32, 2013.
111. S. R. Braam, L. Tertoolen, A. van de Stolpe, T. Meyer, R. Passier, and C. L. Mummery, Prediction of drug-induced cardiotoxicity using human embryonic stem cell-derived cardiomyocytes, *Stem Cell Res*, vol. 4, pp. 107–16, 2010.
112. R. P. Davis, C. W. van den Berg, S. Casini, S. R. Braam, and C. L. Mummery, Pluripotent stem cell models of cardiac disease and their implication for drug discovery and development, *Trends Mol Med*, vol. 17, pp. 475–84, 2011.
113. S. N. Bhatia, G. H. Underhill, K. S. Zaret, and I Fox, Cell and tissue engineering for liver disease, *Sci Transl Med*, vol. 6, p. 245sr2, 16 2014.
114. S. Ng, R. E. Schwartz, S. March, A. Galstian, N. Gural, J. Shan, et al., Human iPSC-derived hepatocyte-like cells support plasmodium liver-stage infection *in vitro*, *Stem Cell Reports*, vol. 4, pp. 348–59, 2015.
115. T. M. Desrochers, E. Palma, and D. L. Kaplan, Tissue-engineered kidney disease models, *Adv Drug Deliv Rev*, vol. 69–70, pp. 67–80, Apr 2014.
116. J. P. Morgan, P. F. Delnero, Y. Zheng, S. S. Verbridge, J. Chen, M. Craven, et al., Formation of microvascular networks *in vitro*, *Nat Protoc*, vol. 8, pp. 1820–36, 2013.
117. D. Huh, Y. S. Torisawa, G. A. Hamilton, H. J. Kim, and D. E. Ingber, Microengineered physiological biomimicry: Organs-on-chips, *Lab Chip*, vol. 12, pp. 2156–64, 2012.
118. G. Vunjak-Novakovic, S. Bhatia, C. Chen, and K. Hirschi, HeLiVa platform: Integrated heart-liver-vascular systems for drug testing in human health and disease, *Stem Cell Res Ther*, vol. 4 Suppl 1, p. S8, 2013.
119. K. K. Jain, *Textbook of Personalized Medicine*. Springer-Verlag, New York, 2009.
120. FDA. *FDA Regulation of Human Cells, Tissues, and Cellular and Tissue-Based Products (HCT/Ps) Product List*. 2009. Available from: http://www.fda.gov/BiologicsBloodVaccines/TissueTissueProducts/RegulationofTissues/ucm150485.htm (Accessed on August 27, 2015).

121. FDA. *Classify Your Medical Device.* 2014. Available from: http://www.fda.gov/MedicalDevices/DeviceRegulationandGuidance/Overview/ClassifyYourDevice/ (Accessed on August 27, 2015).
122. FDA. *PMA Review Process.* 2015. Available from: http://www.fda.gov/MedicalDevices/DeviceRegulationandGuidance/HowtoMarketYourDevice/PremarketSubmissions/PremarketApprovalPMA/ucm047991.htm (Accessed on August 27, 2015).
123. FDA. *Investigational New Drug (IND) Application.* 2014. Available from: http://www.fda.gov/drugs/developmentapprovalprocess/howdrugsaredevelopedandapproved/approvalapplications/investigationalnewdrugindapplication/default.htm (Accessed on August 27, 2015).
124. FDA. *Guidance for Industry: Good Laboratory Practices Questions and Answers,* USDoHaH Services 2007. Available from: http://www.fda.gov/downloads/ICECI/EnforcementActions/BioresearchMonitoring/UCM133748.pdf (Accessed on August 27, 2015).
125. FDA. *The Drug Development Process > Step 3: Clinical Research.* 2015. Available from: http://www.fda.gov/ForPatients/Approvals/Drugs/ucm405622.htm#Clinical_Research_Phase_Studies (Accessed on August 27, 2015).
126. FDA. *Biologics License Applications (BLA) Process (CBER).* 2014 Available from: http://www.fda.gov/biologicsbloodvaccines/developmentapprovalprocess/biologicslicenseapplicationsblaprocess/default.htm (Accessed on August 27, 2015).
127. C. P. Adams and V. V. Brantner, Estimating the cost of new drug development: Is it really 802 million dollars? *Health Aff (Millwood),* vol. 25, pp. 420–8, 2006.
128. FDA. *Summary for basis of approval: Autologous Cultured Chondrocytes, Carticel.* 1997.
129. FDA. *Cellular, Tissue, and Gene Therapies Advisory Committee, Apligraf® (Oral).* 2011. Available from: http://www.fda.gov/downloads/AdvisoryCommittees/CommitteesMeetingMaterials/BloodVaccinesandOtherBiologics/CellularTissueandGeneTherapiesAdvisoryCommittee/UCM284480.pdf (Accessed on August 27, 2015).
130. FDA. Center for Biologics Evaluation and Research Office of Biostatistics and Epidemiology. *Final Review Memorandum: GINTUIT.* 2012. Available from: http://www.fda.gov/downloads/BiologicsBloodVaccines/CellularGeneTherapyProducts/ApprovedProducts/UCM297443.pdf (Accessed on August 27, 2015).
131. D. J. Herzyk, Development of biosimilars. In: L. Plitnick and D. Herzyk, (Eds.) *Nonclinical Development of Novel Biologics, Biosimilars, Vaccines and Specialty Biologics.* Academic Press, London, pp. 141–206, 2013.
132. S. Lam, *Market Insight: A Review of the Trends and Developments in Orphan Drug R&D.* 2013. Available from: http://lsconnect.thomsonreuters.com/market-insight-a-review-of-the-trends-and-developments-in-orphan-drug-rd/
133. BIO and NORD, Impact of the orphan drug tax credit on treatments for rare diseases prepared for the Biotechnology Industry Organization and the National Organization for Rare Disorders, 2015.
134. B. Obama, *FACT SHEET: President Obama's Precision Medicine Initiative,* TWHOotP Secretary, 2015.
135. M. Herper, *How A $440,000 Drug Is Turning Alexion Into Biotech's New Innovation Powerhouse.* 2012. Available from: http://www.forbes.com/sites/matthewherper/2012/09/05/how-a-440000-drug-is-turning-alexion-into-biotechs-new-innovation-powerhouse/ (Accessed on August 27, 2015).
136. New York State Assembly, *An Act to Amend the Public Health Law, in Relation to Establishing the Pharmaceutical Cost Transparency.* 2015. Available from: http://assembly.state.ny.us/leg/?sh=printbill& bn=S5338& term=2015 (Accessed on August 27, 2015).

137. FDA. Developing Products for Rare Diseases & Conditions. Available from: http://www.fda.gov/ForIndustry/DevelopingProductsforRareDiseasesConditions/ucm2005525.htm (Accessed on August 27, 2015).
138. IFMSA, *The International Federation of Medical Students Associations Policy Statement Access to Orphan Drugs*. 2014. Available from: http://ifmsa.org/wp-content/uploads/2015/05/SecGen_2014AM_PS_Access_to_Orphan_Drugs.pdf (Accessed on August 27, 2015).
139. Institute of Medicine Roundtable on Translating Genomic-Based Research, The National Academies Collection: Reports funded by National Institutes of Health, in *The Value of Genetic and Genomic Technologies: Workshop Summary*, National Academies Press (US). National Academy of Sciences, Washington (DC), pp. 19–34, 2010.
140. R. Turna, *Study Provides 'Strong Evidence' Backing Warfarin Initiation PGx Testing; Adoption Impact Unclear*. 2015. Available from: https://www.genomeweb.com/genetic-research/study-provides-strong-evidence-backing-warfarin-initiation-pgx-testing-adoption (Accessed on August 27, 2015).
141. M. G. Aspinall and R. G. Hamermesh, *Realizing the Promise of Personalized Medicine*. 2007. Available from: https://hbr.org/2007/10/realizing-the-promise-of-personalized-medicine (Accessed on August 27, 2015).
142. IBISWorld, *Health & Medical Insurance in the US: Market Research Report*, 2015.
143. L. Setar, *Top Sectors for Regulatory Change*, 2013.
144. C. N. Saha and S. Bhattacharya, Intellectual property rights: An overview and implications in pharmaceutical industry, *J Adv Pharm Technol Res*, vol. 2, pp. 88–93, 2011.
145. R. Patel, *Developing a Patent Strategy a Checklist for Getting Started*. 2008. Available from: https://www.fenwick.com/FenwickDocuments/Patent_Checklist.pdf (Accessed on August 27, 2015).
146. R. Davis, *AIA Reviews are a Game-Changer for Pharma, Survey Shows, Law* 3602015.
147. R. Anderson, *Pharmaceutical Industry Gets High on Fat Profits*. 2014. Available from: http://www.bbc.com/news/business-28212223 (Accessed on August 27, 2015).
148. T. Wizemann, *Public Health Effectiveness of the FDA 510(k), Clearance Process: Balancing Patient Safety and Innovation: Workshop Report*. National Academies Press (US), Washington (DC).
149. D. M. Smith, Assessing commercial opportunities for autologous and allogeneic cell-based products, *Regen Med*, vol. 7, pp. 721–32, 2012.
150. H. Ledford, First biosimilar drug set to enter US market, *Nature*, vol. 517, pp. 253–4, 2015.
151. B. S. Bal, An introduction to medical malpractice in the United States, *Clin Orthop Relat Res*, vol. 467, pp. 339–47, 2009.
152. Deloitte, *2015 Global Life Sciences Outlook. Adapting in an Era of Transformation*. Available from: http://www2.deloitte.com/content/dam/Deloitte/global/Documents/Life-Sciences-Health-Care/gx-lshc-2015-life-sciences-report.pdf (Accessed on August 27, 2015).
153. GPhA. *Biosimilars in the EU*. Available from: http://www.gphaonline.org/gpha-media/gpha-resources/biosimilars-in-the-eu (Accessed on August 27, 2015).
154. A. Antoniou, P. D. Pharoah, S. Narod, H. A. Risch, J. E. Eyfjord, J. L. Hopper, et al., Average risks of breast and ovarian cancer associated with BRCA1 or BRCA2 mutations detected in case series unselected for family history: A combined analysis of 22 studies, *Am J Hum Genet*, vol. 72, pp. 1117–30, 2003.
155. Supreme Court, *Association for Molecular Pathology et al. v. Myriad Genetics, INC.*, et al., SCotU States: Report of Decisions, 2013.
156. R. A, *Who Owns the Biggest Biotech Discovery of the Century?* 2014. Available from: http://www.technologyreview.com/featuredstory/532796/who-owns-the-biggest-biotech-discovery-of-the-century (Accessed on August 27, 2015).

157. P. Knoepfler, Putting the IP in iPS cells: patent war looming? *Knoepfler Lab Stem Cell Blog*, 2013. Available from: http://www.ipscell.com/2013/05/putting-the-ip-in-ips-cells-patent-war-looming/ (Accessed on August 27, 2015).
158. C. A. Minami, J. W. Chung, J. L. Holl, and K. Y. Bilimoria, Impact of medical malpractice environment on surgical quality and outcomes, *J Am Coll Surg*, vol. 218, pp. 271–8. e1–9, 2014.
159. E. Feess, Malpractice liability, technology choice and negative defensive medicine, *Eur J Health Econ*, vol. 13, pp. 157–67, 2012.
160. P. J. Bixenstine, A. D. Shore, W. T. Mehtsun, A. M. Ibrahim, J. A. Freischlag, and M. A. Makary, Catastrophic medical malpractice payouts in the United States, *J Healthc Qual*, vol. 36, pp. 43–53, 2014.
161. G. E. Marchant, R. J. Milligan, and B. Wilhelmi, Legal pressures and incentives for personalized medicine, *Pers Med*, vol. 3, pp. 391–7, 2006.
162. G. E. Marchant, D. E. Campos-Outcalt, and R. A. Lindor, Physician liability: The next big thing for personalized medicine? *Pers Med*, vol. 8, pp. 457–67, 2011.
163. B. S. Bal and L. H. Brenner, Corporate malfeasance, off-label use, and surgeon liability, *Clin Orthop Relat Res*, vol. 471, pp. 4–8, 2013.
164. E. G. Lasker, S. A. Klein, and T. Fishman-Barago, Taking the product out of product liability: Litigation risks and business implications of innovator and co-promoter liability, *Defense Counsel J*, vol. 82, pp. 295–308, 2015.
165. D. Bornstein, Helping new drugs out of research's 'valley of death', *New York Times*, 2011. Available from: http://opinionator.blogs.nytimes.com/2011/05/02/helping-new-drugs-out-of-academias-valley-of-death/?_r=2 (Accessed on August 27, 2015).
166. E. R. Dorsey, J. P. Thompson, M. Carrasco, J. de Roulet, P. Vitticore, S. Nicholson, et al., Financing of U.S. biomedical research and new drug approvals across therapeutic areas, *PLoS One*, vol. 4, p. e7015, 2009.
167. FDA, *New Molecular Entity Approvals for 2010*, 2010.
168. K. D. Harrison, N. S. Kadaba, R. B. Kelly, and D. Crawford, Building a life sciences innovation ecosystem, *Sci Transl Med*, vol. 4, p. 157fs37, 2012.
169. A. Fobes and J. Lawless, *CMS Actuary Finds Health Care Spending Outpacing GDP Growth, TUSSCo Finance*, 2015.
170. S. Rockey, *What are the Chances of Getting Funded?* 2015. Available from: http://nexus.od.nih.gov/all/2015/06/29/what-are-the-chances-of-getting-funded/ (Accessed on August 27, 2015).
171. D. Butler, Translational research: Crossing the valley of death, *Nature*, vol. 453, pp. 840–2, 2008.
172. A. Forman, *Building New York City's Innovation Economy*. 2015, Available from: https://nycfuture.org/data/info/nycs-tech-profile (Accessed on August 27, 2015).
173. BIO, *BIO's Venture Funding of Therapeutic Innovation Report*. 2015. Available from: https://www.bio.org/biovcstudy (Accessed on August 27, 2015).
174. B. Booth, *Where Does All That Biotech Venture Capital Go?* 2015. Available from: http://www.forbes.com/sites/brucebooth/2015/02/09/where-does-all-that-biotech-venture-capital-go/ (Accessed on August 27, 2015).
175. FDA, *Novel New Drugs 2014 Summary*, 2015. Available from: http://www.fda.gov/downloads/drugs/developmentapprovalprocess/druginnovation/ucm430299.pdf (Accessed on August 27, 2015).
176. *The New Role of Academia in Drug Development*, 2010. Available from: http://www.kauffman.org/what-we-do/research/2011/05/the-new-role-of-academia-in-drug-development (Accessed on August 27, 2015).
177. JLL, *Global Competition Heats up as Small and Mid-tier Companies Dictate Demand*, 2014. Available from: http://www.jll.com/Research/2014-global-life-sciences-report-JLL.pdf?654be919-aef1-45a0-bef3-ab01d0a4ece6 (Accessed on August 27, 2015).

178. Etymonline, *Patient*. 2015. Available from: http://www.etymonline.com/index.php?term=patient (Accessed on August 27, 2015).
179. C. Minto, B. Bauce, C. Calore, I. Rigato, F. Folino, N. Soriani, et al., Is Internet use associated with anxiety in patients with and at risk for cardiomyopathy? *Am Heart J*, vol. 170, pp. 87–95, 95.e1-4, 2015.
180. M. J. Mazzini and L. M. Glode, Internet oncology: Increased benefit and risk for patients and oncologists, *Hematol Oncol Clin North Am*, vol. 15, pp. 583–92, 2001.
181. P. Hartzband and J. Groopman, Untangling the Web–patients, doctors, and the Internet, *N Engl J Med*, vol. 362, pp. 1063–6, 2010.
182. NIH, A Pilot Open Label Clinical Trial to Evaluate the Combined Impact of Two Mobile Health Products on Health Outcomes in Overweight Adults with Type 2 Diabetes, *Clinical Trials.gov*, 2015. Available from: https://clinicaltrials.gov/ct2/show/NCT02227303 (Accessed on August 27, 2015).
183. Medidata, *Medidata Completes Study Exploring Potential for Instrumenting Patients with Mobile Health Tools in a Clinical Trial Setting*. 2015. Available from: https://www.mdsol.com/en/newsroom/press-release/medidata-completes-study-exploring-potential-instrumenting-patients-mobile (Accessed on August 27, 2015).
184. P. Wicks and M. Little, The virtuous circle of the quantified self: A human computational approach to improved health outcomes. In: P. Michelucci, (Ed.) *Handbook of Human Computation*. Springer-Verlag, New York, pp. 105–131, 2013.
185. P. Wicks, *Blinding as a Solution to Bias in Biomedical Science and the Courts: A Multidisciplinary Approach*, Elsevier, Cambridge, MA, 2015.
186. P. Wicks, T. E. Vaughan, M. P. Massagli, and J. Heywood, Accelerated clinical discovery using self-reported patient data collected online and a patient-matching algorithm, *Nat Biotechnol*, vol. 29, pp. 411–14, 2011.
187. Etymonline, *Doctor*. 2015. Available from: http://www.etymonline.com/index.php?term=doctor (Accessed on August 27, 2015).
188. K. Zaslav, B. Cole, R. Brewster, T. DeBerardino, J. Farr, P. Fowler, et al., A prospective study of autologous chondrocyte implantation in patients with failed prior treatment for articular cartilage defect of the knee: Results of the Study of the Treatment of Articular Repair (STAR) clinical trial, *Am J Sports Med*, vol. 37, pp. 42–55, 2009.
189. FDA, *Epicel® cultured epidermal autograft (CEA)*, 2007. Available from: http://www.fda.gov/MedicalDevices/ProductsandMedicalProcedures/DeviceApprovalsandClearances/Recently-ApprovedDevices/ucm074878.htm (Accessed on August 27, 2015).
190. Genzyme, *Epicel (cultured epidermal autografts)*, 2007. Available from: http://www.accessdata.fda.gov/cdrh_docs/pdf/H990002d.pdf (Accessed on August 27, 2015).
191. M. Hidalgo, E. Bruckheimer, N. V. Rajeshkumar, I. Garrido-Laguna, E. De Oliveira, B. Rubio-Viqueira, et al., A pilot clinical study of treatment guided by personalized tumorgrafts in patients with advanced cancer, *Mol Cancer Ther*, vol. 10, pp. 1311–16, 2011.
192. P. Malaney, S. V. Nicosia, and V. Dave, One mouse, one patient paradigm: New avatars of personalized cancer therapy, *Cancer Lett*, vol. 344, pp. 1–12, 2014.
193. EpiBone, *EpiBone: Grow Your Own Bone*. 2015. Available from: http://www.epibone.com (Accessed on August 27, 2015).
194. NeoCart, *NeoCart: Autologous Engineered Neocartilage*. 2015. Available from: http://www.neocartimplant.com (Accessed on August 27, 2015).
195. Histogenics, *NeoCart*. 2015. Available from: http://www.histogenics.com/products-platform/neocart (Accessed on August 27, 2015).
196. J. A. Chacon, A. A. Sarnaik, J. Q. Chen, C. Creasy, C. Kale, J. Robinson, et al., Manipulating the tumor microenvironment ex vivo for enhanced expansion of tumor-infiltrating lymphocytes for adoptive cell therapy, *Clin Cancer Res*, vol. 21, pp. 611–21, 2015.

197. I. W. Mak, N. Evaniew, and M. Ghert, Lost in translation: Animal models and clinical trials in cancer treatment, *Am J Transl Res*, vol. 6, pp. 114–18, 2014.
198. Organovo, *Organovo: Structurally and Functionally Accurate Bioprinted Human Tissue Models*. 2015. Available from: http://www.organovo.com (Accessed on August 27, 2015).
199. K. Jakab, C. Norotte, F. Marga, K. Murphy, G. Vunjak-Novakovic, and G. Forgacs, Tissue engineering by self-assembly and bio-printing of living cells, *Biofabrication*, vol. 2, p. 022001, 2010.
200. M. Radisic, H. Park, H. Shing, T. Consi, F. J. Schoen, R. Langer, et al., Functional assembly of engineered myocardium by electrical stimulation of cardiac myocytes cultured on scaffolds, *Proc Natl Acad Sci U S A*, vol. 101, pp. 18129–34, 2004.
201. Z. Miller, *Top Sites for Crowdfunding Scientific Research Best Platforms for Crowdfunding Science*. 2014. Available from: http://crowdfunding.about.com/od/Placeholderrr/tp/Top-Sites-for-Crowdfunding-Scientific-Research.htm
202. P. U. Macneill and B. Ferran, Art and bioethics: Shifts in understanding across genres, *J Bioethical Inquiry*, vol. 8, pp. 71–85, 2011. (Accessed on August 27, 2015).
203. UWA, *SymbioticA*. 2015. Available from: http://www.symbiotica.uwa.edu.au (Accessed on August 27, 2015).
204. Rare Diseases Europe, *Orphan Medicines Regulation*. 2014. Available from: http://www.eurordis.org/content/promoting-orphan-drug-development
205. J. P. Mather. *Stem Cell Culture: Methods in Cell Biology*, vol. 86, Academic Press, 2008.sdfsd

Index

Note: Page numbers ending in "f" and "t" refers to figures and tables, respectively.

E

F

G

H

R

S